PLANETAR

Safeguarding Human Health and the Environment in the Anthropocene

We live in unprecedented times – the Anthropocene – defined by far-reaching human impacts on the natural systems that underpin civilization. *Planetary Health* explores the many environmental changes that threaten to undermine progress in human health, and explains how these changes affect health outcomes, from infectious diseases (including pandemics) to child development to mental health, from chronic diseases to nutrition to injuries. It shows how people can adapt at least partly to those changes that are now unavoidable, through actions that both improve health *and* safeguard the environment. But humanity must do *more* than just adapt: we need *transformative* changes across many sectors – energy, built environment, transport, manufacturing, food, and health care – to reduce the environmental footprints of society. The book discusses specific policies, technologies, and interventions to achieve the scale of change required, and explains how these can be implemented. It presents the *evidence*, builds *hope* in our common future, and aims to motivate *action* by everyone, from the general public to policy-makers to health professionals.

ANDY HAINES is Professor of Environmental Change and Public Health at the Centre for Climate Change and Planetary Health, London School of Hygiene and Tropical Medicine, where he was previously director between 2001 and 2010. He was a member of the Intergovernmental Panel on Climate Change for the Second, Third, and Fifth Assessment Reports. He was chair of the Task Force on Climate Change Mitigation and Public Health in 2008–9, and of the Rockefeller Foundation–*Lancet* Commission on Planetary Health in 2014–15. He is an international member of the US National Academy of Medicine.

HOWARD FRUMKIN is Professor Emeritus of Environmental and Occupational Health Sciences at the University of Washington School of Public Health in Seattle, where he was dean from 2000 through 2006. He was previously head of the Wellcome Trust 'Our Planet, Our Health' initiative, and director of the National Center for Environmental Health at the US Centers for Disease Control and Prevention.

'This book illustrates the dramatic change in humanity's awareness about the realities of our world and our destiny. *Planetary Health* describes the transformation that will move us to a healthier, safer, and more sustainable future, with hope, optimism, and confidence.'

Gro Harlem Brundtland, *former Prime Minister of Norway and former Director-General of the World Health Organization*

'The climate crisis and other environmental challenges pose growing threats to the continued health of humanity. Andy Haines and Howie Frumkin's book cogently and accessibly summarizes these growing threats to human health, and outlines the transformative changes we need to enact to improve health around the world and *at the same time* safeguard the environment for our children and grandchildren. This book is essential reading for every decision-maker in government and NGOs formulating health or environmental policy. The book is also a wonderful overview of planetary health issues for the general reader and student. A more important topic for the future of humanity from two leading experts is hard to imagine. A very timely analysis of the central issues of our time.'

Helen Clark, *former Prime Minister of New Zealand and former Administrator of the United Nations Development Programme*

'A splendid and important piece of work. Andy Haines and Howard Frumkin have, over many decades, led the world in the understanding of the relationships between environment and health. In this crucially important book, they show how our destructive behaviour, on a huge scale, has undermined our environment in ways that foster pandemics and infectious diseases, and which profoundly damage our physical and mental health. They also show, through careful and evidence-based analysis, how we can transform how we live, work, and consume to not only restore our environment but also live in much healthier and more enjoyable ways.'

Professor Lord Nicholas Stern, *London School of Economics*

'Andrew Haines and Howard Frumkin are two of the modern giants in planetary health. Their book highlights the value of cross-cutting approaches to human health and the health of the environment, combining resilience, health, equity, and prosperity. The book highlights the potential for the topic of planetary health to transform higher education, helping us rise to the challenge of multiple global crises. As we emerge from a global pandemic, I cannot imagine a more timely and important topic. An amazing, brilliantly argued compendium.'

Judith Rodin, *former President of the Rockefeller Foundation and former President of the University of Pennsylvania*

'By its very nature, planetary health is a vast and multidisciplinary subject. Our understanding of the field is constantly expanding, our approaches to addressing it rapidly evolving, and the political landscape surrounding it increasingly complex, and yet, Haines and Frumkin masterfully summarize the issue in all of its complexity, leaving the reader with a solid foundation of essential knowledge and – more importantly – hope, and a clearly delineated path forward.'

Michele Barry MD, *Director of the Center for Innovation in Global Health, Stanford University*

'The Anthropocene Era. Our children and our children's children will be astonished that the first fully comprehensive book to describe their everyday life was not written 'til 2021. And as they reflect on what Sir Andy Haines and Howard Frumkin have laid out, either they will ask why did we not follow the ideas for transformative change in this book, or they will stand together in hope saying, humanity took heed, it listened to evidence and common sense, and we were able to tilt the Earth's systems on their axis. This book is a watershed, its integrated science combines separate strands of knowledge, and creates the space for far more equitable participation in and ownership of the actions needed. This book is published at a watershed moment in history, when the health of the planet is in our hands. Can *Planetary Health* become our zeitgeist?'

Liz Grant, *University of Edinburgh*

'Sir Andy Haines and Howard Frumkin's book *Planetary Health: Safeguarding Human Health and the Environment in the Anthropocene* is a fact-filled and thought-provoking volume that explains the origins and proposes sound solutions to the existential crises of the Anthropocene Era. The authors describe the political, economic, technological, and ideological choices, often motivated by greed and science denial that throughout history have brought us to a point where the continued survival of human societies is now under threat. They demonstrate how, after improving the health of human societies, we have in the span of a few decades prepared for its regression, by destroying billions of years of stable climate evolution. Instead of protecting the Earth, we have exploited it wrongly and excessively, polluted it, and created irreversible damage that has increased socioeconomic inequities and fallen most heavily upon the world's poor. This book offers a holistic approach that helps us understand why we must respect, protect, and enforce multi-sectoral frameworks if we are to maintain the integrity of the unique integrated complex system that is the Earth. The book also describes the many actions each of us can choose to take, across sectors and within the finite planetary boundaries and cultural context, to reverse and reduce the current risks to a minimum and to remain healthy. An important read!'

Agnes Binagwaho, *Vice Chancellor of the University of Global Health Equity in Rwanda*

'A brilliant book: planetary health, defined by founders of the field.'

Tony Capon, *Director of the Monash Sustainable Development Institute and Professor of Planetary Health in the School of Public Health and Preventive Medicine at Monash University*

PLANETARY HEALTH

Safeguarding Human Health and the Environment in the Anthropocene

ANDY HAINES

London School of Hygiene and Tropical Medicine

HOWARD FRUMKIN

University of Washington

CAMBRIDGE
UNIVERSITY PRESS

University Printing House, Cambridge CB2 8BS, United Kingdom

One Liberty Plaza, 20th Floor, New York, NY 10006, USA

477 Williamstown Road, Port Melbourne, VIC 3207, Australia

314–321, 3rd Floor, Plot 3, Splendor Forum, Jasola District Centre, New Delhi – 110025, India

103 Penang Road, #05–06/07, Visioncrest Commercial, Singapore 238467

Cambridge University Press is part of the University of Cambridge.

It furthers the University's mission by disseminating knowledge in the pursuit of education, learning, and research at the highest international levels of excellence.

www.cambridge.org
Information on this title: www.cambridge.org/9781108492348
DOI: 10.1017/9781108698054

First published 2021

Printed in the United Kingdom by TJ Books Limited, Padstow Cornwall

A catalogue record for this publication is available from the British Library.

ISBN 978-1-108-49234-8 Hardback
ISBN 978-1-108-72926-0 Paperback

Contents

Preface

Humanity has emerged from the Holocene Epoch, a period of relative climatic stability lasting about 11,500 years, into the Anthropocene Epoch, so called because it is characterized by the dominance of our own species over the global environment. There is a growing number of books about global environmental changes in general, including transformation of our climate, land use, and oceans, and about the Anthropocene Epoch more specifically. So why add to this, sometimes depressing, litany? This book aims to address a major gap in an emerging field – namely how these environmental changes could affect the health of humanity and what can be done to protect and advance the health of today's and future generations in the face of such far-reaching changes.

When we think about medicine and health more generally we often focus on the potential for new cures for currently intractable conditions – cancers, neurological conditions, mental illness, and the pervasive problems of an ageing population. The hope is that through advances in molecular genetics, artificial intelligence, and bioengineering, for example, we might overcome chronic disease and disability to lead long and healthy lives unencumbered by today's afflictions. The biomedical focus of much research can make important contributions to the advancement of health but it also has notable shortcomings, including the growing unaffordability of health care, the large burden of preventable ill-health, and dimensions of human well-being beyond simply the absence of disease. Good health remains far too elusive to poor and disenfranchised people the world over. An overly biomedical approach also fails, in large part, to address the unsustainable basis of today's economy. Our profligate use of energy and materials may improve health statistics in the short run, but over time it cannot be sustained, and it contains the seeds of much human suffering.

This book – whilst acknowledging the importance of biomedicine – also challenges its limited vision, and encompasses a wider view of health based on the integrity of natural systems. This is hardly a new concept. In *On Airs, Waters and Places*, written 2500 years ago, Hippocrates described the importance of considering environmental factors, including climate, in understanding the disease patterns in different locations. A few pioneers, notably the late Tony McMichael, had the vision to see how global environmental change could threaten human societies. Tony's magisterial triad of books published over a quarter century – *Planetary Overload: Global Environmental Change and the Health of the Human Species* (1993), *Human Frontiers, Environments and Disease* (2001), and *Climate Change and the Health of Nations* (published posthumously in 2017), presciently established the foundations of what we now know as Planetary Health (1–3). We both worked and published with Tony, and we honour his memory. Scholarship on the health impacts of planetary changes slowly accumulated despite the failure to attract appreciable funding or attention from much of the health research community for many years; we think Tony would be gratified to know that this approach is now gathering considerable momentum.

In recent years there has been a pronounced shift in interest, particularly from early career researchers and students, in transdisciplinary collaborations to understand how natural systems underpin human health and effective policies to address the interlinked challenges of the Anthropocene Epoch. This book is dedicated both to the early pioneers and to a new generation of informed and open-minded scientists, practitioners, policymakers, and increasingly the wider public from all walks of life. The motivation for writing the book stems in part from our involvement in the Rockefeller Foundation–*Lancet* Commission on Planetary Health, one of us (AH) as Chair. The Commission's work involved synthesizing large amounts of evidence from a diverse range of sources. We have updated the evidence where relevant and we hope the synthesis will be useful to a wide readership. We acknowledge with gratitude the contributions of other members of the Commission and the many colleagues with whom we have worked over the years. Their names feature prominently in the reference lists to many of the chapters.

This book aspires to provide a compelling overview of how human activities impact on natural systems, and the implications of the resulting changes for health, well-being, and even survival both today and in the foreseeable future. We do not stop at describing the challenges; we aim also to provide a critical appraisal of policy options to sustain and improve health while rapidly reducing the environmental footprint of humanity.

Chapter 1 includes a brief history of the Anthropocene Epoch and description of the concept, as well as the concept of finite planetary boundaries within which humanity can flourish. It discusses the 'Great Acceleration' in Earth systems changes over the last century or two, and especially since the end of World War II. It describes how the health, development, and wealth of humanity have improved in recent decades (although inequitably), but at the cost of widespread degradation of natural systems.

The chapter outlines the key trends in the drivers and impacts of environmental change (including climate change, biodiversity loss, ocean acidification, soil degradation, land use change, and pollution), and discusses interactions between local and global trends and between different drivers.

Chapters 2 and 3 describe the impacts of planetary changes on human health. Chapter 2 focuses on perhaps the best recognized of these processes, climate change, which affects health through a range of pathways and mechanisms. These include direct effects such as increased exposure to extreme heat, droughts, and floods; those mediated by changes in ecosystems (such as changes in infectious disease exposure or nutritional effects of reduced food availability and quality); and socially mediated effects such as increased poverty, population displacement, conflict, and violence.

Chapter 3 explores a range of other global environmental changes affecting health including pollution, land use change, declining freshwater availability, and biodiversity loss. The current burden of environmental ill-health and the relationship between environmental causes of ill-health and development are discussed. This chapter also discusses how the drivers and consequences of environmental change are interlinked, such as how the combustion of fossil fuels and solid biomass threaten health by causing air pollution as well as contributing to climate change.

The concepts of risk, exposure, hazard, and vulnerability to the adverse effects of environmental change are discussed in Chapter 4, including how inequities in health and wealth influence vulnerability. It discusses the geographical, personal, socioeconomic, occupational, and other factors that influence vulnerability to global environmental change and the unfinished agenda of poverty-related environmental ill-health. For example, it outlines how increasing heat stress reduces the labour productivity of outdoor workers in tropical and sub-tropical regions and how sea level rise threatens the health and livelihoods of some coastal populations and those living on small islands. It discusses how multiple environmental stressors acting together can increase vulnerability to adverse health outcomes through a range of pathways, and the challenges of assessing vulnerability, including the assessment of individual and community risk factors that predispose to different health outcomes.

Chapter 5 begins our pivot toward solutions. Here we discuss the main conceptual, knowledge, and implementation challenges that need to be addressed to protect health. *Conceptual challenges* include the dominant focus on economic throughput, exemplified by over-reliance on the Gross Domestic Product as the main metric of human progress, and the dominance of biomedicine and curative approaches in health care systems, resulting in lack of attention to 'upstream' preventive approaches. *Knowledge challenges* are exemplified by the relative lack of research on Planetary Health issues and the poor quality of existing evidence to guide policy. *Implementation challenges* refer to our collective failure to put into policy and practice what we know. Among the many causes are the failure to tailor evidence to the needs of policymakers and organized opposition from vested interests.

Even with decisive action now we cannot prevent all environmental change. Chapter 6 addresses both how societies can become more resilient to environmental change and specific adaptation strategies that can help to protect health. It also stresses the importance of understanding the limits to adaptation (e.g. due to extreme heat, droughts, etc.) and how these limits vary by location and other factors. Here we explore the role of ecosystem approaches in adaptation and resilience (such as protecting wetlands or mangroves to prevent flooding), the potential for maladaptation (e.g. increasing exposure to infectious diseases from poorly designed adaptation), and the need to anticipate and manage inevitable trade-offs. The chapter also outlines what is known about national adaptation plans and their evaluation, including how these plans will change in the future depending on our ability to reduce – and ultimately halt – adverse environmental change.

The Sustainable Development Goals (SDGs) which frame the development agenda up to 2030 are discussed in Chapter 7. Although SDG 3 focuses specifically on health, with a strong emphasis on universal coverage with affordable health care, many of the other SDGs have important implications for health, for example by reducing poverty (SDG 1), improving food security and promoting sustainable agriculture (SDG 2), providing sustainable water and sanitation (SDG 6), increasing access to clean energy (SDG 7), and designing and operating sustainable cities (SDG 11). This chapter discusses how the SDGs can contribute to Planetary Health, outlines interactions among the goals that can support progress, and identifies potential trade-offs.

Providing universal access to clean renewable energy sources in the electricity generation, domestic, industrial, and transport sectors can reduce both household and ambient air pollution exposure, thus saving millions of lives in the near term, and at the same time help to decarbonize the economy. Chapter 8 discusses the health co-benefits of policies to provide access to non-polluting, climate-neutral sources of energy, and how valuing these can help offset the modest increase in costs associated with decarbonization. This chapter outlines steps towards a circular economy that reduces harmful emissions and incentivizes re-use, re-manufacturing, and recycling, and minimizes waste. It examines how resource use efficiency can be incentivized and how tax and subsidy reform can help direct the world economy towards a more sustainable course.

The majority of the world's population now lives in cities that account for about 75% of global greenhouse gas emissions and about 85% of global economic activity. The process of urbanization is still evolving rapidly in Africa and Asia. Chapter 9 discusses how appropriate policies can reduce the environmental footprint of cities and make them more pleasant, healthy places to live by reducing inequities and providing opportunities for productive employment. It focuses particularly on housing, transport, water and sanitation, and access to green spaces, describing how intersectoral policies can reduce the environmental impacts of urban living and improve health, and identifying trade-offs and conflicts where relevant. Through examples it shows how the challenges vary by region. For example, the rapidly growing cities of Sub-Saharan Africa can take advantage of innovations in sustainable transport systems, low-cost resilient housing in informal settlements, access to green space and clean energy. In high-income countries existing cities may need re-design to reduce the adverse effects of urban sprawl and road congestion.

A key priority is to reduce the environmental impacts of the food and agricultural system and provide healthy food choices for 9–10 billion people later this century, within environmental limits. Chapter 10 assesses the potential for changing behaviour at the individual and population levels to promote the consumption of healthy, low environmental impact diets and reduce food waste. It outlines potential opportunities and limitations of different policy and technological options, ranging from tax and subsidy reform to cellular agriculture, biofortification, genetically modified crops, and sustainable intensification policies. The chapter acknowledges the important contributions of small farms to food production in many countries, whilst recognizing the need for technological innovation. It gives a brief overview of the need for sustainable fisheries and aquaculture in the face of adverse trends.

Chapter 11 focuses on the health sector. Health systems and health professionals have a key role in addressing the challenges of the Anthropocene Epoch. One mandate is to provide universal health coverage that is resilient to environmental shocks. In addition, health care systems are responsible for substantial emissions of greenhouse gases and pollutants. Policies to reduce the environmental footprint of the health sector can also have a catalytic effect on other sectors. In part this is because health systems are large purchasers of energy, food, equipment, and supplies, so they can exert considerable market pressure. Moreover, health professionals are highly trusted voices, so their advocacy for sustainable practices can be widely influential. Finally, health professionals have a role to play in family planning programmes, which (together with the education of girls and the empowerment of women) are key strategies for slowing population growth.

The final chapter (12) considers why progress is so difficult. Barriers to change are diverse, including vested interests, disinformation, externalities, growing inequities, population growth, and consumerism, but these can and must be addressed. This chapter includes case studies of how local and national governments are addressing the challenges and how to use evidence in the 'post-truth era'. The chapter discusses policies and practices that can address the conceptual, knowledge, and implementation challenges introduced in Chapter 5.

A word on data: Planetary Health is a wide-ranging field, and pertinent data come from a multitude of sources – government agencies, multinational organizations, universities, research institutes, private firms, and others. Methods for collecting and analysing data are sometimes unstandardized, yielding different estimates from different sources. Published data are updated frequently. Some phenomena, such as the number of people affected by 'natural' disasters or the number of species that have gone extinct, are intrinsically difficult to quantify. While recognizing these challenges, we have strived to provide consistent and current data throughout this book. We acknowledge, though, that some inconsistencies invariably remain. We encourage readers who wish to verify quantitative points, or to seek updated figures, to consult the latest versions of the original sources we cite.

A book on Planetary Health is a vast undertaking. While we have ranged broadly across many environmental and social threats, we could not be comprehensive. A notable omission is the challenge of weapons of mass destruction. We are deeply concerned about the potential catastrophic effects of the use of chemical, biological, radiological, and nuclear weapons by accident or design. Nuclear weapons, in particular, have the potential to trigger global-scale changes acutely (4–6). The remedies lie largely in the field of arms control and are outside the scope of this book.

In recent years, it has become common to hear global challenges described as 'existential' and as emergencies. Neither characterization is hyperbolic. We are indeed at a critical juncture. Human societies must shift to a more sustainable trajectory to safeguard the health of today's population and of future generations in the Anthropocene Epoch. We hope that this book will contribute, in a small way, to the needed transition.

References

1. McMichael AJ. *Planetary Overload: Global Environmental Change and the Health of the Human Species*. Cambridge, UK: Cambridge University Press; 1993.
2. McMichael AJ. *Human Frontiers, Environments and Disease: Past Patterns, Uncertain Futures*. Cambridge, UK: Cambridge University Press; 2001.
3. McMichael AJ, Woodward A, Muir C. *Climate Change and the Health of Nations: Famines, Fevers, and the Fate of Populations*. Oxford and New York: Oxford University Press; 2017.
4. Helfand I, Haines A, Ruff T, et al. The growing threat of nuclear war and the role of the medical community. *World Medical Journal*. 2016;62(3):86–94.
5. Robock A, Toon OB. Local nuclear war, global suffering. *Scientific American*. 2010;302(1):74–81. doi: 10.1038/scientificamerican0110-74.
6. Jägermeyr J, Robock A, Elliott J, et al. A regional nuclear conflict would compromise global food security. *Proceedings of the National Academy of Sciences*. 2020;117(13): 7071–81. doi: 10.1073/pnas.1919049117.

Acknowledgements

We gratefully acknowledge the collaboration of many colleagues in research and scholarship to advance understanding of Planetary Health in general and specific environmental threats to health, as well as the potential policy options to address these challenges. Colleagues at the Centre for Climate Change and Planetary Health at the London School of Hygiene and Tropical Medicine have contributed greatly to advancing knowledge and we have drawn extensively on their work in writing this book. We salute the leaders and members of the Planetary Health Alliance which has helped propel the growth of this field in recent years. We thank Alan Dangour, Kristine Belesova, and James Milner for helpful comments on individual chapters and Sarah Lambert and Matt Lloyd at Cambridge University Press for their support and forbearance over the long gestation period for this book. We also acknowledge with gratitude the work of Emanuel Santos in preparing the figures and acquiring permissions, together with Margaret Zou's contributions in the early stages of the drafting process. Our families deserve a special vote of thanks for their unstinting encouragement and support throughout. In addition to the seminal contributions of Tony McMichael we would like to recognize the major contributions of other pioneers such as the late Paul Epstein, Georgina Mace, and Kirk Smith, with whom we had the immense privilege of collaborating and from whom we learned so much.

1

Our Changing Planet

The World is a healthier place than at almost any time in the past. People live longer, they are taller and stronger and their children are less likely to get sick or to die. (1)

A continuing trajectory away from the Holocene could lead, with an uncomfortably high probability, to a very different state of the Earth system, one that is likely to be much less hospitable to the development of human societies. (2)

These two quotes, whilst not necessarily incompatible, illustrate the disconnect between different perspectives of the present: is it a time of unprecedented peril or progress, or is it both? What is the state of our planet, and what does it mean for people around the world? How do we manage threats, build on genuine progress, and assure a healthy, habitable planet for future generations?

This book attempts to address these questions. We do so through the lens of human health and well-being. We hope in this way to fill a gap in the growing roster of publications that explore recent dramatic environmental changes and the driving forces behind them. Our premise is that the health of populations depends on the integrity of natural systems. These systems provide 'ecosystem services' such as freshwater and food, energy and building materials, and storm protection. Natural environments also provide meaning and inspiration to billions of people. Importantly, natural systems maintain planetary conditions within safe limits for humanity. However, many natural systems, both globally and regionally, have been altered and their vital functions compromised by human actions. These changes threaten to undermine progress, with far-reaching consequences for the health of today's and future populations.

To understand these transformational changes in natural systems, we need to analyse more than just their physical manifestations. We need to confront questions of history, politics, economics, and even human nature. Did humanity stumble unknowingly into the future we are creating, or did the current state of the planet result, to some degree, from deliberate decisions by parties who could have pursued a different and more benign development pathway but had strong incentives to exploit short-term benefits over future sustainability? In an era when 'facts' can be manipulated to serve powerful interests on a scale never previously witnessed, this line of inquiry is highly charged. Perceptions about the relative importance of a threat or an opportunity may be filtered by ideologies, coloured by preconceptions, and shaped by forces with private agendas. This book aims not merely to document the burgeoning threats to human health but also to suggest potential actions that could be taken to adapt to environmental change that cannot be prevented, and more importantly to halt, and where possible reverse, the environmental damage that has already occurred. We aspire to provide a clear trail of evidence to support our conclusions, and to be transparent about inevitable uncertainties. In doing so we draw on insights from diverse disciplines and academic traditions, without claiming to be expert in many of them. We invite you, the reader, if in doubt, to consult the original sources and make up your own mind.

The Long Arc of Human History

The story has to begin during the long (in human but not geological terms) period of the last nearly 12,000 years forming the Holocene Epoch, when humanity emerged from hunter-gatherer to agrarian communities and later into growing urban settlements founded on trade, and increasingly manufacturing. The Holocene was notable for its relative climatic stability, which allowed civilization as we know it to emerge. It was interrupted only by little ice ages – significant on human scale but minimally so on a geological scale. Much can be learned from the impacts of relatively modest fluctuations in climate on human society over this period (3). These lessons help us assess the likely effects of rapid climate and other changes on health and development in the future (see Chapter 2).

The industrial revolution occurred during the eighteenth and nineteenth centuries. While technology advanced in many places around the world, the epicentre was in Europe and North America. The steam engine emerged in the early eighteenth century. A vast range of industries, from textiles to iron and steel to shipbuilding, saw rapid innovation, mechanization, and growth. Machines began to replace human and animal labour on farms. An economic transition from agriculture to industry, population shifts from rural to urban, and population growth, all followed. A second wave, sometimes called the second industrial revolution, occurred during the late nineteenth and early twentieth centuries, featuring mass production on assembly lines, transportation advances including the internal combustion engine, and distributed electrical grid systems. While a full account of this history is beyond the scope of this chapter, one underlying theme is important to emphasize: the central role of energy (4). It was unprecedented access to concentrated energy, first from coal and later from petroleum and natural gas, that unlocked the revolutionary changes that ushered in the modern world – both its benefits and some of its greatest challenges.

Progress in health gathered pace during the late nineteenth and early twentieth centuries. Consider life expectancy in England, where good historical data are available. There was little change between 1550 and 1850 (although the life expectancy of ducal families began to rise around 1750) (5). Between 1850 and 1950 average life expectancy increased from 40 to 70 years, with most of the increase occurring from 1900 onwards (apart from a dip in 1918 due to the post-World War I influenza pandemic). The increase was observed in other countries for which there are data. It was due primarily to reduced chances of dying in childhood and predated the advent of many effective treatments for adult diseases.

Improved nutrition was certainly part of the story but not the only explanation. Economic growth alone is also an unlikely candidate because the improvements in child mortality were uniform across Europe despite the highly uneven onset of economic growth (see (1) for discussion). The most likely explanation is the adoption of public health measures, including improved sanitation, led by pioneering researchers such as John Snow, whose epidemiological study of cholera transmission in London showed how it was spread through faecal contamination of drinking water (6, 7).

By the mid-twentieth century, the post-World War II economic expansion brought a further growth in global population, increasingly rapid technological change (such as the development of the synthetic organic chemical and electronics industries), and increasing demands for raw materials, food, and energy. This scaling up of the human enterprise has been called the Great Acceleration (8). It is best captured by the iconic set of graphs that demonstrate human progress as assessed by a range of indicators (population, economic growth, resource use, urbanization, globalization, transport, and communication) (**Figure 1.1**).

Human Thriving in the Great Acceleration

It is obvious from the graphs in Figure 1.1 that this period has been largely a positive experience for much of humanity. The average life expectancy increased worldwide from 47 years in 1950–1955 to 70.9 years in 2010–2015 (9). Death rates in children younger than 5 years of age worldwide decreased substantially from an average of 214 per thousand live births in 1950–1955 to 39 in 2018 (10). According to World Bank estimates, in 2015, 10% of the world's population lived below the international poverty line, currently set at US$1.90 a day (in 2011 purchasing power parity (PPP) dollars), compared with 36% in 1990. Nearly 1.1 billion people have moved out of extreme poverty since 1990. In 2015, 736 million people lived on less than US$1.90 a day, down from 1.9 billion in 1990 (11). These advances have been driven largely by progress in East Asia (China) and the Pacific (Indonesia) as well as South Asia (India), with only modest progress in Sub-Saharan Africa, where most people in absolute poverty now live. This admittedly unequal escape from poverty (1) has been accompanied by unparalleled advances in public health, health care, education, human rights legislation, and technological development that have brought great benefits to multitudes of people.

The second half of the twentieth and early twenty-first centuries brought many advances in public health and health care. Immunization coverage against common infectious diseases expanded, promoted by the World Health Organization (WHO) Expanded Programme on Immunization from 1974. Primary health care workers delivered a growing stock of effective treatments such as oral rehydration therapy for diarrhoeal diseases. Advances in contraceptive technology and availability, and improvements in the education of girls (in some countries), helped to lower fertility, thus reducing the risks of pregnancy and childbirth. Again there is no consistent relationship between economic growth and improvements in health indicators such as infant mortality. The relatively recent declines in infectious diseases such as HIV/AIDS, malaria, and tuberculosis reflect effective disease control

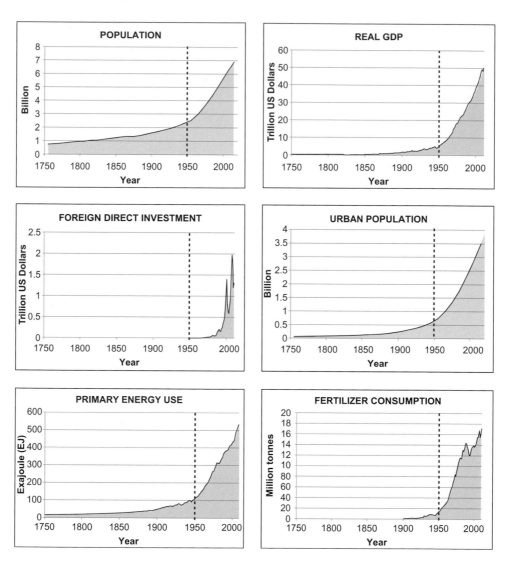

Figure 1.1. The Great Acceleration. These graphs show the rapid upscaling of the human enterprise, using a wide range of indicators. Dramatic growth began in the nineteenth century and accelerated rapidly in the second half of the twentieth century.

Source: International Geosphere–Biosphere Programme. www.igbp.net/news/pressreleases/pressreleases/planetarydashboardshowsgreataccelerationinhumanactivitysince1950.5.950c2fa1495db7081eb42.html.

measures together with investments in research and health care more than they reflect Gross Domestic Product. In high-income countries declines in non-communicable disease mortality (including ischaemic heart disease, stroke, and some types of cancer) reflect declines in smoking and improved prevention and treatment, particularly in primary care settings. The epidemiological transition in many low- and middle-income countries (LMICs), with its growing burden of non-communicable disease, is partly due to declines in infectious disease

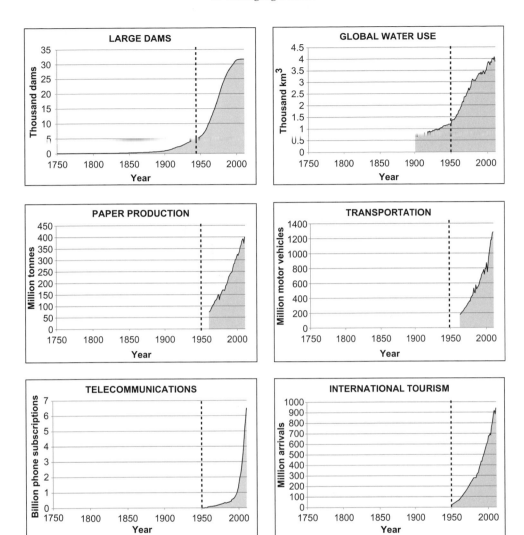

Figure 1.1. (*cont.*)

mortality (12). Pathways linking communicable and non-communicable diseases and the environment are discussed in Chapter 2.

In summary, humanity has benefitted greatly, although inequitably, from advances in the past two to three centuries. But these benefits have come at a cost, which has been borne particularly by the Earth's natural systems.

A Changing Planet

The growing human presence on Earth, depicted in Figure 1.1, is associated with another set of changes, shown in **Figure 1.2**. These are changes to Earth systems; systems that

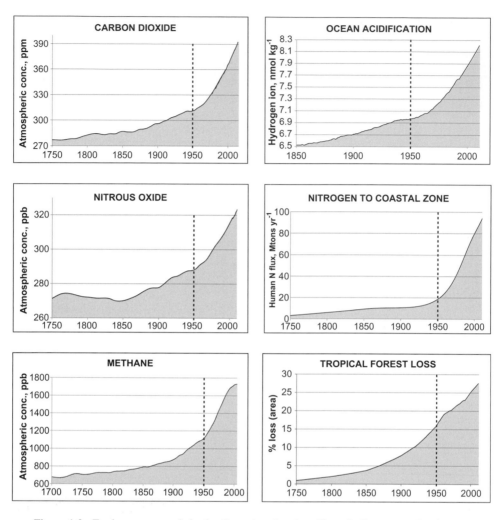

Figure 1.2. Earth system trends in the Great Acceleration. These indicators can be thought of as the planet's vital signs. All show significant and rapid changes as part of the Great Acceleration, especially since the mid-twentieth century.

Source: International Geosphere-Biosphere Programme. www.igbp.net/news/pressreleases/pressreleases/planetarydashboardshowsgreataccelerationinhumanactivitysince1950.5.950c2fa1495db7081eb42.html.

are so intrinsic to the Earth's balance of energy and cycling of materials that they can be likened to an organism's metabolism, and the indicators in Figure 1.2 to rapid changes in vital signs.

Geological epochs ordinarily persist for millions of years but contemporary planetary changes are so rapid and far-reaching that they demarcated a new geological epoch less than 12,000 years into the Holocene. Nobel laureate Paul Crutzen, a Dutch atmospheric chemist, proposed the term Anthropocene to describe these dramatic changes (13). The transition of a geological epoch is marked by a detectable global signal in fossils, rock, or

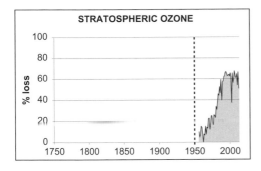
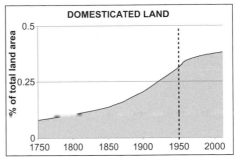
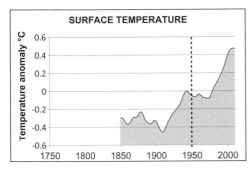
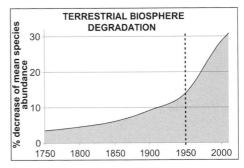

Figure 1.2. (*cont.*)

both, such as the layer of iridium in sediments around the world from 66 million years ago, when a meteorite collided with Earth, wiped out the dinosaurs, and ended the Cretaceous Epoch. Scientists have debated the starting point of the Anthropocene. Crutzen pointed to the early nineteenth century when, during the industrial revolution, large-scale fossil fuel (especially coal) combustion initiated the rise of the atmospheric carbon dioxide concentration from 284 parts per million (ppm), the Holocene maximum (14). (The level is now well above 400 ppm and rising.) Others have suggested inflection points as early as the European conquest of the Americas (because the demographic collapse of Amerindian populations from infectious disease, war, and forced labour left enduring traces in the geosphere and biosphere (15, 16)), or even earlier, back at the dawn of human agriculture. Other commentators have proposed that the point of inflection is as recent as the mid-twentieth century (when multiple changes in the nitrogen, phosphorus, and carbon cycles were observed, and when nuclear weapons testing left indelible traces) (17). Irrespective of the starting point of the Anthropocene Epoch, it is clear that profound changes in Earth systems are underway, and that they have significance not only for those of us alive today, but on geological timescales. A brief summary follows; the Rockefeller Foundation–*Lancet* Commission report (18) provides a more complete discussion of trends and primary data sources.

Land use changes and soil degradation (18, 19): About one-third of the Earth's ice-free and desert-free land surface has been converted to cropland or pasture and 2.3 million km^2 of primary forest have been cut down since 2000, particularly in tropical regions. (In temperate regions much of the extensive deforestation occurred historically and in some places reforestation is occurring.) Roads are being built throughout the world's remaining forests, and habitats are being fragmented – so much so that, by one estimate, 70% of remaining forest is within 1 km of a forest edge (20). In addition, urbanized land is expanding in both large and small–medium conurbations around the world, much of it in prime locations such as along rivers where it replaces cultivated farmland or natural vegetation (21). Soil degradation, which renders land unfit for cultivation, affects about 1–2.9 million hectares (10,000–29,000 km^2) of agricultural land annually, with some estimates suggesting much larger affected areas. This also reduces soil carbon, increases susceptibility to flooding, and reduces soil microbial diversity. More than 50% of desertification is due to underlying soil degradation from human activities.

Biodiversity loss: Overall, biodiversity loss is occurring at vastly increased rates compared with pre-human times, amounting to as much as a 1000-fold increased rate of extinctions, which has not been equalled for 66 million years (22). Species abundance of vertebrates declined by 58% between 1970 and 2012, with the largest recorded decline (81%) in freshwater systems. A recent census of biomass on Earth (23) shows that the biomass of humans (0.06 gigatonnes of carbon, Gt C) is an order of magnitude greater than that of all wild mammals combined (≈ 0.007 Gt C), and the biomass of livestock exceeds them both (≈ 0.1 Gt C, dominated by cattle and pigs). This also applies to wild and domesticated birds, for which the biomass of domesticated poultry (≈ 0.005 Gt C, dominated by chickens) is about three-fold higher than that of wild birds (≈ 0.002 Gt C).

Freshwater appropriation: Humans exploit approximately 50% of all accessible freshwater annually. In the relentless search for energy and water for irrigation and other uses, over 60% of the world's rivers have been dammed, amounting to more than 0.5 million km in total length, and many more dams are planned (24). Aquifers, reserves of freshwater built up over millennia that cannot be replenished in the foreseeable future, are being depleted rapidly in many regions. For example, the Arab world has experienced a 75% decrease in per capita freshwater availability since 1962 and the total demand is 16% higher than available renewable freshwater resources. These trends imply that increasing numbers of countries will become water stressed over coming decades (**Figure 1.3**) with serious implications for the irrigation of crops, industrial production, and energy supply.

Changing oceans: Oceans too are subject to rapid and extensive pressures with about 90% of wild-catch fisheries fully or overexploited. According to the Food and Agriculture Organization (FAO) data, global fish catches peaked in 1996 and have declined slowly since, but independent estimates suggest a higher peak catch and subsequent faster decline, reflecting overfishing rather than reductions in catch to allow restocking of fish populations (25). The oceans act as a sink for both carbon dioxide and for increasing heat – reducing the rate of atmospheric warming – and as a result are undergoing profound changes. As carbon dioxide rises in the atmosphere and dissolves in seawater, the oceans become more acidic. The pH of ocean surface water has decreased by 0.1 since the beginning of the industrial era, equivalent to a 26% increase in hydrogen ion concentration (26). Increasing acidity has many, but still incompletely understood, implications for ecosystems and for humanity. It results in thinning of the shells of crustaceans as they

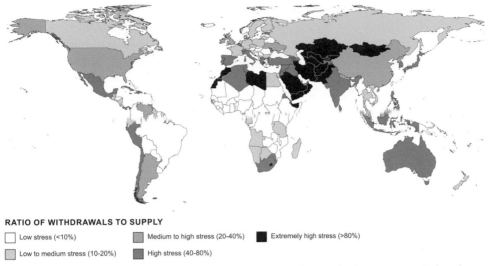

RATIO OF WITHDRAWALS TO SUPPLY

☐ Low stress (<10%)   ▨ Medium to high stress (20-40%)   ■ Extremely high stress (>80%)

▨ Low to medium stress (10-20%)   ▨ High stress (40-80%)

Figure 1.3. Projected water stress by country in 2040. These projections assume continued high greenhouse gas emissions (RCP8.5) and a 'middle of the road' development pathway that largely continues historical patterns (SSP2).

Source: World Resources Institute. www.wri.org/resources/charts-graphs/water-stress-country.

find it more difficult to lay down calcium carbonate, and contributes to the degradation of coral reefs which are subject to multiple environmental stressors. Heating is contributing to coral bleaching (27) and changing the distribution of fish stocks, resulting in movement of some fish species from equatorial to higher latitudes (28). Many coastal ecosystems are also subject to high loading from nitrogen, phosphorus, and carbon run-off from agricultural land and sewage outflows, contributing to local acidification. In some cases these inflows lead to 'dead zones' where high levels of nutrients result in eutrophication, characterized by reduced oxygen levels, the growth of toxic algae, and loss of biodiversity (29).

Climate: Our changing climate is the most prominent and widely known transformation that confronts humanity in the Anthropocene Epoch. Climate change is due to the accumulation of greenhouse gases (GHGs) in the atmosphere driven by human activities, notably the burning of fossil fuels. (In general the term climate change is preferred to global warming because the effects are wider than just temperature increases and include changes in rainfall patterns and sea level rise. The term 'climate change' is being superseded by 'climate crisis' or 'climate emergency' in many non-technical publications because of the growing risks, and 'global warming' may be replaced by 'global heating' for similar reasons.) Carbon dioxide (CO_2) is the most important GHG. According to the World Meteorological Organization (WMO) (30), the last time the Earth experienced a comparable atmospheric concentration of CO_2 was 3 to 5 million years ago when global mean temperatures were 2 to 3 °C higher than today and sea level was 10 to 20 metres higher. The WMO concludes that the rate of increase of atmospheric CO_2 over the past 70 years is nearly 100 times larger than that at the end of the last ice age and the abrupt changes in the atmosphere witnessed in the past 70 years are without precedent. Carbon dioxide persists for long periods in the atmosphere, with around 20% remaining for 1000 years or more – a troubling legacy for future generations. Short-lived climate pollutants (SLCPs) include methane, black carbon (a type of fine particle from incomplete burning of solid fuels), and hydrofluorocarbons, or HFCs (synthetic compounds used for several purposes including air conditioning, refrigerants, solvents; replacing the ozone-depleting chemicals chlorofluoro-carbons, or CFCs). The SLCPs are important because reductions in emissions can deliver immediate reductions in climate forcing, in contrast to long-lived GHGs such as CO_2, whose presence, and impact, will linger for centuries.

In 2015, at the annual conference of the United Nations Framework Convention on Climate Change (UNFCCC) in Paris, the nations of the world reached agreement to limit global heating to well under 2 °C global average temperature rise above pre-industrial levels (with the further aspiration to limit it to 1.5 °C) and pledged to set goals to reduce GHG emissions. While the Paris Agreement marked an important step forward, the Nationally Determined Contributions (NDCs) to which nations committed are, in the aggregate, insuffi-cient to prevent warming exceeding 2 °C. Recent estimates (31) suggest that even if the NDCs are implemented (an unlikely outcome given current trends) the ensuing increase in global average temperature will reach around 3.2 °C.

Consumption patterns and population growth: The human presence on Earth, unlike temperature, pH, and the other domains just discussed, is not a biophysical measure. But it underlies each of the other planetary changes. A key driver is human population growth (**Figure 1.4**). At the turn of the twentieth century, the human population was 1.6 billion. In 1950, shortly after the end of World War II, it was 2.5 billion. By 2000, it had reached 6.1 billion, and by 2020, it surpassed 7.7 billion (adding as many people in 20 years as had lived on the planet in 1900). According to UN mid-level projections (9), the global population will continue to grow, to about 8.5 billion in 2030, 9.7 billion in 2050, and 10.9 billion in 2100. Over half this growth will be in Sub-Saharan Africa and India, much of it in cities. (As discussed in Chapter 4 and shown in Figure 4.4, different assumptions about socioeconomic and population growth yield different planetary scenarios.)

Importantly, population explains only part of humanity's demands on planetary resources. The per capita use of these resources – mediated by affluence, technology, efficiency, and waste – is a major driving factor. Consumption varies widely, as reflected in differences in per capita CO_2 emissions that may vary by over 100-fold between countries (32). Fifty years ago, biologist Paul Ehrlich and environmental scientist John P. Holdren expressed this idea in simple terms with the equation $I = PAT$, in which environmental impacts (I) are a function of population (P), the intensity of individual consumption patterns (A for affluence), and the technologies used to produce and consume (T) (33). Some critics have dismissed this equation as overly simplistic, because it omits such phenomena as interactions among the factors, environmental resilience, and tipping points. But it conveys a powerful point: that addressing both consumption and population is essential to addressing the challenges of the Anthropocene.

With the Great Acceleration, global prosperity has dramatically increased. In the half-century following World War II, the number of people living in extreme poverty began to decline for the first time in human history, while the number of people not in extreme

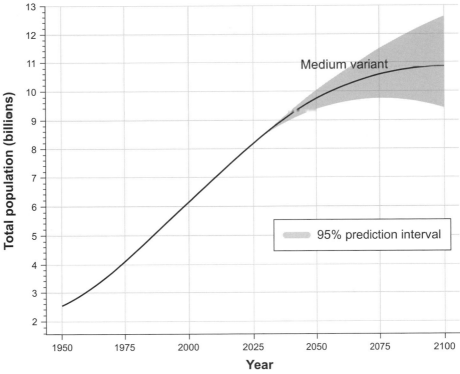

Figure 1.4. Projected global population, in billions.
Source: United Nations, Department of Economic and Social Affairs, Population Division; 2019. *World Population Prospects 2019*. https://population.un.org/wpp/.

poverty grew rapidly (**Figure 1.5**). Poverty is highly inequitably distributed geographically, as discussed below. But with billions of people emerging from poverty, seeking to eat more animal-source protein, fly to distant destinations, drive cars, utilize mobile phones, and store data, the combination of population and prosperity represents a phase change for the planet, with our collective footprint far exceeding the capacity of the planet to support us (**Box 1.1**).

Great care is taken to ensure that scientific research is conducted under ethical principles including informed consent and minimization of risks. However, heedless disruptions of Earth systems are essentially multiple poorly regulated experiments at regional and global scales, the outcomes of which we are now beginning to experience. These 'experiments' are not seen as 'research' but as 'development', with the onus of regulating potentially damaging approaches falling on frequently inadequate systems of governance and accountability.

The advent of the Anthropocene Epoch marks a discontinuity in human history and complicates attempts to predict and project future trends in human development. The dramatic changes seen in the Earth's vital signs raise the possibility that, in achieving (admittedly inequitable) socioeconomic advances for humanity, we have mortgaged the future. These trends challenge conventional thinking and point to the need to live within

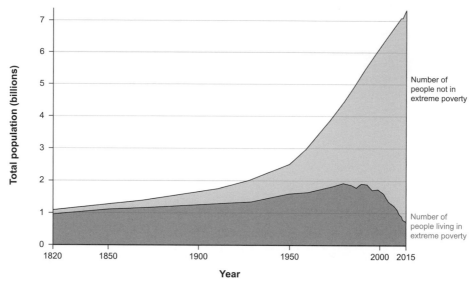

Figure 1.5. Global population living in extreme poverty, 1820–2015. Extreme poverty is defined as living on less than US$1.90 in international dollars per day, adjusted for price differences among countries and for inflation over time.
Source: Our World in Data. https://ourworldindata.org/extreme-poverty.

Box 1.1. **The Ecological Footprint and Biocapacity**

Two useful concepts for considering humanity's use of planetary resources are the ecological footprint and biocapacity. The ecological footprint measures the environmental resources a population requires to produce the goods and services it uses and to absorb its waste (including carbon emissions). Examples of what people need to produce goods and services include metal and stone, plant-based food and fibre products, livestock and fish, timber and other forest products, and space for urban infrastructure. These in turn are drawn from productive surface areas, six of which the ecological footprint tabulates: cropland, grazing land, fishing grounds, built-up land, forest area, and area needed to absorb CO_2 emissions from energy use. Population size and affluence are the main drivers of environmental footprint, while urbanization, age distribution, and economic structure (a country's position on the development continuum from extractive industries and manufacturing towards a service economy) have less effect (34).

While ecological footprint is a demand-side concept, reflecting human consumption patterns, biocapacity is a supply-side concept. Biocapacity measures nature's ability to absorb our waste and generate new resources – to support us by providing 'ecosystem services'. This corresponds to the productivity of ecological assets – cropland and forests. The biocapacity of a place can be enhanced or degraded, depending upon how it is managed.

Both ecological footprint and biocapacity are expressed in units of hectares, so they can be compared with each other. If a population's ecological footprint exceeds the biocapacity of the region it inhabits, that population is not living within sustainable limits, and is said to run an ecological deficit. This requires importing from other regions, depleting the region's ecological assets (e.g. overfishing), and/or emitting waste (e.g. CO_2) into the atmosphere. If on the other hand a region's biocapacity exceeds its ecological footprint, it is said to have an ecological reserve.

The biocapacity needed to produce the natural resources and services that humanity consumed in 2016 is equivalent to that provided by 1.7 planet Earths – a powerful testament to the unsustainable trajectory of current development pathways. As of 2020 human demand on the Earth's ecosystems is estimated to exceed the regenerative capacity of nature by 75%.

Still another way to express the idea of resource use relative to limits is to calculate a country's 'Overshoot Day' – the day when its use of resources exceeds the amount it would use in one year if it were living within ecological limits (see **Figure 1.8**, below). A country that reaches Overshoot Day early in the year is using resources well in excess of what is sustainable, while a country that reaches Overshoot Day near the end of the year is living within its ecological means. There is also a global Overshot Day, which varies over time with global resource use. During the decade 2010–2019, that day varied within a few days of 3 August, but in 2020 it was delayed until 22 August by the COVID-19-related global economic slowdown. While the concept can be criticized on the grounds that it combines different impacts, it provides a readily understandable measure of sustainability as a single figure.

Sources:
- Wackernagel M, Beyers B. *Ecological Footprint: Managing Our Biocapacity Budget.* Gabriola Island, BC: New Society Publishers; 2019.
- Global Footprint Network. www.footprintnetwork.org/our-work/ecological-footprint.
- Earth Overshot Day. www.overshootday.org/.

finite constraints of a small planet at a time of increasing demand for resources from an expanding world population. The first step towards that end is to define as far as possible the safe environmental limits within which humanity has a reasonable prospect of flourishing.

Planetary Boundaries

Growing understanding of the pervasive changes in Earth's natural systems has led to attempts to define a safe operating space within which humanity can flourish. A team based at the Stockholm Resilience Institute proposed nine planetary boundaries in 2009 (35), and has continued to refine and improve the concept since then (2) (**Figure 1.6** and **Table 1.1**). The current concept proposes two core boundaries – climate and biosphere integrity – that cut across all the others, providing connections among them and operating at the level of the whole Earth system, having co-evolved over 4 billion years or so. They are both regulated by the other processes and provide a broad framework within which the other boundaries operate. Ecosystems (terrestrial, marine, and freshwater) and their biota, which constitute the biosphere, regulate the flows of energy and materials, thus determining the responses of the Earth system to changes in a range of processes.

Several aspects of the planetary boundaries are important. First, they are based on our evolving understanding of the conditions that allowed humanity to thrive over the course of the Holocene Epoch. Second, the various boundaries, and the driving forces that threaten to

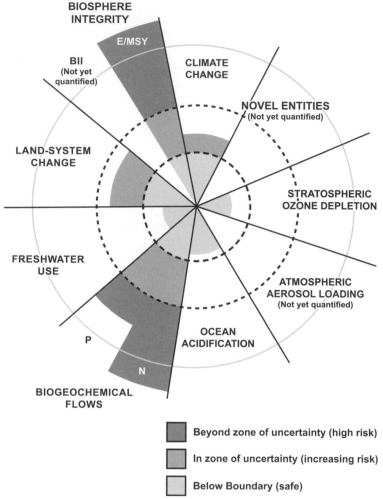

Figure 1.6. Planetary boundaries. The nine processes shown in the diagram regulate the Earth's stability and resilience. If humanity operates within boundaries for each domain, we can continue to develop and thrive for generations to come. But transgressing the boundaries – as has occurred with biosphere integrity and biogeochemical flows – increases the risk of large-scale abrupt or irreversible environmental changes.

Key: Under biosphere integrity, BII is the biodiversity intactness index, representing functional biodiversity, and E/MSY is extinctions per million species-years, representing genetic biodiversity. Novel entities are chemicals such as organic pollutants, radioactive materials, nanomaterials, and microplastics that are anthropogenic, generally persistent, and harmful.

Source: Stockholm Resilience Institute. www.stockholmresilience.org/research/planetary-boundaries.html.

breach them, interact, reflecting the concept that the Earth is essentially a single integrated complex system. For example, climate change accelerates biodiversity loss because many species cannot adapt quickly enough to changes in temperature and precipitation, including by being unable to migrate quickly enough to keep pace with rapid changes. Similarly,

Table 1.1. *The nine planetary boundaries: risks of exceeding each boundary, control variables, proposed values and zones of uncertainty, and current status*

Earth system process	Risk of exceeding boundary	Control variable(s)	Planetary boundary (zone of uncertainty)	Current status
Climate change	Loss of polar ice sheets. Regional climate disruptions. Loss of glacial freshwater supplies. Weakening of carbon sinks.	Atmospheric CO_2 concentration, ppm; Energy imbalance at top of atmosphere, W/m^2	350 ppm CO_2 (350–550 ppm) +1.0 W/m^2 (+1.0–1.5 W/m^2)	415 ppm CO_2 (2020) 2.3 W/m^2 (1.1–3.3 W/m^2)
Change in biosphere integrity (biodiversity)	Disrupted ecosystem functioning at continental and ocean basin scales. Impact on many other boundaries – C storage, freshwater, N and P cycles, land systems.	Genetic diversity: Extinction rate; Functional diversity: Biodiversity Intactness Index (BII)	<10 E/MSY (10–100 E/MSY) with an aspirational goal of ~1 E/MSY (the background rate of extinction loss). Maintain BII at 90% (90–30%) or above, assessed geographically by biomes/large regional areas (e.g. southern Africa), major marine ecosystems (e.g. coral reefs), or large functional groups.	100–1000 E/MSY 84%, analysed for southern Africa only.
Stratospheric ozone depletion	Severe and irreversible UV-B radiation effects on human health and ecosystems.	Stratospheric ozone concentration, DU	<5% reduction from pre-industrial level of 290 DU (5–10%).	Only transgressed over Antarctica in Austral spring (~200 DU); currently improving.
Atmospheric aerosol loading	Disruption of monsoon systems. Human health effects. Interacts with climate change and freshwater boundaries.	Global: AOD (but with much regional variation); Regional: AOD as a seasonal average over a region	Global: To be determined. Regional (South Asian Monsoon as a case study): AOD over Indian subcontinent of 0.25 (0.25–0.50), absorbing (warming) AOD <10% of total AOD.	0.30 AOD over South Asia

15

Table 1.1. (*cont.*)

Earth system process	Risk of exceeding boundary	Control variable(s)	Planetary boundary (zone of uncertainty)	Current status
Ocean acidification	Conversion of coral reefs to algal-dominated systems. Regional loss of some marine biota.	Carbonate ion concentration, Ω_{arag}	Sustain ≥80% (≥80%–≥70%) of the pre-industrial mean surface ocean Ω_{arag}	~84% of the pre-industrial aragonite saturation state
Biogeochemical flows (phosphorus and nitrogen cycles)	P: Oceanic anoxic events with impacts on marine ecosystems. N: Reduced resilience of ecosystems via acidification of terrestrial ecosystems and eutrophication of coastal and freshwater systems.	P global: P flows from freshwater systems into oceans. P regional: P flows from fertilizers to erodible soils. N global: industrial and intentional biological fixation of N	11 Tg P/year (11–100 Tg P/year). 6.2 Tg P/year (6.2–11.2) mined and applied to erodible (agricultural) soils (global average that varies regionally). 62 (62–82) Tg N/year globally; regional distribution of fertilizer N is critical.	~22 Tg P/year ~14 Tg P/year ~150 Tg N/year
Land-system change	Trigger of irreversible and widespread conversion of biomes to undesired states. Reduced carbon storage and resilience via changes in biodiversity and landscape heterogeneity.	Global: forested land area as percentage of original forested area. Biome: forested land area as percentage of potential forest	Global: 75% (75–54) (a weighted average of tropical, boreal, and temperate forests) Biome-specific: • Tropical: 85% (85–60) • Temperate: 50% (50–30) • Boreal: 85% (85–60)	62%
Global freshwater use	Could affect regional climate patterns (e.g. monsoon behaviour) by affecting moisture feedback, biomass production, carbon uptake by terrestrial systems and reducing biodiversity.	Consumptive blue water use (km³/year)	Global: 4000 km³/year (4000–6000 km³/year)	~2600 km³/year

| Chemical pollution ('Novel entities' see Figure 1.6 above) | Thresholds leading to unacceptable impacts on human health and ecosystem function possible but largely unknown. | Emissions, concentrations, or effects on ecosystem and Earth system function of such entities as persistent organic pollutants (POPs), plastics, endocrine disruptors, heavy metals, and nuclear waste. | To be determined. |

Key: ppm: parts per million, W/m^2: watts per square metre; E/MSY: extinctions per million species-years; DU: Dobson unit, defined as 0.01 mm thickness at standard temperature and pressure; AOD: aerosol optical depth; Ω_{arag}: average global surface ocean aragonite saturation state. (Aragonite is a form of calcium carbonate formed by many marine organisms. Acidification lowers Ω_{arag}; at $\Omega_{arag} < 1$, aragonite dissolves, threatening the integrity of these organisms.)

Source: Adapted from Steffen W, Richardson K, Rockstrom J, et al. Planetary boundaries: guiding human development on a changing planet. *Science.* 2015;347(6223):1259855, and Stockholm Resilience Centre. www.stockholmresilience.org/research/planetary-boundaries/planetary-boundaries/about-the-research/quantitative-evolution-of-boundaries.html.

ocean acidification depends directly on the level of CO_2 in the atmosphere; there would be no risk of exceeding the boundary if CO_2 levels had not exceeded 350 ppm. Third, different boundaries operate on different spatial scales – continental or global in some cases, regional or local in others. Some regional changes have global implications; for example, the saturation or degradation of regional carbon sinks (such as forests or wetlands) increases CO_2 levels in the atmosphere and accelerates climate change. Fourth, while 'control variables' are proposed for each boundary, there are inevitable uncertainties about the exact quantification of boundaries. Nevertheless, there is value to the notion of quantitative boundaries, set at levels that account for uncertainties and that allow for inevitable lags in implementing protective policies before reaching dangerous tipping points (Box 1.2).

The climate boundary: There is a risk that, even at the 2 °C target of heating agreed in Paris, feedbacks could come into play which would amplify warming and set the Earth on a highly dangerous course (38). These feedbacks can include permafrost thawing releasing methane, Amazon and boreal forest dieback and, most significantly, weakening of the carbon sinks on land and in the oceans leading to reduced ability to take up carbon. These feedbacks combined could lead to an estimated mean additional 0.47 °C (0.24–0.66) of heating (above that due to greenhouse gas emissions) by the end of the century and could become self-perpetuating. There is potential for both continuous change, e.g. melting of large masses of ice, and for sudden tipping points (**Box 1.2**). The number of abrupt shifts, particularly in terrestrial systems, is projected to increase in high GHG emission compared with low-emission scenarios (39). Cascading tipping points may occur at different levels of heating; irreversible changes to the Greenland and West Antarctic ice sheets, Arctic summer sea-ice, Alpine glaciers, and coral reefs may occur at 1–3 °C heating above pre-industrial levels (indeed, may be underway now), while tipping points in the great ocean currents, Amazon and boreal forests, the Sahel, Indian summer monsoons, and the El Niño Southern Oscillation are unlikely to occur until 3–5 °C warming is reached, and irreversible loss of Arctic winter sea-ice, Northern Hemisphere permafrost, and the East Antarctic ice

---

Box 1.2. **Tipping Points**

A **tipping point** in an Earth system is a point at which sudden, non-linear, shifts may occur, with high risk of adverse effects. Tipping points represent a sudden change from one state to another, which may be difficult or impossible to reverse and result in major reductions in ecosystem services important for human societies. Examples are collected by the Regime Shifts database (http://regimeshifts.org/) which also provides evidence that the shift has occurred and for the mechanisms behind the shift as well as the potential for reversibility. Some of these shifts are predominantly local phenomena but many are linked to underlying global processes – for example the move to an ice-free Arctic in summer months due to climate change, or coastal eutrophication from increased inflows of nitrogen and phosphorus, which leads to murky, high nutrient waters with lower oxygen levels and reduced biodiversity. Tipping points may cascade, meaning that one may trigger another (36) (**Figure 1.7**). Because approaching tipping points leads to high risks of serious, and potentially catastrophic, consequences, sensible public policy entails a precautionary approach that aims to avoid reaching them (37).

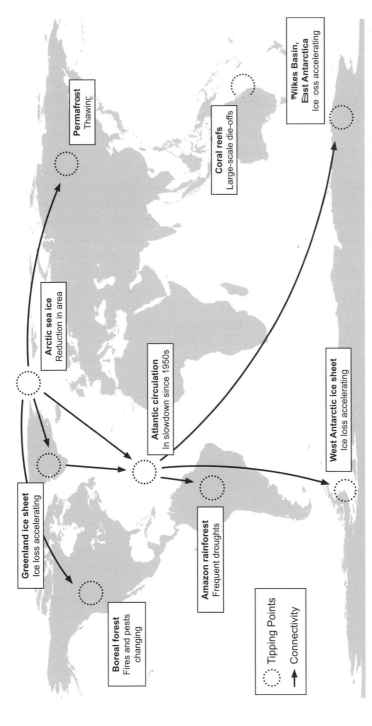

Figure 1.7. Potential tipping cascades from climate change. Many planetary systems are subject to irreversible changes, or tipping points. Potential interactions among the tipping elements that could generate cascading domino effects are shown by arrows.

Source: Lenton TM, Rockström J, Gaffney O, et al. Climate tipping points – too risky to bet against. *Nature.* 2019;575:592–5.

sheet are unlikely until more than 5 °C warming has occurred (38). The various tipping points are inter-related; for instance, collapse of ocean circulation would hasten loss of Antarctic ice, contributing to rising sea levels (**Figure 1.7**). The implication of this analysis is that we may approach a planetary threshold above which there are unacceptable risks of cascading effects even at temperature rises compatible with the Paris Agreement.

The biodiversity boundary reflects two key roles of the biosphere within the overall Earth system. The first is genetic diversity, which determines the capacity for further evolution to adapt to changing conditions, and the second is functional diversity, which is 'the value, range, distribution and relative abundance of the functional traits of the organisms present in an ecosystem or biota' (40). The appropriate metric for genetic diversity would be phylogenetic species variability (PSV), the 'information bank' of genetic variability that represents the capacity for continued evolutionary adaptation. However, global PSV data are not available, so the metric of genetic diversity is the global extinction rate. This metric has serious limitations: our inadequate understanding of baseline extinction rates, the difficulty of measuring extinction rates across all taxa, and the time lag it entails. The target is set at ten extinctions per million species-years (E/MSY), a ten-fold increase over the presumed background rate (1 E/MSY) – a level that may or may not reflect a level of loss that can be sustained without significant irreversible damage to the Earth system. Ideally the extinction rate would be kept to the background rate. The metric of functional biodiversity is the Biodiversity Intactness Index (BII), which assesses population abundance across a wide range of key taxa or functional groups in a given biome or ecosystem, as a percentage of the pre-industrial abundance. The BII has only been applied to a few countries and there is much uncertainty (90–30%) about the appropriate boundary level, but the upper limit is suggested as the boundary to reduce risks.

Nitrogen and phosphorus cycling: Nitrogen and phosphorus are key elements in the biochemical cycles that support life – components of DNA and RNA, of the energy-carrier adenosine triphosphate (ATP), of proteins, and more. Both elements cycle through biotic and abiotic systems naturally, but human activities – principally the manufacturing and distribution of fertilizers – have distorted these cycles significantly and now dominate global N and P cycling (41, 42). The run-off of fertilizer, human and animal waste, and other N and P sources into waterways can lead to overabundance of these nutrients – eutrophication – which in turn drives overgrowth of phytoplankton (cyanobacteria, dinoflagellates, and especially algae). As these organisms die and decompose, oxygen in the water is depleted, creating dead zones. Substantial dead zones are found in the Baltic and Black Seas, the Gulf of Mexico, the Bay of Bengal, the US Great Lakes, and numerous other coastal and inland waterways. Since the main contributions arise from a limited number of agricultural regions, redistribution from areas of excessive use to areas in which fertilizers are underused could increase global crop yield and prevent exceedance of N and P boundaries at the regional scale.

Land system changes: Of the many land types – forest, woodland, savannah, grassland, tundra, and so on – the focus of the land system boundary is forests, because of the critical impacts of forests on global climate. Forests help regulate climate through changes in albedo (reflectance of sunlight) and cooling from evapotranspiration, and act as carbon sinks. The boundary is set as a percentage of original forest cover remaining. Tropical and boreal forests play a more vital role than temperate forests in climate regulation, and are therefore disproportionately weighted in the land use boundary. Forest protection also plays a key role in biodiversity conservation and vice versa, such that if the BII boundary of 90% were respected, the achievement of the forest cover boundary would be assured.

Stratospheric ozone depletion is a qualified success story. The Montreal Protocol on the regulation of ozone-depleting substances such as chlorofluorocarbons (CFCs) has led to the progressive elimination of these substances with the aim of limiting the damage that they cause to the Earth's ozone layer (43). The ozone layer in the stratosphere is a shield against damaging levels of ultraviolet radiation reaching the Earth's surface. The Montreal Protocol was the first treaty in the history of the United Nations to achieve universal ratification, making it arguably the most successful environmental global action, one which was agreed in advance of incontrovertible proof of the cause of the ozone hole because of the consensus that it was wise to act in a precautionary manner. The sustained damage to the ozone layer is due to the long lifetimes in the atmosphere of ozone-depleting chemicals. Happily, recent evidence shows that during the first two decades of the present century there was a decline of about 0.8% annually in reactive chlorine species that are responsible for the ozone depletion (44). But recent evidence of illegal CFC emissions from eastern China reinforce the need for vigilance (45). This concern is further heightened by evidence that the Protocol's success is being undermined by new, unexpected emissions, not only of several CFCs but also of carbon tetrachloride and hydrofluorocarbons (46). The latter were introduced as non-ozone-depleting replacements for CFCs but are potent greenhouse gases.

Atmospheric aerosol loading: Atmospheric aerosols represent a wide range of liquid, solid, or mixed particles, including particles that are directly emitted into the atmosphere and gaseous molecules that are converted into particulate-phase species. They derive from natural sources such as volcanoes and deserts, and from human activities such as fossil fuel combustion and agriculture (linking them closely to climate change). Chemically, atmospheric aerosols are divided into inorganic species (salts, metals), carbonaceous compounds, and water. There are direct human health impacts, including millions of deaths each year (see Chapter 2), as well as regional and global environmental impacts. The boundary for aerosol atmospheric loading is based on the regional impacts of aerosols on ocean–atmosphere circulation – specifically, on the threshold at which aerosol loading could cause the Indian monsoon to switch to a drier state, with potentially far-reaching effects on the regional economy and agricultural production.

Novel entities: Finally, the introduction of novel entities (i.e. new substances, new forms of existing substances, novel life forms) is proposed as a boundary because of the potential for unwanted effects on the environment and human health. The risks were exemplified by the introduction of CFCs, discussed above, which were thought to be unreactive but later found to be ozone-depleting. The releases of over 100,000 synthetic chemicals into the environment by the burgeoning global chemical industry, whose economic output rose from US$171 billion in 1970 to over US$5 trillion in 2019 (47), are likely to increase the risks of untoward events in future, but at present no global boundary can be defined. Major sources of chemical contamination include: pesticides from agricultural run-off; dioxins associated with combustion and electronics recycling; mercury and other heavy metals from mining and coal combustion; butyl tins, heavy metals, and asbestos released during ship-breaking; dyes, heavy metals, and other pollutants associated with textile production; toxic metals, solvents, polymers, and flame retardants used in electronics manufacturing and released during waste disposal; and drug or pharmaceutical pollution through excretion in urine and inadequate disposal (18, 47, 48). Many of these sources may pose serious local threats to human health and the environment. In order for a chemical to pose a threat to the Earth system, three conditions need to be fulfilled, according to Persson et al. (49): (a) the chemical has a disruptive effect on a vital Earth system process; (b) the disruptive effect is not discovered until it is a problem at the global scale; and (c) the effect is not readily reversible. This approach highlights the central role of uncertainty in managing the risk of novel entities. When risks are uncertain but there is sufficient evidence to give grounds for concern (e.g. because of effects on reproductive performance of some species), precautionary actions may be taken to reduce risks (although these actions may be opposed by vested interests which can use the presence of uncertainty as a rationale for inaction). When little or nothing is known about the risks the challenge is to address gaps in knowledge. Because of the diversity of novel entities, their complex geographical spread, multiple interacting exposures, and the consequently major gaps in knowledge, accompanying often lax systems of regulation, this boundary is likely to be the most problematic in terms of defining appropriate metrics and credible boundaries. Ultimately it can most effectively be addressed through re-designing our economy based on careful regulation of toxic (and potentially toxic) substances together with the principles of circularity, including re-use, recycling and re-manufacturing, as outlined in Chapter 10.

Interactions among the key processes corresponding to planetary boundaries are more the norm than the exception, although much remains to be learned about them. As noted above, two of the boundaries – global climate and biodiversity – serve as cross-cutting constructs that interact with each other and integrate the interactive effects of the other seven. One example is the oceans. Carbon dioxide drives both climate change and ocean acidification. In addition, through varied oceanographic, ecological, and physiological processes, climate change exacerbates the hypoxic conditions that drive ocean dead zones. Warmer water holds less dissolved oxygen, and warmer surface waters amplify water column stratification, preventing more oxygenated surface waters from mixing with more hypoxic deep layers (50). Chemical contaminants offer another example. Climate-related meteorological factors such as temperature, wind speed, and precipitation increase the mobilization and transport of many environmental contaminants, including mercury, polycyclic aromatic hydrocarbons (PAHs), and other persistent organic pollutants (POPs) (51, 52). And climate change is a principal threat to biosphere integrity, contributing to biomass reductions and species loss in settings ranging from fisheries to pollinators (53, 54).

As so much of our current economic system is based on ignorance of the biophysical limits within which humanity can flourish, the concept of finite limits is likely to be unwelcome to many and poses a serious threat to current paradigms of development. The challenge is amplified by the pronounced inequities in consumption, and thus environmental footprints, between and within countries, both historically and contemporaneously.

Inequities in Greenhouse Gas Emissions and Environmental Footprints

Countries vary greatly in their use of resources and in their emissions of GHGs (55), as do people at varying levels of prosperity within countries. In 2010 for example, median per capita emissions in high-income countries (13 tCO_2eq/person/year) were almost ten times those in low-income countries (1.4 tCO_2eq/person/year). Recent data from the Global Carbon Atlas show the dramatic differences even more starkly – from 38 tonnes/person/year for Qatar to less than 0.1 tonne for some Sub-Saharan African countries. This means that the carbon footprint of one Qatari is equivalent to those of over 380 citizens of Malawi or Chad, and highlights the importance of consumption patterns in global carbon emissions (32).

Within broad categories of countries there is considerable variability; per capita emissions in high-income countries range from 8.2 to 21 tCO_2eq/person/year, for the (population weighted) 10th and 90th percentiles. Amongst low-income countries the average is pulled up (to 4.3 tCO_2eq/person/year) by a very few countries with high emissions from land use change (55). While CO_2 emissions from fossil fuel burning are fairly reliably quantified, methane emissions and emissions from land use change have greater uncertainty. About half of cumulative fossil fuel CO_2 emissions to 2010 were from high-income economies (as designated by the Organisation for Economic Co-operation and Development, OECD), 20% from the economies in OECD's 'transition region', 15% from the Asia region, and the remainder from Latin America, the Middle East, Africa, and international shipping. Thus, high-income countries have benefitted greatly from fossil fuel-derived energy which has powered their development and led to the advent of the consumer society.

Corresponding to the inequities in GHG emissions, there are pronounced inequalities in transgressions of other planetary boundaries. For example 33% of the world's sustainable nitrogen budget is used to produce meat for people in the EU – just 7% of the world's population (56). Several studies have compared environmental footprints across countries (34, 57). **Figure 1.8** shows results of such analysis, illustrating the dramatic differences between, say, Qatar and Vietnam. More than 80% of the world's population lives in countries that are using more resources than can be supported by the Earth's natural systems.

Whilst population growth has clearly contributed importantly to increasing GHG emissions, the large gap in per capita emissions between the highest and lowest emitting nations reveals profound and sustained inequalities. The variation between countries shows that, at least in principle, some countries can achieve high levels of human development at substantially lower GHG emissions than others, although no high-income countries have yet achieved the very low emission levels required to meet or exceed the mitigation targets agreed in Paris.

Globalization, Development, and Changing Environmental Footprints

In recent decades, the movement of manufacturing to countries such as China has contributed to the plateauing, and in some cases decline, of energy-related GHG emissions in high-income countries. Climate negotiations focus on production-related rather than consumption-related emissions; however, when countries such as China produce goods for

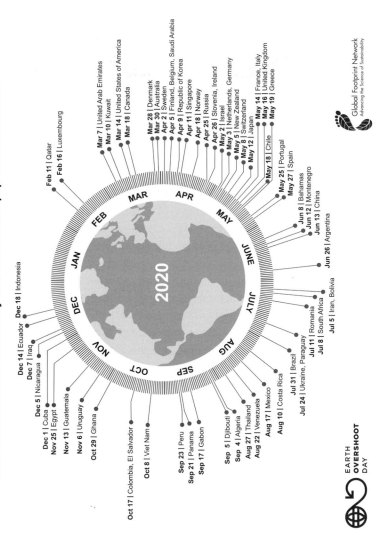

COUNTRY OVERSHOOT DAYS 2020

When would Earth Overshoot Day land if the world's population lived like...

Figure 1.8. Overshoot Days. This diagram shows the day in the year on which the Earth would reach its sustainable limit if everyone consumed to the same extent as the population of the country in question. The only nations that can be said to be approaching sustainability are those whose Overshoot Days occur in the last few months of the year. However, even those countries that appear to be approaching sustainability face significant challenges. Indonesia, for example, has experienced extensive forest fires, including peat burning, and in 2016 ranked fifth in the world for coal production (www.worldometers.info/coal/indonesia-coal/).

Source: Global Footprint Network. www.overshootday.org/content/uploads/2020/02/GFN-Country-Overshoot-Day-2020.pdf.

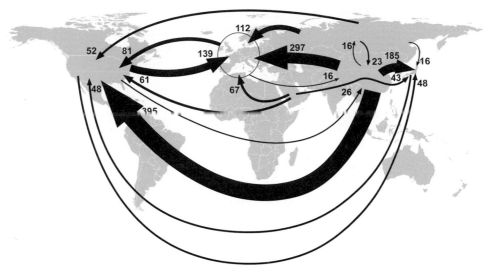

Figure 1.9. Major inter-regional fluxes of emissions embodied in trade (Mt CO_2/year) from dominant net exporting countries to dominant net importing countries. Countries of Western Europe are aggregated. The largest flux is embodied in the flow of manufactured goods from China to the USA.

Source: Adapted from Davis SJ, Caldeira K. Consumption-based accounting of CO_2 emissions. *Proceedings of the National Academy of Sciences.* 2010;107(12):5687–92.

consumption in wealthy countries, the responsibility for associated CO_2 emissions arguably lies with those who consume the goods produced. The concept of embodied emissions – emissions intrinsic to products that enter global trade – is therefore pertinent to a full understanding of national contributions to climate change (58) (**Figure 1.9**). According to one estimate of the supply chain of global CO_2 emissions, 37% of global emissions are from fossil fuels traded internationally and 23% of global emissions are embodied in traded goods (59).

That said, GHG emissions in LMICs are growing substantially as rising prosperity alters consumption patterns. Global growth in GHG emissions is increasingly driven by the energy and industry sectors in upper-middle-income countries which, according to the IPCC, accounted for 60% of the rise in global GHG emissions between 2000 and 2010. Despite a global trend away from coal use, new coal-fired power plants are planned or under construction in China, India, Vietnam, Indonesia, Bangladesh, South Africa, Egypt, the Philippines, Pakistan, and Zimbabwe (60) – a process that has been called 'carbonization' (61). Animal product (meat and dairy) consumption is rising rapidly in many LMICs (with the possible exception of India), a function of shifts in dietary preferences with growing prosperity and urbanization (62). Rapid urbanization, especially in Africa and Asia, has been accompanied by dramatic increases in private car and truck traffic, often consisting of poorly maintained, ageing vehicles that emit high levels of GHGs as well as other air pollutants (63, 64). Overall, the increase in fossil fuel CO_2 emissions between 1970 and 2010 can be attributed to changes in population (+87%), per capita GDP adjusted for Purchasing Power Parity (PPP) (+103%), energy intensity of GDP (−35%) and CO_2 intensity of energy (−15%). The increases in per capita production and consumption are major drivers worldwide of rising GHG emissions and increases in both energy efficiency and decarbonization have not been sufficient to offset the effects of GDP growth.

The Roots of the Anthropocene

How did we arrive at this point? How did the trajectory of humanity diverge so drastically from the constraints of the natural world?

Human separation from nature has a long history, at least in Western culture. In ancient Greece, some thinkers abstracted human learning from nature. In the Platonic dialogue *Phaedrus*, Socrates finds himself strolling outside the city walls, and grumbles to his companion, 'I'm a lover of learning, and trees and open country won't teach me anything, whereas men in town do' (65, p. 479). Even earlier, the book of Genesis included the well-known divine mandate, 'Be fertile and multiply. Fill the land and conquer it. Dominate the fish of the sea, the birds of the sky, and every beast that walks the land.' (Some contemporary thinkers have imputed a gentler meaning to this passage, emphasizing stewardship rather than conquest – but the long-standing standard interpretation speaks volumes about conventional attitudes.) Modern economies often evolved with little thought of protecting nature, the ultimate source of food, raw materials and energy, or reducing hazardous waste and pollution, ignoring them altogether or classifying them at best as externalities. These externalities can be difficult to quantify and even when they can be costed they are often not reflected in the current prices we pay for consumer goods or energy. This failure to internalize true costs forfeits opportunities to prevent ongoing irreversible damage to planetary life-support systems.

There is perhaps no more chilling illustration of this callous and even brutal attitude toward the natural world than these words from French political theorist Henri de Saint-Simon (1760–1825):

The object of industry is the exploitation of the globe, that is to say, the appropriation of its products for the needs of man; and by accomplishing this task, it modifies the globe and transforms it, gradually changing the conditions of its existence... From this point of view, Industry becomes religion. (65, p. 291, cited in 67, p. xii)

But there were opposing views too, embedding humanity within the fabric of nature. This world-view had long characterized indigenous cultures (68). It has emerged with regularity in Western culture, from the New England transcendentalists to the British and German Romantic painters of the eighteenth and nineteenth centuries. As a scientific framework, the interdependence of humans and the natural world was foreshadowed by such writers as John Muir, George Perkins Marsh, and Aldo Leopold. The rise of evolutionary biology in the nineteenth and twentieth centuries made clear that humans were subject to a complex framework of natural laws, and the rise of ecology as a scientific discipline in the twentieth century provided much empirical evidence.

With the Anthropocene, a third approach has emerged – humans not as *exploiters* of the natural world, nor as *citizens* of the natural world, but as *managers* of the natural world. In some cases, fear of the future is replaced by the hubris of audacious schemes – giant mirrors in space orbit, stratospheric sulfate particles to reflect sunlight back into space, ocean fertilization with iron to stimulate CO_2-absorbing algae. Such schemes are represented as planetary rescue plans, despite profound questions of safety, governance, and ethics, and the possibility that such efforts will distract from or undermine efforts to achieve sustainable practices (69).

What is clear is that humanity did not stumble completely unaware or unwittingly into the Anthropocene. The basic principles of global heating have been clear for well over a century. In 1857, French lawyer and scientist Eugène Huzar, observing the explosive growth of technology powered by fossil fuels, wrote:

In one or two hundred years, criss-crossed by railways and steamships, covered with factories and workshops, the world will emit billions of cubic meters of carbonic acid and carbon oxide, and, since the forests will have been destroyed, these hundreds of billions of carbonic acid and carbon oxide may indeed disturb the harmony of the world. (70, p. 106, cited in 67, p. xii)

In 1896, Swedish scientist Svante Arrhenius calculated the effect of a doubling of atmospheric CO_2 on Earth's surface temperature, with estimates that proved prescient. The problem was not a novel one; 'A great deal has been written,' Arrhenius noted in his paper, 'on the influence of the absorption of the atmosphere upon the climate' (71). In the 1930s, British steam engineer Guy Callendar collected several decades of temperature and CO_2 measurements, and concluded that global heating was underway. He wrote:

Few of those familiar with the natural heat exchanges of the atmosphere, which go into the making of our climates and weather, would be prepared to admit that the activities of man could have any influence upon phenomena of so vast a scale. In the following paper I hope to show that such influence is not only possible, but is actually occurring at the present time. (72)

In 1957, Roger Revelle and Hans Suess of the Scripps Institution correctly suggested: '. . . human beings are now carrying out a large scale geophysical experiment. . . Within a few centuries we are returning to the atmosphere and oceans the concentrated organic carbon stored in the sedimentary rocks over hundreds of millions of years' (73). Nor were such observations confined to arcane scientific journals. Bonneuil and Fressoz, in *The Shock of the Anthropocene*, show that the significance of the Great Acceleration was widely appreciated by mid-century (67). Books entitled *Road to Survival* by William Vogt and *Our Plundered Planet* by Henry Fairfield Osborn detailing the profound human implications of environmental destruction sold 20–30 million copies in the immediate post-World War II period. Rachel Carson's *Silent Spring*, published in 1962, called attention to the ravages of pesticides through ecosystems, and became an immediate best-seller. Archival research has documented that fossil fuel companies were not only aware of their contributions to climate change from the 1970s or earlier; they were actively engaged in research to investigate the phenomenon (74). Despite claims by the World Economic Forum in 2019 that, 'in relation to the environment ... the world is most clearly sleepwalking into catastrophe' (75), sleepwalking hardly seems the right metaphor for a process undertaken against a background of well-founded growing awareness.

Instead, the Anthropocene is the result of particular choices – political, economic, technological, and ideological choices, frequently driven by powerful interests – so much so that some have suggested that 'Capitalocene' is a more apt term than Anthropocene (76). Clearly, there have been deliberate attempts to bake in aspects of the modern economy that reinforce dangerous trends. With regard to energy, giant fossil fuel corporations have won subsidies, suppressed alternative power sources, battled labour unions, and manipulated markets to assure the dominance first of coal, and later of petroleum (77–79). With regard to the built environment (Chapter 9), passive house technologies requiring little external energy were developed in the 1930s, and affordable solar homes were promoted in 1945 (80), but these were eclipsed by the development of suburbs in post-war America and the insistence by developers and utility companies on connections to the fossil-powered electric grid. In the USA, suburbanization – supported by Federal and local policies, mortgage lending practices, and real estate developers (81) – also fostered mass ownership of private cars. With regard to food and agriculture (Chapter 10), it has been argued that the Green Revolution of the 1960s, which took place against the backdrop of Cold War ideological struggles, helped to ensure dominance of an agro-industrial approach to food production (82). This approach was successful in improving food security by increasing yields of wheat, rice, and maize but also led to heavy dependence on high inputs of chemical fertilizers, pesticides, and energy whilst depleting water tables, polluting water sources, and leading to salination of soils.

However, we cannot exclusively blame a capitalist system that thrived on externalizing the full costs of economic activities. There is ample evidence of environmental destruction in the Soviet Union (83), East Germany (84), the People's Republic of China (85), and other non-capitalist economies. Greenhouse gas emissions declined sharply after the fall of Soviet Communism (86). A substantial contribution was made by deep reductions in beef consumption when subsidies for livestock were removed and by subsequent carbon sequestration on abandoned cropland. However, it is likely that in the absence of policy action emissions will rebound. Both centrally controlled and market-driven economies have therefore contributed to unsustainable patterns of development, but now the consumer economy reigns largely supreme and depends on the pursuit of economic growth irrespective of the impacts on the natural systems that sustain human civilization.

The Path Forward

If the Anthropocene is the result of human action, then reversing course, and achieving a sustainable world to bequeath to future generations, can also flow from human action. Many of the needed technologies are already available. Many of the needed policies are clear. The institutions that need to be transformed (or sunsetted) are well documented (87, 88). They include fossil fuel industries and other companies whose business model depends on degrading the natural systems that underpin health. And a growing global movement of concerned citizens is calling for change.

During the long sweep of human history the environment has been exploited to benefit societies and powerful interests within them. Many indicators of human health and well-being have improved over the past century, but persistent inequities and adverse trends in indicators of the health of natural systems underpinning human progress are deeply concerning. In the absence of transformative change, current practices threaten to undermine the prospects for today's and future generations. Subsequent chapters of this book aim to describe threats to human health in the Anthropocene Epoch, to discuss the extent to which we can adapt to the changes unleashed by human activities, and to suggest which societal choices can help to safeguard health, and human progress more generally, at much lower levels of environmental impact.

References

1. Deaton A. *The Great Escape: Health, Wealth, and the Origins of Inequality*. Princeton, NJ and Oxford: Princeton University Press; 2013.
2. Steffen W, Richardson K, Rockström J, et al. Planetary boundaries: guiding human development on a changing planet. *Science*. 2015;347(6223):1259855. DOI: 10.1126/science.1259855.
3. McMichael AJ, Woodward A, Muir C. *Climate Change and the Health of Nations: Famines, Fevers, and the Fate of Populations*. Oxford and New York: Oxford University Press; 2017.
4. Smil V. *Energy and Civilization: A History*. Cambridge, MA: MIT Press; 2017.
5. Harris B. Public health, nutrition, and the decline of mortality: the McKeown thesis revisited. *Social History of Medicine*. 2004;17(3):379–407.
6. Rosen G. *A History of Public Health*. Revised expanded edition. Baltimore: Johns Hopkins University Press; 2015.
7. Melosi MV. *The Sanitary City: Urban Infrastructure in America from Colonial Times to the Present*. Baltimore: Johns Hopkins University Press; 2000.
8. Steffen W, Crutzen PJ, McNeill JR. The Anthropocene: are humans now overwhelming the great forces of nature? *AMBIO: A Journal of the Human Environment*. 2007;36(8):614–21. doi: 10.1579/0044-7447(2007)36[614:TAAHNO]2.0.CO;2.
9. UN Department of Economic and Social Affairs Population Division. *World Population Prospects 2019. Volume I: Comprehensive Tables*. New York: United Nations; 2019. Contract No. ST/ESA/SER.A/426.
10. UN Inter-agency Group for Child Mortality Estimation. *Levels & Trends in Child Mortality: Report 2019*. New York: UNICEF, World Health Organization, and World Bank; 2019.
11. World Bank. *Piecing Together the Poverty Puzzle*. World Bank; 2018.
12. Omran AR. The epidemiologic transition. A theory of the epidemiology of population change. *The Milbank Memorial Fund Quarterly*. 1971;49(4):509–38.
13. Crutzen PJ. Geology of mankind. *Nature*. 2002;415(6867):23. https://doi.org/10.1038/415023a.
14. Crutzen PJ, Steffen W. How long have we been in the Anthropocene era? *Climatic Change*. 2003;61(3):251–7. doi: 10.1023/B:CLIM.0000004708.74871.62.
15. Lewis SL, Maslin MA. Defining the Anthropocene. *Nature*. 2015;519(7542):171–80. doi: 10.1038/nature14258.
16. Lewis SL, Maslin M. *The Human Planet: How We Created the Anthropocene*. London: Pelican; 2018.

17. Zalasiewicz J, Waters CN, Williams M, et al. When did the Anthropocene begin? A mid-twentieth century boundary level is stratigraphically optimal. *Quaternary International*. 2015;383:196–203.
18. Whitmee S, Haines A, Beyrer C, et al. Safeguarding human health in the Anthropocene Epoch: report of The Rockefeller Foundation–*Lancet* Commission on Planetary Health. *The Lancet*. 2015;386(10007):1973–2028. https://doi.org/10.1016/S0140-6736(15)60901-1.
19. IPBES. *Global Assessment Report on Biodiversity and Ecosystem Services*. Bonn, Germany: Intergovernmental Science-Policy Platform on Biodiversity and Ecosystem Services; 2019. https://doi.org/10.5281/zenodo.3553579.
20. Haddad NM, Brudvig LA, Clobert J, et al. Habitat fragmentation and its lasting impact on Earth's ecosystems. *Science Advances*. 2015;1(2):e1500052. DOI: 10.1126/sciadv.1500052.
21. Güneralp B, Reba M, Hales BU, Wentz EA, Seto KC. Trends in urban land expansion, density, and land transitions from 1970 to 2010: a global synthesis. *Environmental Research Letters*. 2020;15(4):044015. https://doi.org/10.1088/1748-9326/ab6669.
22. Pimm SL, Jenkins CN, Abell R, et al. The biodiversity of species and their rates of extinction, distribution, and protection. *Science*. 2014;344(6187). DOI: 10.1126/science.1246752.
23. Bar-On YM, Phillips R, Milo R. The biomass distribution on Earth. *Proceedings of the National Academy of Sciences*. 2018;115(25):6506–11. https://doi.org/10.1073/pnas.1711842115.
24. Zarfl C, Lumsdon AE, Berlekamp J, Tydecks L, Tockner K. A global boom in hydropower dam construction. *Aquatic Sciences*. 2015;77(1):161–70. DOI: 10.1007/s00027-014-0377-0.
25. Pauly D, Zeller D. Catch reconstructions reveal that global marine fisheries catches are higher than reported and declining. *Nature Communications*. 2016;7:10244. https://doi.org/10.1038/ncomms10244.
26. Rhein M, Rintoul SR, Aoki S, et al. Observations: Oceans. In Stocker TF, Qin D, Plattner G-K, et al., editors. *Climate Change 2013: The Physical Science Basis. Contribution of Working Group I to the Fifth Assessment Report of the Intergovernmental Panel on Climate Change*. Cambridge, UK: Cambridge University Press; 2013. pp. 255–315.
27. Hughes TP, Anderson KD, Connolly SR, et al. Spatial and temporal patterns of mass bleaching of corals in the Anthropocene. *Science*. 2018;359(6371):80–3. doi: 10.1126/science.aan8048.
28. Frainer A, Primicerio R, Kortsch S, et al. Climate-driven changes in functional biogeography of Arctic marine fish communities. *Proceedings of the National Academy of Sciences*. 2017;114(46):12202–7. https://doi.org/10.1073/pnas.1706080114.
29. Breitburg D, Levin LA, Oschlies A, et al. Declining oxygen in the global ocean and coastal waters. *Science*. 2018;359(6371):eaam7240. DOI: 10.1126/science.aam7240.
30. WMO. Greenhouse Gas Bulletin: the state of greenhouse gases in the atmosphere based on global observations through 2016. Geneva: World Meteorological Organization; 2017.
31. UNEP. Emissions Gap Report 2019. Nairobi: United Nations Environment Programme; 2019.
32. Global Carbon Atlas. CO_2 emissions 2020. Available from www.globalcarbonatlas.org/en/CO2-emissions.
33. Ehrlich PR, Holdren JP. Impact of population growth. *Science*. 1971;171:1212–17. doi: 10.1126/science.171.3977.1212.

34. Dietz T, Rosa EA, York R. Driving the human ecological footprint. *Frontiers in Ecology and the Environment*. 2007;5(1):13–18. doi: 10.1890/1540-9295(2007)5[13:Dthef]2.0. Co;2.

35. Rockström J, Steffen W, Noone K, et al. Planetary boundaries: exploring the safe operating space for humanity. *Ecology and Society*. 2009;14(2).

36. Cai Y, Lenton TM, Lontzek TS. Risk of multiple interacting tipping points should encourage rapid CO_2 emission reduction. *Nature Climate Change*. 2016;6(5):520–5. https://doi.org/10.1038/nclimate2964.

37. Lenton TM, Rockström J, Gaffney O, et al. Climate tipping points – too risky to bet against. *Nature*. 2019;575:592–5. https://doi.org/10.1038/d41586-019-03595-0.

38. Steffen W, Rockström J, Richardson K, et al. Trajectories of the Earth system in the Anthropocene. *Proceedings of the National Academy of Sciences*. 2018;115(33):8252–9. https://doi.org/10.1073/pnas.1810141115.

39. Drijfhout S, Bathiany S, Beaulieu C, et al. Catalogue of abrupt shifts in Intergovernmental Panel on Climate Change climate models. *Proceedings of the National Academy of Sciences*. 2015;112(43):E5777–86. doi: 10.1073/pnas.1511451112.

40. Mace GM, Reyers B, Alkemade R, et al. Approaches to defining a planetary boundary for biodiversity. *Global Environmental Change*. 2014;28:289–97. https://doi.org/10.1016/j.gloenvcha.2014.07.009.

41. Fowler D, Coyle M, Skiba U, et al. The global nitrogen cycle in the twenty-first century. *Philosophical Transactions of the Royal Society B: Biological Sciences*. 2013;368(1621):20130164. https://doi.org/10.1098/rstb.2013.0164.

42. Yuan Z, Jiang S, Sheng H, et al. Human perturbation of the global phosphorus cycle: changes and consequences. *Environmental Science & Technology*. 2018;52(5):2438–50. doi: 10.1021/acs.est.7b03910.

43. Gonzalez M, Taddonio KN, Sherman NJ. The Montreal Protocol: how today's successes offer a pathway to the future. *Journal of Environmental Studies and Sciences*. 2015;5(2):122–9. DOI: 10.1007/s13412-014-0208-6.

44. Strahan SE, Douglass AR. Decline in Antarctic ozone depletion and lower stratospheric chlorine determined from Aura Microwave Limb Sounder observations. *Geophysical Research Letters*. 2018;45(1):382–90. https://doi.org/10.1002/2017GL074830.

45. Rigby M, Park S, Saito T, et al. Increase in CFC-11 emissions from eastern China based on atmospheric observations. *Nature*. 2019;569(7757):546–50. doi: 10.1038/s41586-019-1193-4.

46. Solomon S, Alcamo J, Ravishankara AR. Unfinished business after five decades of ozone-layer science and policy. *Nature Communications*. 2020;11(1):4272.

47. UNEP. Global Chemicals Outlook II – from legacies to innovative solutions: implementing the 2030 Agenda for Sustainable Development. Geneva: United Nations Environment Programme; 2019.

48. Landrigan PJ, Fuller R, Acosta NJR, et al. The Lancet Commission on Pollution and Health. *The Lancet*. 2018;391:462–512. doi: 10.1016/S0140-6736(17)32345-0.

49. Persson LM, Breitholtz M, Cousins IT, et al. Confronting unknown planetary boundary threats from chemical pollution. *Environmental Science & Technology*. 2013;47(22):12619–22. https://doi.org/10.1021/es402501c.

50. Altieri AH, Gedan KB. Climate change and dead zones. *Global Change Biology*. 2015;21(4):1395–406. https://doi.org/10.1111/gcb.12754.

51. Obrist D, Kirk JL, Zhang L, et al. A review of global environmental mercury processes in response to human and natural perturbations: changes of emissions, climate, and land use. *Ambio*. 2018;47(2):116–40. doi: 10.1007/s13280-017-1004-9.

52. Nadal M, Marquès M, Mari M, Domingo JL. Climate change and environmental concentrations of POPs: a review. *Environmental Research.* 2015;143:177–85. doi: 10.1016/j.envres.2015.10.012.
53. Barange M, Bahri T, Beveridge MCM, et al. Impacts of Climate Change on Fisheries and Aquaculture: Synthesis of Current Knowledge, Adaptation and Mitigation Options. Rome: Food and Agriculture Organization of the United Nations; 2018. Contract No. 627. ISBN 978-92-5-130607-9.
54. Marshman J, Blay-Palmer A, Landman K. Anthropocene crisis: climate change, pollinators, and food security. *Environments.* 2019;6(2). https://doi.org/10.3390/ environments6020022.
55. Stocker TF, Qin D, Plattner G-K, et al., editors. *Climate Change 2013: The Physical Science Basis. Contribution of Working Group I to the Fifth Assessment Report of the Intergovernmental Panel on Climate Change.* Cambridge, UK and New York: Cambridge University Press; 2013.
56. Sutton MA, Oenema O, Erisman JW, et al. Too much of a good thing. *Nature.* 2011;472(7342):159–61. doi: 10.1038/472159a.
57. Global Footprint Network. Ecological footprint 2019. Available from www .footprintnetwork.org/our-work/ecological-footprint/.
58. Davis SJ, Caldeira K. Consumption-based accounting of CO_2 emissions. *Proceedings of the National Academy of Sciences.* 2010;107(12):5687–92. https://doi.org/10.1073/ pnas.0906974107.
59. Davis SJ, Peters GP, Caldeira K. The supply chain of CO_2 emissions. *Proceedings of the National Academy of Sciences.* 2011;108(45):18554–9. https://doi.org/10.1073/ pnas.1107409108.
60. Shearer C, Mathew-Shah N, Myllyvirta L, Yu A, Nace T. *Boom and Bust 2019: Tracking the Global Coal Plant Pipeline.* San Francisco: Global Energy Monitor, Sierra Club, and Greenpeace; 2019.
61. Steckel JC, Hilaire J, Jakob M, Edenhofer O. Coal and carbonization in Sub-Saharan Africa. *Nature Climate Change.* 2020;10(1):83–8. https://doi.org/10.1038/s41558- 019-0649-8.
62. Zhai FY, Du SF, Wang ZH, et al. Dynamics of the Chinese diet and the role of urbanicity, 1991–2011. *Obesity Reviews.* 2014;15(Suppl. 1):16–26. DOI: 10.1111/ obr.12124.
63. Kutzbach MJ. Motorization in developing countries: causes, consequences, and effectiveness of policy options. *Journal of Urban Economics.* 2009;65(2):154–66.
64. Doumbia M, Toure NDE, Silue S, et al. Emissions from the road traffic of West African cities: assessment of vehicle fleet and fuel consumption. *Energies.* 2018;11(9):2300. https://doi.org/10.3390/en11092300.
65. Hamilton E, Cairns H, editors. *Plato: The Collected Dialogues.* Princeton, NJ: Princeton University Press; 1961.
66. Saint-Simon H. *Doctrine de Saint-Simon, Vol 2.* Paris: Aux Bureaux de l'Organisateur; 1830.
67. Bonneuil C, Fressoz J-B. *The Shock of the Anthropocene.* London and New York: Verso; 2016.
68. Gomez-Baggethun E, Corbera E, Reyes-Garcia V. Traditional ecological knowledge and global environmental change: research findings and policy implications. *Ecology and Society.* 2013;18(4):72.
69. Hamilton C. *Earthmasters: The Dawn of the Age of Climate Engineering.* New Haven, CT: Yale University Press; 2014.
70. Huzar E. *L'Arbre de la science.* Paris: Dentu; 1857.

71. Arrhenius S. On the influence of carbonic acid in the air upon the temperature of the ground. *Philosophical Magazine and Journal of Science*. 1896;41:237–76.
72. Callendar GS. The artificial production of carbon dioxide and its influence on temperature. *Quarterly Journal of the Royal Meteorological Society*. 1938;64(275): 223–40.
73. Revelle R, Suess HE. Carbon dioxide exchange between atmosphere and ocean and the question of an increase of atmospheric CO_2 during the past decades. *Tellus*. 1957;9(1):18–27.
74. Supran G, Oreskes N. Assessing ExxonMobil's climate change communications (1977–2014). *Environmental Research Letters*. 2017;12(8):084019. DOI: 10.1088/ 1748-9326/aa815f.
75. WEF. *The Global Risks Report 2019*. Geneva: World Economic Forum; 2019.
76. Moore JW, editor. *Anthropocene or Capitalocene? Nature, History, and the Crisis of Capitalism*. Oakland, CA: PM Press/Kairos; 2016.
77. Freese B. *Coal: A Human History*. Cambridge, MA: Perseus; 2003.
78. Yergin D. *The Prize: The Epic Quest for Oil, Money, and Power*. New York: Simon & Schuster; 1991.
79. Auzanneau M. *Oil, Power, and War: A Dark History*. White River Junction, VT: Chelsea Green Publishing; 2015.
80. Barber DA. Tomorrow's house: solar housing in 1940s America. *Technology and Culture*. 2014;55(1):1–39.
81. Jackson KT. *Crabgrass Frontier: The Suburbanization of the United States*. New York and Oxford: Oxford University Press; 1985.
82. Cullather NT. *The Hungry World: America's Cold War Battle Against Poverty in Asia*. Cambridge, MA: Harvard University Press; 2010.
83. Peterson DJ. *Troubled Lands: The Legacy of Soviet Environmental Destruction*. Washington, DC: Westview Press; 1993.
84. Dominick R. Capitalism, communism, and environmental protection: lessons from the German experience. *Environmental History*. 1998;3(3):311–32.
85. Shapiro J. *Mao's War Against Nature: Politics and the Environment in Revolutionary China*. Cambridge, UK: Cambridge University Press; 2001.
86. Schierhorn F, Kastner T, Kuemmerle T, et al. Large greenhouse gas savings due to changes in the post-Soviet food systems. *Environmental Research Letters*. 2019;14(6):065009. https://doi.org/10.1088/1748-9326/ab1cf1.
87. Heede R. Tracing anthropogenic carbon dioxide and methane emissions to fossil fuel and cement producers, 1854–2010. *Climatic Change*. 2014;122(1):229–41. DOI 10.1007/s10584-013-0986-y.
88. Griffin PM. *The Carbon Majors Database: CDP Carbon Majors 2017*. London: Carbon Disclosure Project; 2017.

2

Climate Change

Environmental changes, and the driving forces that dominate the Anthropocene outlined in Chapter 1, can have wide-ranging and pervasive effects on health through a range of direct and indirect pathways. One of the best recognized of these pathways is climate change – a relatively anodyne term that, thanks to growing awareness and alarm in recent years, is often now replaced by 'global heating', the 'climate crisis', or even the 'climate emergency'.

The signature feature of climate change is the increase in global mean temperature. This is driven principally by the emission of greenhouse gases and other contaminants that alter the atmosphere and therefore the planet's energy balance, increasing the solar energy the Earth's atmosphere retains. The principal greenhouse gas (GHG) is carbon dioxide (CO_2), a product of fossil fuel combustion and other sources; other GHGs include methane, nitrous oxide, water vapour, and fluorinated gases. Black carbon, another product of combustion, also warms the Earth by absorbing solar energy, by reducing the albedo (or reflectivity) of light surfaces, and through complex interactions with clouds. These climate pollutants remain in the atmosphere for highly variable periods of time – days to weeks for black carbon, about 12 years for methane and more than 100 years for CO_2 (with perhaps 15–20% remaining in the atmosphere for 1000 years or more).

But a hotter Earth is not the only important feature of climate change. In many mid-latitude and sub-tropical dry regions, precipitation will decrease, whereas in many mid-latitude wet regions, precipitation is projected to increase, over coming decades. Extreme precipitation events over most of the mid-latitude land masses and over wet tropical regions will very likely become more intense and more frequent by the end of this century, as global mean surface temperature increases. Precipitation variability is increasing (1) (giving rise to yet another term, 'climate chaos'). Increases in the most intense tropical cyclones are likely. Because water expands as it warms, and because land-based ice masses are melting, sea levels are rising. Warmer water is less able to dissolve oxygen, and the oceans are absorbing carbon dioxide, leading to reduced oxygen levels and reduced pH (more acidity) in the oceans. Climate change therefore entails a wide range of Earth system changes (2). Many of these changes are already evident; they are not only predictions, but contemporary observations, and for more and more people, lived experience.

These linked and cascading phenomena affect human health in many ways. Some pathways are direct: excessive heat exposure, or the increased frequency and intensity of extreme events such as floods and droughts. Other pathways are mediated through natural systems, such as changes in the distribution of vector-borne diseases, notably dengue or malaria. Still other pathways operate through social systems, such as by increasing poverty, undernutrition, or forced migration. These pathways are not always distinct; in fact, they typically interact. For example, increasing heat exposure directly reduces physical labour capacity, and heat and drought both reduce crop productivity; together, these pathways can reduce the food available in vulnerable areas, deepen poverty and undernutrition, and potentially drive migration.

In this chapter we first consider the sources of insight about the impacts of climate on human health and well-being – how we know what we know. These sources include historical analysis of changes in climate; studies of a recurring weather phenomenon, the El Niño Southern Oscillation, which provides a model for climate impacts; contemporary epidemiological and other health studies; and modelled projections of future effects. An important challenge is causal attribution – how we determine that climate change is responsible for observed weather events and their human impacts. Next, we turn to the various specific pathways through which climate affects health: heat; severe weather; infectious diseases; threats to the food supply; effects on poverty, migration, and conflict; and grief and other mental health effects.

The Foundations of Knowledge about Climate and Health

Historical Perspective

While public awareness of the modern concept of climate change from accumulation of GHGs is just a few decades old, humans have lived with changes in climate since the dawn of the species. Nowhere is this legacy better recounted than in the magisterial *Climate Change and the Health of Nations*, by the Australian physician, epidemiologist, and writer Tony McMichael. McMichael's writing provides a panoramic view of the effects of a changing climate on health over millennia (3). As the Pleistocene transitioned to the Holocene (**Figure 2.1**) only one of the five existing *Homo* species endured: *Homo sapiens.* We were well placed to exploit the opportunities presented by a warmer world.

With the end of the glacial period between 19,000 and 11,000 years ago, ice sheets retreated. But about 12,800 years ago, a sudden cooling occurred, lasting about 1000 years: the Younger Dryas period. This precipitous cooling may have been caused by the sudden release of cold freshwater into the Atlantic from extensive lakes in Canada fed by the melting of the vast Laurentian ice sheet. This in turn could have disrupted the Gulf Stream which conveys warm water to Europe and the Middle East. There is also a (somewhat controversial) theory that a massive comet impact contributed by throwing clouds of dust into the atmosphere, partly blocking the sun's rays (4, 5). No matter what the trigger, the impacts of the Younger Dryas confirmed the importance of maintaining an optimum and stable climate for the development of humanity. The rapid cooling was accompanied by dramatic declines in human settlements in the Nile Valley (modern day Egypt), and there is evidence of mass violence, presumably due to competition for scarce food and resources. In southern Africa the number of stable settlements fell by half during this period. In the

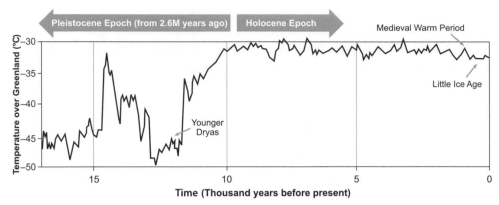

Figure 2.1. Global temperature during the late Pleistocene and Holocene Epochs, as represented by Greenland. This figure is on a scale of thousands of years. Compare with the temperature tracing in Figure 1.2 (which depicts less than 200 years) to see how rapid and drastic the recent temperature increase has been, and what a departure it is from the long-term stability of the Holocene.

northern Fertile Crescent (modern day Israel, Palestine, Jordan, Lebanon, and most of Syria) Natufian settlements that had emerged over the previous millennium disbanded due to food shortages. The surviving settlements turned to cultivation of cold-tolerant rye grass which may have been a forerunner of modern agriculture.

As the Younger Dryas transitioned to the early Holocene, agriculture evolved over a period of 2000 years or so, in a number of locations that were not in direct contact with each other. This suggests that the new more stable and warmer climatic conditions favoured the development of cultivation and harvesting of crops over the hunter-gatherer lifestyle that had prevailed for perhaps the preceding 2 million years.

Agriculture provided greater access to food but also brought some penalties, because agrarian diets, particularly in the early days, were of limited nutritional diversity and quality. The intake of carbohydrates rose, while that of protein fell. Early farming communities without access to marine resources suffered from declines in physical health, including reduced stature and poor dentition. The domestication of herd animals emerged, probably about 10,500 years ago around the Tigris River, initially led by the herding of sheep and goats, followed by pigs and cattle (the former descended from wild boar and the latter from aurochs). These herds provided a higher quality and more dependable supply of protein than from cultivation alone. From these tentative beginnings today's human population now dominates the global landscape in its quest to provide food for over 7 billion people, accounting for one-quarter of the planet's primary photosynthetic product to feed itself and using about 40% of the world's available land (6).

McMichael's encyclopaedic review of the historical links between climatic fluctuations and human disease and survival illustrates how societies depend on stable climates to flourish. Amongst the notable examples he cites, the Great Famine, which affected Northern Europe between 1315 and 1322, was probably the single worst recorded subsistence crisis experienced by the continent (3). Climate – specifically, the El Niño phenomenon (see below for an explanation) – likely played a key role. During this period of transition between the Medieval Warm Period and the Little Ice Age, European weather became cooler and wetter, whilst in low-latitude Asia conditions were warmer and drier, with weakened monsoon activity. Most of Europe's grain crops failed during this period, leading for example to devastating declines in crop yields in England between 1315 and 1317, accompanied by tripling of the grain price. An important lesson from this tragic event is that it resulted from two consecutive harvest failures; a single failure had relatively little effect on society but recurrent failure overwhelmed coping mechanisms, resulting in a doubling of death rates and a five-fold escalation in crime rates. Outbreaks of ergotism (St Anthony's Fire) were frequent, caused by spoilage of rye by a fungus producing a mycotoxin that caused burning pains, hallucinations, psychosis, and gangrene, often resulting in death. Throughout Europe, famine was often accompanied by spikes in infectious diseases in animals and humans. Whilst the exact nature of these is often unclear, there is no doubt that they contributed greatly to the increase in death rates. A complex combination of climatic, socioeconomic, and demographic factors was probably responsible for the wide-ranging catastrophic effects on European society of the famine and accompanying disease outbreaks (7). These in turn resulted in long-lasting damage to the fabric of society and may have contributed to the devastating impacts of the Black Death three decades later, possibly through a legacy of chronic undernutrition and increased susceptibility to some infectious diseases.

The study of history yields many other telling examples of societies that were unable to withstand sudden shocks or adapt to gradual changes in the environment. A compelling case is that of the collapse of Mayan civilization between AD 750 and 950 – what has been called 'a demographic disaster as profound as any other in human history' (8). This coincided with climatic changes, including the southern displacement of the Intertropical Convergence Zone, triggering a prolonged dry period punctuated by a series a of severe multi-year droughts. These in turn led to declines in food production, causing intractable stresses on the social and political fabric of Mayan society. Skeletal remains from the late eighth century show increasing proportions of infant, child, and female adult deaths, many with signs of undernutrition including stunting (9). Many Mayan centres became fragmented and ultimately collapsed.

These droughts also had severe impacts further north on the ancestral Pueblo (sometimes called Anasazi) population in the San Juan Basin, in today's southwestern USA. These settlements experienced three droughts between the early ninth century and late thirteenth century, the first linked to the drought further south that dealt Mayan civilization a fatal blow (10). The megadrought commencing in the twelfth century proved pivotal in determining the future of several regional populations, including the ancestral Pueblo. Despite their expertise in water conservation and knowledge of wild plants as backup food sources, they were vulnerable because their diets were nutritionally deficient in some essential amino acids and because population growth in preceding centuries had brought them perilously close to the limits of carrying capacity of the land they inhabited. Their sophisticated irrigation systems were ultimately unable to cope as intensifying drought combined with local environmental degradation, including soil erosion and deforestation, leading to a loss of vital sources of food such as piñon nuts. Their society collapsed in violent conflict, including invasions by neighbouring groups, which finally led to their out-migration southward along traditional trading routes (11).

An important consideration arising from historical profiles such as these is the extent to which the direction of departure – excessive heat or cold – from long-term stable climatic conditions determines the impacts on society. It is probably the magnitude and rapidity of change that matter, because crops and human societies more generally are adapted to prevailing climatic conditions, and incapable of rapid change. When temperatures are outside the optimal zone – either too low or too high – then indicators such as mortality (12,13) and crop failure increase, creating a U-shaped curve.

Whilst historical analysis of past events provides salutary lessons for contemporary society we cannot apply them uncritically to inform our current predicament. First, current societies are interconnected, with supply chains stretching around the world. This implies that a stressor such as a regional collapse in food production could be addressed by importing food from elsewhere (providing that the importing nation has resources to pay). Second, contemporary governance mechanisms and political structures may be more responsive to citizens' needs than in the past, although we cannot be complacent given the persistence, and in some cases strengthening, of autocratic governments. Nevertheless, even in most autocracies, access to the news media is better than in previous eras and it is commensurately more difficult to conceal incompetence and mismanagement. Third, our access to high-quality information is much greater than in the past, although again we cannot be complacent when objective data collection and analysis are under attack in some quarters because they offend powerful interests. Fourth, we have access to diverse techno-logical innovations that could aid adaptation to environmental change to avert crises.

Against these advantages, compared with our ancestors, we face unprecedented chal-lenges, emerging over what is in historical terms the blink of an eye, with a much larger and high-consuming world population driving further environmental change. These trends are taking us farther and farther from the conditions of the Holocene in which civilization as we know it arose. In a world of unprecedented challenges, to paraphrase Antonio in Shakespeare's *The Tempest*, 'the past may no longer be prologue'. We therefore need to supplement historical study with other methods in order fully to understand and anticipate the effects of climate on humanity.

The El Niño Southern Oscillation

The El Niño Southern Oscillation (ENSO) phenomenon is an ocean–atmosphere cycle, based in the tropical Pacific Ocean, that occurs every 2–7 years, usually lasting 9–12 months but sometimes lasting several years. ENSO has two phases: the more frequent El Niño phase, when the central or eastern Pacific (off the South American coast) warms while Indonesia and the western Pacific cool, and the La Niña phase, when the reverse pattern prevails. El Niño typically occurs towards the end of the year, around Christmas (hence the term El Niño, which refers to the Christ child). When a large area of the tropical Pacific warms, moist air rises over the warm area, not only increasing regional rainfall, but affecting weather over far-flung parts of the planet through 'teleconnections'. El Niño was recognized by fishermen off the South American coast as early as the seventeenth century, but modern meteorology has greatly clarified its characteristics, permitting a better understanding of human impacts. This natural phenomenon is a powerful model for learning about climate impacts on human health.

The short-term changes in temperature and precipitation arising from the ENSO cycle affect human health and well-being in many ways. For example, between 1964 and 1993 the number of people experiencing climate-related disasters worldwide (especially drought, often leading to food shortages) increased dramatically during and immediately following El Niño years compared with the pre-El Niño years, with particularly strong associations in South Asia. The increased exposure to such disasters with El Niño years can affect as much as 3% of the world's population (14). A more recent study using data from 1964–2017 showed that the annual number of people affected by drought globally increased sharply in El Niño years (relative to neutral years) (15). However, the disaster type most strongly associated with El Niño is flooding, with the largest affected populations in Asia, especially India. Associations between El Niño and disasters may be changing over time, with the La Niña phase more likely to be associated with drought events over the recent 20-year period. So far there is no evidence that climate change is affecting the El Niño cycle.

There is strong evidence that ENSO is associated with malaria epidemics in parts of Latin America and South Asia and with increased cholera incidence in Bangladesh (16). In Southeast Asia during strong El Niño years, an increased number of fires contribute to massively increased annual average air pollution exposure near fire sources and the pollution is transported large distances downwind. These fires contribute an additional 200 days per year that exceed the World Health Organization (WHO) target for fine particulate matter (PM) air pollution exposure (50 $\mu g/m^3$ 24-hr $PM_{2.5}$), and account for an estimated 10,800 (6800–14,300) yearly excess cardiovascular deaths in the region during El Niño years (17).

According to the WHO, the 2015–2016 El Niño subjected more than 60 million people, particularly in eastern and southern Africa, the Horn of Africa, Latin America and the Caribbean, and the Asia-Pacific region, to severe drought and associated food insecurity, flooding, rains, and temperature increases. These resulted in a wide range of health problems, including vector-borne and water-borne disease outbreaks, undernutrition, heat stress, and respiratory diseases (18). The ENSO phenomenon illustrates the profound sensitivity of human health and well-being to perturbations in climate.

Contemporary Health Studies

A range of epidemiological and other public health study methods has also been applied to understanding the association between climate and health. Often these studies entail linking meteorological data with health data, which can be challenging given differing data owners, temporal and spatial measurement scales, and data architecture. Another challenge is the paucity of reliable health and meteorological data in areas of the world most vulnerable to the effects of climate change. Still another challenge is the multifactorial nature of the health outcomes affected by climate change; climate change is often an effect modifier (a threat amplifier), or one of many causal factors operating through intricate webs, so complex conceptual and statistical models are needed.

Nevertheless, the body of modern evidence linking climate change with human health is substantial and growing. For example, studies have shown rising mortality risk due to extreme heat exposure in many countries (19) and epidemiological surveillance has linked increased incidence of malaria in high-altitude areas of East Africa to rising temperatures (20). Authoritative reviews describe the many pathways by which climate change can impact health, the vulnerability of specific populations and subgroups, estimates of numbers affected, and finally the potential for solutions that protect health through adaptation, or more importantly through policies to cut GHG emissions. Key sources include the Intergovernmental Panel on Climate Change (IPCC), most recently in its Fifth Assessment Report (21), the World Health Organization (WHO) (22), the US (23, 24), UK (25), and other national governments' assessment reports, the *Lancet* Countdown on Health and Climate Change (26), original articles in academic journals (27–30), and books (31–33). The evidence generated by this research is summarized below.

Modelling and Forecasting

In the setting of rapidly evolving changes, evidence describing associations to date do not suffice to give us a clear view of the future. We know that the planet will be very different in 50 years than ever before in human history. Accordingly, a key part of the evidence base comes from projections of future health impacts. These are necessarily based on assumptions about future trends in greenhouse gas emissions, economic development, population growth, technology innovation and uptake, and other factors. Despite the inevitable uncertainties, forecasting can provide valuable insights about potential impacts and the effect of actions to address climate change.

A starting point is climate system models such as general circulation models (also called global climate models, or GCMs). The GCMs apply physical principles to describe and predict changes in the Earth's climate based on climate forcings such as CO_2 levels, and assumptions about the sensitivity of Earth systems to these forcings. The models are computationally intensive; they divide the planet into horizontal and vertical grids, and account for many processes such as cloud formation, ocean currents, and jet streams (**Figure 2.2**). Most GCMs have resolutions at the scale of hundreds of kilometres, too coarse to provide localized insights, so GCMs are typically supplemented with downscaled regional climate models (RCMs) to provide finer resolution (34).

Scientists evaluate their models retrospectively, by comparing what models would have predicted at some point in the past with what actually ensued. In general, models developed over recent decades have been highly accurate, and have improved over time (35). This gives confidence in predictions of future climate.

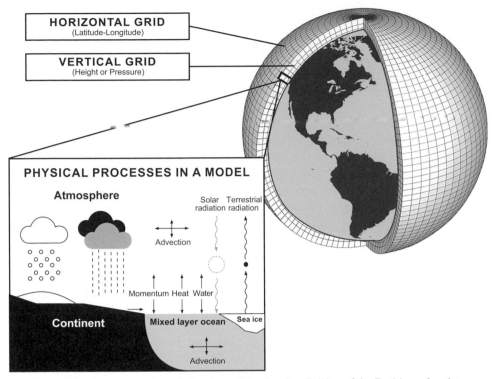

Figure 2.2. A representation of climate models, showing division of the Earth's surface into grids, and the physical processes that are incorporated.

Source: NOAA. Adapted from www.climate.gov/file/atmosphericmodelschematicpng.

The next step was to develop sets of future trajectories of GHG concentrations based on model outputs for use by the climate research community. Representative Concentration Pathways (RCPs) were adopted for climate modelling and research by the IPCC (36). There are four RCPs, each designated by a level of radiative forcing (the difference between incoming solar energy absorbed by the Earth, and energy radiated back into space, expressed in watts per square metre (W/m^2)): RCP8.5, RCP6, RCP4.5, and RCP2.6. RCP8.5 is a high-emissions pathway, accompanied by destructive land use change, while RCP2.6 is the most benign pathway, predicated on rapid and deep cuts in emissions. The resulting simulations of future global temperature changes, and other climate impacts, vary accordingly.

A set of scenarios that illustrate different potential futures, called shared socioeconomic pathways (SSPs), was intended to complement the RCPs (37, 38). These replaced earlier attempts dating from the 1990s that did not reflect recent large-scale socioeconomic changes. The SSPs are based on narratives about societal trajectories, and include assumptions about demographics, human development, economy and lifestyle, policies and institutions, technology, and environment and natural resources. The five SSPs are outlined in **Table 2.1**. **Figure 2.3** depicts how they differ in terms of their adaptation and mitigation challenges. As the names and descriptions suggest, the five scenarios lead to widely

Table 2.1. *Shared socioeconomic pathways (SSPs)*

SSP name	Population growth	Urbanization	Education & health investments	Equity	Social cohesion & participation
1 Sustainability: Taking the green road	Low	High, well managed	High	High	High
2 Middle of the road	Medium	Medium, continuing historical patterns	Medium	Medium	Medium
3 Regional rivalry: A rocky road	Low in HICs High in LICs	Low, poorly managed	Low	Low	Low
4 Inequality: A road divided	Low in HICs High in LICs	Mixed	Variable & inequitable across regions	Medium	Low
5 Fossil-fuelled development: Taking the highway	Low	High, some sprawl	High	High	High

Source: O'Neill BC, Kriegler E, Ebi KL, et al. The roads ahead: narratives for shared socioeconomic pathways describing world futures in the 21st century. *Global Environmental Change.* 2017;42:169–80.

diverging projections by the year 2100: global energy consumption ranging from 400 to 1200 EJ, and land use dynamics ranging from reduced cropland area to a massive expansion by more than 700 million hectares.

Based on constructs such as the RCPs and the SSPs, investigators can project future climate conditions and impacts, using integrated assessment models (IAMs). These models, as the name suggests, integrate projections about climate with projections about a range of social, economic, and other variables. They can also support projections of health impacts; SSP1, the sustainability pathway, is likely to yield far lower health burdens than SSP5, the fossil fuel pathway (39), but more work is needed to quantify these differences in outcome and the relative contributions of environmental and socioeconomic factors.

But Was Climate Change the Cause?

A key question is the extent to which climate change – alone or interacting with other environmental and social processes – causes extreme events such as severe storms, or gradual changes such as sea level rise and increasing heat exposure, that in turn lead to such human impacts as the spread of infectious disease or armed conflict. The historical analyses described above have triggered vigorous debate about the causal role of climate drivers, with some observers arguing that quantitative associations do not necessarily

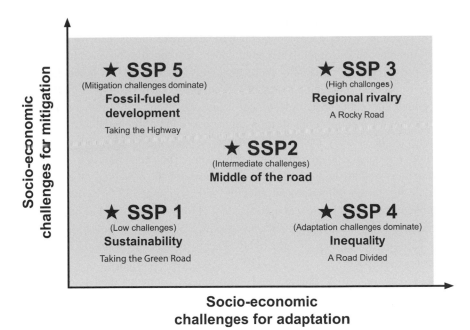

Figure 2.3. The five shared socioeconomic pathways (SSPs), showing differing challenges to mitigation and adaptation.
Source: O'Neill BC, Kriegler E, Ebi KL, et al. The roads ahead: narratives for shared socioeconomic pathways describing world futures in the 21st century. *Global Environmental Change*. 2017;42:169–80.

indicate causation and that many analyses ignore historical complexities. Investigators in Hong Kong and China for example developed a robust method to test historical causality using data from a long period that included times of harmonious development as well as times of great social unrest and population decline (40). This approach considers whether there is a plausible explanation for a causal relationship, assesses whether the strength, consistency, and timing of associations suggest a causal link, and evaluates whether the proposed causal variable can reliably predict the outcome. The investigators collected and tabulated all available historical data about climate, agroecology, economy, society, human ecology, and demography in Europe between AD 1500 and 1800, using 14 relevant variables, to assess causal relationships. They were able to simulate 'golden' and 'dark' periods in Europe, and the Northern Hemisphere more widely. They found that cooling was a major driver of socioeconomic decline, which in turn caused widespread crisis. In AD 1650, the European population fell to its lowest point (105 million) in the period due to war and famine. The changing climate was found to play a causal role: declines of temperature were followed by peaks of social disturbance, such as rebellions, revolutions, and political reforms, with a time lag varying between 1 and 15 years.

That said, the role of climate must be placed in context. Whilst climatic and other environmental changes can play key roles in triggering crises, many social scientists consider that the effectiveness of governance mechanisms and the resilience of socioeconomic systems determine the ultimate impact of these stressors on human societies. 'Undue attention to stressors,' writes geographer, ecologist, and archaeologist Karl W. Butzer, 'risks underestimating the intricate interplay of environmental, political and sociocultural resilience in limiting the damages of collapse or facilitating reconstruction' (41). **Figure 2.4** summarizes Butzer's proposed conceptual model of societal breakdown.

It is perhaps inevitable that, in historical analyses, different disciplinary perspectives yield somewhat different conceptualizations of the role of environmental change in societal crises, either as the primary cause or as a trigger subject to modulation by political and sociocultural responses and governance mechanisms. This may reflect the diversity of events studied as well as the methods used. In some cases the effects of an environmental trigger could have been minimized by effective and prompt action or by having resilient and equitable mechanisms to prevent upheaval and ultimately collapse. For other events the environmental drivers are predominant and so powerful that even well-ordered societies would be hard-pressed to respond effectively. This is increasingly likely to be the case in the Anthropocene Epoch.

In recent years, improved datasets and more refined computer models and analytical approaches have enabled investigators to attribute events to climate change with increasing confidence – an emerging discipline known as attribution science. Such attribution is an exercise in probabilistic thinking: an event that would have been X% likely to occur without climate change became Y% likely to occur in the setting of climate change, so based on the excess of Y relative to X, one attributes the event to climate change with defined levels of probability and confidence (42, 43). For example, a study published after Hurricane Harvey in 2017 estimated that human-induced climate change likely increased the probability of the most catastrophic levels of rain in Houston by a factor of at least 3.5, and that total rainfall in the worst affected areas was likely increased by ∼38% (with a lower bound of ∼19%) (44). Analogous thinking helps in attributing human health outcomes to climate change. For example, a study of the 2003 European heat wave (discussed further below), using computer simulations of weather that year both with and (counterfactually) without anthropogenic climate change, found that climate change increased the risk of heat-related mortality by ∼70% in Central Paris and by ∼20% in London, where peak temperatures were lower. Of the estimated ∼735 and ∼315 summer deaths attributed to the heat wave in Central Paris and Greater London, respectively, the investigators attributed 506 (±51) in Paris and 64 (±3) in London to anthropogenic climate change (45). Such analyses are becoming increasingly robust, and are applied to other health outcomes such as Lyme disease and Vibrio emergence (46). Results increasingly help provide evidence-based answer to the question 'Is it causal?'

The Impacts of Climate Change on Human Health and Well-Being

The diverse range of pathways by which climate change can affect health is shown in **Figure 2.5**. The complexity of potential linkages means that effects are likely to be pervasive and far-reaching but also introduces uncertainties in estimating numbers of people

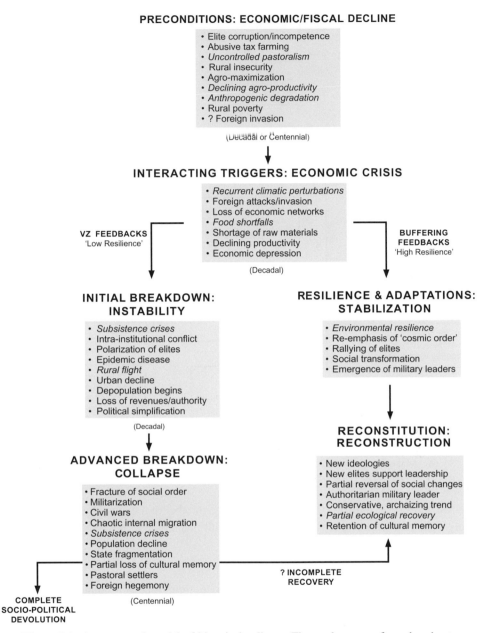

Figure 2.4. A conceptual model of historical collapse. Timescales range from decades to centuries. Alternate pathways point to important qualities of resilience. Environmental components (italicized) are secondary to sociopolitical factors.

Source: Adapted from Butzer KW. Collapse, environment, and society. *Proceedings of the National Academy of Sciences.* 2012;109(10):3632–9.

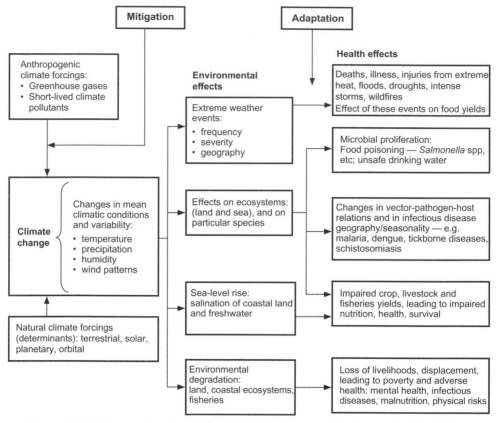

Figure 2.5. Pathways through which climate change can affect human health. (Major examples shown, but figure is not exhaustive.)

Source: Adapted from McMichael AJ, Woodruff RE, Hales S. Climate change and human health: present and future risks. *The Lancet*. 2006;367(9513):859–69.

affected. Estimates of effects that focus on a limited number of the most direct pathways are likely to be most reliable, but underestimate the full magnitude of the impacts.

Heat

The medical consequences of extreme heat range from relatively minor, self-limiting conditions, such as heat rash and cramping, to severe and sometimes fatal outcomes such as heat stroke (47). Physiological adaptation to heat is possible, but only to a point. Even for healthy young adults, there are limits to the ability to tolerate extreme heat exposure. When the core body temperature exceeds 38 °C there is a risk of heat exhaustion with impaired cognitive function and reduced ability to undertake physical labour, and above 40.6 °C heat stroke is likely, with risks of loss of consciousness, organ failure and death.

More consequentially from a population point of view, mortality rates rise during periods of heat, especially amongst the elderly, mostly due to increases in cardiovascular deaths (48). The threshold temperature at which death rates rise with heat varies according to location, being higher in places that are acclimatized to hot climates. A notable example is the European heat wave of 2003, which resulted in about 15,000 deaths in France. Deaths across 16 European countries brought the total to about 70,000 (49). In France 80% of the heat-related deaths occurred in people aged 75 and over. A distinctive feature of the 2003 event was that, unlike smaller heat waves where a significant proportion of the deaths is merely brought forward by a short period and followed subsequently by a dip in deaths (called mortality displacement, or rather unkindly, the harvesting effect), only a small proportion of the total deaths appeared due to short-term displacement. This implies that the event was severe enough to kill large numbers of people who were not expected to die in the near future. Such events place substantial burdens on health care systems, although there have been few efforts to quantify the costs and consequences (50). Many heat-related deaths also occur outside defined heat waves, when temperatures are nevertheless above the local threshold, but prolonged exposure to extreme heat is particularly hazardous.

It has been estimated that the number of vulnerable people exposed to heat-wave events increased by about 125 million between 2000 and 2016, compared to the 1986–2008 average, with a record 175 million more people exposed to heat waves in 2015 (51). This is largely due to an increase in the elderly population in areas experiencing increasing heat waves. There is evidence, in some parts of the world, that the mortality response to extreme heat has decreased over recent years, indicating some adaptation (ranging from behavioural responses to air conditioning) (52, 53). However, exposure to heat stress increases non-linearly with climate change, which means that a given level of future warming could have larger societal impacts than experienced historically. This is particularly so when using metrics that take into account increasing humidity. In South Asia, for example, a study showed that when global average temperature increases by only 1.5 °C above pre-industrial levels, twice as many megacities could experience similar levels of heat exposure to those in the extreme heat event of 2015 (54). This would expose more than 350 million additional people to 'deadly heat' by 2050 under a mid-range population growth scenario. If temperature increases reach 2 °C, the 2015 extreme heat event could become a yearly occurrence for major cities such as Karachi and Kolkata.

In many parts of the world cold-related deaths are more numerous than heat-related deaths (13), which suggests the potential for reduced mortality as global average temperatures rise. However, recent research (55) shows that with continued heating under high-emission scenarios, the rapid increase in heat-related deaths in regions such as South and Southeast Asia and southern Europe will exceed any decline in cold-related deaths. (Few data are available from African locations but they are likely to face similar threats.) Some studies have cast doubt on whether cold-related mortality will in fact decrease substantially in a warmer world because of the effects of potential increases in temperature variability, as exemplified by the dozens of deaths in Texas during an unaccustomed severe cold snap in February 2021 (21).

Heat has wide-ranging effects short of mortality. These range from increased risk of kidney stones (56) and possibly chronic renal disease (57) to impaired sleep (58) to pregnancy complications and adverse birth outcomes (59, 60), and from increased workplace injuries (61) to increased violence (62) and possibly suicide (63).

The effects of heat on working people are a special concern, because of the impact of reduced labour productivity on poverty and development (64, 65). Increasingly hot weather affects a range of occupations in tropical and sub-tropical regions, particularly where access to reliable air conditioning is infeasible for economic or logistical reasons. As heat stress increases it becomes progressively more difficult and potentially hazardous to undertake physical labour. A commonly used metric of heat exposure is the wet bulb globe temperature (WBGT) which takes into account humidity as well as temperature and includes solar radiation, which adds substantially to heat exposure when people are working outdoors without shade. The US National Institute of Occupational Safety and Health (NIOSH) recommends that no physical labour be undertaken above WBGT 34 °C to minimize risks to health (66).

As heat stress increases, productivity falls. **Figure 2.6** shows the steep decline of labour capacity with increasing heat at different levels of work output (67). According to one estimate, global labour capacity in populations exposed to increased temperatures decreased by 5.3% from 2000 to 2016 (51). This trend is projected to continue, with a predicted 1 billion or more workers exposed to WBGT levels above the NIOSH safe work threshold, and about 20 million pushing the limits of survivability, at a global temperature increase of 2.5 °C above pre-industrial levels (68). According to another projection, later this century, in some areas, 30–40% of annual daylight hours will become too hot for work to be carried out. The social and economic impacts could include global Gross Domestic Product (GDP) losses greater than 20% by 2100 as a result (69). Taken together these and other sources of evidence suggest the potential for severe disruption of work patterns later this century as a result of increasing heat, particularly for those least able to adjust their working patterns such as day labourers, construction workers, and subsistence farmers (see Chapter 3 for further discussion).

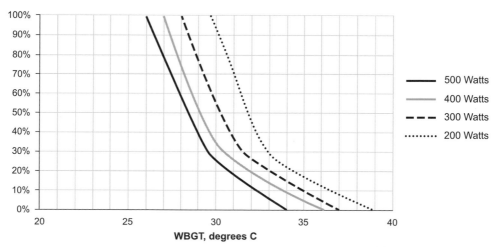

Figure 2.6. The association between work capacity and wet bulb globe temperature (WGBT) for four work intensities.

Source: Kjellstrom T, Kovats RS, Lloyd SJ, Holt T, Tol RS. The direct impact of climate change on regional labor productivity. *Archives of Environmental & Occupational Health.* 2009;64(4):217–27.

Extreme Weather

Floods: Climate change increases the frequency and intensity of floods through a range of mechanisms including heavy rainfall, sea level rise and melting of glaciers. Floods can cause a range of health effects ranging from short-term deaths and injuries to infectious diseases such as leptospirosis, diarrhoeal diseases, vector-borne diseases and loss of long-term medication in people with chronic diseases when supply chains are disrupted, particularly in low-income countries (70, 71). Flooding and heavy rainfall are also associated with an increase in snake bites, particularly when flood waters recede (72). The mental health effects of floods are increasingly recognized and can be of long duration. Common mental disorders (anxiety and depression) were two to five times more prevalent in people reporting flood water entering their home than in non-flooded comparison populations following the 2007 floods in the UK (73); leading risk factors for mental health problems after a flood include being displaced and suffering financial insecurity (74, 75).

A related problem is saltwater intrusion into coastal aquifers. In Bangladesh, sea level rise compounded by local factors including shrimp farming and damming of rivers have led to salination of freshwater supplies. Studies in that country show associated increases in pregnancy-related hypertension and pre-eclampsia (76) and in blood pressure in non-pregnant adults (77). Some evidence, however, suggests that for those consuming water with modest levels of salinity there was a paradoxical beneficial effect on blood pressure, perhaps due to displacement of sodium by high levels of calcium and magnesium in the drinking water (78).

Drought: Defining the presence and the start and end dates of a drought is often challenging (79). There are 150 definitions of drought in the literature, although most feature a deficit in water availability compared with the local norm. Different categories of drought include meteorological (as a result of reduced rainfall), hydrological (reflecting the effect of lack of rain on the stocks of surface- or groundwater), agricultural (indicating the effects of inadequate water availability on crop growth), and socioeconomic (when the shortfall in water causes an imbalance between supply and demand of goods and therefore on key economic indicators such as income, affordability of food, etc.). These types of drought are interlinked and reflect a focus on different processes and impacts. **Figure 2.7** summarizes the growing evidence about the direct and indirect health effects of droughts.

Weather disasters: A disaster is an unforeseen and often sudden event that causes great damage, destruction, and human suffering. The world's most comprehensive disaster database, the Emergency Events Database (EM-DAT) at the Catholic University of Louvain (emdat.be), applies a further criterion in its operational definition: that the situation or event overwhelms local capacity, requiring a request to national or international authorities for assistance. According to EM-DAT estimates, over 2 billion people worldwide were affected by floods between 1998 and 2017; the next most common disaster was droughts, which affected about 1.5 billion people. It is likely that these numbers are underestimates, particularly in low-income countries where data collection is challenging. This makes it difficult to follow disaster trends in relation to climate change and other drivers accurately. Between 2007 and 2016, the EM-DAT recorded an annual mean of 306 weather-related

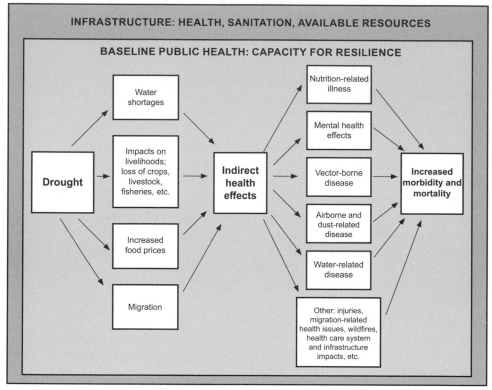

Figure 2.7. Health impacts of drought and factors that influence them.
Source: Stanke C, Kerac M, Prudhomme C, Medlock J, Murray V. Health effects of drought: a systematic review of the evidence. *PLoS Currents Disasters*. 2013;5.

disasters, an increase of 46% from the 1990–1999 average. However, this apparent increase may have been partly due to more complete reporting in recent years. Thus far, there is no evidence that death rates or numbers of people affected have increased, probably due to poverty reduction and improved disaster preparedness. The number of people affected by floods and storms in Africa has increased significantly over recent decades, although the lethality of these events has not changed significantly. Death rates from disasters are much higher in low-income countries than in wealthy countries. There is considerable national variability in trends however, and some countries have witnessed increases in weather-related disaster deaths (see (51) for discussion) and costs. In the USA, the annual average of billion-dollar disasters was 6.5 between 1980 and 2019, but that average more than doubled, to 13.8, during the final five years (2015–2019) (80), with losses reaching at least US$400 billion per year. Some of this increase is attributable to population growth and the growing concentration of valuable infrastructure in disaster-prone areas, but some reflects rising storm intensity. A key question for the future is for how long will poverty reduction and adaptation strategies be able to hold the global numbers of affected people in check. The high death toll of about 3000 in Puerto Rico following Hurricane Maria in 2018 is a reminder that powerful storms can have major impacts even in developed settings (81).

Infectious Diseases

Climatic influences determine the distribution and incidence of many infectious diseases, notably vector-borne, food-borne, and water-borne diseases. At the same time, there are advances in disease control, availability of clean water and sanitation, and food safety that can counterbalance the effect of climate change to varying extents, depending on location, available resources, and the magnitude of environmental change (see Chapters 3 and 4).

Vector-borne diseases: The two major vector-borne threats to health globally are malaria and dengue, while other vector-borne diseases such as those caused by West Nile and Zika viruses have much smaller global impact (although this is not to minimize their local importance). Quantifying the role of climate can often be challenging; for example, the northward movement of the ticks that transmit Lyme disease in North America with climate change has not yet been accompanied by major increases in human infection in the affected locations. In Europe changes in the distribution of tick-borne encephalitis cannot be solely ascribed to climate change and in some locations changes in recreational and agricultural activities are likely to be responsible (see (21) for discussion).

Malaria is caused by five distinct species of the *Plasmodium* parasite transmitted by different *Anopheline* mosquito vectors, which vary in their responses to climatic factors. Over the past century the dominant factors affecting the geographical extent and endemicity of malaria have been socioeconomic development and control strategies, which have greatly outweighed the effects of the global warming experienced so far (82). An estimated 228 million cases of malaria occurred worldwide in 2018 (93% of these in Africa), compared with 237 million cases in 2010. The incidence rate of malaria declined globally between 2010 and 2018, from 71 to 57 cases per 1000 population at risk, but the decline slowed dramatically from 2014 to 2018, and incidence actually rose in the Eastern Mediterranean and Americas. Malaria deaths also declined globally over this period, from 585,000 in 2010 to 405,000 in 2018, as a result of improved diagnosis, treatment, and vector control, but again, this trend did not hold in the Eastern Mediterranean and Americas regions, where the mortality rates increased after 2014 (83). *Plasmodium falciparum* is responsible for most of the deaths. According to the WHO (84, p. 68), the challenges to achieving and maintaining progress include 'the lack of robust, predictable and sustained international and domestic financing; the risks posed by conflict and other complex situations; the emergence of parasite resistance to antimalarial medicines and of mosquito resistance to insecticides; and the inadequate performance of health systems'.

In some regions, such as the highlands of East Africa, increasing temperatures may have contributed to malaria spread at higher altitudes. The response to temperature is non-linear so even small increases may have a disproportionate effect on transmission in suitable locations. Research based on records from highland regions of Ethiopia and Colombia suggests that future climate warming will substantially increase the risks of malaria transmission in densely populated regions of Africa and South America, if disease monitoring and control efforts are not enhanced (84). On the other hand, climate change will limit malaria propagation in some places, as extremely high temperatures exceed the limits at which vectors can survive and impair the abilities of parasites to reproduce in the mosquito.

Dengue fever is another important climate-sensitive vector-borne disease. It is caused by one of four types of dengue virus (called serotypes), and is the commonest vector-borne disease worldwide, responsible for an estimated 390 million (range 284–528) infections each year, of which only one in four causes symptoms (85). There has been a 30-fold increase in the dengue disease burden over the past 50 years, due to a combination of factors including trade patterns, travel, urbanization, inadequate domestic water supplies leading to peri-domestic water storage and the changing climate. There is evidence that vectorial capacity (the capacity to transmit infections) of the two common mosquito vectors for dengue, *Aedes aegypti* and *Aedes albopictus*, has increased since 1950 by 9.4% and 11.1%, respectively, as a result of climate trends (51). Although the fatality rate is low, dengue can incapacitate large numbers of people at a time. In a small proportion of cases, probably those involving re-infection with a different strain of dengue, dengue haemorrhagic fever can develop as a result of abnormal blood clotting, in some cases leading to circulatory failure, dengue shock syndrome, and a high risk of death. Modelling studies suggest the potential for expansion of the populations where transmission can occur due to climate change. One study, using projected changes in vapour pressure, a measure of humidity, estimated that with a high greenhouse gas emission scenario, by 2085 about 5–6 billion people (50–60% of the projected global population) would be at risk of dengue transmission, compared with 3.5 billion people, or 35% of the population, in the absence of climate change (86). The storage of water in containers can provide breeding sites for mosquito vectors providing an indirect pathway by which climate variability and change can influence transmission. The two dengue vectors also carry other important emerging or re-emerging arboviruses, including Yellow Fever, Chikungunya, Mayaro, and Zika viruses, and these infections may, therefore, also be affected by climate change. The incidence of vector-borne diseases depends of course not just on environmentally influenced transmission patterns, but also on the effectiveness of disease control systems and on development more generally, for example through improved housing or water storage (see Chapter 3).

Food- and water-borne illnesses: Food safety is likely to be compromised by climate change through a range of mechanisms and pathways including increases in bacteriological and viral contamination, and increases in algal biotoxins and fungal toxins (mycotoxins). The total burden of food-borne diseases is substantial. The WHO estimates that there were between 420 and 960 million cases of food-borne illness worldwide in 2010 and about 420,000 (95% uncertainty interval: 310,000–600,000) deaths as a result (87). The largest category was diarrhoeal disease, particularly from (non-typhoidal) *Salmonella* organisms, which caused the death of about 230,000 people. Microbiological contamination of food and water become more likely in hot weather, resulting in increases in gastrointestinal disease (88–90). Higher temperatures generally encourage bacterial growth. For example, a study in European countries found that above a threshold of 6 °C, there was a linear relationship between ambient temperature and the number of reported cases of salmonellosis. The strongest associations were with *Salmonella enteridis* (a serotype of *Salmonella*) infection in adults and with temperatures one week before the onset of the illness (91). In

England and Wales, food poisoning due to campylobacteriosis, salmonellosis, *Salmonella typhimurium* and *Salmonella enteritidis* infections were associated with temperature in the current and previous week. There were significant reductions in the impact of temperature on food-borne illnesses from the 1970s to the early 2000s, probably reflecting reduced pathogen concentrations in food and improved food hygiene (92). In low-income countries, the burden is likely to be much higher and control systems less effective. An estimated 48,000 additional deaths from climate change among children under the age of 15 resulting from diarrhoeal illness are projected by 2030 (93).

Mycotoxins are produced by a large range of fungi and can cause a range of acute and chronic health effects, including increased risk of liver cancer and probably of stunting, particularly in Sub-Saharan Africa. About 25% of the world's crop production annually is contaminated by mycotoxins. Increasing moisture levels, temperature and carbon dioxide levels may further increase contamination (94) – a trend that has been documented in European maize and wheat when temperature increases exceed 2 °C (95, 96). Ingestion may occur directly through consumption of affected crops or by consuming livestock products from animals that have consumed such crops.

Water-related illnesses: The distribution of schistosomiasis, a water-related parasitic disease for which aquatic snails are intermediate hosts, is influenced by climate throughout its complicated ecology, including reproduction, survival, and fecundity of the parasite and the snail hosts and transmission to humans (97, 98). In China, the increased incidence of schistosomiasis over the decade preceding 2005 may have partly reflected the observed warming trend. The critical 'freeze line' which limits the survival of the intermediate water snail host, and hence the transmission of the parasite *Schistosoma japonicum*, moved northwards, exposing an additional 20.7 million people to the risk of schistosomiasis (99). By 2050, an additional 784,000 km^2 could come within the transmission zone with further northward movement under climate change (100). There is less certainty about the impact of climate change on schistosomiasis distribution in Africa (101), where the predominant species of parasite are *S. haematobium* and *S. mansoni*.

The native marine genus *Vibrio* includes several organisms that cause human disease, most notably *V. cholerae*, the causative agent of cholera, an infection known and feared throughout history because of its ability to cause pandemics affecting many thousands of people. It was once considered 'America's greatest scourge' (102) after it ravaged communities between New Orleans and New York City in 1849, but virtually disappeared with improved sanitation and is now responsible for only sporadic cases in the USA. The organisms can be present in riverine, coastal, and estuarine environments where they have a symbiotic relationship with plankton, particularly copepods (103). By metabolizing the insoluble chitin coat of the copepod and returning polysaccharides to the ecosystem in a soluble form, Vibrios help to support the carbon and nitrogen cycles. By adhering to the copepods their metabolic activity is enhanced; adherence to eggs, which are widely dispersed in water, facilitates the spread of the organisms. Vibrios can also enter a dormant state when conditions are unfavourable for active growth, allowing them to survive until conditions improve. **Figure 2.8** summarizes the complex environmental relationships that influence *V. cholerae* transmission and shows how climatic factors, particularly temperature

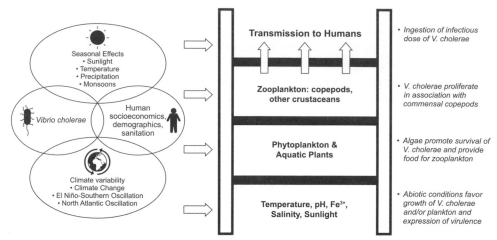

Figure 2.8. A hierarchical model for environmental cholera transmission.
Source: Adapted from Lipp EK, Huq A, Colwell RR. Effects of global climate on infectious disease: the cholera model. *Clinical Microbiology Reviews*. 2002;15(4):757–70. Figure redrawn by Emanuel Santos.

and rainfall, can interact with local environmental factors such as salinity and nutrient supply. During algal blooms *V. cholerae* are more likely to attach to copepods. In the Indian Ocean during August and September, after the monsoon season, levels of nutrients significantly increase and blooms of phytoplankton followed by zooplankton (copepods) occur, leading to increased production of copepod eggs. Another peak occurs between February and April. Both of these correlate well with the incidence of diarrhoea and cholera cases, particularly in India and Bangladesh (103).

A strong relationship between *Vibrio* numbers and sea surface temperature in the North Sea has been established by DNA analyses of samples collected over a 44-year period; increased sea surface temperature explained 45% of the variance in *Vibrio* data (104). In 2018, 107 days were suitable for transmission of pathogenic *Vibrio* (non-cholera) in the Baltic Sea, the highest since records began, and twice the number in the early 1980s (26). Levels of *V. vulnificus* and *V. parahaemolyticus* in shellfish are also influenced by sea surface temperature and by rainfall affecting salinity levels (105).

A symptom of altered coastal systems is harmful algal blooms, which are becoming more frequent and extensive in many parts of the world (106, 107). Even small 1–2 °C increases in ocean temperatures may yield relatively large increases in growth rates of common dinoflagellate organisms, such as *Gambierdiscus* and *Fukuyoa* species, which produce substances that are toxic to humans. People absorb these toxins primarily by eating contaminated fish and shellfish. Health impacts include various forms of shellfish poisoning (amnesic, neurotoxic, paralytic, and diarrhetic) and ciguatera fish poisoning, with a range of gastrointestinal, neurological, and cardiovascular symptoms. These health impacts can be compounded by economic impacts on fishing, recreation, and tourism during large algal blooms. The combination of nutrient loading, hypoxic water, and algal blooms can trigger major fish die-offs, as has occurred periodically along the Philippine coast (108, 109). Such events can threaten local nutrition and livelihoods.

Threats to the Food Supply

One of the major challenges for humanity in the Anthropocene Epoch is to safeguard the food supply for a growing world population in the face of climate and other environmental changes. There is scientific consensus that climate change is likely to reduce the yield of staple crops, particularly at low latitudes, in coming decades. A systematic review that screened over 1100 studies and used data from 52 of them to assess future trends in crop yield under climate change in Africa and South Asia concluded that climate change posed risks to food security, particularly in countries that were already food insecure. By the 2050s the mean projected yield decline was −8% across all crops studied. Average yield changes of −17% (wheat), −5% (maize), −15% (sorghum), and −10% (millet) across Africa and −16% (maize) and −11% (sorghum) across South Asia were projected. No mean change in yield was detected for rice (110). Inevitably research syntheses such as this integrate a range of assumptions across different models, but by systematically assessing and combining the results of different studies in an unbiased way, they provide the best available evidence of likely impacts.

A subsequent meta-analysis assessed the potential of adaptation strategies to offset the yield losses. Consistent with the meta-analysis just described, this one also projected reduced production of wheat, rice, and maize in both temperate and tropical regions at a 2 °C increase in local temperature in the absence of adaptation (111). It also projected that adaptation could raise yields by 8–15% for wheat and rice relative to no adaptation, largely countering the yield losses, with much less benefit for maize. Some of the adaptations considered were incremental changes to existing cropping systems, such as changes in varieties, planting times, irrigation, and residue management. Others were more systemic changes such as crop species or grazing integration, and still others were transformational options such as crop relocation or shifting from irrigated to dryland systems. Taking into account the potential for adaptation there are major projected declines in wheat and maize yield in tropical regions increasing after mid-century. Although there is less consistent evidence for temperate regions, even modest increases in temperature can cause yield declines in some locations.

Consistent with the results of that meta-analysis, the IPCC also concluded that the response of crops to climate varies according the prevailing climate. In temperate regions increasing temperatures may increase yield, at least for a time, but in warmer regions declines in yield are more likely (112). Positive trends are already evident in some high-latitude regions but negative trends are more common. Around daytime temperatures of 30 °C many crops show pronounced declines in yield. Similarly, climate change is reducing the harvesting of aquatic species, both freshwater and marine, in some tropical developing countries, but other regions are becoming more favourable for exploitation.

Current projections suggest that effects become more negative from mid-century. However, many such assessments do not consider the full range of environmental challenges facing the food system, including the effects of ozone pollution, salination of soils, prolonged exposure to high temperatures, reduced labour productivity from increased heat stress, and increases in pests and weeds – all of which could contribute additionally to reduced yields. For example, yield losses from insect pests for the three most important grain crops – namely wheat, rice, and maize – will probably increase by 10–25% for each degree C warming, particularly in temperate regions (113). This will make it more difficult to compensate for decreases in crop yield in tropical and sub-tropical regions by increasing yields elsewhere.

Much of the work on crop yields has been undertaken on staple crops, which are essential for food security. There has been less attention to fruit, vegetables, nuts, and seeds, which are important for preventing non-communicable diseases. A recent systematic review and meta-analysis of studies of the effects of environmental factors on vegetable and legume yield showed substantial negative impacts of increasing temperatures above $20\,^{\circ}\text{C}$, as well as from salination of soils and exposure to tropospheric ozone (114). High levels of carbon dioxide increased yield but this is likely to be more than offset by the combined negative effects of the other environmental exposures due to climate change.

Increases in CO_2 concentration tend to have larger effects on some plants than others, depending upon the photosynthetic pathway used. While not a function of climate change itself, we discuss the effects here because CO_2 is a major driver of climate change. In C3 plants including most food crops – wheat, rice, rye, barley, soy, sugar beets, potatoes and yams, cassava, spinach, tomatoes, apples, and peaches – the first product of photosynthesis is a 3-carbon compound. In C4 plants, including maize, sorghum, and sugarcane, the first product of photosynthesis is a 4-carbon compound. Rates of photosynthesis in C3 crops are more responsive to increases in ambient CO_2. Increased CO_2 may, however, have indirect effects on the growth of C4 crops, particularly under conditions of water restriction, because it increases the efficiency with which plants use water by closing the stomata and thus reducing water loss. The effects of CO_2 on growth vary according to genotype, availability of nutrients and temperature, with low and high temperatures limiting the response. Fertilization by CO_2 decreases the concentrations of micronutrients important for human health and reduces protein in a range of crops. For example, wheat, barley, potatoes. and rice grown in high CO_2 conditions have lower protein levels and concentrations of zinc and iron are also decreased (115, 116).

In a multi-year, multi-location free-air CO_2 enrichment (FACE) study of 18 genetically diverse rice lines currently grown throughout Asia, crops were exposed to varying elevated levels of CO_2, including levels expected to be reached later in this century. The results confirmed the expected declines in protein, iron, and zinc in the rice, and also found consistent declines in vitamins B1, B2, B5, and B9 with higher levels of CO_2, but conversely, an increase in vitamin E (117). Rice is the primary food source for more than 2 billion people and these findings suggest a net negative nutritional effect with important implications for public health. Another study analysed data from FACE and open-top chamber experiments to model dietary zinc intake at elevated CO_2 concentrations of 550 ppm, the concentration expected by 2050. It estimated that an additional 138 million (95% CI: 120–156 million) people could be placed at new risk of zinc deficiency by 2050. The most affected populations are projected to live in Africa and South Asia, with nearly 48 million (95% CI: 32–64 million) living in India (118). A more recent analysis using dietary data for 151 countries estimated that (assuming 2050 population and CO_2 projections) elevated CO_2 could cause an additional 175 million people to be zinc deficient and an additional 122 million people to be protein deficient as well as worsening the situation of those with existing nutritional deficiencies. In addition, 1.4 billion women of childbearing age and children under 5 living in countries with greater than 20% anaemia prevalence would lose more than 4% of dietary iron, putting them at increased risk of iron-deficiency anaemia. South and Southeast Asia, Africa, and the Middle East are projected to be at highest risk (119). As with so many health impacts of climate change, the nations at highest risk are not those responsible for most CO_2 emissions, raising important questions of justice and fairness (120).

Emissions of CO_2 are often accompanied by ozone precursors (e.g. volatile organic compounds from road transport or methane from gas leaks) and the resulting increases in tropospheric ozone have reduced crop yield, in the case of wheat and soybean for example by about 10%. The impacts are particularly severe over India and China (121, 122).

Food prices are a major determinant of food security and, based on studies cited by IPCC (123), it is very likely that climate change will lead to increased food prices by 2050, with estimated increases ranging from 3 to 84%. These estimates do not take into account CO_2 fertilization or the negative impacts of several other factors on yield such as ozone, pests, and freshwater declines, together with the projected declines in nutritional quality. A systematic review of studies from 162 countries showed that increases in food prices led to reduced consumption and that these reductions were greater in low-income than high-income countries. Within countries at all levels of development, poorer households suffered the most adverse effects on food consumption (124).

A modelling study of the effect of climate change on nutrition-related deaths compared a reference 'middle-of-the-road' development scenario (SSP2, as explained in Table 2.1) both without climate change and with the RCP8.5 emissions pathway. Without climate change the reference scenario projected improvements in per capita food availability, including in fruit and vegetable consumption, between 2010 and 2050 and thus reductions in nutrition-related deaths. In contrast, the climate change scenario (with a 2 °C increase in global mean surface temperature by 2046–2065 compared to 1986–2005) projected mean per-person reductions of 3.2% in global food availability, 4.0% in fruit and vegetable consumption, and 0.7% in red meat consumption by 2050. This translated to 529,000 climate change-related deaths worldwide (95% CI: 314,000–736,000) as a result of decreased food availability by 2050, most in South and East Asia. This mortality burden was estimated to be double that from climate-related increases in the prevalence of underweight. Adoption of climate change mitigation policies would reduce the number of nutrition-related deaths by 29–71%, depending on their effectiveness in reducing emissions (125).

The effects of climate change on child nutrition are of particular concern because of the potential impacts on children's life prospects. Stunting (lower than expected height for age) is due to inadequate nutrition in the first 1000 days of life and leaves a long-term legacy by increasing the risk of death in childhood (126) (about four-fold in the case of severe stunting), as well as impairing cognitive development and thus reducing educational attainment and income in adulthood. About 149 million (22%) of all children aged under 5 worldwide are stunted (127). Climate change is likely to increase the risk of stunting. For example, a high-emission scenario resulted in a projected relative increase in severe stunting of 23% in Sub-Saharan Africa and 62% in South Asia in the 2050s (128). In such global-level studies, climate change increases the risk of undernutrition by reducing crop yield, which affects post-trade national calorie availability. However, the pace of economic advances may be more important than climate change in influencing the risk of under-nutrition, at least in the relatively near term. About two-thirds of the reduction in stunting between 1970 and 2012 was estimated to be due to improvements in female education, gender equality, and access to clean water and sanitation services (129). Thus, climate change can also affect stunting through its impacts on the economic and development prospects of the poor.

A study of the interactions between economic development, climate change, and stunting aimed to test the effects of different assumptions about poverty reduction and stunting in the presence of climate change (130). It showed that in the absence of climate change, under a 'prosperity scenario' broadly consistent with achieving the Sustainable Development Goals, the number of stunted children aged under 5 fell to 83 million in 2030, while under a 'poverty scenario' with the fraction of poor people held constant, the number was 110 million. Estimates of climate change-attributable stunting ranged from 570,000 children under the prosperity/low climate change scenario to 1 million under the poverty/high climate change scenario, with larger effects in rural than in urban areas. The relatively small impact is due to the short time horizon (changes in policy now would have relatively little impact by 2030) and would no doubt be much greater in subsequent decades. The investigators projected that rising prices would tend to increase stunting in countries with high levels of poverty and relatively high food prices. In contrast, in countries with higher incomes and relatively low food prices, rising prices would tend to decrease stunting. These findings have important implications for policy, suggesting that food prices that provide decent incomes to poor farmers together with high employment at living wages will reduce undernutrition and vulnerability to climate change.

The climate-related impacts on health, mediated by agricultural losses, do not operate only through nutritional pathways. A study of growing season temperatures and suicide rates in India found that temperatures above an optimum for crop yield of 20 °C were associated with an increase in suicide rates, consistent with an estimate that warming over the last 30 years is responsible for 59,300 suicides in India, accounting for 6.8% of the total upward trend (131). Although the findings and their interpretation were criticized by a number of respondents, the author defended her findings robustly (132).

There are few empirical data on nutrition, crop yield, and climate change from Sub-Saharan Africa where vulnerability is high. A study of 44,616 children aged under 5 in a subsistence farming community in rural Burkina Faso showed that survival was appreciably worse in children born in years with low crop yield (full-adjustment hazard ratio = 1.11 (95% CI: 1.02, 1.20) for a 90th- to 10th-centile decrease in annual crop yield). The effect appeared to persist up to the age of 5, suggesting a long-term adverse consequence of reduced food availability around the time of, or shortly after, birth (133). Low per capita household crop production was also associated with poorer nutritional status of children in this Burkina Faso study. Climate change is likely to increase child deaths in such communities.

According to the FAO (127), after many years of decline the trend reversed in the middle of the second decade of this century, with the number of undernourished people increasing from about 785 million in 2015 to 822 million in 2018, a number not seen since 2010. The FAO attributed food crises to climate change in ten countries in Africa, Latin America, and the Caribbean (and Pakistan), to conflict in seven countries in Africa and the Middle East (and Ukraine), and to both climate change and conflict in sixteen countries, predominantly in Africa. Only time will tell whether this is the beginning of a long-term trend influenced by climate change but this seems likely.

While agricultural crops dominate the literature on climate change, food, and nutrition, both livestock and fish are important considerations as well. Livestock provides important sources of nutrients, particularly to some pastoral and subsistence farming communities and of course provide livelihoods for farmers in countries at all levels of development. In livestock, as productivity increases, the capacity to tolerate heat decreases because of increased metabolic heat production, although it is possible to select for animals with higher heat tolerance through breeding. Heat stress can cause increased mortality in cows and interferes with reproductive efficiency in pigs. Higher temperatures would lead to declining dairy production, reduced weight gain and lower feed conversion efficiency, particularly in sub-tropical and tropical developing regions but with substantial negative effects projected also in parts of the USA, Southern Europe, Latin America, and Australasia (134). Blue tongue virus, a common vector-borne disease of livestock, has been shown to be highly climate sensitive and in Europe has spread northward with climate change. Tropical vector-borne diseases of livestock, such as Rift Valley fever and African horse sickness, may also spread to more temperate regions.

Fish also represent a substantial source of dietary protein for many populations, but unfortunately we cannot look to the oceans to make up any qualitative and quantitative shortfalls in land-based food production. Global fisheries, already compromised by over-fishing (135), are threatened by climate change, especially at low to mid-latitudes (136, 137), although fish catches may increase at higher latitudes. Estimates suggest that declining fish catches will leave 845 million people vulnerable to dietary deficiencies of iron, zinc, and vitamin A and 1.4 billion people vulnerable to deficiencies of vitamin B_{12} and omega-3 long-chain polyunsaturated fatty acids (138). Aquaculture will play a growing role in supplying fish, but climate change and ocean acidification threaten production, particularly of bivalves such as oysters, mussels, and scallops (139, 140).

Effects on Poverty, Migration, and Conflict

Poverty: The indirect effects of climate change through impacts on socioeconomic systems are likely to be pervasive and problematic to address through current governance mechanisms. The prediction that climate change will increase poverty is probably the least controversial and supported by extensive evidence marshalled by the World Bank (141). Currently, many of the factors that push households into poverty are related to climate variability, including food price shocks, natural disasters, and ill-health from climate-sensitive conditions (141). Small changes in the balance between people escaping poverty and returning to poverty can greatly affect the rate of poverty reduction over time. Poor people are more likely to be exposed to climate events, they tend to lose proportionally more of their income or assets for a given exposure, and they receive less support from social networks or formal financial mechanisms such as loans or insurance payouts. For example, although food price spikes may benefit farmers who can charge more for the food they sell (142), that benefit disappears if the cause of the price spike is the inability to produce food. In any case, net consumers of food, including most urban residents, suffer economically when food prices rise; the 2010–11 food price spike drove 44 million into poverty (143). Climate events can also adversely affect employment prospects because of the degradation of natural systems on which poor people depend (e.g. artisanal fisheries), reduced labour productivity from extreme heat, or destruction of local businesses (e.g. following floods) (141). Poor families are less likely to provide adequate nutritious food to their children or pay for needed health care after being exposed to extreme events. Chapter 4 describes in more detail the vulnerability of poor populations to climate change and subsequent chapters discuss policies to reduce the impacts amongst the poor.

A World Bank report (141) compared the effects of two distinct scenarios on poverty as well as on the effects of climate change. The first of these, 'Prosperity', assumes that the World Bank's goals of extreme poverty eradication and shared prosperity are met by 2030 (less than 3% of the world population remains in extreme poverty), and that virtually universal access to basic services is achieved. The second scenario, 'Poverty', is much more pessimistic; 11% of the world population remains in extreme poverty. The Bank's findings are probably underestimates because they do not take into account interactions between sectors, they are relatively short term, and they omit some potentially important pathways, for example, the loss of ecosystem services or potential migration after disasters. Nevertheless, the findings are striking and illustrative of the potential importance of economic and social policies in ameliorating the impacts of climate change. By 2030, between 3 and 16 million additional people in the 'prosperity' scenario and between 35 and 122 million additional people in the 'poverty' scenario would be in poverty because of climate change, relative to scenarios with no climate change effects (141).

Conflict: Conflict is an enormously complex phenomenon. Many factors – poor governance, ethnic rivalry, imperial ambition, economic pressure, resource scarcity – may contribute. For example, about two-thirds of all civil wars since 1946 have been fought along ethnic lines (144), possibly related to differential access to power and resources as well as the capacity to rapidly mobilize combatants along ethnic divisions (145). A study of conflicts between 1980 and 2010 showed that about 23% of conflict outbreaks in ethnically highly fractionalized countries robustly coincide with climate-related disasters (146). There was no direct evidence that the climate events triggered the conflicts but the presence of ethnic fractionalization may have caused the societal stresses which led to a greater likelihood of the event culminating in violence compared with more ethnically homogeneous societies. These results support the view that climate change likely contributes to conflict by aggravating underlying risk factors.

The study of climate as a contributor to conflict has sometimes sparked controversy because of the heterogeneity between studies. The IPCC (147) concluded that there was *medium confidence* that climate variability had an effect on conflict and *low confidence* that climate change has an effect. Nevertheless, it concluded that climate change has the potential to become a key risk because factors such as poverty and economic shocks that increase conflict risks are themselves sensitive to climate change (148).

Our understanding of these issues has been advanced by research linking conflict event data from Africa and Asia to remote sensing data on agricultural land use and on settlements of different ethnic groups (149). This approach enables quantification of drought conditions during the growing season of locally dominant crops and assessment of socioeconomic, livelihood, and political vulnerabilities. In very poor countries, local growing-season drought was found to increase the likelihood of sustained conflict for agriculturally dependent and politically marginalized groups, with conflict clusters in West Africa, the African Great Lakes region, and the Horn, as well as in South Asia (149).

A landmark paper using 45 different conflict datasets from 60 primary studies across all major world regions spanning 10,000 BC to the present found that deviations from mild temperatures and from normal precipitation patterns increased the risk of conflict in a highly statistically significant fashion (150). Each one standard deviation increase in temperature or extreme (high or low) rainfall was associated with a 14% increase in intergroup conflict (political violence, ethnic conflicts, and other forms of collective violence) and a 4% increase in interpersonal violence (such as murder, assault, and domestic violence). The investigators noted that the planet is projected to warm by 2–4 standard deviations by mid-century, signalling the potential for substantial increases in conflict. In a subsequent analysis (151), the same research team confirmed the findings. They noted that early studies suggested a U-shaped relationship between temperature and conflict, with both cold and heat extremes increasing conflict risks, but studies since 1950 during a relatively warm period in human history demonstrated a dominant relationship between higher temperatures and conflict. This analysis showed that temperature had a larger effect than rainfall. It is important to recognize, though, that there is considerable diversity between studies, particularly in the temperature findings, which may represent contextual differences and/or differences in sampling of climate variables.

Migration: Migration has long been a feature of human life. Households make decisions about migration for complex reasons (152), including:

- *Economic* factors including poverty, income volatility, and wage differentials; all might induce people to relocate from an area of low economic opportunity to one with more opportunity.
- *Political* factors including poor or dysfunctional governance, instability, and conflict.
- *Social* factors: family and cultural norms such as the need to acquire funds for dowries, seek an education, or rejoin diaspora communities settled abroad.
- *Demographic* factors such as population growth that reduces opportunity in a place of origin, and/or an ageing population in a receiving area that creates employment opportunities.
- *Environmental* factors including threats such as rising sea levels or severe storms, and loss of ecosystem services such as drought or soil degradation; these may arise suddenly or gradually.

Adding to the complexity is the fact that the drivers of migration interact with each other. For example, environmental degradation may lead to poverty, which in turn may drive migration.

Given this complexity, disentangling the drivers of migration (and specifically the role of climate change) is challenging. A number of theoretical models are in use, and different disciplines – demography, public health, geography, economics, and others – vary in their approaches (153). Estimates of the future role of climate in driving migration during this century vary widely, from a few hundred million to as high as 1 billion (154–156). Sea level rise alone, according to projections based on a high-emission scenario, could cause the forced displacement of up to 187 million people over the century (up to 2.4% of global population) (157). Temperature gradients are small in tropical regions, so to escape rising temperatures, people would have to move large distances – an alteration of the more typical pattern of short-distance migration (158). As a result, by the end of this century, 12.5% of the global population, mostly in tropical zones, would have to migrate more than 1000 kilometres to maintain stable temperatures – meaning large-scale migration to the edges of the tropics or beyond. Reflecting the uncertainty in projections of migration, the IPCC has classified migration as an emergent climate risk (147).

It is clear that migration is a growing phenomenon. The number of international migrants globally grew from 173 million in 2000 to 220 million in 2010, and to 258 million in 2017. The majority of movement is south-to-south, much of it between countries in the same region. In Africa and Asia, 80% of all international migrants were born in their region of residence (159). Among immigrants in Europe, those originating in Europe exceed the combined number of those from Africa and Asia. Data do not support the perception in some wealthy countries (and promulgated by some populist politicians) of a mass migration from the Global South to the Global North up to now (160).

Sea level rise (making some coastal settlements uninhabitable), severe storms (destroying communities), extreme heat (compromising liveability), and drought (curtailing agricultural production) are climate-related drivers of migration. Research suggests that low crop yields in Mexico drive migration to the USA (161), that weather anomalies drive rural to urban migration in Sub-Saharan Africa (162), that drought contributed to conflict and migration in Syria (163), and that heat stress drives people to depart their villages in Pakistan (164) and in Indonesia (165). Recent assessments concluded that climate change is driving resettlement in Vietnam's Mekong River delta, along the Limpopo River of Mozambique, on the Alaskan coast, in the Inner Mongolia Autonomous Region, and from the Carteret Islands to Bougainville Island in Papua New Guinea (165, 166). A study of the effects of climate change on migration into the EU using data from 103 source countries showed that deviation from an optimum temperature of about 20 °C in the growing season was associated with an increase in asylum applications (167). In the recent past such applications averaged about 350,000 annually from the source countries studied; the investigators projected that under a high-emissions pathway an increase of about 660,000 was projected by the end of this century.

In the Anthropocene, climate-related environmental pressures may operate alongside other drivers, such as pollinator loss and soil loss. But the environment is not working in isolation as a driver of migration and the role of other factors must be considered. For example, the decision to relocate may be influenced by social networks in the receiving area, government policies towards migrants in the receiving area, and levels of aid available to adapt or rebuild locally.

Reviews of empirical findings on climate-related migration have identified several patterns (153, 168). One is labour migration – what might be called 'ecological-economic migrants' who confront diminished livelihoods at home due to environmental changes, and seek better opportunities elsewhere. This kind of migration often involves just one member of a household, is mostly temporary (either seasonal or long term), and often features remittances sent back home. Forced migration consists of people fleeing disasters or circumstances that are no longer liveable. Such migrants generally move to the nearest place where life is bearable, often where aid is provided or where relatives live, especially after rapid-onset environmental changes. They tend to remain anchored to their place of origin (169, 170), and to return to their place of origin as soon as conditions permit (159). Long-term displacement, or resettlement, consists of permanently displaced populations, such as people leaving coastal areas or small island nations that can no longer sustain their way of life. People's social networks play a large role in the decision to migrate (171). Only where local social and cultural forces are positively disposed towards migration is it likely that people experiencing adverse effects from environmental change will begin to consider out-migration (172).

The alternative to all of these forms of movement is to remain in place, sometimes because of deprivation, sometimes because of a judgement that the risk of moving exceeds the risk of staying (152). Level of wealth is an important factor; people in poverty are less able to relocate than people with access to resources. Poverty increases vulnerability to acute and long-term environmental stressors, and limits a community's ability to adapt. But it also limits the ability to migrate in response to these changes, trapping some populations in place (173).

Poverty may also limit people's ability to return to their place of origin following forced migration. For example, when Hurricane Katrina hit New Orleans, many African-Americans lived in lower-lying areas of the city. Facing two challenges – greater levels of damage to their homes, and fewer resources with which to rebuild – many were unable to return to the city following the disaster (174).

Migration may be framed as a last resort – as a kind of failure – or more positively as an adaptive solution. In particular, planned and managed relocation can transition people from vulnerable situations to safer places (173, 175, 176). The deliberate relocation of populations is known as 'managed retreat' (although this term is controversial as it may connote defeat; less tendentious equivalent terms include 'planned relocation' and 'managed realignment') (177). Small coastal populations and some island populations have implemented this process, but managed retreat is rarely – some would say too rarely – contemplated in the context of larger settlements such as coastal cities (178).

While migration can be a viable form of adaptation, it can also be maladaptive. This can happen in at least two ways. First, when economically active young people depart an environmentally pressured area – say, at a time of drought – their community may be left less resilient, with a deficit of skills and knowledge, and less capable of carrying out activities such as planting and harvesting (179). Second, a disturbing possibility is that as people migrate from rural to urban areas, driven at least in part by environmental pressures, they may relocate to areas of *increased* risk (180). Many cities are highly vulnerable to such hazards as rising sea levels, severe weather events, and heat.

Grief and Other Mental Health Effects

In addition to the mental health impacts of extreme events such as floods and droughts (see above), climate change may threaten mental health in more pervasive ways, through the loss of familiar and beloved places, culturally significant features, or long-standing sources of employment. This has been termed 'ecological grief' (181) or 'global mourning', or, in the context of destruction of Australia's Great Barrier Reef, 'reef grief' (182). A closely related concept is 'solastalgia', defined as 'the pain experienced when there is recognition that the place where one resides and that one loves is under immediate assault' (183, p. 48). Research suggests that this response is more likely when it touches people's 'sense of place' – their place-attachment and place-based identity – than when their connection with place is simply one of aesthetic appreciation (182, 184).

Grief is a normal human response to loss of a loved one; the subsequent mourning processes help us to recover over time to varying degrees and to resume our personal and social lives as far as possible. The idea that grief responses can extend to environmental loss is a relatively recent development but one that has a growing evidence base, from observations in such climate-impacted places as Northern Canada and the Australian wheat belt (181). But in the context of climate change, the grief is not only place-based; similar reactions have been described in university students learning about climate change (185) and in climate scientists who regularly confront grim data (186). Grief may also be an anticipatory response to projected losses of familiar environments, particularly those identified with concepts of 'home', signifying profound emotional bonds with a specific location. This has been termed 'pre-traumatic stress syndrome' (187).

In Australia, research has shown how individuals and rural communities have lost confidence in their ability to predict and cope with seasonal variations in weather because of climatic change (181). This causes deep concerns about the future of their farming communities and the capacity of the land to sustain the livelihoods of their descendants (188, 189). Amongst the Inuit of Nunatsiavut, a community in Northern Canada, middle-aged and older people express sadness that their traditional knowledge of the migration routes of animals, on safe travel routes, and on weather patterns from generations of observations is becoming increasingly unreliable as the climate changes (190, 191). These deep feelings of loss extend beyond individual grief to the loss of cultural assets that have accrued over generations. Both the Australian farmers and the Inuit may express guilt that they have been unable to ensure good stewardship of 'their' land (189), leading to feelings of helplessness and impotence to prevent change (190, 191). Ways of reducing the suffering associated with loss may include deliberate anticipation of coming losses, collective action to prevent further loss, and memorializing what has been lost; the most effective approaches need to be clarified through what has been called a 'science of loss' (192, 193).

Nor is grief the only mental health response to the awareness of climate change. There is increasing recognition that concerns about climate change can manifest as anxiety and depression (194–196), particularly in children (197). Much of this surfaces on social media, especially in the context of the youth-led climate activism that blossomed around the world beginning in 2018. 'Terrified of climate change? You might have eco-anxiety,' read a 2019 headline in *Time* magazine (198). '"Overwhelming and terrifying": the rise of climate anxiety,' read another, a few months later, in the *Guardian* (199). 'Young people seek help for anxiety over climate change,' announced *The Times* (200) in early 2020, as the *Washington Post*, in a story entitled 'The environmental burden of generation Z', reported that 'Kids are terrified, anxious and depressed about climate change' (201). The full dimensions of these problems, and the optimal approaches to treatment, remain to be defined. However climate activism, and personal choices to live in climate-sensitive ways, are increasingly recognized as potential strategies for overcoming feelings of powerlessness and despair (202) (see Chapter 12).

Conclusions

Overall, climate change poses many current threats to health, which will intensify going forward, particularly from mid-century onwards, if we fail to achieve the emissions reductions proposed under the Paris Agreement. Current trends are off course to limit warming to 2 °C above pre-industrial temperatures and the future of humanity hangs in the balance. Humanity can undoubtedly adapt to some of the projected impacts, such as those affecting infectious disease distribution. Others, such as extreme heat exposure will be much more difficult, particularly for vulnerable populations. Corrupt, autocratic governments can in all probability amplify the effects of climate change through their failure to mount effective policy responses. The impacts of climate change on health will depend not just on whether humanity is able to achieve the climate goals agreed in Paris but also on wider development trends and particularly whether socioeconomic inequalities are addressed. These challenges are discussed in subsequent chapters. As Chapter 1 has shown, climate change is only one of several environmental trends posing unprecedented challenges to humanity. Chapter 3 describes the effects of other environmental changes on health.

References

1. Pendergrass AG, Knutti R, Lehner F, Deser C, Sanderson BM. Precipitation variability increases in a warmer climate. *Scientific Reports*. 2017;7(1):17966. https://doi.org/10.1038/s41598-017-17966-y.
2. Stocker TF, Qin D, Plattner G-K, et al., editors. *Climate Change 2013: The Physical Science Basis. Contribution of Working Group I to the Fifth Assessment Report of*

the Intergovernmental Panel on Climate Change. Cambridge, UK and New York: Cambridge University Press; 2013.

3. McMichael AJ, Woodward A, Muir C. Climate Change and the Health of Nations: Famines, Fevers, and the Fate of Populations. Oxford and New York: Oxford University Press; 2017.

4. Firestone RB, West A, Kennett JP, et al. Evidence for an extraterrestrial impact 12,900 years ago that contributed to the megafaunal extinctions and the Younger Dryas cooling. Proceedings of the National Academy of Sciences. 2007;104(41): 16016–21. https://doi.org/10.1073/pnas.0706977104.

5. Pinter N, Scott AC, Daulton TL, et al. The Younger Dryas impact hypothesis: a requiem. Earth-Science Reviews. 2011;106(3):247–64. DOI: 10.1002/jqs.2724.

6. Whitmee S, Haines A, Beyrer C, et al. Safeguarding human health in the Anthropocene Epoch: report of The Rockefeller Foundation–Lancet Commission on Planetary Health. The Lancet. 2015;386(10007):1973–2028.

7. Jordan WC. The Great Famine: Northern Europe in the Early Fourteenth Century. Princeton, NJ: Princeton University Press; 1996.

8. Haug GH, Günther D, Peterson LC, et al. Climate and the collapse of Maya civilization. Science. 2003;299(5613):1731–5. doi: 10.1126/science.1080444.

9. Wright LE, White CD. Human biology in the classic Maya collapse: evidence from paleopathology and paleodiet. Journal of World Prehistory. 1996;10(2):147–98. https://www.jstor.org/stable/25801093?seq=1.

10. Benson LV, Berry MS, Jolie EA, et al. Possible impacts of early-11th-, middle-12th-, and late-13th-century droughts on western Native Americans and the Mississippian Cahokians. Quaternary Science Reviews. 2007;26(3–4):336–50. doi: 10.1016/j. quascirev.2006.08.001.

11. Fagan B. Chaco Canyon: Archeologists Explore the Lives of an Ancient Society. Oxford: Oxford University Press; 2005.

12. Gasparrini A, Guo Y, Hashizume M, et al. Mortality risk attributable to high and low ambient temperature: a multicountry observational study. The Lancet. 2015; 386:369–75. doi: 10.1016/S0140-6736(14)62114-0.

13. Hatfield JL, Prueger JH. Temperature extremes: effect on plant growth and development. Weather and Climate Extremes. 2015; 10:4–10.

14. Bouma MJ, Kovats RS, Goubet SA, Cox JS, Haines A. Global assessment of El Niño's disaster burden. The Lancet. 1997;350(9089):1435–8. https://doi.org/10.1016/ s0140-6736(97)04509-1.

15. Lam CH, Haines A, McGregor G, Chan YE, Hajat S. Time-series study of associations between rates of people affected by disasters and the El Niño Southern Oscillation (ENSO) cycle. International Journal of Environmental Research and Public Health. 2019;16(17). doi: 10.3390/ijerph16173146.

16. Kovats RS, Bouma MJ, Hajat S, Worrall E, Haines A. El Niño and health. The Lancet. 2003;362(9394):1481–9. DOI: 10.1016/S0140-6736(03)14695-8.

17. Marlier ME, DeFries RS, Voulgarakis A, et al. El Niño and health risks from landscape fire emissions in southeast Asia. Nature Climate Change. 2013;3(2):131–6. https://doi. org/10.1038/nclimate1658.

18. WHO. El Niño affects more than 60 million people. Geneva: World Health Organization; 2016. Available from www.who.int/news-room/feature-stories/detail/ el-ni%C3%B1o-affects-more-than-60-million-people.

19. Ahmadalipour A, Moradkhani H. Escalating heat-stress mortality risk due to global warming in the Middle East and North Africa (MENA). Environment International. 2018;117:215–25. https://doi.org/10.1016/j.envint.2018.05.014.

20. Alonso D, Bouma MJ, Pascual M. Epidemic malaria and warmer temperatures in recent decades in an East African highland. *Proceedings of the Royal Society B: Biological Sciences.* 2011;278(1712):1661–9. doi: 10.1098/rspb.2010.2020.

21. Smith KR, Woodward A, Campbell-Lendrum D, et al. Human health: impacts, adaptation, and co-benefits. In Field CB, Barros VR, Dokken DJ, et al., editors. *Climate Change 2014: Impacts, Adaptation, and Vulnerability Part A: Global and Sectoral Aspects. Contribution of Working Group II to the Fifth Assessment Report of the Intergovernmental Panel of Climate Change* Cambridge, UK and New York: Cambridge University Press; 2014. pp. 709–54.

22. WHO. Health & Climate Change. COP24 Special Report. Geneva: World Health Organization; 2018.

23. Crimmins A, Balbus J, Gamble JL, et al. *The Impacts of Climate Change on Human Health in the United States: A Scientific Assessment.* Washington, DC: US Global Climate Research Program (USGCRP); 2016.

24. Ebi KL, Balbus JM, Luber G, et al. Human health. In Reidmiller DR, Avery CW, Easterling DR, et al., editors. *Impacts, Risks, and Adaptation in the United States: Fourth National Climate Assessment. II.* Washington, DC: US Global Change Research Program; 2018. pp. 539–71.

25. Vardoulakis S, Heaviside C, editors. *Health Effects of Climate Change in the UK 2012: Current Evidence, Recommendations and Research Gaps.* Health Protection Agency, Public Health England; 2012.

26. Watts N, Amann M, Arnell N, et al. The 2019 report of The *Lancet* Countdown on Health and Climate Change: ensuring that the health of a child born today is not defined by a changing climate. *The Lancet.* 2019;394(10211):1836–78.

27. McMichael AJ. Globalization, climate change, and human health. *The New England Journal of Medicine.* 2013;368(14):1335–43. DOI: 10.1056/NEJMra1109341.

28. Patz JA, Frumkin H, Holloway T, Vimont DJ, Haines A. Climate change: challenges and opportunities for global health. *JAMA.* 2014;312(15):1565–80.

29. Semenza JC. Climate change and human health. *International Journal of Environmental Research and Public Health.* 2014;11(7):7347–53.

30. Haines A, Ebi K. The imperative for climate action to protect health. *New England Journal of Medicine.* 2019;380(3):263–73. DOI: 10.1056/NEJMra1807873.

31. Luber G, Lemery J. *Global Climate Change and Human Health: From Science to Practice.* San Francisco: Jossey-Bass; 2015.

32. Levy BS, Patz JA. editors, *Climate Change and Public Health.* New York: Oxford University Press; 2015.

33. Butler CD. *Climate Change and Global Health.* Wallingford, UK and Boston, MA: CABI; 2014.

34. Holmes KJ, Wender BA, Weisenmiller R, Doughman P, Kerxhalli-Kleinfield M. Climate assessment moves local. *Earth's Future.* 2020;8(2):e2019EF001402.

35. Hausfather Z, Drake HF, Abbott T, Schmidt GA. Evaluating the performance of past climate model projections. *Geophysical Research Letters.* 2019;47(1).

36. van Vuuren DP, Edmonds J, Kainuma M, et al. The representative concentration pathways: an overview. *Climatic Change.* 2011;109(1):5.

37. O'Neill BC, Kriegler E, Ebi KL, et al. The roads ahead: narratives for shared socioeconomic pathways describing world futures in the 21st century. *Global Environmental Change.* 2017;42:169–80. doi: 10.1016/j.gloenvcha.2015.01.004.

38. Riahi K, van Vuuren DP, Kriegler E, et al. The Shared Socioeconomic Pathways and their energy, land use, and greenhouse gas emissions implications: an overview. *Global Environmental Change.* 2017;42:153–68. doi: 10.1016/j.gloenvcha.2016.05.009.

39. Sellers S, Ebi KL. Climate change and health under the Shared Socioeconomic Pathway Framework. *International Journal of Environmental Research and Public Health*. 2017;15(1). doi: 10.3390/ijerph15010003.
40. Zhang DD, Lee HF, Wang C, et al. The causality analysis of climate change and large-scale human crisis. *Proceedings of the National Academy of Sciences*. 2011;108(42):17296–301. doi: 10.1073/pnas.1104268108.
41. Butzer KW. Collapse, environment, and society. *Proceedings of the National Academy of Sciences*. 2012;109(10):3632–9. doi: 10.1073/pnas.1114845109.
42. Committee on Extreme Weather Events and Climate Change Attribution. *Attribution of Extreme Weather Events in the Context of Climate Change*. Washington, DC: The National Academies Press; 2016.
43. Ornes S. Core concept: how does climate change influence extreme weather? Impact attribution research seeks answers. *Proceedings of the National Academy of Sciences*. 2018;115(33):8232–5. https://doi.org/10.1073/pnas.1811393115.
44. Risser MD, Wehner MF. Attributable human-induced changes in the likelihood and magnitude of the observed extreme precipitation during Hurricane Harvey. *Geophysical Research Letters*. 2017;44(24):12,457–64.
45. Mitchell D, Heaviside C, Vardoulakis S, et al. Attributing human mortality during extreme heat waves to anthropogenic climate change. *Environmental Research Letters*. 2016;11(7):074006. https://doi.org/10.1088/1748-9326/11/7/074006.
46. Ebi KL, Ogden NH, Semenza JC, Woodward A. Detecting and attributing health burdens to climate change. *Environmental Health Perspectives*. 2017;125(8): 085004.
47. Gauer R, Meyers BK. Heat-related illnesses. *American Family Physician*. 2019; 99(8):482–9.
48. Kovats RS, Hajat S. Heat stress and public health: a critical review. *Annual Review of Public Health*. 2008;29:41–55. doi:10.1146/annurev.publhealth.29.020907.090843.
49. Robine JM, Cheung SL, Le Roy S, et al. Death toll exceeded 70,000 in Europe during the summer of 2003. *Comptes Rendus Biologies*. 2008;331(2):171–8.
50. Wondmagegn BY, Xiang J, Williams S, Pisaniello D, Bi P. What do we know about the healthcare costs of extreme heat exposure? A comprehensive literature review. *Science of The Total Environment*. 2019;657:608–18. doi: 10.1016/j.scitotenv.2018.11.479.
51. Watts N, Amann M, Ayeb-Karlsson S, et al. The *Lancet* Countdown on Health and Climate Change: from 25 years of inaction to a global transformation for public health. *The Lancet*. 2017;391:581–630. doi:10.1016/S0140-6736(16)32124-9.
52. Kinney PL. Temporal trends in heat-related mortality: implications for future projections. *Atmosphere*. 2018;9(10):409. https://doi.org/10.3390/atmos9100409.
53. Arbuthnott K, Hajat S, Heaviside C, Vardoulakis S. Changes in population susceptibility to heat and cold over time: assessing adaptation to climate change. *Environmental Health*. 2016;15:S33. https://doi.org/10.1186/s12940-016-0102-7.
54. Matthews TKR, Wilby RL, Murphy C. Communicating the deadly consequences of global warming for human heat stress. *Proceedings of the National Academy of Sciences*. 2017;114(15):3861–6. doi: 10.1073/pnas.1617526114.
55. Gasparrini A, Guo Y, Sera F, et al. Projections of temperature-related excess mortality under climate change scenarios. *The Lancet Planetary Health*. 2017;1(9):E360–7.
56. Tasian GE, Pulido JE, Gasparrini A, et al. Daily mean temperature and clinical kidney stone presentation in five U.S. metropolitan areas: a time-series analysis. *Environmental Health Perspectives*. 2014;122(10):1081–7.
57. Sorensen C, Garcia-Trabanino R. A new era of climate medicine: addressing heat-triggered renal disease. *New England Journal of Medicine*. 2019;381(8):693–6.

58. Obradovich N, Migliorini R, Mednick SC, Fowler JH. Nighttime temperature and human sleep loss in a changing climate. *Science Advances*. 2017;3(5).
59. Kuehn L, McCormick S. Heat exposure and maternal health in the face of climate change. *International Journal of Environmental Research and Public Health*. 2017;14(8). doi:10.3390/ijerph14080853.
60. Zhang Y, Yu C, Wang L. Temperature exposure during pregnancy and birth outcomes: an updated systematic review of epidemiological evidence. *Environmental Pollution*. 2017;225.700–12. https://doi.org/10.1016/j.envpol.2017.02.066.
61. Spector JT, Masuda YJ, Wolff NH, Calkins M, Seixas N. Heat exposure and occupational injuries: review of the literature and implications. *Current Environmental Health Reports*. 2019;6(4):286–96. doi: 10.1007/s40572-019-00250-8.
62. Mares DM, Moffett KW. Climate change and interpersonal violence: a 'global' estimate and regional inequities. *Climatic Change*. 2015;135(2):297–310.
63. Burke M, González F, Baylis P, et al. Higher temperatures increase suicide rates in the United States and Mexico. *Nature Climate Change*. 2018;8(8):723–9.
64. Kjellstrom T, Briggs D, Freyberg C, et al. Heat, human performance, and occupational health: a key issue for the assessment of global climate change impacts. *Annual Review of Public Health*. 2016;37:97–112. doi:10.1146/annurev-publhealth-032315-021740.
65. Dunne JP, Stouffer RJ, John JG. Reductions in labour capacity from heat stress under climate warming. *Nature Climate Change*. 2013;3(6):563–6.
66. Jacklitsch B, Williams J, Musolin K, Coca A, Kim J-H, Turner N. *Occupational Exposure to Heat and Hot Environments: Criteria for a Recommended Standard, Revised 2016*. Cincinnati, OH: US National Institute for Occupational Safety and Health; 2016. Available from www.cdc.gov/niosh/docs/2016-106.
67. Kjellstrom T, Kovats RS, Lloyd SJ, Holt T, Tol RS. The direct impact of climate change on regional labor productivity. *Archives of Environmental & Occupational Health*. 2009;64(4):217–27.
68. Andrews O, Le Quéré C, Kjellstrom T, Lemke B, Haines A. Implications for workability and survivability in populations exposed to extreme heat under climate change: a modelling study. *The Lancet Planetary Health*. 2018;2(12):e540–7.
69. Kjellstrom T. Impact of climate conditions on occupational health and related economic losses: a new feature of global and urban health in the context of climate change. *Asia Pacific Journal of Public Health*. 2016;28(2 Suppl.):28s–37s.
70. Saulnier DD, Brolin Ribacke K, von Schreeb J. No calm after the storm: a systematic review of human health following flood and storm disasters. *Prehospital and Disaster Medicine*. 2017;32(5):568–79. https://doi.org/10.1017/s1049023x17006574.
71. Paterson DL, Wright H, Harris PNA. Health risks of flood disasters. *Clinical Infectious Diseases*. 2018;67(9):1450–4. doi: 10.1093/cid/ciy227.
72. Alirol E, Sharma SK, Bawaskar HS, Kuch U, Chappuis F. Snake bite in South Asia: a review. *PLoS Neglected Tropical Diseases*. 2010;4(1):e603.
73. Paranjothy S, Gallacher J, Amlot R, et al. Psychosocial impact of the summer 2007 floods in England. *BMC Public Health*. 2011;11:145.
74. Lamond JE, Joseph RD, Proverbs DG. An exploration of factors affecting the long term psychological impact and deterioration of mental health in flooded households. *Environmental Research*. 2015;140:325–34.
75. Tong S. Flooding-related displacement and mental health. *The Lancet Planetary Health*. 2017;1(4):e124–5. https://doi.org/10.1016/S2542-5196(17)30062-1.
76. Khan AE, Scheelbeek PF, Shilpi AB, et al. Salinity in drinking water and the risk of (pre)eclampsia and gestational hypertension in coastal Bangladesh: a case-control

study. *PLoS One*. 2014;9(9):e108715. https://doi.org/10.1016/S2542-5196(17) 30062-1.

77. Scheelbeek PFD, Chowdhury MAH, Haines A, Vineis P. Drinking water salinity and raised blood pressure: evidence from a cohort study in coastal Bangladesh. *Environmental Health Perspectives*. 2017;125(5):057007.

78. Naser AM, Rahman M, Unicomb L, et al. Drinking water salinity, urinary macro-mineral excretions, and blood pressure in the southwest coastal population of Bangladesh. *Journal of the American Heart Association*. 2019;8(9):e012007.

79. Slette IJ, Post AK, Awad M, et al. How ecologists define drought, and why we should do better. *Global Change Biology*. 2019;25(10):3193–200.

80. Smith AB. 2010–2019: A landmark decade of U.S. billion-dollar weather and climate disasters. Washington, DC: NOAA; 2020. Available from www.climate.gov/news-features/blogs/beyond-data/2010-2019-landmark-decade-us-billion-dollar-weather-and-climate.

81. Andrade E, Barrett N, Colon-Ramos U, et al. Ascertainment of the Estimated Excess Mortality from Hurricane María in Puerto Rico. Washington, DC: George Washington University; 2018.

82. Gething PW, Smith DL, Patil AP, et al. Climate change and the global malaria recession. *Nature*. 2010;465(7296):342–5. doi: 10.1038/nature09098.

83. WHO. World Malaria Report 2019. Geneva: World Health Organization; 2019.

84. Siraj AS, Santos-Vega M, Bouma MJ, et al. Altitudinal changes in malaria incidence in highlands of Ethiopia and Colombia. *Science*. 2014;343(6175):1154–8.

85. Bhatt S, Gething PW, Brady OJ, et al. The global distribution and burden of dengue. *Nature*. 2013;496(7446):504–7. https://doi.org/10.1038/nature12060.

86. Hales S, de Wet N, Maindonald J, Woodward A. Potential effect of population and climate changes on global distribution of dengue fever: an empirical model. *The Lancet*. 2002;360(9336):830–4. https://doi.org/10.1016/S0140-6736(02)09964-6.

87. WHO. WHO Estimates of the Global Burden of Foodborne Diseases. Geneva: World Health Organization; 2015.

88. Ghazani M, FitzGerald G, Hu W, Toloo GS, Xu Z. Temperature variability and gastrointestinal infections: a review of impacts and future perspectives. *International Journal of Environmental Research and Public Health*. 2018;15(4).

89. Lake IR, Barker GC. Climate change, foodborne pathogens and illness in higher-income countries. *Current Environmental Health Reports*. 2018;5(1):187–96.

90. Cisse G. Food-borne and water-borne diseases under climate change in low- and middle-income countries: further efforts needed for reducing environmental health exposure risks. *Acta Tropica*. 2019;194:181–8.

91. Kovats RS, Edwards SJ, Hajat S, et al. The effect of temperature on food poisoning: a time-series analysis of salmonellosis in ten European countries. *Epidemiology and Infection*. 2004;132(3):443–53. doi: 10.1017/s0950268804001992.

92. Lake IR, Gillespie IA, Bentham G, et al. A re-evaluation of the impact of temperature and climate change on foodborne illness. *Epidemiology and Infection*. 2009;137(11): 1538–47. doi: 10.1017/S0950268809002477.

93. Hales A, Kovats S, Lloyd S, Campbell-Lendrum D. Quantitative Risk Assessment of the Effects of Climate Change on Selected Causes of Death, 2030s and 2050s. Geneva: World Health Organization; 2014.

94. Cotty PJ, Jaime-Garcia R. Influences of climate on aflatoxin producing fungi and aflatoxin contamination. *International Journal of Food Microbiology*. 2007;119(1–2): 109–15. doi: 10.1016/j.ijfoodmicro.2007.07.060.

95. Battilani P, Toscano P, Van der Fels-Klerx HJ, et al. Aflatoxin B(1) contamination in maize in Europe increases due to climate change. *Scientific Reports*. 2016;6:24328.

96. Assunção R, Martins C, Viegas S, et al. Climate change and the health impact of aflatoxins exposure in Portugal – an overview. *Food Additives & Contaminants: Part A*. 2018;35(8):1610–21. doi: 10.1080/19440049.2018.1447691.

97. Adekiya TA, Aruleba RT, Oyinloye BE, Okosun KO, Kappo AP. The effect of climate change and the snail-schistosome cycle in transmission and bio-control of schistosomiasis in Sub-Saharan Africa. *International Journal of Environmental Research and Public Health*. 2019;17(1). https://dx.doi.org/10.3390% 2Fijerph17010181

98. Kalinda C, Chimbari M, Mukaratirwa S. Implications of changing temperatures on the growth, fecundity and survival of intermediate host snails of schistosomiasis: a systematic review. *International Journal of Environmental Research and Public Health*. 2017;14(1):80. https://dx.doi.org/10.3390%2Fijerph14010080.

99. Zheng J, Gu XG, Xu YL, et al. Relationship between the transmission of schistosomiasis japonica and the construction of the Three Gorge Reservoir. *Acta Tropica*. 2002;82(2):147–56.

100. Zhou XN, Yang GJ, Yang K, et al. Potential impact of climate change on schistosomiasis transmission in China. *American Journal of Tropical Medicine and Hygiene*. 2008;78(2):188–94.

101. Stensgaard AS, Vounatsou P, Sengupta ME, Utzinger J. Schistosomes, snails and climate change: current trends and future expectations. *Acta Tropica*. 2019;190: 257–68. doi: 10.1016/j.actatropica.

102. Chambers JS. *The Conquest of Cholera*. New York: MacMillan; 1938.

103. Constantin de Magny G, Colwell RR. Cholera and climate: a demonstrated relationship. *Transactions of the American Clinical and Climatological Association*. 2009;120:119–28.

104. Vezzulli L, Brettar I, Pezzati E, et al. Long-term effects of ocean warming on the prokaryotic community: evidence from the vibrios. *ISME J*. 2012;6(1):21–30.

105. Baker-Austin C, Trinanes J, Gonzalez-Escalona N, Martinez-Urtaza J. Non-cholera vibrios: the microbial barometer of climate change. *Trends in Microbiology*. 2017;25(1):76–84. doi: 10.1016/j.tim.2016.09.008.

106. Gobler CJ, Doherty OM, Hattenrath-Lehmann TK, et al. Ocean warming since 1982 has expanded the niche of toxic algal blooms in the North Atlantic and North Pacific oceans. *Proceedings of the National Academy of Sciences*. 2017;114(19): 4975–80. https://doi.org/10.1073/pnas.1619575114.

107. Wells ML, Trainer VL, Smayda TJ, et al. Harmful algal blooms and climate change: learning from the past and present to forecast the future. *Harmful Algae*. 2015; 49:68–93. https://dx.doi.org/10.1016%2Fj.hal.2015.07.009.

108. Escobar MTL, Sotto LPA, Jacinto GS, et al. Eutrophic conditions during the 2010 fish kill in Bolinao and Anda, Pangasinan, Philippines. *Journal of Environmental Science and Management*. 2013;35(Special Issue No. 1).

109. San Diego-McGlone ML, Azanza RV, Villanoy CL, Jacinto GS. Eutrophic waters, algal bloom and fish kill in fish farming areas in Bolinao, Pangasinan, Philippines. *Marine Pollution Bulletin*. 2008;57(6):295–301.

110. Knox J, Hess T, Daccache A, Wheeler T. Climate change impacts on crop productivity in Africa and South Asia. *Environmental Research Letters*. 2012;7(3):034032.

111. Challinor AJ, Watson J, Lobell DB, et al. A meta-analysis of crop yield under climate change and adaptation. *Nature Climate Change*. 2014;4(4):287–91.

112. Porter JR, Xie L, Challinor AJ, et al. Food security and food production systems. In Field CB, Barros VR, Dokken DJ, et al., editors. *Climate Change 2014: Impacts, Adaptation and Vulnerability Part A: Global and Sectoral Aspects. Contribution of Working Group II to the Fifth Assessment Report of the Intergovernmental Panel*

on Climate Change. Cambridge, UK and New York: Cambridge University Press; 2014. pp. 485–533.

113. Deutsch CA, Tewksbury JJ, Tigchelaar M, et al. Increase in crop losses to insect pests in a warming climate. *Science*. 2018;361(6405):916–19. DOI: 10.1126/science. aat3466.

114. Scheelbeek PFD, Bird FA, Tuomisto HL, et al. Effect of environmental changes on vegetable and legume yields and nutritional quality. *Proceedings of the National Academy of Sciences*. 2018;115(26):6804–9. doi: 10.1073/pnas.1800442115.

115. Beach RH, Sulser TB, Crimmins A, et al. Combining the effects of increased atmospheric carbon dioxide on protein, iron, and zinc availability and projected climate change on global diets: a modelling study. *The Lancet Planetary Health*. 2019;3(7): e307–17. https://doi.org/10.1016/S2542-5196(19)30094-4.

116. Myers SS, Smith MR, Guth S, et al. Climate change and global food systems: potential impacts on food security and undernutrition. *Annual Review of Public Health*. 2017;38:259–77. https://doi.org/10.1146/annurev-publhealth-031816-044356.

117. Zhu C, Kobayashi K, Loladze I, et al. Carbon dioxide (CO_2) levels this century will alter the protein, micronutrients, and vitamin content of rice grains with potential health consequences for the poorest rice-dependent countries. *Science Advances*. 2018;4(5).

118. Myers SS, Wessells KR, Kloog I, Zanobetti A, Schwartz J. Effect of increased concentrations of atmospheric carbon dioxide on the global threat of zinc deficiency: a modelling study. *The Lancet Global Health*. 2015;3(10):e639–45.

119. Smith MR, Myers SS. Impact of anthropogenic CO_2 emissions on global human nutrition. *Nature Climate Change*. 2018;8(9):834–9.

120. Moore ERH, Smith MR, Humphries D, Dubrow R, Myers SS. The mismatch between anthropogenic CO_2 emissions and their consequences for human zinc and protein sufficiency highlights important environmental justice issues. *Challenges*. 2020; 11(1):4. https://doi.org/10.3390/challe11010004.

121. Van Dingenen R, Dentener FJ, Raes F, et al. The global impact of ozone on agricultural crop yields under current and future air quality legislation. *Atmospheric Environment*. 2009;43(3):604–18. https://doi.org/10.1016/j.atmosenv.2008.10.033.

122. Ainsworth EA, Lemonnier P, Wedow JM. The influence of rising tropospheric carbon dioxide and ozone on plant productivity. *Plant Biology*. 2019. https://doi .org/10.1111/plb.12973.

123. IPCC. *Climate Change and Land. An IPCC Special Report on Climate Change, Desertification, Land Degradation, Sustainable Land Management, Food Security, and Greenhouse Gas Fluxes in Terrestrial Ecosystems*; 2019. Available from www .ipcc.ch/srccl/.

124. Green R, Cornelsen L, Dangour AD, et al. The effect of rising food prices on food consumption: systematic review with meta-regression. *BMJ*. 2013;346:f3703.

125. Springmann M, Mason-D'Croz D, Robinson S, et al. Global and regional health effects of future food production under climate change: a modelling study. *The Lancet*. 2016;387(10031):1937–46. https://doi.org/10.1016/s0140-6736(15)01156-3.

126. Black RE, Allen LH, Bhutta ZA, et al. Maternal and child undernutrition: global and regional exposures and health consequences. *The Lancet*. 2008;371(9608):243–60.

127. FAO, IFAD, UNICEF, WFP, WHO. *The State of Food Security and Nutrition in the World 2019: Safeguarding Against Economic Slowdowns and Downturns*. Rome: FAO; 2019.

128. Lloyd SJ, Kovats RS, Chalabi Z. Climate change, crop yields, and undernutrition: development of a model to quantify the impact of climate scenarios on child undernutrition. *Environmental Health Perspectives*. 2011;119(12):1817–23.

129. Smith LC, Haddad L. Reducing child undernutrition: past drivers and priorities for the post-MDG era. *World Development*. 2015;68:180–204.

130. Lloyd SJ, Bangalore M, Chalabi Z, et al. A global-level model of the potential impacts of climate change on child stunting via income and food price in 2030. *Environmental Health Perspectives*. 2018;126(9):097007.

131. Carleton TA. Crop-damaging temperatures increase suicide rates in India. *Proceedings of the National Academy of Sciences*. 2017;114(33):8746–51.

132. Carleton TA. Reply to Plewis, Murari et al. and Das: The suicide temperature link in India and the evidence of an agricultural channel are robust. *Proceedings of the National Academy of Sciences*. 2018;115(2):E118–21.

133. Belesova K, Gasparrini A, Sié A, Sauerborn R, Wilkinson P. Annual crop-yield variation, child survival, and nutrition among subsistence farmers in Burkina Faso. *American Journal of Epidemiology*. 2017;187(2):242–50.

134. Rojas-Downing MM, Nejadhashemi AP, Harrigan T, Woznicki SA. Climate change and livestock: impacts, adaptation, and mitigation. *Climate Risk Management*. 2017;16:145–63. https://doi.org/10.1016/j.crm.2017.02.001.

135. Pauly D, Zeller D. Catch reconstructions reveal that global marine fisheries catches are higher than reported and declining. *Nature Communications*. 2016;7:10244.

136. Pörtner H-O, Karl DM, Boyd PW, et al. Ocean systems. In Field CB, Barros VR, Dokken DJ, et al., editors. *Climate Change 2014: Impacts, Adaptation, and Vulnerability Part A: Global and Sectoral Aspects. Contribution of Working Group II to the Fifth Assessment Report of the Intergovernmental Panel on Climate Change*, Cambridge, UK and New York: Cambridge University Press; 2014. pp. 411–84.

137. Barange M, Bahri T, Beveridge MCM, et al. *Impacts of Climate Change on Fisheries and Aquaculture: Synthesis of Current Knowledge, Adaptation and Mitigation Options*. Rome: Food and Agriculture Organization of the United Nations; 2018. Contract No. 627.

138. Golden CD, Allison EH, Cheung WW, et al. Nutrition: fall in fish catch threatens human health. *Nature*. 2016;534(7607):317–20. doi: 10.1038/534317a.

139. Clements JC, Chopin T. Ocean acidification and marine aquaculture in North America: potential impacts and mitigation strategies. *Reviews in Aquaculture*. 2016;9(4):326–41. https://doi.org/10.1111/raq.12140.

140. Froehlich HE, Gentry RR, Halpern BS. Global change in marine aquaculture production potential under climate change. *Nature Ecology & Evolution*. 2018;2(11): 1745–50. https://doi.org/10.1038/s41559-018-0669-1.

141. Hallegate S, Bangalore M, Bonzanigo L, et al., editors. *Shock Waves: Managing the Impacts of Climate Change on Poverty*. Washington, DC: World Bank; 2016.

142. Headey DD. Food prices and poverty. *The World Bank Economic Review*. 2016;32(3):676–91.

143. Ivanic M, Martin W, Zaman H. Estimating the short-run poverty impacts of the 2010–11 surge in food prices. *World Development*. 2012;40(11):2302–17.

144. Denny EK, Walter BF. Ethnicity and civil war. *Journal of Peace Research*. 2014; 51(2):199–212. https://doi.org/10.1177%2F0022343313512853.

145. Franck R, Rainer I. Does the leader's ethnicity matter? Ethnic favoritism, education, and health in Sub-Saharan Africa. *American Political Science Review*. 2012;106(2): 294–325. doi:10.1017/S0003055412000172.

146. Schleussner C-F, Donges JF, Donner RV, Schellnhuber HJ. Armed-conflict risks enhanced by climate-related disasters in ethnically fractionalized countries. *Proceedings of the National Academy of Sciences*. 2016;113(33):9216–21.

147. IPCC. *Climate Change 2014: Synthesis Report. Contribution of Working Groups I, II and III to the Fifth Assessment Report of the Intergovernmental Panel on Climate Change* [Core Writing Team, RK Pachauri and LA Meyer (editors)]. Geneva: Intergovernmental Panel on Climate Change; 2014.

148. Oppenheimer M, Campos M, Warren R, et al. Emergent risks and key vulnerabilities. In Field CB, Barros VR, Dokken DJ, et al., editors. *Climate Change 2014: Impacts, Adaptation, and Vulnerability Part A: Global and Sectoral Aspects. Contribution of Working Group II to the Fifth Assessment Report of the Intergovernmental Panel on Climate Change*. Cambridge, UK and New York: Cambridge University Press; 2014. pp. 1039–99.

149. von Uexkull N, Croicu M, Fjelde H, Buhaug H. Civil conflict sensitivity to growing-season drought. *Proceedings of the National Academy of Sciences*. 2016;113(44): 12391–6. https://doi.org/10.1073/pnas.1607542113.

150. Hsiang SM, Burke M, Miguel E. Quantifying the influence of climate on human conflict. *Science*. 2013;341(6151):1235367. DOI: 10.1126/science.1235367.

151. Burke M, Hsiang SM, Miguel E. Climate and conflict. *Annual Review of Economics*. 2015;7(1):577–617. https://doi.org/10.1146/annurev-economics-080614-115430.

152. Black R, Adger WN, Arnell NW, et al. The effect of environmental change on human migration. *Global Environmental Change*. 2011;21:S3–11.

153. Kaczan DJ, Orgill-Meyer J. The impact of climate change on migration: a synthesis of recent empirical insights. *Climatic Change*. 2020;158(3):281–300.

154. Brown O. *Migration and Climate Change*. Geneva: International Organization for Migration; 2008. Contract No. 31.

155. Baird R, Migiro K, Nutt D, et al. *Human Tide: The Real Migration Crisis*. London: Christian Aid; 2007.

156. Rigaud KK, de Sherbinin A, Jones B, et al. *Groundswell: Preparing for Internal Climate Migration*. Washington, DC: World Bank; 2018.

157. Nicholls RJ, Marinova N, Lowe JA, et al. Sea-level rise and its possible impacts given a 'beyond 4°C world' in the twenty-first century. *Philosophical Transactions of the Royal Society A: Mathematical, Physical and Engineering Sciences*. 2011;369(1934): 161–81. https://doi.org/10.1098/rsta.2010.0291.

158. Hsiang SM, Sobel AH. Potentially extreme population displacement and concentration in the tropics under non-extreme warming. *Scientific Reports*. 2016;6: 25697.

159. UNDESA. International Migration Report 2017. New York: UNDESA. Contract No. ST/ESA/SER.A/403.

160. Findlay AM. Migrant destinations in an era of environmental change. *Global Environmental Change: Human and Policy Dimensions*. 2011;21:S50–8.

161. Feng S, Krueger AB, Oppenheimer M. Linkages among climate change, crop yields and Mexico–US cross-border migration. *Proceedings of the National Academy of Sciences*. 2010;107(32):14257–62. doi: 10.1073/pnas.1002632107.

162. Marchiori L, Maystadt J-F, Schumacher I. The impact of weather anomalies on migration in Sub-Saharan Africa. *Journal of Environmental Economics and Management*. 2012;63(3):355–74. DOI: 10.1016/j.jeem.2012.02.001.

163. Kelley CP, Mohtadi S, Cane MA, Seager R, Kushnir Y. Climate change in the Fertile Crescent and implications of the recent Syrian drought. *Proceedings of the National Academy of Sciences*. 2015;112(11):3241–6. https://doi.org/10.1073/pnas.1421533112.

164. Mueller V, Gray C, Kosec K. Heat stress increases long-term human migration in rural Pakistan. *Nature Climate Change*. 2014;4(3):182–5. https://dx.doi.org/10.1038%2Fnclimate2103.

165. Bohra-Mishra P, Oppenheimer M, Hsiang SM. Nonlinear permanent migration response to climatic variations but minimal response to disasters. *Proceedings of the National Academy of Sciences.* 2014;111(27):9780–5.

166. de Sherbinin A, Castro M, Gemenne F, et al. Preparing for resettlement associated with climate change. *Science.* 2011;334(6055):456–7. DOI: 10.1126/science. 1208821.

167. Missirian A, Schlenker W. Asylum applications respond to temperature fluctuations. *Science.* 2017;358(6370):1610–4. doi: 10.1126/science.aao0432.

168. Borderon M, Sakdapolrak P, Muttarak R, et al. Migration influenced by environmental change in Africa: a systematic review of empirical evidence. *Demographic Research.* 2019;41:491–544. https://dx.doi.org/10.4054/DemRes.2019.41.18.

169. Renaud FG, Bogardi JJ, Dun O, Warner K. *Control, Adapt or Flee: How to Face Environmental Migration?* Bonn, Germany: United Nations University Institute for Environment and Human Security; 2007.

170. Kniveton D, Schmidt-Verkerk K, Smith C, Black R. *Climate Change and Migration.* IOM Migration Research Series: UN; 2008.

171. Warner K, Erhart C, de Sherbinin A, Adamo SB, Chai-Onn TC. In Search of Shelter: Mapping the Effects of Climate Change on Human Migration and Displacement. A policy paper prepared for the 2009 Climate Negotiations. Bonn, Germany: United Nations University, CARE, CIESN Columbia University; 2009.

172. Piguet E, Pécoud A, de Guchteneire, P, editors. *Migration and Climate Change.* Cambridge, UK and New York: Cambridge University Press; 2011.

173. Black R, Bennett SRG, Thomas SM, Beddington JR. Migration as adaptation. *Nature.* 2011;478(7370):447–9. https://doi.org/10.1038/478477a.

174. Fussell E, Sastry N, VanLandingham M. Race, socioeconomic status, and return migration to New Orleans after Hurricane Katrina. *Population and Environment.* 2009;31(1–3):20–42. doi: 10.1007/s11111-009-0092-2.

175. Naik A. Migration and natural disasters. In Laczko F, Aghazarm A, editors. *Migration, Environment and Climate Change: Assessing the Evidence.* Geneva: International Organization for Migration; 2009. pp. 245–317.

176. Government Office for Science. *Foresight: Migration and Global Environmental Change: Future Challenges and Opportunities.* London: Government Office for Science; 2011.

177. Dannenberg AL, Frumkin H, Hess JJ, Ebi KL. Managed retreat as a strategy for climate change adaptation in small communities: public health implications. *Climatic Change.* 2019;153(1–2):1–14. DOI: 10.1007/s10584-019-02382-0.

178. Koslov L. The case for retreat. *Public Culture.* 2016;28(2 79):359–87.

179. Jacobson C, Crevello S, Chea C, Jarihani B. When is migration a maladaptive response to climate change? *Regional Environmental Change.* 2018;19(1):101–12.

180. Geddes A, Adger WN, Arnell NW, Black R, Thomas DSG. Migration, environmental change, and the challenges of governance. *Environment and Planning C: Government and Policy.* 2012;30(6):951–67.

181. Cunsolo A, Ellis NR. Ecological grief as a mental health response to climate change-related loss. *Nature Climate Change.* 2018;8(4):275–81. https://doi.org/10.1038/s41558-018-0092-2.

182. Marshall N, Adger WN, Benham C, et al. Reef grief: investigating the relationship between place meanings and place change on the Great Barrier Reef, Australia. *Sustainability Science.* 2019;14(3):579–87. https://doi.org/10.1007/s11625-019-00666-z.

183. Albrecht GA. 'Solastalgia': a new concept in health and identity. *PAN: Philosophy Activism Nature.* 2005;3:44–59. https://doi.org/10.4225/03/584f410704696.

184. Nicolosi E, Corbett JB. Engagement with climate change and the environment: a review of the role of relationships to place. *Local Environment*. 2018;23(1):77–99.
185. Verlie B. Bearing worlds: learning to live-with climate change. *Environmental Education Research*. 2019;25(5):751–66. https://doi.org/10.1080/13504622.2019.1637823.
186. Head L, Harada T. Keeping the heart a long way from the brain: the emotional labour of climate scientists. *Emotion, Space and Society*. 2017;24:34–41.
187. Colino S. Fearing the future: pre-traumatic stress reactions. US News and World Report. 24 May 2017.
188. Ng FY, Wilson LA, Veitch C. Climate adversity and resilience: the voice of rural Australia. *Rural and Remote Health*. 2015;15(4):3071.
189. Ellis NR, Albrecht GA. Climate change threats to family farmers' sense of place and mental wellbeing: a case study from the Western Australian Wheatbelt. *Social Science & Medicine*. 2017;175:161–8.
190. Cunsolo Willox A, Harper S, Ford J, et al. Climate change and mental health: an exploratory case study from Rigolet, Nunatsiavut, Canada. *Climatic Change*. 2013;121(2):255–70. DOI: 10.1007/s10584-013-0875-4.
191. Cunsolo Willox A, Harper SL, Ford JD, et al. 'From this place and of this place': climate change, sense of place, and health in Nunatsiavut, Canada. *Social Science & Medicine*. 2012;75(3):538–47. doi: 10.1016/j.socscimed.2012.03.043.
192. Barnett J, Adger WN. Climate change, human security and violent conflict. *Political Geography*. 2007;26(6):639–55.
193. Barnett J, Tschakert P, Head L, Adger WN. A science of loss. *Nature Climate Change*. 2016;6(11):976–8. doi: 10.1038/nclimate3140.
194. Fritze JG, Blashki GA, Burke S, Wiseman J. Hope, despair and transformation: climate change and the promotion of mental health and wellbeing. *International Journal of Mental Health Systems*. 2008;2(1):13. https://doi.org/10.1186/1752-4458-2-13.
195. Searle K, Gow K. Do concerns about climate change lead to distress? *International Journal of Climate Change Strategies and Management*. 2010;2(4):362–79.
196. Clayton S, Manning C, Krygsman K, Speiser M. *Mental Health and Our Changing Climate: Impacts, Implications, and Guidance*. American Psychological Association and EcoAmerica; 2017.
197. Burke SEL, Sanson AV, Van Hoorn J. The psychological effects of climate change on children. *Current Psychiatry Reports*. 2018;20(5):35. https://doi.org/10.1007/s11920-018-0896-9.
198. Nugent C. Terrified of climate change? You might have eco-anxiety. *Time*. 21 November 2019.
199. Taylor M, Murray J. 'Overwhelming and terrifying': the rise of climate anxiety. *Guardian*. 10 February 2020.
200. Burgess K. Young people seek help for anxiety over climate change. *The Times*. 4 January 2020.
201. Plautz J. The environmental burden of generation Z. *Washington Post*. 3 February 2020.
202. Busby E. Climate change activism 'reducing mental health symptoms among young people'. *Independent*. 28 November 2017.

3

Pollution, Land Use, Biodiversity, and Health

While climate change is a vitally important environmental change confronting humanity, the planet is changing in other unprecedented ways. Many of these changes – pollution, biodiversity loss, land use changes, and others – correspond to the planetary boundaries introduced in Chapter 1. Like climate change, these planetary changes also have implications for human health and well-being – the subject of this chapter. We turn first to pollution, a broad category that includes air and water pollution by substances including metals, pesticides, plastics, and pharmaceuticals. Next we consider land use and biodiversity loss – two closely intertwined processes. After land we turn to freshwater – exploring the many ways in which humans have altered the planet's hydrology. Finally, we explore how these many changes interact with each other in complex ways.

Pollution: A Multidimensional Challenge

Pollution has often been seen as primarily a local issue, with individuals or communities exposed to a nearby source – workers exposed to endocrine disruptors on the job, a neighbourhood near a contaminated former industrial site, a city choking on air pollution. Increasingly, however, pollution is recognized as a threat to health on a planetary scale.

Pollution interacts with many of the planetary boundaries (1) that frame Planetary Health. One is climate change; a major proportion of fine particulate and ozone air pollution originates from sources that also emit carbon dioxide, the major greenhouse gas, and short-lived climate pollutants, including methane and black carbon. Carbon dioxide also leads to ocean acidification, another of the planetary boundaries. The boundary related to biogeochemical flows is defined largely by nitrogen and phosphorus flows, most of which are related to fertilizer use. As pollutants, these cause eutrophication of aquatic systems. Through such mechanisms pollution also contributes to biodiversity loss, yet another planetary boundary. Stratospheric ozone depletion is slowly reversing following the phaseout of ozone-depleting substances – CFCs – under the Montreal Protocol, a rare relative success story (although even this is being partly undermined by leakage of CFCs from sources such as refrigerators, air conditioners, and insulation foam blowing (2), and possibly from illicit production in China (3)) (see Chapter 1). For atmospheric aerosol loading, the boundary was initially set based on the impact on the Indian monsoons, but this is also a Planetary Health issue, as aerosols are a component of health-damaging air pollution.

Another reason to consider pollution as more than a local threat is that it spreads. Consider air pollution, which, while generally dominated by local sources, can reflect more distant sources as well. For example, on days when there are strong westerly winds, 12–24% of sulfates, 4–6% of carbon monoxide, and up to 11% of black carbon pollution detected in the air of the western USA is of Chinese origin (4). Another class of pollutants, the persistent organic pollutants or POPs, circulate on a global scale, and are routinely found in biota, including human tissues, thousands of kilometres from where they were made and used (5, 6). Still another class of pollutants, the plastics, discarded in waterways, flow to oceans and accumulate in massive ocean gyres, cluttering the beaches of remote islands (7, 8). So the interconnections of planetary processes, and the fact that pollutants know no boundaries, mark pollution as a Planetary Health problem – and one that, as discussed below, requires global cooperation to address. There is a proposed planetary boundary related to the introduction of novel entities, defined as 'new substances, new forms of existing substances and modified life-forms that have the potential for unwanted geophysical and/or biological effects' (1), particularly when they are persistent, widely distributed, and potentially harmful to planetary processes or systems. Specifying a single boundary for a complex mixture of novel entities is, however, a challenge and is currently beyond the capabilities of the science community.

The Global Burden of Pollution

The *Lancet* Commission on Pollution and Health showed compellingly that pollution is a neglected cause of a large burden of ill-health (9). In 2018, the Commission estimated that 9 million deaths were attributed to pollution, amounting to 16% of premature deaths worldwide, three times as many as malaria, tuberculosis, and AIDS combined. Ninety two per cent of pollution-related deaths occur in low- and middle-income countries (LMICs) with the highest burden in rapidly industrializing countries.

Estimates of the disease burden from pollution, such as those of the *Lancet* Commission, are likely to understate the true burden, as depicted in **Figure 3.1** (9). Pollutants in Zone 2 of the figure have adverse effects that are not fully characterized, precluding reliable estimates of disease burden; examples include soil pollution by heavy metals and toxic chemicals at contaminated industrial and mining sites. Zone 3 encompasses emerging chemical threats whose effects are only now coming into focus. These include developmental neurotoxicants; endocrine disruptors; new classes of pesticides such as the neonicotinoids; chemical herbicides such as glyphosate; and pharmaceutical wastes. As more is known about the effects of individual chemicals some will move from Zone 3 to Zones 2 and 1.

Air Pollution

One of the sentinel events that raised public and professional awareness of the health effects of air pollution was the London smog of 1952. Between the 5th and 9th of December 1952, London was shrouded in a dense fog of air pollution from the burning of coal, and from December 1952 through March 1953, there were over 13,500 more deaths than normal. The

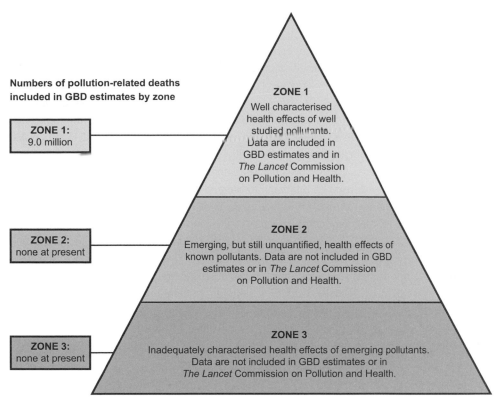

Figure 3.1. The 'pollutome', the totality of all forms of pollution with the potential to harm human health. GBD refers to Global Burden of Disease; for explanation see source.
Source: Landrigan PJ, Fuller R, Acosta NJR, et al. The *Lancet* Commission on Pollution and Health. *The Lancet.* 2018;391:462–512.

relative importance of air pollution and influenza, which affected London immediately afterwards, has been a matter of debate, with subsequent estimates suggesting that about 12,000 deaths could have been due to air pollution (10). The serious public health consequences of air pollution led to the Clean Air Acts of 1956 and 1968, which introduced measures to dramatically reduce industrial and domestic fossil fuel emissions with great effect. Nevertheless, it is striking that over 65 years later many parts of the world are still struggling to effectively control air pollution. Even in London, despite considerable improvements in air quality and the banning of domestic coal burning, air pollution continues to claim up to 9400 premature deaths per year, reducing average life expectancy by about a year (11).

What Is Air Pollution?

Air pollution is a complex mixture of components, from a variety of sources. As such it varies from place to place. The principal components of air pollution are solid particles, suspended liquids, and gases. The particles ('particulate matter', or PM) are often considered by reference to their size ('mean aerodynamic diameter') in microns (μm), or millionths of a metre, because size has a bearing on health impact. The PM may be coarse or fine; the fraction smaller than 10 μm in diameter is referred to as PM_{10}, and the even finer particles, smaller than 2.5 μm, as $PM_{2.5}$. Finer particles penetrate more deeply into the lungs and are more closely associated with some adverse health outcomes. Ultrafine particles ($PM_{0.1}$) pass through the alveolar-capillary membrane and are carried around the body via the bloodstream to virtually all tissues (12). The gases in air pollution include oxides of nitrogen (NO_x) and oxides of sulfur (combustion by-products), hydrocarbons, carbon monoxide, and ozone. Ozone is known as a secondary pollutant; rather than being emitted from smokestacks and tailpipes, it forms in the air through complex atmospheric chemical reactions from precursors including volatile organic compounds, methane, and oxides of nitrogen.

Many other terms are used in reference to air pollution, including aerosols (suspended particles and/or liquids), smoke (a visible suspension of particles and gases formed by burning), haze (a visible suspension of water vapour and particles), and smog (a portmanteau of smoke and fog, a suspension of particles, oxides of nitrogen and sulfur, and ozone that is typically formed from combustion emissions and the photochemical reactions that result). While the most troublesome air pollution generally results from combustion, some occurs naturally; sources include vegetation (which releases hydrocarbons such as terpenes), lightning (which can generate oxides of nitrogen), and wind-blown dust (called 'aeolian dust' after Aeolus, the Greek God of wind). It is also increasingly recognized that agricultural ammonia emissions strongly contribute to fine particulate air pollution ($PM_{2.5}$).

How Are People Exposed to Air Pollution?

Typically, air pollution is divided into two principal categories: household air pollution and outdoor, or ambient, air pollution. Household air pollution is most severe in places where people use solid fuels, such as wood, charcoal, coal, and dung, for cooking and heating. Ambient air pollution comes from power generation, agriculture, household sources, industry, and vehicles, and is generally most severe in cities in LMICs. Smoking tobacco also entrains a high concentration of particles and gases directly into the airways, and second-hand exposure to environmental tobacco smoke is a well-recognized cause of ill-health, although usually considered separately from ambient and household air pollution.

The pollution from power generation, industry, and transportation generally comes from fossil fuel combustion. Coal – used in power generation and industry – is the most polluting fossil fuel, and coal combustion is an important cause of both pollution and climate change. Each stage in the extraction, transport, processing, and combustion of coal generates a waste stream, and thus the full life cycle of coal results in multiple hazards for health and the environment. These costs are not borne by the coal industry and are thus often considered 'externalities', many of which are cumulative. One study focusing on Appalachia estimated that the life cycle costs of coal to the US public amounted to between a third and half a trillion dollars annually (13). If these costs were fully taken into account in pricing, even using conservative assessments, the cost of coal would double or triple, making it economically non-viable against low-carbon alternatives and energy

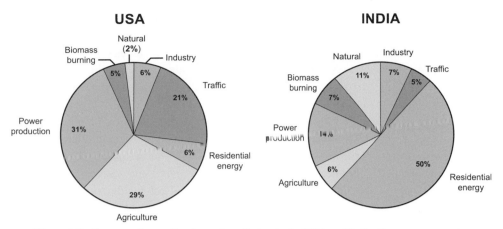

Figure 3.2. Sources of mortality from air pollution in the USA and India. Percentages are the proportions of the deaths attributable to ambient air pollution. Excess deaths attributable to air pollution in 2015 were estimated at 120,000 (95% CI: 81,000–156,000) in the USA and 967,000 (753,000–1,150,000) in India. Natural refers to natural sources of air pollution, predominantly aeolian dust.

Source: Lelieveld J, Haines A, Pozzer A. Age-dependent health risk from ambient air pollution: a modelling and data analysis of childhood mortality in middle-income and low-income countries. *The Lancet Planetary Health*. 2018;2(7):e292–e300.

conservation. It will be necessary to leave at least a third of global oil reserves, half of gas reserves, and over 80% of current coal reserves unused from 2010 to 2050 in order to meet the Paris climate target of 2 °C warming above pre-industrial temperatures (14).

The relative contribution of different air pollution sources varies across different countries. **Figure 3.2** compares the USA and India, using mortality rather than emissions as an indicator. India shows a much larger relative contribution from household sources and lower contribution from transport and other sectors compared with the USA (15). The distribution of sources is likely to change as India moves to less-polluting household fuels such as LPG and private vehicle use increases.

How Does Air Pollution Affect Health?

While air pollution is often treated as a single phenomenon, each pollutant has its own epidemiological features. One of the challenges faced by air pollution researchers is to disentangle the effects of different pollutants, particularly where these are often found together and concentrations are therefore strongly correlated. This is why multi-pollutant models are used to separate the effects of different pollutants on health. The strongest evidence linking air pollution and ill-health is for fine particulates and ozone (in the troposphere or lower atmosphere, rather than stratospheric ozone which shields the Earth's surface against harmful UV rays), but NO_x is also recognized as toxic.

Particulate matter is perhaps best recognized as a risk factor for cardiopulmonary mortality, following both short-term and long-term exposure (16–19). Recent research has helped establish a mechanistic basis for this association, demonstrating links between PM exposure and cardiometabolic risk factors including systemic inflammation (20), arrhythmias (21), high blood pressure (22), obesity (23), and diabetes (24, 25). Exposure to PM is also associated with a remarkably broad set of other risks, ranging from adverse birth outcomes (26, 27) to neuropsychiatric disorders (28–30), and it increases the risk of respiratory infections including tuberculosis (31) and pneumonia (32). Whilst evidence is still emerging, and there are no widely accepted disease burden estimates for some of these outcomes (see also (9)), there are plausible pathways for many of them and the weight of evidence is growing over time.

Particulate matter is chemically different from place to place and source to source, and little is known about how different chemical forms of PM affect health risk. It is generally assumed that fine particles have similar effects irrespective of their source but this may be an oversimplification given their chemical diversity. An exception is black carbon (BC), a component of fine particulate matter emitted by combustion engines (notably diesel), residential burning of wood and coal, power stations using heavy oil or coal, open burning of agricultural wastes, and forest and vegetation fires. Black carbon seems to be more harmful at a given level of concentration than fine particles in general, perhaps because toxins are adsorbed onto the surface of BC particles. However, when expressed across the interquartile range the effects are similar to PM in general (33). Black carbon is also a powerful short-lived climate pollutant, and when deposited onto snow and ice it accelerates melting and may also affect rainfall patterns.

Ozone was long recognized as a trigger of respiratory symptoms, especially in people with asthma and chronic obstructive lung disease, but recent research has confirmed that it also causes premature deaths: an estimated 1.04–1.23 million respiratory deaths annually in adults according to one study that used updated relative risk estimate and exposure parameters (34). Increases in estimated attributable mortality were larger in northern India, southeast China, and Pakistan than in other locations. Like BC, ozone has environmental impacts in addition to its human impacts. In the stratosphere it protects the Earth against ultraviolet (UV) radiation. In the lower atmosphere (the troposphere), it functions as a GHG, and its toxicity extends to plants, so higher levels of ozone reduce agricultural output (35).

The effects of NO_x have been more difficult to disentangle from the effects of PM and ozone. The Committee on the Medical Effects of Air Pollution in the UK undertook an assessment of the independent contribution of NO_2 to premature deaths from air pollution (36). They were unable to achieve complete consensus but a majority of the Committee, having considered the evidence from meta-analyses of seven available cohort studies including measurements of two (six studies) and three (one study) pollutants concluded that NO_2 probably has an independent effect on mortality in addition to that of PM. Studies of the toxicology of NO_2 and chamber studies, in which volunteers are exposed to different concentrations of NO_2, also provide evidence of adverse effects on the respiratory tract. Overall the uncertainties do not allow robust global estimates of the independent effects of NO_2 on premature deaths, but it is clear that NO_2 is hazardous to human health. There is increasing concern that NO_2 contributes to the burden of asthma in children. A study of the NO_2-attributable burden of asthma incidence in children aged 1–18 years in 194 countries

Table 3.1. *Percentage of disability-adjusted life-years (DALYs) attributable to air pollution (household plus ambient) by disease and country income group*

	Lower respiratory infections	Tracheal, bronchial and lung cancer	Ischaemic heart disease	Ischaemic stroke	Haemorrhagic stroke	Chronic obstructive pulmonary disease	Cataracts
High income	12%	8%	13%	9%	11%	16%	1%
Upper middle income	34%	30%	24%	20%	24%	41%	14%
Lower middle income	57%	38%	35%	28%	31%	52%	25%
Low income	64%	48%	43%	36%	22%	51%	35%
Global	53%	24%	28%	37%	27%	44%	19%

Source: Landrigan PJ, Fuller R, Acosta NJR, et al. The *Lancet* Commission on Pollution and Health. *The Lancet*. 2018;391:462–512.

and 125 major cities estimated that 4.0 million (95% uncertainty interval [UI] 1.8–5.2) of new paediatric asthma cases could be attributable to NO_2 pollution annually, accounting for 13% (6–16) of global incidence (37). This NO_2 exposure is largely a result of traffic-related air pollution (TRAP) and about two-thirds of these cases occur in cities.

What Is the Health Burden of Air Pollution?

There are varying estimates of deaths from ambient air pollution depending on how exposures are estimated and the assumed relationship between exposure and health outcome. The WHO estimated that 4.2 million premature deaths worldwide in 2016 could be attributed to ambient air pollution from $PM_{2.5}$ (38). More recent research, based on a novel Global Exposure Mortality Model and using data from dozens of cohort studies, suggests that ambient air pollution accounts for nearly 9 million premature deaths globally each year, substantially higher than previous estimates (39, 40). This is mainly from exposure to $PM_{2.5}$ although exposure to tropospheric ozone is also included in this estimate. Furthermore, this recent research shows that even levels of air pollution below the WHO guideline levels could have adverse effects on health. Since 91% of the world's population is exposed to pollution levels above the guideline level of 10 $\mu g/m^3$ annual mean, this implies that virtually all of us are breathing polluted air at levels sufficient to harm our health. This rivals the burden of disease from cigarette smoking, and averaged across the global population it translates into 2.9 years of life lost per person, with the impact greatest in China, India, and parts of Africa.

Some of the reduced life expectancy is due to the effects of pollution on children (15). A study combining data from nearly 1 million births in Sub-Saharan Africa with satellite-based measurements of exposure to $PM_{2.5}$ suggested that $PM_{2.5}$ concentrations above minimum exposure levels were responsible for 22% (95% CI: 9–35) of infant deaths in the 30 study countries, resulting in a total of 449,000 (95% CI: 194,000–709,000) additional deaths of infants in 2015. This figure is about three times higher than previous estimates (41). Such research is difficult in low-income countries where exposures are often high, as air pollution measurements are often lacking.

Table 3.1 shows the impact of air pollution, expressed in disability-adjusted life years (DALYs), on key health outcomes in countries at different income levels. Overall, air pollution accounts for large proportions of respiratory and cardiovascular disease and lung cancer.

Not all air pollution is caused by human activities; anthropogenic (i.e. human-related) air pollution is thought to cause the premature deaths of about 5.5 million people every year (see Figure 8.5) (42). This suggests the potential for large public health gains by addressing the preventable causes of air pollution, for example by phasing out the use of fossil fuels and other sources of pollution from human activities. Many of these policies can also reduce GHG emissions and therefore benefit the climate as well as health (see Chapters 8 and 12).

Air Pollution Trends

Overall, trends in ambient air pollution present a mixed picture. Global deaths just from fine PM are estimated to have increased by about 20% between 1990 and 2015 as a result of population growth, ageing populations, and increasing levels of air pollution in LMICs (9). More recently there have been perceptible improvements in China. During the 2020 COVID-19 pandemic, economic activity and therefore air pollution fell dramatically in locations such as Italy, China, and India (43, 44) – a fortuitous development, as air pollution emerged as a possible risk factor for COVID-19 mortality (45). People with conditions to which air pollution contributes, such as heart disease, stroke, chronic obstructive pulmonary disease, and diabetes are also at an increased risk of death from COVID 19. However, we should not interpret those short-term falls in air pollution as an unalloyed positive development because the economic contraction was catastrophic for people's livelihoods and health. Increased poverty related to the COVID-19 economic depression will probably force more people to burn cheap but polluting solid fuels in the home. Economic recovery will be accompanied by large increases in air pollution and GHG emissions in the absence of active policies to prevent that happening (see Chapter 12).

In all countries, the poor and marginalized tend to be exposed to higher levels of air pollution than wealthier groups. Under a business as usual scenario in which no new pollution controls are implemented, ambient air pollution-related deaths are projected to increase by about 50% worldwide by 2050, with the ageing population (older adults being more susceptible to the effects of air pollution than younger adults) being a major factor. The growing cities of South and Southeast Asia are particularly likely to experience large increases in air pollution-related deaths. In high-income countries air pollution levels have declined but populations are still exposed to unacceptably high levels responsible for large numbers of deaths and a high disease burden. Household air pollution, which also contributes substantially to ambient air pollution particularly in some LMICs, is declining in many parts of the world as a result of reductions in poverty and access to cleaner fuels.

Microbial Contamination of Water

Water pollution is also a major killer although key sources differ in their estimates of the total disease burden. This likely reflects, in part, different definitions of 'safe water'; the WHO considers only access to an improved water source whereas the Global Burden of Disease Study (GBD) requires safe water at both point of access and point of use (9). Under the Millennium Development Goals, which shaped the global development agenda from 2000 to 2015, considerable progress was made in water and sanitation, particularly in urban settings. From a baseline of 1990, 2.6 billion people achieved access to improved drinking water sources and 2.1 billion people gained access to improved sanitation by 2015. As a result of these and other improvements, for example in effective hand hygiene and treatment with oral rehydration, worldwide child deaths from diarrhoeal disease fell by about 60%. Nevertheless, over 2 billion people still have unimproved sanitation, with nearly a billion practising open defecation (9). These populations represent large numbers of people who have not benefitted substantially so far from the advances of humanity in the Anthropocene and are at increased vulnerability to the consequent dramatic changes in the global environment to which they have contributed little. The numbers quoted above do not

include those affected by chemical contamination of water supplies for which no comprehensive assessments have been published on a global scale.

Planetary trends may contribute to microbial contamination of water at the local level.
For example, water-borne diseases such as those caused by *E. coli*, salmonella, shigella,
and campylobacter are more common during warm weather (46–48) – a risk that is likely to
rise with continued heating (49). Severe rainfall events are also associated with increases
in diarrhoea (50). As meat consumption rises globally, and with it industrial livestock
production, increased microbial contamination of surface and groundwater from animal
feed operations may occur (51, 52). Planetary trends, from temperature to rainfall to food
system dynamics, are reflected in water quality in many places.

Chemical Pollution

Humans have used chemicals for thousands of years: combining alkali and limestone to make
glass, sulfur and saltpetre to make explosives, and sulfur and bitumen to make poison gas;
smelting copper, tin, and arsenic to make bronze; using asphalt as building mortar and ship
caulking and lead to make pipes. But the Anthropocene has featured a massive scale-up of
chemical manufacturing and use. Naturally occurring chemicals have continued to be used,
but an entire industry has grown up around synthetic chemicals as well (**Figure 3.3**). Only a
minority of these have been adequately tested for toxicity to humans. The volumes of synthetic
chemicals produced are impressive; the most recent values, for example, are 6×10^6 tonnes
of pesticides globally, and 0.23×10^6 tonnes of US pesticides (53).

Persistent Organic Pollutants and Endocrine Disruption

Emblematic of the challenges of modern chemical exposure are persistent organic pollutants
(POPs). As the name implies, these are synthetic organic chemicals that are toxic to humans
and wildlife and remain for many years in the environment. They are widely distributed,
including in areas far from their original use. They also bioaccumulate, particularly in fatty
tissue where they can be 70,000 times more concentrated than the background levels, and
concentrations increase at higher levels in the food web.

One of the most important health impacts of POPs is interference with hormone action
(54, 55). A wide range of products, including some that are not POPs, are implicated as
endocrine-disrupting chemicals (EDCs); these include pharmaceuticals, personal care products, and commercial chemicals. The EDCs are suspected of causing wide-ranging adverse
health effects in humans and animals (56). Reasons for concern include the increasing
prevalence of endocrine disorders in humans, together with endocrine-related effects in
wildlife and evidence from laboratory experiments that exposure to such chemicals increases
the risk of endocrine abnormalities. Amongst the observations causing concern are:

• Up to 40% of young men in some countries have poor quality semen with low sperm
  counts which may impair fertility (57, 58).
• Rates of genital malformations including undescended testicles and penile malformations
  (hypospadias) in baby boys are rising, and in some cases have plateaued at high levels (59, 60).

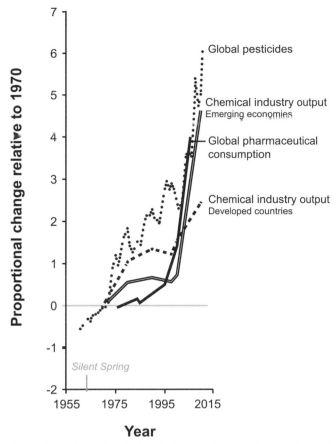

Figure 3.3. Upscaling of the chemical industry over the past half century. Trends in the global trade value of synthetic chemicals (in both developed and emerging economies), and for the pesticide and pharmaceutical chemical sectors individually. Note: All trends are shown relative to 1970 values, except pharmaceutical consumption, where the earliest data reported are from 1975. *Silent Spring* refers to the book by Rachel Carson, published in 1962 by Houghton Mifflin, which raised public awareness of the pervasive environmental effects of the indiscriminate use of pesticides (see p. 92).

Source: Adapted from Bernhardt ES, Rosi EJ, Gessner MO. Synthetic chemicals as agents of global change. *Frontiers in Ecology and the Environment*. 2017;15(2):84–90.

- Endocrine-related cancers (including breast, endometrial, prostate, ovarian, testicular) are rising globally, increases that are not fully explained by more complete detection and diagnosis (61).
- There are global increases of neuro-behavioural disorders, in some cases linked with thyroid abnormalities, with evidence that EDCs contribute (62–64).
- Metabolic disorders related to endocrine disruption have been found among amphibians, which may help explain declines in amphibian and other wildlife populations (56, 65).

About 800 chemicals are known or thought to disrupt endocrine function by interfering with hormone receptors, hormone synthesis or conversion. However, only a small proportion has been rigorously tested in intact organisms and in some cases no internationally recognized and validated tests exist. Children may have higher exposures than adults because of their hand-to-mouth behaviours and high metabolic rates, and are especially susceptible because of their developmental stage (66). Both humans and animals are exposed to multiple EDCs simultaneously; this further complicates assessment of causal relationships and raises additional concerns about the potential for additive or synergistic effects. The EDCs can also interfere with fat storage and metabolism, bone development, and the immune system, suggesting the potential for even more pervasive effects than the above examples.

Using rigorous approaches based on those developed by the IPCC and WHO criteria for assessing the strength of evidence, an international panel of scientists achieved consensus at least for probable (>20%) causation by EDCs for a range of health outcomes (54): IQ loss and intellectual disability, autism, attention-deficit hyperactivity disorder, childhood and adult obesity, adult (type 2) diabetes, undescended testes, male infertility, and mortality associated with reduced testosterone. Accounting for probability of causation, for which they used the midpoint of the range for each outcome, they estimated a median cost of €157 billion in Europe, corresponding to 1.23% of EU GDP. The EU has better regulation than many other regions of the world so it is likely that the burden is now or will become even higher in other regions, if such chemicals are allowed to be widely used and disseminated.

Plastics

Plastics are malleable materials made from synthetic organic compounds, typically polymers, that can be formed into a wide range of objects, from packaging to consumer goods to building components. The monomers used to form plastics include ethylene, propylene, styrene, urethane – all familiar terms when preceded by 'poly'. In addition to the polymer itself, plastic products typically contain chemical additives such as plasticizers (to achieve desired texture), flame retardants, stabilizers, and colourants (67). While some plastic products are relatively durable, others are intended to be used just once and disposed of.

Plastics present both opportunities and challenges for health. Benefits include protection of foods from bacteriological contamination, water storage, and reduced transport emissions due to the light weight of components and containers. The challenges are increasingly apparent; they arise from the relative indestructability of plastic molecules and fragments, the potential for wide dispersion in terrestrial and marine environments, the potential to contaminate soil, water, and air, the migration of additives and other compounds from plastics into the food chain, and the toxicity of some of these compounds. While some monomers are relatively innocuous, others are more worrisome. For example, bisphenol A (BPA), the building block of polycarbonates, acts as an oestrogen disruptor (68, 69), and vinyl chloride, used to make PVC for everything from plumbing pipes to electrical cables, from beverage bottles to credit cards, causes both hepatocellular carcinoma and angiosarcoma of the liver at high exposures (70). Similarly, some plastic additives are toxic. For example, the plasticizers known as phthalates act as endocrine disruptors, and have been associated with impaired neurodevelopment in children and ailments as diverse as reduced sperm counts, diabetes, and obesity across the lifespan (71–73). The burden of disease from such pollution is currently unknown but there are legitimate grounds for concern.

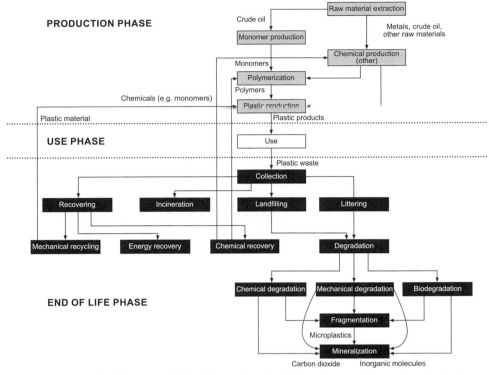

Figure 3.4. The life cycle of plastic products (excluding energy input and emissions) created in STAN (subSTance flow ANalysis) software.

Source: Adapted from Hahladakis JN, Velis CA, Weber R, Iacovidou E, Purnell P. An overview of chemical additives present in plastics: migration, release, fate and environmental impact during their use, disposal and recycling. *Journal of Hazardous Materials*. 2018;344:179–99.

Migration of additives leading to human exposure can occur during the three life-cycle phases of plastics – production, use, and end of life (**Figure 3.4**). For example, uncontrolled burning of plastics can release metals such as antimony and bromine (74) as well as POPs including dioxin, and disposal in landfill is inefficient whilst potentially leading to local contamination of soil and water. Unregulated recycling can result in potentially hazardous levels of contaminants in children's toys and food packaging. The effects of some emerging practices, such as the incorporation of plastic waste into bitumen road surfaces, which may only last 4–6 years, are unknown because the risks have not been adequately assessed.

A large quantity of plastic ends up in waterways, and then flows into oceans. Estimates of the total volume of plastics entering the oceans range from 4.8 to 12.7 million tonnes (Mt) per year (75). Marine plastic pollution is a growing problem with immediate impacts on a range of marine ecosystems and potential implications for human health, considering that the seas provide food for billions of people. Plastics can cause direct damage to a range of species by entanglement, smothering, and ingestion and may allow invasive species to spread to new locations by providing a resilient platform capable of travelling great distances (76). Plastic and other debris are increasingly found in fish and shellfish for human consumption. In Indonesian markets, for example, 28% of individual fish and 55% of all species contained plastic debris, and in US markets, anthropogenic debris was found in 25% of individual fish and in 67% of all species but was primarily synthetic fibres (77).

In December 2017 China, which had been at the centre of the global recycling trade in plastic waste, having imported a cumulative total of 45% of plastic waste since 1992 (78), dramatically curtailed non-industrial plastic waste imports with little notice, having recognized the costs in terms of ill-health and environmental damage of often poor-quality waste. The ban affected eight types of plastic waste in the commercial recycling stream, including polyethylene (PE), polystyrene (PS), polyvinyl chloride (PVC), polyethylene terephthalate (PET), and polypropylene (PP), as well as bales of PET plastic bottles, aluminium plastic film, and compact/digital video disks. Whilst this led to increases in imports to other Asian countries such as Vietnam and Thailand, they will not be able to make up for the loss of Chinese imports and are likely to increasingly regulate their own imports. Most of the exports are from high-income countries, with the EU being collectively the largest exporter. In the medium to long run this restriction on imported waste is likely to increase pressures for the development of a circular economy (see Chapter 12), although it will also increasingly pose threats to health in many countries with inadequate regulation.

Pharmaceutical Pollution

Worldwide about 600 pharmaceutical compounds have been detected in the environment – particularly in aquatic systems, notably surface waters (79, 80). There is growing pollution of rivers by pharmaceutical agents, which in some studies dissipate very slowly after discharge from sources such as wastewater treatment plants. A study that measured the concentrations of five commonly used pharmaceutical agents in UK rivers showed negligible degradation of these agents 5 km downstream from the source, suggesting that freshwater ecosystems are widely exposed to substantial concentrations (81). Pharmaceuticals used in humans and animals may affect ecosystems even at low exposure levels; for example, antidepressants affect the growth and feeding behaviour of fish (82–84) and diclofenac (a non-steroidal anti-inflammatory drug) has caused kidney failure in vultures that ate the carcasses of domestic animals given the drug, resulting in population declines of this important scavenger (85). Hormonal medications may affect the endocrine systems of fish, molluscs, invertebrates, and birds, and antiparasitic agents may affect worms, aquatic invertebrates, and insects. Environmental risk assessments in Europe show that about 10% of pharmaceutical agents pose risks (80). But many compounds have not yet been adequately assessed.

A key priority is to address the spread of antimicrobial resistance (AMR), which threatens the progress made since the middle of the last century in reducing the deaths and illness from many infectious diseases. **Figure 3.5** summarizes the drivers of antibiotic resistance and environmental hotspots from which spread is likely. The contamination of freshwater sources with antimicrobial agents from unregulated pharmaceutical company effluent in India has been well documented and is associated with the selection and dissemination of antibiotic-resistant organisms such as carbapenemase-producing pathogens (86). A large majority

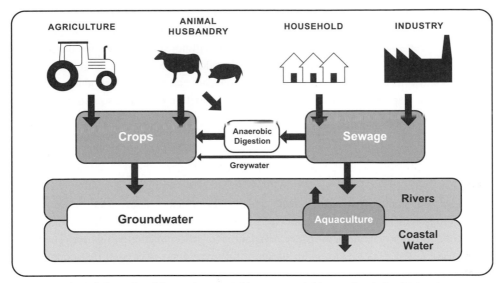

Figure 3.5. Schematic of the environmental hotspots and drivers of antimicrobial resistance (AMR).

Source: Adapted from Singer AC, Shaw H, Rhodes V, Hart A. Review of antimicrobial resistance in the environment and its relevance to environmental regulators. *Frontiers in Microbiology.* 2016;7:1728.

(86%) of bacterial strains sampled from a wastewater treatment plant serving bulk production pharmaceutical companies in India were resistant to 20 or more antibiotics and showed a high prevalence of integrons that permit the exchange of genetic material with other organisms and could therefore facilitate the transmission of antibiotic resistance to pathogenic bacteria (87). Other sources of antibiotic pollution in the environment include from use in animals, aquaculture, agricultural run-off (from the spreading of manure and sewage sludge) and from municipal and industrial wastewater. Importantly, other compounds such as heavy metals, biocides (i.e. disinfectants and surfactants), natural substances (e.g. plant-derived) and xenobiotics (e.g. solvents such as octanol, hexane, and toluene) can also select for resistance genes (88). A complex combination of factors determines the prevalence of resistance genes in the environment that reflect a dynamic balance between fitness costs and benefits. Evidence suggests that antimicrobial resistance in common pathogens increases with local temperature and population density (89). It is therefore plausible to suggest modest increases in the threat of antimicrobial resistance as a result of climate change.

There is an extraordinary range of antibiotic use in animals among European countries, from a low level of 3.1 mg/PCU (population correction unit) in Norway to a high of 423.1 mg/PCU in Cyprus (90), with levels declining in many countries since 2010. This suggests that profligate antibiotic use may not be necessary to achieve highly productive agricultural systems and that current patterns of use in some countries may reflect differences in regulatory mechanisms and their enforcement rather than an objective and comprehensive assessment of their benefits and risks. A policy imperative is to reduce the veterinary use of antimicrobial agents that are needed to treat infections in humans.

However, the drivers of antimicrobial pollution share many of the characteristics of other pollution sources, such as vested economic interests, and merely focusing on promoting more appropriate veterinary and medical prescribing will not suffice to resolve the pressing challenges to public health and to ecosystems. In addition to reducing antibiotic misuse in human and veterinary use, biocide use in personal care and household products will need to decline; and capture, re-use and recycling of metals within the waste stream will need to be promoted.

Pesticides

Pesticides – including insecticides, fungicides, and herbicides – are used worldwide to protect crops against insects, fungi, and weeds, and are also used by public health authorities in vector control programmes to reduce populations of insect vectors of disease. The range of chemicals used as pesticides is broad and a full exposition is beyond the scope of this chapter, but is available elsewhere (91).

While pesticides have played an important role in public health, both in protecting crops from pests and therefore in raising agricultural output, and in controlling vectors of infectious diseases, pesticides have also had unintended harmful consequences for both humans and the planet. It was DDT, an organochlorine pesticide that found wide use following World War II, that inspired Rachel Carson's 1962 classic book, *Silent Spring*, which helped launch the modern environmental movement.

Pesticides are toxic: this is intrinsic to their control of unwanted pest species. But in practice, the toxic effects of pesticides often manifest beyond the target species, and beyond the desired time and place of action. Many classes of pesticides are toxic to humans. The acute toxicity has long been recognized; in fact, organophosphates have been used not only as insecticides but also as chemical warfare agents, precisely because of their human toxicity. The research community was slower to recognize chronic effects of pesticides, but these are now well established, and range from cancer to neurotoxicity (92). Recent years have seen a growing awareness of toxic effects following very low dose exposures, including endocrine disruption (93).

People are exposed to pesticides through various routes, including occupational exposures, ingestion of pesticides in food, and intentional ingestion. Ingestion of foods contaminated with pesticides affects the largest number of people. The risk of such exposures is assessed and managed in standard ways by the World Health Organization and many national governments. It begins with hazard identification, which classifies substances according to their adverse effects on health (94). A well-known example is the International Agency for Research on Cancer classification of substances with respect to carcinogenicity. Next, the level of risk is assessed, and an Acceptable Daily Intake (ADI) is established accordingly. For pesticide residues in food, this is carried out jointly by two UN bodies: the Food and Agriculture Organization and the World Health Organization (95). The ADIs, in turn, support the setting of maximum residue limits (MRLs) in food – the levels of pesticides, or their residues, that if consumed over a lifetime are expected not to have adverse effects, and that are therefore permitted in foods. National authorities are expected to enforce adherence to MRLs. In the UK, for example, the Expert Committee on Pesticide Residues in Food monitors such residues. In 2018, it oversaw the analysis of 3385 samples of 40 different types of food, of which 45% contained a residue (96). However, only five samples were referred to the Food Standards Agency because of concerns about threats to human health and 21 were referred to the Health and Safety Executive because they contained pesticides either prohibited for use in the UK or for the crop in question. The concern must, therefore, particularly be for countries where no such mechanisms are in place or where they do not function effectively. In these circumstances, older pesticides that persist in the environment for many years may continue to be used and newer pesticides may be used at levels that exceed the threshold for unacceptable risks to human health.

Suicide by pesticide is a particular tragedy. Pesticides are often the most readily available means of suicide, especially in rural areas. A systematic review of the evidence conservatively estimated that there are 258,234 (plausible range 233,997 to 325,907) deaths from pesticide self-poisoning worldwide each year, responsible for 30% (range 27% to 37%) of suicides globally (97). Furthermore, the authors documented wide geographical differences in the relative contribution from 4% in the European region to 50% in the western Pacific region. These differences were related not to total volume of pesticides sold but to differences in the toxicity and patterns of use. Global trends signal the need for ongoing concern. Rural communities in India and elsewhere confront increasing stress as a result of climate change and other environmental stressors, increasing suicide risks for farmers (98, 99). At the same time, pesticide use will likely increase as a result of increases in pests from climate change (100). Solutions must include reducing the toxicity of pesticides, developing climate-resilient agricultural practices, and robust social and mental health safety nets for rural communities.

Pesticides also have environmental impacts beyond those intended. They can contaminate surface and groundwater, soil, and biota. While some pesticides are degraded relatively rapidly, others can persist in these environmental media for prolonged periods, bioaccumulating and bioconcentrating. Other species than the targeted pests may be poisoned, ranging from insects (including some that are beneficial) to a variety of plants, birds, and mammals. These can lead to feedback loops and second-order effects. For example, the neonicotinoid class of pesticides is implicated in reducing pollinator populations, which can in turn reduce the productivity of crops and other plants (101). Soil fertility can be compromised as the result of disruptions of microbial communities (102). Importantly, pesticide resistance develops among many target species, an entirely unsurprising manifestation of evolutionary principles, emphasizing the need for reducing reliance on pesticides in favour of multifaceted approaches to agricultural productivity (103) (see Chapter 10).

Exposure to Lead and Other Metals

The effects of lead exposure have probably been more widely studied than those of all toxic pollutants. People began to mobilize lead from natural geological reservoirs during the Bronze Age (as early as 5000 years ago), leaving enough stratigraphic evidence that some have suggested this legacy as a marker for the early Anthropocene (104, 105). Lead was widely used in pipes and food and water vessels in Roman times (106), and Pliny accurately described the signs and symptoms of lead poisoning (107). In modern times, lead was best recognized as an occupational hazard. But in the last two centuries, as lead was used for water piping, incorporated into products such as paint, and added to petrol (gasoline) to improve engine performance, general population exposure became widespread. For example in the USA, where lead was widely used in petrol until the 1980s, childhood exposure is estimated to have reduced population intelligence significantly, such that the number of people with superior intelligence (an IQ over 130) fell by half and the number with an IQ below 70 doubled (108). **Figure 3.6** illustrates how a 5-point reduction of IQ from population exposure to a pollutant can greatly affect the proportion of people with very low and high IQs. Lead affects lifetime prospects; with a negative impact on IQ, an increased risk of attention-deficit hyperactivity disorder (109), and an association with antisocial and criminal behaviours (110, 111), it is little wonder that childhood lead exposure is associated with reduced socioeconomic circumstances later in life (112). According to one analysis, reducing blood lead levels to less than 1 μg/dL among all US children between birth and age 6 years would increase timely secondary school graduation rates and reduce crime, yielding net societal benefits of US$50,000 (SD, US$14,000) per child annually at a discount rate of 3% and overall estimated savings of about US$1.2 trillion (SD, US$341 billion) (113). The socioeconomic impacts are also large in LMICs; foregone earnings are lower in less affluent economies, but the impact at the individual level potentially larger because of more severe exposures (114).

Long-term exposure to lead also increases the risks of cardiovascular diseases, including hypertension, ischaemic heart disease, stroke, cardiac arrhythmias, and peripheral arterial disease, which account for the majority of deaths attributable to lead in adults (see (9) for overview). Further, this relationship holds at comparatively low levels of lead in the blood. Although lead-free petrol is now used in 175 countries, widespread contamination has left a long-term legacy that is still being felt. Global estimates of deaths and disability due to lead, which amount to 0.5 million premature deaths and 9.3 million life-years lost (DALYs) in 2015 according to the GBD study, do not take into account exposure at lead battery recycling sites (see p. 96) (9). The WHO estimates somewhat higher impacts on DALYs and that 0.6 million children suffer from mild to moderate mental retardation as a consequence of lead poisoning (see (9) for more detail).

Lead is only one of a group of metals of hydrogeological origin that human activity has helped disseminate, threatening human health. Others include arsenic, cadmium, mercury, and copper. Arsenic and cadmium are classified by the International Agency for Research on Cancer as group 1 carcinogens. A systematic review of 37 studies including nearly 350,000 participants looked for evidence that these metals affect the risk of cardiovascular disease, an important contributor to the global burden of disease. Arsenic, lead, cadmium, and copper were all independently associated with substantial and highly significant increases in cardiovascular disease risk, after adjustment for confounders such as smoking. The only exception was mercury (largely from fish consumption) for which there was no increase in risk (115). Falling levels of environmental metals in the USA as a result of reductions in

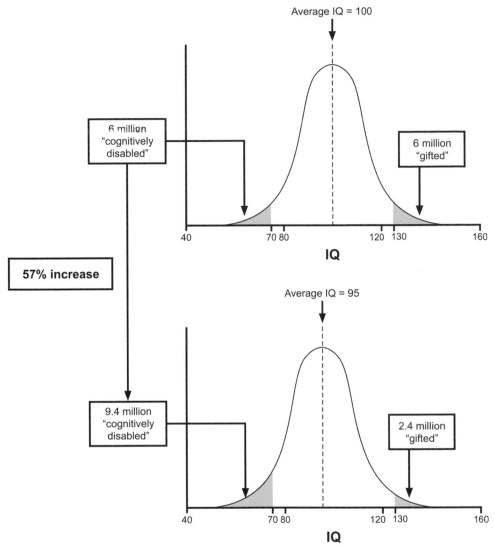

Figure 3.6. Model of intelligence losses associated with a mean 5-point drop in IQ of a population of 100 million.
Source: Landrigan PJ, Fuller R, Acosta NJR, et al. The *Lancet* Commission on Pollution and Health. *The Lancet*. 2018;391:462–512.

smoking and air pollution and tighter regulations may have contributed to declines in cardiovascular disease over recent decades, but in other parts of the world exposures to a range of pollutants may be increasing, for example due to increased trade in electronic waste.

Mercury exemplifies the human disruption of natural systems, with resulting health consequences, in several ways. First, humans have come to dominate the cycling of this metal. Anthropogenic emissions of mercury have been higher than natural emissions for about 200 years, and currently account for about 90% of the flux of mercury – about one-third of this from current anthropogenic emissions and about two-thirds from 're-emissions' of previously released mercury (116). Second, mercury emissions are large and consequential. Global mercury emissions total about 2000 tonnes per annum. Without improved controls, emissions of mercury are projected to rise. The main sources are coal burning and artisanal mining followed by the production of ferrous and non-ferrous metals and of cement. Third, emissions are not uniformly distributed. Almost half the mercury emissions arise from Asia as a result of growing industrialization, and considerable emissions occur in deprived areas of Africa, Latin America, and Asia, where artisanal gold mining accounts for about 20% of the global gold supply (117). The *Lancet* Commission on Pollution and Health estimated that between 14 and 19 million artisanal and small-scale miners were exposed to hazardous levels of mercury. Mercury is mixed with gold-containing materials and then heated, vaporizing the mercury to yield gold. Fourth, this human activity alters ecosystems. Anthropogenic emissions have doubled mercury levels in the top 100 metres of the ocean over the past 100 years, with slower increase in the deep layers of the ocean. However, in parts of the Atlantic levels are decreasing, showing that reduced emissions from North America and Europe are starting to have an effect and reinforcing the need for better regulatory policies elsewhere. Fifth, global-scale processes such as organification and bioconcentration affect the fate and transport of emitted mercury. Micro-organisms in soil and water convert inorganic mercury to a highly toxic organic form, methylmercury, which is recycled in the biosphere for decades or centuries. Concentrations of methylmercury in plankton are as high as 10,000 times the concentration in seawater, and rise as methylmercury is biomagnified up the food web. In some species of Arctic marine animals, average mercury content has increased by 12 times since the pre-industrial period. Finally, mercury is toxic to humans, principally through effects on the nervous system in adults and impaired neurological development in children, but with widespread effects on other organs as well (118). Some indigenous populations consume large amounts of top marine predators – fish, whales, and seals – and are, therefore, exposed to high levels of mercury. Among populations that consume seafood regularly, pregnant women and young children are particularly at risk because exposure in utero is associated with developmental neurotoxicity (e.g. deficits in fine motor skills, language, and memory). As in the case of lead, this could have serious implications for future educational and economic prospects and employment (119).

Waste Dumping

A particularly disturbing aspect of global chemical contamination is the transfer of waste from producer countries to receiving countries, known as dumping. Many low-income countries lack the governance and regulatory systems needed to detect, interdict, and prevent potentially hazardous imports. The Basel Convention on the Control of Transboundary Movements of Hazardous Wastes and Their Disposal, which entered into force in 1992, was intended to control such dumping, but it has had only limited success. The *Lancet* Pollution Commission reported that 61 million people in the 49 countries surveyed to date are exposed to heavy metals and toxic chemicals at contaminated sites. Although clearly an underestimate because there may be hundreds of thousands of sites globally, this illustrates the scale of the potential threat to health (9). The Commission estimated that between 6 million and 16 million people were exposed to hazardous levels of lead at sites where lead batteries were recycled. These activities take place in a range of low-income settings around the world – particularly in Sub-Saharan Africa – and illustrate the dangers of exposure to toxic hazards in countries that lack the capability and/or political will for adequate regulation.

Electronic waste, or e-waste, has emerged as a major form of international waste transfer. The global quantity of e-waste generated from end-of-life electronic equipment in 2016 was estimated to be about 44.7 Mt, or 6.1 kg per capita (120). The problem is increasing; the global value of trade in e-waste is estimated to have risen from US$9.8 billion in 2012 to US$41.4 billion in 2019. Only about 20% of e-waste generated is documented for collection and recycling and the fate of the majority of e-waste is unrecorded. Europe generates the second highest per capita levels of e-waste (16.6 kg/person), after Oceania which contributes little in absolute terms, but also has relatively high recycling rates (35%). The Americas generate 11.6 kg/person but with lower recycling rates (17%). Asia generates only 4.2 kg/person but because of its vast population and low recycling rates is responsible for about 40% of the total global e-waste. Of the waste exported much ends up in countries that are ill-equipped to recycle the waste safely. Studies by the Basel Action Network, placing GPS trackers in exported e-waste, have shown that 87% of e-waste exported from the USA ended up in Asia, the majority in Hong Kong (121), while most exported e-waste from Europe went to Nigeria, Ghana, and Tanzania (122).

Reclamation of valuable elements such as gold and copper from e-waste has become an income-generating activity in some poor communities but exposes people, including children, to a range of toxic pollutants including lead, cadmium, chromium, brominated flame retardants, and polychlorinated biphenyls (PCBs) through direct contact, inhalation of toxic fumes, or contamination of soil and water. A 2013 systematic review of studies of the health effects of e-waste exposure found 23 epidemiological studies meeting quality criteria, all in southeast China (123); the literature has continued to expand substantially with later reviews (124, 125) and published accounts focusing on India (126, 127), Pakistan (128), Indonesia (129), and Africa (130). In general, these studies have found adverse health effects both in exposed workers and in nearby communities, including adverse birth outcomes, reduced thyroid function, behavioural changes, decreased lung function, and evidence of genetic damage.

Land Use Change, Biodiversity Loss, and Disease Risk

Land Use

The land use changes described in Chapter 1 threaten human health in direct and indirect ways. Some key examples include water-borne diseases, vector-borne diseases, and zoonotic diseases.

Forest alterations can have a major impact on stream and river flow and on water quality. One study examined watershed quality in relation to diarrhoeal disease risk in children. Among nearly 300,000 children under 5 years of age in 35 countries, after controlling for socioeconomic factors, the presence of improved water and sanitation, and other potential confounders, more intact tree cover in the upper watershed reduced the probability of diarrhoea in children in downstream communities. Thirty per cent more tree cover offered roughly the same protection as improved sanitation infrastructure (but not as much as wealth, education, or an improved water supply) (131). Similar results emerged from studies in Fiji, which showed that forest clearing and road construction through forested areas were associated with increased risk of typhoid fever in nearby communities. A key factor seemed to be the risk of soil erosion on altered land surfaces; erosion and associated run-off delivered contaminated water to streams and rivers, and undermined latrines and septic systems (132).

Land use change in forested regions is associated not only with water-borne disease, but also with vector-borne disease. This relationship has been extensively studied in the Amazon, where road-building and deforestation are associated with increased risk of malaria. Probable mechanisms include some combination of altered microclimates, increases in mosquito breeding sites, selective favouring of competent mosquito species, and increased human exposure through in-migration and poor housing (133–136).

Similar findings have emerged from the other side of the world. In Malaysian Borneo, clearing of forests for palm oil plantations leads to forest fragmentation, concentration of surviving macaque monkeys in the remaining habitat, and probably increased transmission of zoonotic malaria, *Plasmodium knowlesi*, within the monkey populations (137). With plantations abutting the remaining forests, mosquitoes are able to transmit the malaria from the macaques to plantation workers.

The relationship between deforestation and malaria risk is complex, featuring feedback loops and second-order effects (138). For example, increased malaria in a particular location may be associated with decreased subsequent deforestation (133), while secondary forest growth may be associated with increased malaria transmission (135). There are also regional differences; for example, in India, transmission decreases with deforestation (see (139) for discussion). This complexity emphasizes the need to consider unintended adverse health consequences of strategies to reverse land degradation.

Nor is the link between land use change and vector-borne disease confined to the tropics. Lyme disease has expanded its range considerably in North America and Europe in recent years. One contributor is climate change, which has expanded the range of the tick vector, *Ixodes* species (140). Also important, however, is forest fragmentation, as occurs when cities sprawl into rural areas. This creates more edge habitats favoured by deer that carry the ticks, and more contact between people and ticks, increasing the risk of disease (141). Of note, this association has not been demonstrated in all locations tested, and may depend on local circumstances (142).

The problem of zoonotic disease spillover to humans, and the contributions of ecosystem disruption, were brought into sharp relief over recent decades by emerging viral diseases, including HIV, Ebola, SARS, MERS, and COVID-19. While the circumstances of each of these diseases vary, contributing factors include habitat destruction, human incursion into wildlife habitat, the trade in bushmeat including in wet markets, and globalization including mass air travel which facilitates rapid disease spread. Some features of habitat destruction and biodiversity loss are especially conducive to zoonotic disease emergence (143). One recent study showed that the abundance of zoonotic viruses in wild mammalian species is associated with the global abundance of the species (144). With changes in land use, some species are 'losers' – specialists with highly specific requirements – while others, the more adaptable generalists, are 'winners'. These winners are often smaller animals with fast, short lives, and they are more likely to harbour pathogens than are the losers (145, 146). Continued land use changes, biodiversity loss, and increased human–animal contact – what has been termed 'human–animal promiscuity' – are likely to pose continuing risk of zoonotic diseases.

Another pathway from land use change to disease runs not through infection, but through air quality. It stems from the growing demand for biofuels in Europe, and food, cooking oil, and wood in India, Indonesia, and China. In parts of South Asia, notably Indonesia, native tropical forests are being replaced by palm oil and timber production. To clear the native forests and peatlands, fire is commonly used; the resulting smoke blows in defined ways, affecting populations in Indonesia and the Malay peninsula (147). This smoke, containing fine particulate matter, is an established risk factor for cardiovascular mortality (148). Studies combining data on land types, land use, fire occurrence, wind patterns, smoke composition, and health outcomes across the region reveal that Indonesian fires cause an average of approximately 11,000 excess regional deaths in an average year, but in a pattern that varies considerably with such factors as El Niño (149). In an especially bad year, 2015, the toll was a full order of magnitude higher, estimated at just over 100,000 excess deaths (150). In this case, the forest alterations not only affect local ecosystems and communities, but the use of fire to clear forests threatens regional cardiovascular health.

Still another example of the impact of land use on health pertains to nutrition. In a study of children in Malawi, children living in communities with higher percentages of forest cover were more likely to consume vitamin A-rich foods and less likely to experience diarrhoea (151) compared to those living near less forest cover. Using data from the 2010 Demographic and Health Survey (DHS) linked to satellite remote sensing data on forest cover, researchers showed that a net gain in forest cover over the ten-year period prior to the survey was associated with a statistically significant 34% decrease in the odds of children experiencing diarrhoea, and an increase in consumption of vitamin A-rich foods compared to children living in areas with less forest cover. Whilst this study design could not fully account for a range of potential explanatory factors, it supports the view that intact forests benefit health in nearby communities.

A final example of land use changes affecting health pertains not to forests, but to arid drylands, at the border of Kazakhstan and Uzbekistan. Following World War II, the Soviet government sought to convert this region to agriculture. The massive irrigation schemes required diverting the rivers that fed the Aral Sea. Within decades, the surface area of one of the world's largest inland lakes had dropped by more than 90% (152). The former seabed surface, dry and impregnated with salt, contributed to regional dust storms, a phenomenon that is likely worsening with climate change. Moreover, depleted farmland in the region – permeated with fertilizers and pesticides that had been used in an ultimately unsuccessful attempt to prop up production – also gave rise to dust as it dried out. This dust creates respiratory health risks across large downwind areas (153–156). In this case, the conversion of dryland to agriculture, with a complex of first- and second-order effects, threatens health on a large scale.

Biodiversity and Ecosystem Services

Biodiversity encompasses not only diversity between and within species and ecosystems but also the complex inter-relationships and biological structures that sustain ecosystems. More biodiverse ecosystems are more productive, stable and resilient to environmental threats. For this reason, it is prudent to assume that maximizing species, functional and phylogenetic diversity is likely to enhance an ecosystem's value over the long term (157, 158). The complex linkages between biodiversity and human health have been extensively described by Eric Chivian and Aaron Bernstein in their encyclopaedic volume *Sustaining Life* (159). The profound loss of biodiversity afflicting the planet could negatively affect human health through several direct and indirect pathways, operating at interacting local, regional, and global scales (160, 161). An important framework for understanding the benefits of biodiversity is the concept of ecosystem services (**Box 3.1**).

Biodiversity is key to the delivery of a wide range of ecosystem services (**Table 3.2**). Two of the most important pathways are provisioning services – food and pharmaceuticals – and a third is a regulating function – protection from infectious diseases.

Biodiversity and food: Pollinators represent an especially important form of biodiversity, supporting one-third of the global food supply. Key pollinators include certain species of bees and wasps, butterflies and moths, and flies, beetles, and mosquitoes. The numbers and diversity of many of these species have been declining in recent years, the result of land use change, pesticide use, climate change, and other factors, and part of a larger pattern of loss of 'entomofauna' (165). This presents threats to food security (166). Using a database of supplies of 224 types of food in 156 countries, researchers estimated that loss of all pollinator services could reduce global supplies of fruit by about 23%, vegetables by about 16%, and nuts and seeds by 22%, with significant differences between countries. This could aggravate vitamin A deficiency in the 2.2 billion people whose intake is currently below required levels, as well as consign an additional 71 million people to vitamin A deficiency. There would also be major declines in the availability of folic acid (a nutritionally important B vitamin). Whilst such extreme scenarios are implausible, and in some settings wild pollinators could be replaced by commercially available pollinators, this and other studies illustrate the potential links between pollinators and human health (167).

Biodiversity supports dietary diversity, and food security, in many LMICs, where bushmeat and foraging are important components of the food system (168). Downsides include the threat to hunted species (169) and the risk of zoonotic disease transmission to those who hunt, handle, and consume bushmeat (170). Indigenous communities often make

Box 3.1. **Ecosystem Services**

Ecosystem services are the benefits provided by ecosystems, including provisioning, regulating, cultural, and supporting services (**Figure 3.7**). These benefits flow from biodiversity, from air, water, from soil – indeed, from almost every domain of the natural world.

The contribution of ecosystem services to human well-being and health is non-linear such that when a service is scarce a small decline can have disproportionately large effects. The Millennium Ecosystem Assessment (162) concluded that 15 of 24 ecosystem services assessed were in decline, the majority of which were regulating and supporting services. Examples of declining ecosystem services include pollination, the capacity of the atmosphere to remove pollutants, the control of agricultural pests, supplies of natural medicines, and freshwater and marine fisheries, all of which benefit health directly or indirectly.

Mangrove forests exemplify habitats that provide a variety of ecosystem services, from storm protection (a 'regulating' function) to nutrition (a 'provisioning' function). A study of several hundred villages in Orissa, India, impacted by a super cyclone in 1999, found that villages with wider expanses of mangroves separating them from the coast experienced significantly fewer deaths than ones with narrower or no mangroves (163). Riverine mangrove forests are especially

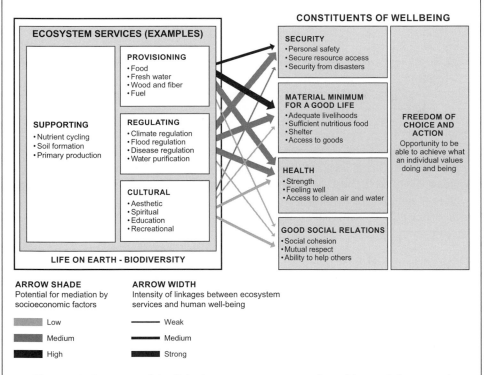

Figure 3.7. Summary of the links between ecosystem services, drivers of change, and human well-being.

Source: Millennium Ecosystem Assessment Program. *Ecosystems and Human Well-Being: Synthesis*. Washington, DC: Island Press; 2005.

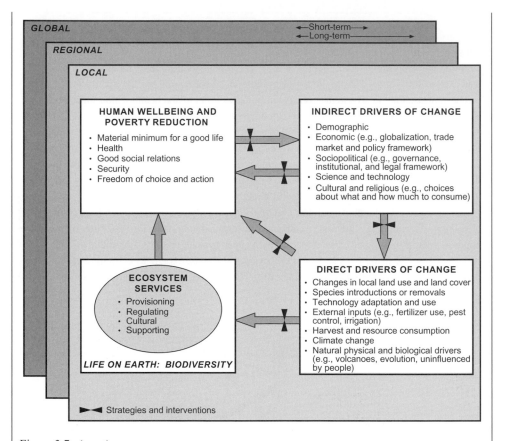

Figure 3.7. (*cont.*)

important for supporting plant and animal productivity, and basin mangroves – the interior stretches that lie behind tidal and riverine systems – enhance nutrient concentrations, providing sources of essential products such as wood (164). These ecosystem services depend on the maintenance of biodiversity in the mangrove forests.

extensive use of wild foods; in both industrialized and developing countries, they use an average of 120 wild species per community (171). In many cases, wild species are actively managed, suggesting that the distinction between hunter-gatherer and agriculturalist may be an oversimplification because many communities show features of both (see also Chapter 10). The FAO estimates that about 1 billion people use wild foods in their diet and 300 million use non-timber forest products (172). In these circumstances, decreases in biodiversity and degradation of local ecosystems can directly threaten nutrition and well-being. For example, a study in northeastern Madagascar, a biodiversity hotspot, showed that children who consumed more local wildlife had significantly higher haemoglobin concentrations, and that loss of access to wild foods could increase the prevalence of anaemia by 29% overall and as much as three-fold in the poorest households (173). Either

Table 3.2. *Biodiversity components affect ecosystem services in multiple and complex ways. The level of certainty, and the importance of the effect, vary across the components listed.*

Ecosystem services	Main components of diversity involved and mechanisms that produce the effect
Amount of biomass produced by plants considered important by humans	*** Functional composition–Faster-growing, bigger, more locally adapted plants produce more biomass, irrespective of the number of species present; in species-poor systems, coexisting plants with different resources use strategies or that facilitate each other's performance may take up more resources. ** Number of species–Within a constant resources and disturbance regime, a large species pool is more likely to contain groups of complementary or facilitating species and highly productive species, both of which could lead to higher productivity of the community.
Stability of biomass production by plants considered important by humans	*** Genetic diversity–Large genetic variability within a crop species buffers production against losses due to diseases and environmental change. *** Number of species–Cultivation of more than one species in the same plot or landscape maintains production over a broader range of conditions. *** Functional composition–Life history characteristics and resource use strategy of dominant plants determine the capacity of ecosystem processes to remain unchanged or return to their initial state in the face of perturbations.
Preservation of the fertility of soils that sustain the production of plants and animals considered important by humans	*** Functional composition–Fast-growing, nutrient-rich plants enhance soil fertility; dense root systems prevent soil erosion.
Regulation of quantity and quality of water available to humans, domestic animals, and crops	*** Arrangement and size of landscape units–Intact riparian corridors and extensive areas with dense vegetation cover reduce erosion and improve water quality. *** Functional composition–Vegetation dominated by large, fast-growing, big-leafed, deep-rooted plants has high transpiration rate, reducing stream flow.
Pollination essential for the immediate production of fruits by, and the perpetuation of, important plant species	*** Functional composition of pollinator assemblage–Loss of specialized pollinators leads to genetic impoverishment and lower number and quality of fruits. ** Number of species of pollinator assemblage–Lower number of pollinator species leads to genetic impoverishment of plant species. ** Arrangement and size of landscape units–Large and/or well-connected landscape units allow movement of pollinators among plants of the same species, thus maintaining plant genetic pool.

Table 3.2. (*cont.*)

Ecosystem services	Main components of diversity involved and mechanisms that produce the effect
Resistance to invasive organisms that have negative ecological, economic, and/or cultural impacts	*** Functional composition–Some key native species are very competitive or can act as biological controls to the spread of aliens. *** Arrangement of landscape units–Landscape corridors (e.g., roads, rivers, and extensive crops) can facilitate the spread of aliens; size and nature of suitable corridors are likely to be different for different organisms. ** Number of species–All else being equal, species-rich communities are more likely to contain highly competitive species and to contain less unused resources, and therefore be more resistant to invasions.
Pest and disease control in agricultural systems	*** Genetic diversity of crops–High intraspecific genetic diversity reduces density of hosts for specialist pests and, thus, their ability to spread. ** Number of crop, weed, and invertebrate species–High number of species acts similarly as genetic diversity and also increases habitat for natural enemies of pest species. ** Spatial distribution of landscape units–Natural vegetation patches intermingled with crops provide habitat for natural enemies of insect pests.
Regulation through biophysical feedbacks of climatic conditions suitable for humans and the animals and plants they consider important	*** Arrangement and size of landscape units–Size and spatial arrangement of landscape units over large areas influence local-to-regional climate by lateral movement of air masses of different temperature and moisture; the threshold for effect is patch size of about 10 km diameter, depending on wind speed and topography. ** Functional composition–Height, structural diversity, architecture, and phenology modify albedo, heat absorption, and mechanical turbulence, thus changing local air temperature and circulation patterns.
Regulation through carbon sequestration in the biosphere of climatic conditions suitable for humans and the animals and plants they consider important	*** Arrangement and size of landscape units–Carbon loss is higher at forest edges, therefore as forest fragments decline in size or area/perimeter ratio a larger proportion of the total landscape is losing carbon. ** Functional composition–Small, fast-growing, fast-decomposing, short-lived plants retain less carbon in their biomass than large, slow-growing, slow-decomposing, long-lived plants. * Number of species–High number of species can slow down the spread of pests and pathogens, which are important agents of carbon loss from ecosystems.

Table 3.2. (*cont.*)

Ecosystem services	Main components of diversity involved and mechanisms that produce the effect
Protection against natural hazards (storms, floods, hurricanes. fires) that cause damage to humans and the animal production systems that they depend on	*** Arrangement and size of landscape units–Large patches of structurally complex vegetation or small, close-by patches are likely to offer more shelter to nearby ecosystems, and buffer them against flooding, sea intrusion, and wind. *** Functional composition–Deep-rooted plants are less susceptible to uprooting by hurricanes; extensive, mat-forming, superficial root systems protect soil against erosion by floods and storms; deciduous canopy types decrease flammability.

Asterisks indicate importance and/or degree or certainty (*** > ** > *) of the link between the ecosystem service in question and different components of biodiversity. Biodiversity components refer to plant assemblages unless otherwise specified. The putative mechanisms have been empirically tested in some cases, but remain speculative in others. The list of ecosystem services is illustrative, rather than exhaustive.
Source: Díaz S, Fargione J, Chapin FS, III, Tilman D. Biodiversity loss threatens human well-being. *PLOS Biology.* 2006;4(8):e277. doi: 10.1371/journal.pbio.0040277.t001.

unsustainable exploitation of wildlife or overzealous application of conservation measures could have deleterious consequences for health.

Wild foods have played a traditionally important role in many Arctic and sub-Arctic communities. This dietary pattern is changing in complex ways. Climate change threatens the availability of some mammalian and fish species. Moreover, many of the fish and marine mammals consumed by these communities concentrate POPs and other toxins through biomagnification (174) – a process that climate change may be accelerating (175). When store-bought processed foods replace wild foods, this brings with it risks of obesity and diabetes due the excessive consumption of refined carbohydrates, fat, and energy (176).

Biodiversity pressures interact across different ecosystems, with implications for human health and well-being. For example, depletion of wild-catch fisheries can increase the demand for land-based wild foods (177). A study using three decades of data from Ghana documented how bushmeat demand in villages rose during years of poor fish supply (177), resulting in declines in 41 species of wild carnivores, primates, and herbivores in six of the nation's nature reserves. Fish declines were driven by a number of factors including the increasing presence of EU fishing boats off the coast of West Africa, with EU fish harvests there increasing by a factor of 20 from 1950 to 2001 (178).

The global food system relies largely on a tiny number of key species with rice, maize, and wheat providing 60% of the human diet (179). Twelve crops together with five animal species provide about 75% of the energy intake (180). This contrasts with the 30,000 edible plants of which about 7000 have been cultivated or collected for food throughout history (181, 182). This increasing homogenization has allowed the provision of energy-dense foods to an increasing population but also brings risks as it renders food supplies more vulnerable to pests and diseases affecting a single species (see Chapter 10). Lack of dietary diversity has adverse effects on nutrition; there is wide acceptance that increasing dietary diversity improves micronutrient intake (183–185). An additional concern is the unplanned spread of novel genes from genetically modified organisms into wild crops or traditional varieties, which could have unanticipated effects on their resilience and thus long-term survival.

Biodiversity and medications: By destroying biodiversity at an unprecedented rate we are also losing the potential of plant-derived pharmaceutical agents. These formed the mainstay of medical treatment for much of recorded history. The first written records come from Mesopotamia in 2600 BC. Early remedies included liquorice (*Glycyrrhiza glabra*) and the opium poppy (*Papaver somniferum*). The first written record of Egyptian medicines describes 700 drugs, many derived from plants. Documentation of Indian Ayurvedic plant-based medicines dates from about 1000 BC and Chinese medicines comprised 850 different prescriptions by AD 659. The ancient Greek and Roman texts were preserved and built on by physicians such as Avicenna and subsequently Ibn al-Baytar from Andalusia, who in the thirteenth century described the medicinal uses of 1400 plants in widely influential books. In 1597, John Gerrard, the curator of the Physic Garden of the Royal College of Physicians of London, published a book that included 2200 woodcut images of medicinal plants; it was a standard reference for at least two centuries (159).

Many examples of plant-based medicines are relevant to contemporary medicine. Artemisinin, a potent antimalarial, was isolated by Chinese scientists from the sweet wormwood plant *(Artemisia annua)*, which had been used for over 2000 years to treat fevers. Vinca alkaloids extracted from the Madagascan periwinkle (*Catharanthus roseus*) have been used to treat childhood leukaemia (vincristine) and Hodgkins disease (vinblastine) with many related compounds of potential value having been synthesized. Aspirin (acetylsalicylic acid), widely used for pain relief, as an anti-inflammatory agent, and for its ability to reduce platelet stickiness and thus reduce risk of heart attacks and strokes, originally came in the form of salicylic acid from the white willow (*Salix alba vulgaris*). An acetyl group was added by chemists to make it less irritating to the stomach. Sweet clover (*Melilotus* species) produces a powerful anticoagulant, dicoumarol, when the coumarin in the clover is metabolized by various moulds. It was originally discovered when a farmer, whose cattle had died of unstoppable haemorrhages after consuming spoiled sweet clover, drove through a snowstorm with a dead heifer, a milk can filled with unclotted blood, and a large sample of spoiled sweet clover to an agricultural research laboratory. By chance the agricultural experimental station he was aiming for was closed and he ended up at the laboratory of Karl Paul Link at the University of Wisconsin, who had been researching 'sweet clover disease'. Six years later one of Link's colleagues synthesized the active compound, dicoumarol. The resulting anticoagulant was called warfarin after the Wisconsin Alumni Research Foundation, which funded Link's research (186). Such discoveries continue (187), now reinforced by genomic technologies (188).

Herbal remedies are widely used in high- and low-income countries, with nearly 5000 medicinal plant species thought to be in use in China, 3000 in India, and thousands more in other countries, totalling over 50,000. Some of these are cultivated and many are wild. Demand is increasing because of the prohibitive cost of conventional pharmaceuticals to many low-income populations. However, an estimated 15,000 medicinal plant species worldwide are endangered because of overharvesting, land use changes, and other factors (189).

Animals are the source of many other medically important compounds such as anticoagulants or antihypertensives. Chivian and Bernstein identify several groups of animals that are valuable to human medicine and that are threatened: amphibians, bears, non-human primates, cone snails, sharks, and horseshoe crabs. They document compellingly how invasive species, pollution, hunting, harvesting, habitat loss, climate change, and emerging disease threaten these organisms in different ways with potentially irreversible consequences. Amphibians are probably the most threatened group, in part due to climate change, which may alter the timing of migration and reproduction and thus disrupt their relationship with other species on which they depend for food, or disadvantage them in comparison to predators or competitors. Amphibians produce an array of chemicals isolated from more than 500 species and including alkaloid toxins, antimicrobial peptides (potentially significant in view of growing antimicrobial resistance), bradykinins and other bioactive peptides, and novel compounds or mixtures such as biological glue that can bind animal tissue. By destroying this biological treasury we lose the potential to study and make use of a range of active substances produced by living organisms.

Biodiversity and infectious disease risk: Biodiversity loss is associated with increased likelihood of disease transmission. For example, the transmission of West Nile virus by mosquitoes is influenced by the diversity of passerine birds which act as hosts (190). Thus, in the USA, the incidence of West Nile encephalitis is higher where the number of bird species is lower because these areas tend to be dominated by species that amplify the virus, whereas in more diverse locations there are more species that are less competent hosts.

Links between biodiversity loss and disease risk may be direct or indirect (191). One example is schistosomiasis, which affects about 200 million people worldwide, causing serious effects on liver and urinary tract depending on the causative species. This disease is caused by five main species of blood fluke, a parasite whose host is freshwater snails. In Lake Malawi, for example, overfishing and the use of fine mesh seine nets appears to have caused a decline in a fish called the snailcrusher hap (*Trematocranus placodon*), the snails' natural predator, with a consequent increase in schistosomiasis (191, 192).

In the case of Lyme disease, transmitted by infected ticks, the level of host diversity strongly influences the risks to humans (191). In areas with low biodiversity, where there are large numbers of white-footed mice that can host infected ticks, transmission to humans is likely, but in areas where other hosts less likely to support transmission predominate, the risks are lower (193). Virginia opossums (*Didelphis virginiana*) are poor hosts for the pathogen and tend to kill the ticks that feed on them so they act as a host with a strong buffering effect. The opossum tends to disappear from forest fragments where biodiversity is lost and the white-footed mouse, which amplifies transmission, tends to remain (190). A similar protective effect of biodiversity appears to operate for hantavirus pulmonary syndrome, a condition with a high mortality rate transmitted by aerosolized urine or faeces of infected rodents. In Panama, for example, where the rice rat is a host, more diverse rodent populations reduce the risk of transmission to humans (194). Resilience of a specific species to environmental change may be linked to its competence as a host for infections, possibly reflecting species differences in immune function. For example, plants that are classified as 'weeds' can grow rapidly as species diversity declines but may also be more susceptible to pathogens and pests; a similar pattern may also occur in vertebrates (190).

Thus, for pathogens whose transmission is already established, biodiversity loss is often associated with increased transmission. A different situation may occur in the case of emerging diseases caused by new pathogens, which may jump from wildlife to humans as a result of humans venturing into close proximity with biodiverse environments (195, 196). The transmission of zoonotic diseases is influenced by a range of environmental changes. Zoonotic diseases may also pass from domestic animals to humans. For example, in Malaysia, Nipah virus, which causes a severe febrile encephalitis in humans with case fatality rates of 40–75%, was transmitted from wild fruit bats to domestic pigs and then to humans (197). The transmission of zoonotic infections from animals to humans probably accelerated as humans transitioned from hunter-gatherers to agrarian communities when changing patterns of land use and domestication of animals brought humans and animals into closer proximity. Many emerging disease risks are related to land use change or agricultural industry change such as intensive production of livestock, and exacerbated by international air travel which facilitates rapid spread of diseases (117). Land use change is a major driver of biodiversity loss.

The relationships between disease transmission and biodiversity are likely mechanism- and context-dependent. A global assessment that accounted for increased reporting in more densely populated locations showed that zoonotic emerging infectious disease risk is elevated in forested tropical regions experiencing land use changes and where there is high wildlife biodiversity (as indicated by mammal species richness). Although the overall trend is for heightened risk of disease emergence with higher mammalian richness, this neither excludes nor confirms the possibility of a dilution effect for specific diseases and is consistent with the conclusion that the relationship between biodiversity and disease risk is complex and context-specific (198).

An example is the Sin Nombre hantavirus, which is fatal to about 35% of infected humans. The reservoir host is the North American deermouse (*Peromyscus maniculatus*). A high diversity of small mammals lowers the prevalence of the virus in deermouse populations, because the lower density of the host results in less spread among host individuals (dilution effect). However, at any given level of deermouse density, more diversity is associated with greater transmission (amplification effect). Thus, both effects operate, with dilution generally predominating (199). A meta-analysis of the relevant literature also suggests that disease risk is probably a local phenomenon depending on the specific reservoir of hosts and vectors, and their ecological characteristics, rather than solely related to patterns of species biodiversity (200).

Future research will need to focus on understanding how biodiversity affects individual mechanisms separately, and their net effects when considered in aggregate, in order to make generalizable predictions.

Freshwater Resources

Human-induced changes in water resources affect health in various ways. As noted in Chapter 1, these changes include less rainfall in areas that are already dry (a result of climate change), depletion of groundwater, and/or damming of rivers.

Water scarcity may be episodic – known as drought – or long-standing, as in arid parts of the world. The most water-stressed parts of the world are the Middle East and North Africa, although droughts have become more severe in places from Australia to North America in recent decades (see Figure 1.3). Freshwater scarcity is predicted to increase in coming decades due to climate change (201–204), saline contamination of coastal groundwater (205), and depletion of groundwater aquifers (206, 207) (supply shortfalls) and growing populations (demand increases). Water scarcity has a range of health effects (208, 209).

Water scarcity threatens food production, and thus nutrition and health. In rural areas during periods of drought, as food production drops off, undernutrition and child stunting increase, with lifelong implications for affected children (210, 211). The nutritional impacts of water scarcity play out on a global scale, as a substantial proportion of the global trade in staple crops involves exports from countries that deplete groundwater to produce these crops. This unsustainable situation highlights the risk of coming food shortfalls in importing countries (212). Other pathways from water scarcity to health include impacts on mental health, including an increased risk of suicide in agricultural communities (213), increased risk of some vector-borne diseases as people store water in containers that serve as mosquito breeding sites (214), violent conflict, particularly among agriculturally dependent groups and politically excluded groups in very poor countries (215), and impacts that follow dislocation and migration.

On the other hand, health benefits could flow from efforts to address water scarcity. One modelling study showed that, in India, shifting diets to meet local constraints on water availability, including a reduction of wheat, dairy, and poultry, and an increase in legumes, could reduce the risk of non-communicable diseases (216).

Rivers change when they are dammed. Impacts may include changes in water flow, temperature, and sedimentation, with resulting habitat changes, reduced water quality, loss of wetlands, disruption of fish migration, and even species extinction (217). While damming offers the health benefits that derive from clean electricity, irrigation, and flood control, it may threaten health in at least three ways. First, dam construction may displace people living in riverside towns and villages that are flooded (218). Construction of China's Three Gorges Dam, for example, completely or partially flooded 13 cities and towns, 365 townships, and 1711 villages, inundated about 26,000 hectares of farmland, and displaced at least 1.3 million residents (219). Impoverishment, shattered social support networks, homelessness, and unemployment, and consequent health impacts including depression and poor self-rated health, followed (220). Second, dams can alter river ecology in ways that increase the risk of infectious disease in nearby populations; this has been best documented for schistosomiasis (221) and malaria (222). Third, dam collapses can lead to catastrophic damage to people and property downstream. Like any ecosystem alteration intended to yield particular benefits, careful adaptive management is needed to identify trade-offs and manage unwanted consequences (223).

Multiple Drivers of Disease Risk

Much research focuses on a single environmental factor acting alone, but increasingly we need to assess the effect of multiple environmental and social factors acting in concert as we live in a dynamic world where changes may be additive or synergistic. A study of infectious disease threat events (IDTEs) in Europe, where disease surveillance systems are more robust than in many regions, showed that of 116 IDTEs detected between 2008 and 2013 by the European Centre for Disease Prevention and Control, most resulted from a combination of two or more drivers, with 25% resulting from three or more drivers (224). The top five individual drivers of IDTEs were travel and tourism, food and water quality, natural environment, global trade, and climate. The natural environment drivers included changes in land use and land cover, waterways, oceans, coastlines, water resources, habitats, and biodiversity. Climate and the natural environment were particularly important drivers of vector-borne infections. For example, a large outbreak of over 2000 cases of dengue in Madeira was driven by climate changes, changes in the natural environment, and travel and tourism (225). A large outbreak of hantavirus infection in Germany in 2010 was attributed to increases in bank vole (*Clethrionomys glareolus*) populations, due to high levels of seed production the previous year (226). Increased outdoor activities in the warm and dry summer months of 2010 resulted in exposure to dust contaminated by rodent excreta. Better diagnosis and reporting may also have contributed to the large numbers of notified cases.

While the history of COVID-19 has not yet been written, it seems very likely, given the similarity of SARS-CoV-2 to bat SARS-CoV-like coronaviruses, that bats were reservoir hosts for the virus. Malayan pangolins (*Manis javanica*) are hosts to coronaviruses similar to SARS-CoV-2 and were illegally imported into Guangdong province as a luxury food or for use in traditional Chinese medicine. Many early cases of COVID-19 were linked to the Huanan market in Wuhan where live animals were traded. Although neither bats nor pangolins had, as of early 2021, yielded the direct progenitor of SARS-CoV-2, it is highly plausible that it originated in one or both of them. After making the jump into humans SARS-CoV-2 could have acquired the genomic features required for human transmission through adaptation during initially undetected human-to-human transmission (227). Rapid spread was facilitated by mass air travel and delayed, and in some cases, inconsistent public health responses, particularly in Europe and North America.

In aquatic systems, degraded water quality from increased nutrient flows – nitrogen and phosphorus – can cause eutrophication that allows algal blooms to develop and persist (228). The high biomass of algal blooms can cause deoxygenation of water, killing large numbers of fish and reducing biodiversity. Climatic factors may also influence the distribution of algae (see also Chapter 2). Suitable habitat for many algae species may extend further northward with climate change to the central and northern regions of the North Sea, the North Pacific Ocean, and lakes in northern locations (229, 230). An additional factor is ocean acidification, which has been shown to increase the abundance of the toxic microalga *Vicitus globosus* by stimulating growth and decreasing the loss due to grazing because of its increased toxicity (231).

The operation of these complex, interacting drivers, and the resulting algal blooms, can threaten human health. A principal pathway is the formation of phycotoxins by certain species of algae. Bivalves (molluscs such as clams, oysters, mussels, and scallops) filter and accumulate the phycotoxins formed by these blooms, and consumption of affected bivalves can trigger such conditions as paralytic, diarrhetic, and amnesic shellfish poisoning. Other pathways operate as well: the loss of income from tourism and fishing, and the loss of recreational opportunities.

The combination of climate change, nutrient loading, and ocean acidification is likely to increase the probability of harmful blooms. The magnitude of this threat, the species specificity, the vulnerable times and locations, and the optimal adaptive responses are not fully understood at this point. But the concern must be that such interactions threaten the ocean food web, increasing mortality of farmed and wild fish with implications for the economic welfare of coastal communities, and the nutrition of those populations depending substantially on fish consumption.

Multiple environmental stressors can also affect land-based food systems but are rarely taken into account in future projections. Climate change, salination, and tropospheric ozone concentrations all reduce vegetable and legume yield, an impact that is unlikely to be offset by the increased growth resulting from carbon dioxide fertilization (232). Decreased labour productivity will also reduce work output of subsistence farmers who are unlikely to access mechanization at a time when their crops and livestock are also increasingly affected by climate change. Food trade depends on exports from a small number of countries to a larger number of import-dependent countries. During periods of environmental shocks such as intense heat or droughts affecting critical regions, some exporting countries suspend exports to protect their own populations (233), at times abruptly, which can trigger rapid increases in food prices and political disturbance. The number of unstable countries unable to feed themselves without recourse to global trade has increased in recent decades. For the most vulnerable countries even a small perturbation may create major instability. New bioenergy policies and increased demand for animal products intensify the risks and overexploitation of finite supplies of freshwater, including in food exporting nations, which further exacerbates the threat level (234).

Conclusions

Environmental change in the Anthropocene Epoch threatens to undermine the dramatic progress in health achieved in recent history. Climate change poses many such threats, as described in Chapter 2; so do pollution, land use change, and biodiversity loss as described in this chapter. Moreover, these pathways can interact in unpredictable ways with impacts on a range of health outcomes, both mental and physical. Vulnerability to these adverse effects depends not just on exposure to a given environmental hazard but also on the susceptibility of individuals and communities to a given exposure. Humans have been adapting to changes in their environments throughout history with varying degrees of success. In Chapter 4 we examine what is known about how humans can adapt to emerging challenges and create more resilient societies able to withstand shocks and recreate themselves to reach a new equilibrium with their environment. At the same time there are limits to adaptation, as shown by historical case studies of collapse of whole societies faced by overwhelming rapid changes or slow to react to gradually emerging challenges. Exploring the potential for adaptation and its limits will help to inform us about additional policies needed to reduce the risks to humanity in the Anthropocene Epoch.

References

1. Steffen W, Richardson K, Rockstrom J, et al. Planetary boundaries: guiding human development on a changing planet. *Science*. 2015;347(6223):1259855. https://science.sciencemag.org/content/347/6223/1259855.
2. Lickley M, Solomon S, Fletcher S, et al. Quantifying contributions of chlorofluorocarbon banks to emissions and impacts on the ozone layer and climate. *Nature Communications*. 2020;11(1):1380. https://doi.org/10.1038/s41467-020-15162-7.
3. Rigby M, Park S, Saito T, et al. Increase in CFC-11 emissions from eastern China based on atmospheric observations. *Nature*. 2019;569(7757):546–50. doi: 10.1038/s41586-019-1193-4.
4. Lin J, Pan D, Davis SJ, et al. China's international trade and air pollution in the United States. *Proceedings of the National Academy of Sciences*. 2014;111(5): 1736–41. https://doi.org/10.1073/pnas.1312860111.
5. Ashraf MA. Persistent organic pollutants (POPs): a global issue, a global challenge. *Environmental Science and Pollution Research*. 2017;24(5):4223–7. https://doi.org/10.1007/s11356-015-5225-9.
6. Scheringer M. Long-range transport of organic chemicals in the environment. *Environmental Toxicology and Chemistry*. 2009;28(4):677–90. doi: 10.1897/08-324R.1.
7. Almroth BC, Eggert H. Marine plastic pollution: sources, impacts, and policy issues. *Review of Environmental Economics and Policy*. 2019;13(2):317–26. doi: 10.1093/reep/rez012.
8. Villarrubia-Gómez P, Cornell SE, Fabres J. Marine plastic pollution as a planetary boundary threat: the drifting piece in the sustainability puzzle. *Marine Policy*. 2018;96:213–20. https://doi.org/10.1016/j.marpol.2017.11.035.
9. Landrigan PJ, Fuller R, Acosta NJR, et al. The *Lancet* Commission on Pollution and Health. *The Lancet*. 2018;391:462–512. doi: 10.1016/S0140-6736(17)32345-0.
10. Bell ML, Davis DL, Fletcher T. A retrospective assessment of mortality from the London smog episode of 1952: the role of influenza and pollution. *Environmental Health Perspectives*. 2004;112(1):6–8. https://dx.doi.org/10.1289%2Fehp.6539.
11. London Air Quality Network. Air pollution research in London 2020. Available from www.londonair.org.uk/LondonAir/General/research.aspx.
12. Schraufnagel DE, Balmes JR, Cowl CT, et al. Air pollution and noncommunicable diseases: a review by the Forum of International Respiratory Societies' Environmental Committee, Part 2: Air pollution and organ systems. *Chest*. 2019; 155(2):417–26. doi: 10.1016/j.chest.2018.10.041.

13. Epstein PR, Buonocore JJ, Eckerle K, et al. Full cost accounting for the life cycle of coal. *Annals of the New York Academy of Sciences*. 2011;1219(1):73–98. doi: 10.1111/j.1749-6632.2010.05890.x.

14. McGlade C, Ekins P. The geographical distribution of fossil fuels unused when limiting global warming to 2°C. *Nature*. 2015;517(7533):187–90. https://doi.org/10.1038/nature14016.

15. Lelieveld J, Haines A, Pozzer A. Age-dependent health risk from ambient air pollution: a modelling and data analysis of childhood mortality in middle income and low-income countries. *The Lancet Planetary Health*. 2018;2(7):e292–300. DOI: https://doi.org/10.1016/S2542-5196(18)30147-5.

16. Requia WJ, Adams MD, Arain A, et al. Global association of air pollution and cardiorespiratory diseases: a systematic review, meta-analysis, and investigation of modifier variables. *American Journal of Public Health*. 2018;108(S2):S123–30. doi: 10.2105/AJPH.2017.303839.

17. Yang Y, Ruan Z, Wang X, et al. Short-term and long-term exposures to fine particulate matter constituents and health: a systematic review and meta-analysis. *Environmental Pollution*. 2019;247:874–82. doi: 10.1016/j.envpol.2018.12.060.

18. Newell K, Kartsonaki C, Lam KBH, Kurmi OP. Cardiorespiratory health effects of particulate ambient air pollution exposure in low-income and middle-income countries: a systematic review and meta-analysis. *The Lancet Planetary Health*. 2017;1(9):e368–80.

19. Jaganathan S, Jaacks LM, Magsumbol M, et al. Association of long-term exposure to fine particulate matter and cardio-metabolic diseases in low- and middle-income countries: a systematic review. *International Journal of Environmental Research and Public Health*. 2019;16(14). https://doi.org/10.3390/ijerph16142541.

20. Liu Q, Gu X, Deng F, et al. Ambient particulate air pollution and circulating C-reactive protein level: a systematic review and meta-analysis. *International Journal of Hygiene and Environmental Health*. 2019;222(5):756–64.

21. Yang HJ, Liu X, Qu C, et al. Main air pollutants and ventricular arrhythmias in patients with implantable cardioverter-defibrillators: a systematic review and meta-analysis. *Chronic Diseases and Translational Medicine*. 2017;3(4):242–51.

22. Yang BY, Qian Z, Howard SW, et al. Global association between ambient air pollution and blood pressure: a systematic review and meta-analysis. *Environmental Pollution*. 2018;235:576–88. doi: 10.1016/j.envpol.2018.01.001.

23. An R, Ji M, Yan H, Guan C. Impact of ambient air pollution on obesity: a systematic review. *International Journal of Obesity*. 2018;42(6):1112–26.

24. Ma R, Zhang Y, Sun Z, Xu D, Li T. Effects of ambient particulate matter on fasting blood glucose: a systematic review and meta-analysis. *Environmental Pollution*. 2020;258:113589. doi: 10.1016/j.envpol.2019.

25. Yang BY, Fan S, Thiering E, et al. Ambient air pollution and diabetes: a systematic review and meta-analysis. *Environmental Research*. 2020;180:108817.

26. Yuan L, Zhang Y, Gao Y, Tian Y. Maternal fine particulate matter (PM(2.5)) exposure and adverse birth outcomes: an updated systematic review based on cohort studies. *Environmental Science and Pollution Research International*. 2019;26(14):13963–83.

27. Li X, Huang S, Jiao A, et al. Association between ambient fine particulate matter and preterm birth or term low birth weight: an updated systematic review and meta-analysis. *Environmental Pollution*. 2017;227:596–605.

28. Tsai TL, Lin YT, Hwang BF, et al. Fine particulate matter is a potential determinant of Alzheimer's disease: a systemic review and meta-analysis. *Environmental Research*. 2019;177:108638. https://doi.org/10.1016/j.envres.2019.108638.

29. Liu, Q., Wang, W., Gu, X. et al. Association between particulate matter air pollution and risk of depression and suicide: a systematic review and meta-analysis. Environ Sci Pollut Res 28, 9029–9049 (2021). https://doi.org/10.1007/s11356-021-12357-3.
30. Donzelli G, Llopis-Gonzalez A, Llopis-Morales A et al. Particulate matter exposure and attention-deficit/hyperactivity disorder in children: a systematic review of epidemiological studies. *International Journal of Environmental Research and Public Health*. 2019;17(1). doi: 10.3390/ijerph17010067.
31. Popovic I, Soares Magalhaes RJ, Ge E, et al. A systematic literature review and critical appraisal of epidemiological studies on outdoor air pollution and tuberculosis outcomes. *Environmental Research*. 2019;170:33–45. doi: 10.1016/j.envres.2018.12.011.
32. Nhung NTT, Amini H, Schindler C, et al. Short-term association between ambient air pollution and pneumonia in children: a systematic review and meta-analysis of time-series and case-crossover studies. *Environmental Pollution*. 2017;230:1000–8.
33. Janssen NAH, Hoek G, Simic-Lawson M, et al. Black carbon as an additional indicator of the adverse health effects of airborne particles compared with PM10 and PM2.5. *Environmental Health Perspectives*. 2011;119(12):1691–9.
34. Malley CS, Henze DK, Kuylenstierna JCI, et al. Updated global estimates of respiratory mortality in adults $\geq$ 30 years of age attributable to long-term ozone exposure. *Environmental Health Perspectives*. 2017;125(8):087021.
35. Mills G, Sharps K, Simpson D, et al. Closing the global ozone yield gap: quantification and cobenefits for multistress tolerance. *Global Change Biology*. 2018;24(10): 4869–93. https://doi.org/10.1111/gcb.14381.
36. COMEAP. Associations of Long-Term Average Concentrations of Nitrogen Dioxide with Mortality. London: Committee on the Medical Effects of Air Pollutants; 2018.
37. Achakulwisut P, Brauer M, Hystad P, Anenberg SC. Global, national, and urban burdens of paediatric asthma incidence attributable to ambient NO_2 pollution: estimates from global datasets. *The Lancet Planetary Health*. 2019;3(4):e166–78.
38. WHO. Ambient (Outdoor) Air Pollution. Geneva: World Health Organization; 2018. Available from www.who.int/news-room/fact-sheets/detail/ambient-(outdoor)-air-quality-and-health.
39. Burnett R, Chen H, Szyszkowicz M, et al. Global estimates of mortality associated with long-term exposure to outdoor fine particulate matter. *Proceedings of the National Academy of Sciences*. 2018;115(38):9592–7.
40. Lelieveld J, Pozzer A, Pöschl U, et al. Loss of life expectancy from air pollution compared to other risk factors: a worldwide perspective. *Cardiovascular Research*. 2020;116(11):1910–17. https://doi.org/10.1093/cvr/cvaa025.
41. Heft-Neal S, Burney J, Bendavid E, Burke M. Robust relationship between air quality and infant mortality in Africa. *Nature*. 2018;559(7713):254–8.
42. Lelieveld J, Klingmüller K, Pozzer A, et al. Effects of fossil fuel and total anthropogenic emission removal on public health and climate. *Proceedings of the National Academy of Sciences*. 2019;116(15):7192–7. https://doi.org/10.1073/pnas.1819989116.
43. Liu F, Wang M, Zheng M. Effects of COVID-19 lockdown on global air quality and health. *Science of The Total Environment*. 2021;755:142533. doi: 10.1016/j.scitotenv.2020.142533.
44. Dutheil F, Baker JS, Navel V. COVID-19 as a factor influencing air pollution? *Environmental Pollution*. 2020;263(Pt A):114466.
45. Wu X, Nethery RC, Sabath BM, Braun D, Dominici F. Exposure to air pollution and COVID-19 mortality in the United States: strengths and limitations of an ecological regression analysis. *Science Advances*. 2020;6(45), p.eabd4049.

46. Ghazani M, FitzGerald G, Hu W, Toloo GS, Xu Z. Temperature variability and gastrointestinal infections: a review of impacts and future perspectives. *International Journal of Environmental Research and Public Health*. 2018;15(4).
47. Philipsborn R, Ahmed SM, Brosi BJ, Levy K. Climatic drivers of diarrheagenic *Escherichia coli* incidence: a systematic review and meta-analysis. *The Journal of Infectious Diseases*. 2016;214(1):6–15. https://doi.org/10.1093/infdis/jiw081.
48. Levy K, Woster AP, Goldstein RS, Carlton EJ. Untangling the impacts of climate change on waterborne diseases: a systematic review of relationships between diarrheal diseases and temperature, rainfall, flooding, and drought. *Environmental Science & Technology*. 2016;50(10):4905–22.
49. Alexander K, Carzolio M, Goodin D, Vance E. Climate change is likely to worsen the public health threat of diarrheal disease in Botswana. *International Journal of Environmental Research and Public Health*. 2013;10(4):1202–30.
50. Curriero FC, Patz JA, Rose JB, Lele S. The association between extreme precipitation and waterborne disease outbreaks in the United States, 1948–1994. *American Journal of Public Health*. 2001;91(8):1194–9. https://doi.org/10.2105/ajph.91.8.1194.
51. Lewis DJ, Atwill ER, Lennox MS, et al. Linking on-farm dairy management practices to storm-flow fecal coliform loading for California coastal watersheds. *Environmental Monitoring and Assessment*. 2005;107(1–3):407–25.
52. Anderson ME, Sobsey MD. Detection and occurrence of antimicrobially resistant *E. coli* in groundwater on or near swine farms in eastern North Carolina. *Water Science and Technology*. 2006;54(3):211–18. doi: 10.2166/wst.2006.471.
53. Bernhardt ES, Rosi EJ, Gessner MO. Synthetic chemicals as agents of global change. *Frontiers in Ecology and the Environment*. 2017;15(2):84–90.
54. Trasande L, Zoeller RT, Hass U, et al. Estimating burden and disease costs of exposure to endocrine-disrupting chemicals in the European Union. *The Journal of Clinical Endocrinology and Metabolism*. 2015;100(4):1245–55.
55. Schug TT, Johnson AF, Birnbaum LS, et al. Minireview: endocrine disruptors: past lessons and future directions. *Molecular Endocrinology*. 2016;30(8):833–47.
56. Matthiessen P, Wheeler JR, Weltje L. A review of the evidence for endocrine disrupting effects of current-use chemicals on wildlife populations. *Critical Reviews in Toxicology*. 2018;48(3):195–216. doi: 10.1080/10408444.2017.1397099.
57. Levine H, Jorgensen N, Martino-Andrade A, et al. Temporal trends in sperm count: a systematic review and meta-regression analysis. *Human Reproduction Update*. 2017;23(6):646–59. doi: 10.1093/humupd/dmx022.
58. Virtanen HE, Jørgensen N, Toppari J. Semen quality in the 21st century. *Nature Reviews Urology*. 2017;14(2):120–30. https://doi.org/10.1038/nrurol.2016.261.
59. García J, Ventura MI, Requena M, et al. Association of reproductive disorders and male congenital anomalies with environmental exposure to endocrine active pesticides. *Reproductive Toxicology*. 2017;71:95–100. doi: 10.1016/j.reprotox.2017.04.011.
60. Skakkebaek NE, Rajpert-De Meyts E, Buck Louis GM, et al. Male reproductive disorders and fertility trends: influences of environment and genetic susceptibility. *Physiological Reviews*. 2016;96(1):55–97. doi: 10.1152/physrev.00017.2015.
61. Soto AM, Sonnenschein C. Environmental causes of cancer: endocrine disruptors as carcinogens. *Nature Reviews Endocrinology*. 2010;6(7):363–70.
62. Jurewicz J, Polanska K, Hanke W. Exposure to widespread environmental toxicants and children's cognitive development and behavioral problems. *International Journal of Occupational Medicine and Environmental Health*. 2013;26(2):185–204.
63. Rock KD, Patisaul HB. Environmental mechanisms of neurodevelopmental toxicity. *Current Environmental Health Reports*. 2018;5(1):145–57.

64. Demeneix BA. Evidence for prenatal exposure to thyroid disruptors and adverse effects on brain development. *European Thyroid Journal.* 2019;8(6):283–92.

65. Regnault C, Usal M, Veyrenc S, et al. Unexpected metabolic disorders induced by endocrine disruptors in *Xenopus tropicalis* provide new lead for understanding amphibian decline. *Proceedings of the National Academy of Sciences.* 2018;115(19): E4416–25. https://doi.org/10.1073/pnas.1721267115.

66. Landrigan PJ, Etzel RA, editors. *Textbook of Children's Environmental Health.* Oxford and New York: Oxford University Press; 2014.

67. Hahladakis JN, Velis CA, Weber R, Iacovidou E, Purnell P. An overview of chemical additives present in plastics: migration, release, fate and environmental impact during their use, disposal and recycling. *Journal of Hazardous Materials.* 2018;344:179–99.

68. Abraham A, Chakraborty P. A review on sources and health impacts of bisphenol A. *Reviews on Environmental Health.* 2020;35(2):201–10.

69. Ma Y, Liu H, Wu J, et al. The adverse health effects of bisphenol A and related toxicity mechanisms. *Environmental Research.* 2019;176:108575.

70. IARC. 1,3-Butadiene, Ethylene Oxide and Vinyl Halides (Vinyl Fluoride, Vinyl Chloride and Vinyl Bromide). Lyon, France: International Agency for Research on Cancer; 2008.

71. Radke EG, Braun JM, Meeker JD, Cooper GS. Phthalate exposure and male reproductive outcomes: a systematic review of the human epidemiological evidence. *Environment International.* 2018;121(Pt 1):764–93.

72. Radke EG, Braun JM, Nachman RM, Cooper GS. Phthalate exposure and neurodevelopment: a systematic review and meta-analysis of human epidemiological evidence. *Environment International.* 2020;137:105408.

73. Zhang Q, Chen XZ, Huang X, Wang M, Wu J. The association between prenatal exposure to phthalates and cognition and neurobehavior of children: evidence from birth cohorts. *Neurotoxicology.* 2019;73:199–212.

74. Zhan L, Zhao X, Ahmad Z, Xu Z. Leaching behavior of Sb and Br from E-waste flame retardant plastics. *Chemosphere.* 2020;245:125684.

75. Jambeck JR, Geyer R, Wilcox C, et al. Plastic waste inputs from land into the ocean. *Science.* 2015;347(6223):768–71. DOI: 10.1126/science.1260352.

76. Gregory MR. Environmental implications of plastic debris in marine settings: entanglement, ingestion, smothering, hangers-on, hitch-hiking and alien invasions. *Philosophical Transactions of the Royal Society B: Biological Sciences.* 2009; 364(1526):2013–25. https://doi.org/10.1098/rstb.2008.0265.

77. Rochman CM, Tahir A, Williams SL, et al. Anthropogenic debris in seafood: plastic debris and fibers from textiles in fish and bivalves sold for human consumption. *Scientific Reports.* 2015;5:14340. https://doi.org/10.1038/srep14340.

78. Brooks AL, Wang S, Jambeck JR. The Chinese import ban and its impact on global plastic waste trade. *Science Advances.* 2018;4(6):eaat0131.

79. aus der Beek T, Weber FA, Bergmann A, et al. Pharmaceuticals in the environment: global occurrences and perspectives. *Environmental Toxicology and Chemistry/ SETAC.* 2016;35(4):823–35. https://doi.org/10.1002/etc.3339.

80. Küster A, Adler N. Pharmaceuticals in the environment: scientific evidence of risks and its regulation. *Philosophical Transactions of the Royal Society B: Biological Sciences.* 2014;369(1656):20130587. doi: 10.1098/rstb.2013.0587.

81. Kay P, Hughes SR, Ault JR, Ashcroft AE, Brown LE. Widespread, routine occurrence of pharmaceuticals in sewage effluent, combined sewer overflows and receiving waters. *Environmental Pollution.* 2017;220:1447–55.

82. Martin JM, Saaristo M, Bertram MG, et al. The psychoactive pollutant fluoxetine compromises antipredator behaviour in fish. *Environmental Pollution*. 2017;222: 592–9. doi: 10.1016/j.envpol.2016.10.010.

83. Sehonova P, Svobodova Z, Dolezelova P, Vosmerova P, Faggio C. Effects of water-borne antidepressants on non-target animals living in the aquatic environment: a review. *Science of The Total Environment*. 2018;631–2:789–94.

84. Ford AT, Fong PP. The effects of antidepressants appear to be rapid and at environmentally relevant concentrations. *Environmental Toxicology and Chemistry*. 2016;35(4):794–8, https://doi.org/10.1002/etc.3087.

85. Green RE, Newton IAN, Shultz S, et al. Diclofenac poisoning as a cause of vulture population declines across the Indian subcontinent. *Journal of Applied Ecology*. 2004;41(5):793–800. doi: 10.1111/j.0021-8901.2004.00954.x.

86. Lübbert C, Baars C, Dayakar A, et al. Environmental pollution with antimicrobial agents from bulk drug manufacturing industries in Hyderabad, South India, is associated with dissemination of extended-spectrum beta-lactamase and carbapenemase-producing pathogens. *Infection*. 2017;45(4):479–91. doi: 10.1007/s15010-017-1007-2.

87. Marathe NP, Regina VR, Walujkar SA, et al. A treatment plant receiving waste water from multiple bulk drug manufacturers is a reservoir for highly multi-drug resistant integron-bearing bacteria. *PLoS One*. 2013;8(10):e77310.

88. Singer AC, Shaw H, Rhodes V, Hart A. Review of antimicrobial resistance in the environment and its relevance to environmental regulators. *Frontiers in Microbiology*. 2016;7:1728. https://dx.doi.org/10.3389%2Ffmicb.2016.01728.

89. MacFadden DR, McGough SF, Fisman D, Santillana M, Brownstein JS. Antibiotic resistance increases with local temperature. *Nature Climate Change*. 2018;8(6): 510–14. https://doi.org/10.1038/s41558-018-0161-6.

90. European Medicines Agency. Sales of Veterinary Antimicrobial Agents in 31 European Countries in 2017. Amsterdam: European Medicines Agency, Veterinary Medicines Division; 2019. Contract No. EMA/294674/2019.

91. Pohanish R. *Sittig's Handbook of Pesticides and Agricultural Chemicals*, 2nd ed. Norwich, NY: William Andrew; 2014.

92. Alavanja MCR, Hoppin JA, Kamel F. Health effects of chronic pesticide exposure: cancer and neurotoxicity. *Annual Review of Public Health*. 2004;25(1):155–97.

93. Combarnous Y. Endocrine disruptor compounds (EDCs) and agriculture: the case of pesticides. *Comptes Rendus Biologies*. 2017;340(9):406–9.

94. WHO. *WHO Human Health Risk Assessment Toolkit: Chemical Hazards*. Geneva: World Health Organization, International Programme on Chemical Safety; 2010.

95. FAO and WHO. Pesticide Residues in Food 2019: Joint FAO/WHO Meeting on Pesticide Residues. Rome: Food and Agriculture Organization of the United Nations; 2019.

96. Expert Committee on Pesticide Residues in Food (PRiF). Annual Report 2018. London: Health and Safety Executive and Food Standards Agency; 2018.

97. Gunnell D, Eddleston M, Phillips MR, Konradsen F. The global distribution of fatal pesticide self-poisoning: systematic review. *BMC Public Health*. 2007;7:357.

98. Merriott D. Factors associated with the farmer suicide crisis in India. *Journal of Epidemiology and Global Health*. 2016;6(4):217–27.

99. Carleton TA. Crop-damaging temperatures increase suicide rates in India. *Proceedings of the National Academy of Sciences*. 2017;114(33):8746–51.

100. Delcour I, Spanoghe P, Uyttendaele M. Literature review: impact of climate change on pesticide use. *Food Research International*. 2015;68:7–15.

101. Sanchez-Bayo F, Goulson D, Pennacchio F, et al. Are bee diseases linked to pesticides? A brief review. *Environment International*. 2016;89–90:7–11.
102. Prashar P, Shah S. Impact of fertilizers and pesticides on soil microflora in agriculture. In Lichtfouse E, editor. *Sustainable Agriculture Reviews: Volume 19*. Cham, Switzerland: Springer International; 2016. pp. 331–61.
103. Gould F, Brown ZS, Kuzma J. Wicked evolution: can we address the sociobiological dilemma of pesticide resistance? *Science*. 2018;360(6390):728–32.
104. Wagreich M, Draganits E. Early mining and smelting lead anomalies in geological archives as potential stratigraphic markers for the base of an early Anthropocene. *The Anthropocene Review*. 2018;5(2):177–201.
105. Hong S, Candelone JP, Patterson CC, Boutron CF. Greenland ice evidence of hemispheric lead pollution two millennia ago by Greek and Roman civilizations. *Science*. 1994;265(5180):1841–3.
106. Delile H, Blichert-Toft J, Goiran J-P, Keay S, Albarède F. Lead in ancient Rome's city waters. *Proceedings of the National Academy of Sciences*. 2014;111(18): 6594–9. https://doi.org/10.1073/pnas.1400097111.
107. Waldron HA. Lead poisoning in the ancient world. *Medical History*. 1973;17(4): 391–9. doi: 10.1017/s0025727300019013.
108. Colborn T, Dumanoski D, Myers JP. *Our Stolen Future*. New York: Penguin; 1997.
109. Donzelli G, Carducci A, Llopis-Gonzalez A, et al. The association between lead and attention-deficit/hyperactivity disorder: a systematic review. *International Journal of Environmental Research and Public Health*. 2019;16(3).
110. Needleman HL, Riess JA, Tobin MJ, Biesecker GE, Greenhouse JB. Bone lead levels and delinquent behavior. *JAMA*. 1996;275(5):363–9.
111. Nevin R. Understanding international crime trends: the legacy of preschool lead exposure. *Environmental Research*. 2007;104(3):315–36.
112. Reuben A, Caspi A, Belsky DW, et al. Association of childhood blood lead levels with cognitive function and socioeconomic status at age 38 years and with IQ change and socioeconomic mobility between childhood and adulthood. *JAMA*. 2017; 317(12):1244–51. doi: 10.1001/jama.2017.1712.
113. Muennig P. The social costs of childhood lead exposure in the post-lead regulation era. *Archives of Pediatrics & Adolescent Medicine*. 2009;163(9):844–9.
114. Attina TM, Trasande L. Economic costs of childhood lead exposure in low- and middle-income countries. *Environmental Health Perspectives*. 2013;121(9):1097–102.
115. Chowdhury R, Ramond A, O'Keeffe LM, et al. Environmental toxic metal contaminants and risk of cardiovascular disease: systematic review and meta-analysis. *BMJ*. 2018;362:k3310. doi: https://doi.org/10.1136/bmj.k3310.
116. UNEP. *UNEP Global Mercury Assessment 2013*. Geneva: United Nations Environment Programme; 2013.
117. Steckling N, Tobollik M, Plass D, et al. Global burden of disease of mercury used in artisanal small-scale gold mining. *Annals of Global Health*. 2017;83(2):234–47.
118. Rice KM, Walker EM, Jr, Wu M, Gillette C, Blough ER. Environmental mercury and its toxic effects. *Journal of Preventive Medicine and Public Health*. 2014;47(2):74–83. doi: 10.3961/jpmph.2014.47.2.74.
119. Sheehan MC, Burke TA, Navas-Acien A, et al. Global methylmercury exposure from seafood consumption and risk of developmental neurotoxicity: a systematic review. *Bulletin of the World Health Organization*. 2014;92(4):254–69F.
120. Baldé CP, Forti V, Gray V, Kuehr R, Stegmann P. The Global E-waste Monitor 2017: Quantities, Flows, and Resources. Bonn, Geneva, and Vienna: United Nations

University (UNU), International Telecommunication Union (ITU), and International Solid Waste Association (ISWA); 2017.

121. Hopson E, Puckett J. Scam Recycling: E-Dumping on Asia by US Recyclers. Seattle, WA: Basel Action Network; 2016.

122. Puckett J, Brandt C, Palmer H. Holes in the Circular Economy: WEEE Leakage from Europe. Seattle, WA: Basel Action Network; 2019.

123. Grant K, Goldizen FC, Sly PD, et al. Health consequences of exposure to e-waste: a systematic review. *The Lancet Global Health.* 2013;1(6):e350–61.

124. Okeme JO, Arrandale VH. Electronic waste recycling: occupational exposures and work-related health effects. *Current Environmental Health Reports.* 2019;6(4):256–68.

125. Vaccari M, Vinti G, Cesaro A, et al. WEEE treatment in developing countries: environmental pollution and health consequences – an overview. *International Journal of Environmental Research and Public Health.* 2019;16(9):1595.

126. Mishra S. Perceived and manifested health problems among informal e-waste handlers: a scoping review. *Indian Journal of Occupational and Environmental Medicine.* 2019;23(1):7–14. https://dx.doi.org/10.4103%2Fijoem.IJOEM_231_18.

127. Awasthi AK, Wang M, Wang Z, Awasthi MK, Li J. E-waste management in India: a mini-review. *Waste Management & Research.* 2018;36(5):408–14.

128. Waheed S, Khan MU, Sweetman AJ, et al. Exposure of polychlorinated naphthalenes (PCNs) to Pakistani populations via non-dietary sources from neglected e-waste hubs: a problem of high health concern. *Environmental Pollution.* 2019;259:113838.

129. Soetrisno FN, Delgado-Saborit JM. Chronic exposure to heavy metals from informal e-waste recycling plants and children's attention, executive function and academic performance. *Science of The Total Environment.* 2020;717:137099.

130. Orisakwe OE, Frazzoli C, Ilo CE, Oritsemuelebi B. Public health burden of e-waste in Africa. *Journal of Health and Pollution.* 2019;9(22):190610.

131. Herrera D, Ellis A, Fisher B, et al. Upstream watershed condition predicts rural children's health across 35 developing countries. *Nature Communications.* 2017;8(1):811. https://doi.org/10.1038/s41467-017-00775-2.

132. Jenkins AP, Jupiter S, Mueller U, et al. Health at the sub-catchment scale: typhoid and its environmental determinants in central division, Fiji. *EcoHealth.* 2016;13(4):633–51. doi: 10.1007/s10393-016-1152-6.

133. MacDonald AJ, Mordecai EA. Amazon deforestation drives malaria transmission, and malaria burden reduces forest clearing. *Proceedings of the National Academy of Sciences.* 2019;116(44):22212–8. https://doi.org/10.1073/pnas.1905315116.

134. Burkett-Cadena ND, Vittor AY. Deforestation and vector-borne disease: forest conversion favors important mosquito vectors of human pathogens. *Basic and Applied Ecology.* 2018;26:101–10. https://doi.org/10.1016/j.baae.2017.09.012.

135. Stefani A, Dusfour I, Corrêa APS, et al. Land cover, land use and malaria in the Amazon: a systematic literature review of studies using remotely sensed data. *Malaria Journal.* 2013;12(1):192. https://doi.org/10.1186/1475-2875-12-192.

136. Vittor AY, Pan W, Gilman RH, et al. Linking deforestation to malaria in the Amazon: characterization of the breeding habitat of the principal malaria vector, *Anopheles darlingi. American Journal of Tropical Medicine and Hygiene.* 2009;81(1):5–12.

137. Fornace KM, Abidin TR, Alexander N, et al. Association between landscape factors and spatial patterns of *Plasmodium knowlesi* infections in Sabah, Malaysia. *Emerging Infectious Disease Journal.* 2016;22(2):201.

138. Tucker Lima JM, Vittor A, Rifai S, Valle D. Does deforestation promote or inhibit malaria transmission in the Amazon? A systematic literature review and critical

appraisal of current evidence. *Philosophical Transactions of the Royal Society B: Biological Sciences*. 2017;372(1722):20160125.

139. Whitmee S, Haines A, Beyrer C, et al. Safeguarding human health in the Anthropocene Epoch: report of The Rockefeller Foundation–*Lancet* Commission on Planetary Health. *The Lancet*. 2015;386(10007):1973–2028.

140. Ostfeld RS, Brunner JL. Climate change and Ixodes tick-borne diseases of humans. *Philosophical Transactions of the Royal Society B: Biological Sciences*. 2015; 370(1665). https://doi.org/10.1098/rstb.2014.0051.

141. MacDonald AJ, Larsen AE, Plantinga AJ. Missing the people for the trees: identifying coupled natural–human system feedbacks driving the ecology of Lyme disease. *Journal of Applied Ecology*. 2019;56(2):354–64.

142. Simon JA, Marrotte RR, Desrosiers N, et al. Climate change and habitat fragmentation drive the occurrence of *Borrelia burgdorferi*, the agent of Lyme disease, at the northeastern limit of its distribution. *Evolutionary Applications*. 2014;7(7):750–64.

143. White RJ, Razgour O. Emerging zoonotic diseases originating in mammals: a systematic review of effects of anthropogenic land-use change. *Mammal Review*. 2020;50(4):336–52.

144. Johnson CK, Hitchens PL, Pandit PS, et al. Global shifts in mammalian population trends reveal key predictors of virus spillover risk. *Proceedings of the Royal Society B: Biological Sciences*. 2020;287(1924):20192736.

145. Gibb R, Redding DW, Chin KQ, et al. Zoonotic host diversity increases in human-dominated ecosystems. *Nature*. 2020;584(7821):398–402.

146. Ostfeld RS, Keesing F. Species that can make us ill thrive in human habitats. *Nature*. 2020;584(7821):346–7. doi: 10.1038/d41586-020-02189-5.

147. Spracklen DV, Reddington CL, Gaveau DLA. Industrial concessions, fires and air pollution in Equatorial Asia. *Environmental Research Letters*. 2015;10(9):091001.

148. Marlier ME, DeFries RS, Kim PS, et al. Fire emissions and regional air quality impacts from fires in oil palm, timber, and logging concessions in Indonesia. *Environmental Research Letters*. 2015;10(8):085005.

149. Marlier ME, DeFries RS, Voulgarakis A, et al. El Niño and health risks from landscape fire emissions in southeast Asia. *Nature Climate Change*. 2013;3(2):131–6.

150. Koplitz SN, Mickley LJ, Marlier ME, et al. Public health impacts of the severe haze in Equatorial Asia in September–October 2015: demonstration of a new framework for informing fire management strategies to reduce downwind smoke exposure. *Environmental Research Letters*. 2016;11(9):094023.

151. Johnson KB, Jacob A, Brown ME. Forest cover associated with improved child health and nutrition: evidence from the Malawi Demographic and Health Survey and satellite data. *Global Health: Science and Practice*. 2013;1(2):237–48.

152. Micklin P, Aladin NV, Plotnikov I, editors. *The Aral Sea: The Devastation and Partial Rehabilitation of a Great Lake*. New York: Springer; 2014.

153. Gupta A, Gupta A. Environmental challenges in Aral Sea basin: impact on human health. *International Research Journal of Social Science*. 2016;6:419–40.

154. Whish-Wilson P. The Aral Sea environmental health crisis. *Journal of Rural and Remote Environmental Health*. 2002;1(2):29–34.

155. O'Hara SL, Wiggs GF, Mamedov B, Davidson G, Hubbard RB. Exposure to airborne dust contaminated with pesticide in the Aral Sea region. *The Lancet*. 2000;355(9204): 627–8. doi: 10.1016/S0140-6736(99)04753-4.

156. Sternberg T, Edwards M. Desert dust and health: a Central Asian review and steppe case study. *International Journal of Environmental Research and Public Health*. 2017; 14(11):1342. DOI: 10.3390/ijerph14111342.

157. Seddon AWR, Macias-Fauria M, Long PR, Benz D, Willis KJ. Sensitivity of global terrestrial ecosystems to climate variability. *Nature*. 2016;531(7593):229–32.

158. Seddon N, Mace GM, Naeem S, et al. Biodiversity in the Anthropocene: prospects and policy. *Proceedings of the Royal Society B: Biological Sciences*. 2016; 283(1844):20162094. https://doi.org/10.1098/rspb.2016.2094.

159. Chivian E, Bernstein A, editors. *Sustaining Life: How Human Health Depends on Biodiversity*. Oxford and New York: Oxford University Press; 2008.

160. Millennium Ecosystem Assessment Program *Ecosystems and Human Well-Being: Synthesis*. Washington, DC: Island Press; 2005.

161. Romanelli C, Cooper D, Campbell-Lendrum D, et al. *Connecting Global Priorities: Biodiversity and Human Health. A State of Knowledge Review*. Geneva: UNEP, CBD, and WHO; 2015.

162. Millennium Ecosystem Assessment Program. *Ecosystems and Human Well-Being: Biodiversity Synthesis*. Washington, DC: World Resources Institute; 2005.

163. Das S, Vincent JR. Mangroves protected villages and reduced death toll during Indian super cyclone. *Proceedings of the National Academy of Sciences*. 2009;106(18): 7357–60. https://dx.doi.org/10.1073%2Fpnas.0810440106.

164. Ewel K, Twilley R, Ong JIN. Different kinds of mangrove forests provide different goods and services. *Global Ecology & Biogeography Letters*. 1998;7(1):83–94.

165. Sánchez-Bayo F, Wyckhuys KAG. Worldwide decline of the entomofauna: a review of its drivers. *Biological Conservation*. 2019;232:8–27.

166. Smith MR, Singh GM, Mozaffarian D, Myers SS. Effects of decreases of animal pollinators on human nutrition and global health: a modelling analysis. *The Lancet*. 2015;386(10007):1964–72. DOI: 10.1016/s0140-6736(15)61085-6.

167. Potts SG, Imperatriz-Fonseca V, Ngo HT, et al. Safeguarding pollinators and their values to human well-being. *Nature*. 2016;540(7632):220–9.

168. Friant S, Ayambem WA, Alobi AO, et al. Eating bushmeat improves food security in a biodiversity and infectious disease 'hotspot'. *EcoHealth*. 2020;17(1):125–38.

169. Ripple WJ, Abernethy K, Betts MG, et al. Bushmeat hunting and extinction risk to the world's mammals. *Royal Society Open Science*. 2016;3(10):160498.

170. Wolfe ND, Daszak P, Kilpatrick AM, Burke DS. Bushmeat hunting, deforestation, and prediction of zoonoses emergence. *Emerging Infectious Diseases*. 2005;11(12): 1822–7. doi: 10.3201/eid1112.040789.

171. Bharucha Z, Pretty J. The roles and values of wild foods in agricultural systems. *Philosophical Transactions of the Royal Society B: Biological Sciences*. 2010; 365(1554):2913–26. https://doi.org/10.1098/rstb.2010.0123.

172. Burlingame B. Wild nutrition. *Journal of Food Composition and Analysis*. 2000; 13(2):99–100. https://doi.org/10.1006/jfca.2000.0897.

173. Golden CD, Fernald LCH, Brashares JS, Rasolofoniaina BJR, Kremen C. Benefits of wildlife consumption to child nutrition in a biodiversity hotspot. *Proceedings of the National Academy of Sciences*. 2011;108(49):19653–6.

174. Kelly BC, Ikonomou MG, Blair JD, Morin AE, Gobas FAPC. Food web-specific biomagnification of persistent organic pollutants. *Science*. 2007;317(5835): 236–9. https://doi.org/10.1126/science.1138275.

175. Wang XP, Sun DC, Yao TD. Climate change and global cycling of persistent organic pollutants: a critical review. *Science China Earth Sciences*. 2016;59(10):1899–911.

176. Rosol R, Powell-Hellyer S, Chan HM. Impacts of decline harvest of country food on nutrient intake among Inuit in Arctic Canada: impact of climate change and possible adaptation plan. *International Journal of Circumpolar Health*. 2016;75:31127.

177. Brashares JS, Arcese P, Sam MK, et al. Bushmeat hunting, wildlife declines, and fish supply in West Africa. *Science*. 2004;306(5699):1180–3.
178. Kaczynski VM, Fluharty DL. European policies in West Africa: who benefits from fisheries agreements? *Marine Policy*. 2002;26(2):75–93.
179. Collins WW, Hawtin GC. Conserving and using crop plant biodiversity. In Collins WW, Qualset CO, editors. *Biodiversity in Agroecosystems*. Boca Raton, FL: CRC Press; 1999. pp. 267–82.
180. Bioversity International. *Mainstreaming Agrobiodiversity in Sustainable Food System: Scientific Foundations for an Agrobiodiversity Index*. Rome: Bioversity International; 2017.
181. Garn SM, Leonard WR. What did our ancestors eat? *Nutrition Reviews*. 1989;47(11): 337–45. https://doi.org/10.1111/j.1753-4887.1989.tb02765.x.
182. FAO. *The State of the World's Plant Genetic Resources for Food and Agriculture*. Rome: Food and Agriculture Organization of the United Nations; 1997.
183. Steyn NP, Nel JH, Nantel G, Kennedy G, Labadarios D. Food variety and dietary diversity scores in children: are they good indicators of dietary adequacy? *Public Health Nutrition*. 2006;9(5):644–50. doi: 10.1079/phn2005912.
184. Arimond M, Wiesmann D, Becquey E, et al. Simple food group diversity indicators predict micronutrient adequacy of women's diets in 5 diverse, resource-poor settings. *Journal of Nutrition*. 2010;140(11):2059S–69S. doi: 10.3945/jn.110.123414.
185. Ndung'u J, Nyanchoka AM. Dietary diversity, nutrient intake and nutritional status of pregnant women aged 18–45 years in developing countries. A systematic review. *International Journal of Food Science and Nutrition*. 2018;3(4):217–20.
186. Duxbury BM, Poller L. The oral anticoagulant saga: past, present, and future. *Clinical and Applied Thrombosis/Hemostasis*. 2001;7(4):269–75. doi: 10.1177/107602960100700403.
187. Newman DJ, Cragg GM. Natural products as sources of new drugs from 1981 to 2014. *Journal of Natural Products*. 2016;79(3):629–61.
188. Harvey AL, Edrada-Ebel R, Quinn RJ. The re-emergence of natural products for drug discovery in the genomics era. *Nature Reviews Drug Discovery*. 2015;14(2): 111–29. doi: 10.1038/nrd4510.
189. Schippmann U, Leaman D, Cunningham AB. A comparison of cultivation and wild collection of medicinal and aromatic plants under sustainability aspects. In Bogers RJ, Craker LE, Lange D, editors. *Medicinal and Aromatic Plants*. Heidelberg: Springer; 2006. pp. 75–95.
190. Keesing F, Belden LK, Daszak P, et al. Impacts of biodiversity on the emergence and transmission of infectious diseases. *Nature*. 2010;468(7324):647–52.
191. Pongsiri MJ, Roman J, Ezenwa VO, et al. Biodiversity loss affects global disease ecology. *BioScience*. 2009;59(11):945–54. https://doi.org/10.1525/bio.2009.59.11.6.
192. Evers BN, Madsen H, McKaye KM, Stauffer JR, Jr. The schistosome intermediate host, *Bulinus nyassanus*, is a 'preferred' food for the cichlid fish, *Trematocranus placodon*, at Cape Maclear, Lake Malawi. *Annals of Tropical Medicine and Parasitology*. 2006;100(1):75–85.
193. Ostfeld RS, LoGiudice K. Community disassembly, biodiversity loss, and the erosion of an ecosystem service. *Ecology*. 2003;84(6):1421–7. https://doi.org/10.1890/02-3125.
194. Ruedas LA, Salazar-Bravo J, Tinnin DS, et al. Community ecology of small mammal populations in Panamá following an outbreak of Hantavirus pulmonary syndrome. *Journal of Vector Ecology*. 2004;29(1):177–91.
195. Bird BH, Mazet JAK. Detection of emerging zoonotic pathogens: an integrated one health approach. *Annual Review of Animal Biosciences*. 2018;6:121–39.
196. Yamada A, Kahn LH, Kaplan, B, et al., editors. *Confronting Emerging Zoonoses: The One Health Paradigm*. Heidelberg: Springer; 2014.

197. Epstein JH, Field HE, Luby S, Pulliam JR, Daszak P. Nipah virus: impact, origins, and causes of emergence. *Current Infectious Disease Reports*. 2006;8(1):59–65.

198. Allen T, Murray KA, Zambrana-Torrelio C, et al. Global hotspots and correlates of emerging zoonotic diseases. *Nature Communications*. 2017;8(1):1124.

199. Luis AD, Kuenzi AJ, Mills JN. Species diversity concurrently dilutes and amplifies transmission in a zoonotic host-pathogen system through competing mechanisms. *Proceedings of the National Academy of Sciences*. 2018;115(31):7979–84.

200. Salkeld DJ, Padgett KA, Jones JH. A meta-analysis suggesting that the relationship between biodiversity and risk of zoonotic pathogen transmission is idiosyncratic. *Ecology Letters*. 2013;16(5):679–86. doi: 10.1111/ele.12101.

201. Hao Z, Singh VP, Xia Y. Seasonal drought prediction: advances, challenges, and future prospects. *Reviews of Geophysics*. 2018;56(1):108–41.

202. Carrão H, Naumann G, Barbosa P. Global projections of drought hazard in a warming climate: a prime for disaster risk management. *Climate Dynamics*. 2018;50(5): 2137–55. https://doi.org/10.1007/s00382-017-3740-8.

203. Schewe J, Heinke J, Gerten D, et al. Multimodel assessment of water scarcity under climate change. *Proceedings of the National Academy of Sciences*. 2014;111(9): 3245–50. https://doi.org/10.1073/pnas.1222460110.

204. Liu W, Sun F, Lim WH, et al. Global drought and severe drought-affected populations in 1.5 and 2°C warmer worlds. *Earth System Dynamics*. 2018;9(1): 267–83. https://doi.org/10.5194/esd-9-267-2018.

205. Werner AD, Bakker M, Post VEA, et al. Seawater intrusion processes, investigation and management: recent advances and future challenges. *Advances in Water Resources*. 2013;51:3–26. https://doi.org/10.1016/j.advwatres.2012.03.004.

206. Konikow LF, Kendy E. Groundwater depletion: a global problem. *Hydrogeology Journal*. 2005;13(1):317–20. http://dx.doi.org/10.1007/s10040-004-0411-8.

207. Famiglietti JS. The global groundwater crisis. *Nature Climate Change*. 2014;4:945.

208. Stanke C, Kerac M, Prudhomme C, Medlock J, Murray V. Health effects of drought: a systematic review of the evidence. *PLoS Currents Disasters*. 2013;5.

209. Hunter PR, MacDonald AM, Carter RC. Water supply and health. *PLoS Medicine*. 2010;7(11):e1000361. https://doi.org/10.1371/journal.pmed.1000361.

210. Cooper MW, Brown ME, Hochrainer-Stigler S, et al. Mapping the effects of drought on child stunting. *Proceedings of the National Academy of Sciences*. 2019;116(35): 17219–24. DOI: 10.1073/pnas.1905228116.

211. Bahru BA, Bosch C, Birner R, Zeller M. Drought and child undernutrition in Ethiopia: a longitudinal path analysis. *PLoS One*. 2019;14(6):e0217821.

212. Dalin C, Wada Y, Kastner T, Puma MJ. Groundwater depletion embedded in international food trade. *Nature*. 2017;543:700. https://doi.org/10.1038/nature21403.

213. Vins H, Bell J, Saha S, Hess JJ. The mental health outcomes of drought: a systematic review and causal process diagram. *International Journal of Environmental Research and Public Health*. 2015;12(10):13251–75. doi: 10.3390/ijerph121013251.

214. Pontes RJ, Freeman J, Oliveira-Lima JW, Hodgson JC, Spielman A. Vector densities that potentiate dengue outbreaks in a Brazilian city. *American Journal of Tropical Medicine and Hygiene*. 2000;62(3):378–83. doi: 10.4269/ajtmh.2000.62.378.

215. von Uexkull N, Croicu M, Fjelde H, Buhaug H. Civil conflict sensitivity to growing-season drought. *Proceedings of the National Academy of Sciences*. 2016;113(44): 12391–6. https://doi.org/10.1073/pnas.1607542113.

216. Milner J, Joy EJM, Green R, et al. Projected health effects of realistic dietary changes to address freshwater constraints in India: a modelling study. *The Lancet Planetary Health*. 2017;1(1):e26–32. https://doi.org/10.1016/S2542-5196(17)30001-3.

217. Schmutz S, Moog O. Dams: ecological impacts and management. In Schmutz S, Sendzimir J, editors. *Riverine Ecosystem Management: Science for Governing Towards a Sustainable Future*. Cham, Switzerland: Springer International; 2018. pp. 111–27.

218. McDonald-Wilmsen B, Webber M. Dams and displacement: raising the standards and broadening the research agenda. *Water Alternatives*. 2010;3(2):142.

219. Hwang S, Cao Y, Xi J. The short-term impact of involuntary migration in China's Three Gorges: a prospective study. *Social Indicators Research*. 2011;101(1):73–92.

220. Xi J, Hwang S. Relocation stress, coping, and sense of control among resettlers resulting from China's Three Gorges Dam Project. *Social Indicators Research*. 2011;104(3):507–22.

221. Steinmann P, Keiser J, Bos R, Tanner M, Utzinger J. Schistosomiasis and water resources development: systematic review, meta-analysis, and estimates of people at risk. *The Lancet Infectious Diseases*. 2006;6(7):411–25.

222. Kibret S, Lautze J, McCartney M, Nhamo L, Yan G. Malaria around large dams in Africa: effect of environmental and transmission endemicity factors. *Malaria Journal*. 2019;18(1):303. https://doi.org/10.1186/s12936-019-2933-5.

223. World Commission on Dams. *Dams and Development: A New Framework for Decision-Making*. London: Earthscan; 2000.

224. Semenza JC, Lindgren E, Balkanyi L, et al. Determinants and drivers of infectious disease threat events in Europe. *Emerging Infectious Diseases*. 2016;22(4):581–9.

225. Lourenço J, Recker M. The 2012 Madeira dengue outbreak: epidemiological determinants and future epidemic potential. *PLoS Neglected Tropical Diseases*. 2014;8(8): e3083. https://doi.org/10.1371/journal.pntd.0003083.

226. Reil D, Rosenfeld UM, Imholt C, et al. Puumala hantavirus infections in bank vole populations: host and virus dynamics in Central Europe. *BMC Ecology*. 2017; 17(1):9. https://doi.org/10.1186/s12898-017-0118-z.

227. Andersen KG, Rambaut A, Lipkin WI, Holmes EC, Garry RF. The proximal origin of SARS-CoV-2. *Nature Medicine*. 2020;26(4):450–2. https://doi.org/10.1038/s41591-020-0820-9.

228. Heisler J, Glibert P, Burkholder J, et al. Eutrophication and harmful algal blooms: a scientific consensus. *Harmful Algae*. 2008;8(1):3–13. doi: 10.1016/j.hal.2008.08.006.

229. Wells ML, Karlson B, Wulff A, et al. Future HAB science: directions and challenges in a changing climate. *Harmful Algae*. 2020;91:101632. https://doi.org/10.1016/j.hal.2019.101632.

230. Trainer VL, Moore SK, Hallegraeff G, et al. Pelagic harmful algal blooms and climate change: lessons from nature's experiments with extremes. *Harmful Algae*. 2020;91: 101591. doi: 10.1016/j.hal.2019.03.009.

231. Riebesell U, Aberle-Malzahn N, Achterberg EP, et al. Toxic algal bloom induced by ocean acidification disrupts the pelagic food web. *Nature Climate Change*. 2018; 8(12):1082–6. https://doi.org/10.1038/s41558-018-0344-1.

232. Scheelbeek PFD, Bird FA, Tuomisto HL, et al. Effect of environmental changes on vegetable and legume yields and nutritional quality. *Proceedings of the National Academy of Sciences*. 2018;115(26):6804–9. doi: 10.1073/pnas.1800442115.

233. Suweis S, Carr JA, Maritan A, Rinaldo A, D'Odorico P. Resilience and reactivity of global food security. *Proceedings of the National Academy of Sciences*. 2015; 112(22):6902–7. https://doi.org/10.1073/pnas.1507366112.

234. Suweis S, Rinaldo A, Maritan A, D'Odorico P. Water-controlled wealth of nations. *Proceedings of the National Academy of Sciences*. 2013;110(11):4230–3. https://doi.org/10.1073/pnas.1222452110.

4

Assessing Vulnerability and Risk in the Anthropocene Epoch

Throughout human history, communities and individuals have adapted more or less successfully to changes in their local environments. The Anthropocene Epoch poses adaptation challenges that are new and different, both quantitatively and qualitatively, from those experienced during human history, because of both the scale and pace of the changes and the sheer numbers of humans who will need to adapt. At the same time, humanity has access – albeit inequitably – to the fruits of scientific and technological progress unavailable to previous generations, which facilitate at least some successful adaptation.

Most planned adaptation actions start with an assessment of the risks to which adaptation is needed. This chapter discusses how the risks of planetary changes are assessed, and Chapter 5 discusses how these risks are managed, through adaptation and the pursuit of resilience.

Much of the contemporary risk assessment framework is grounded in chemical and radiological exposures, often centred on cancer as an outcome, and takes a toxicological approach. A seminal 1983 report by the US National Research Council (NRC) (1) defined a four-step process: hazard identification, dose–response assessment, exposure assessment, and risk characterization. A generation later, a subsequent NRC report (2) broadened this paradigm, while retaining its focus on chemical exposures. It called for initial scoping designed to identify and address practical policy questions, more transparency about uncertainty, and greater consideration of complex contextual issues such as background exposures and disease processes, possible vulnerable populations, and modes of action that may affect human dose–response relationships. Meanwhile, a parallel process was evolving with respect to the environment, called ecological risk assessment (3). Initially focused on chemical pollutants, it expanded over time to consider the potential damage to ecosystems of a range of stressors, such as land use changes, invasive species, and climate change (4). There were efforts to apply ecological risk assessment results to environmental policy goals (5). Eventually, there were efforts to integrate health and ecological risk assessment, deploying such concepts as ecosystem services (6, 7).

These developments are highly relevant to Planetary Health. Assessments of risk identify the relevant *hazards*, the intensity of *exposure* (considering the number of people exposed and the amount of exposure they might sustain), and the *vulnerability* (or sensitivity) of those exposed. These factors combine to yield an assessment of risk (as illustrated in **Figures 4.1** and **4.2**, for climate change).

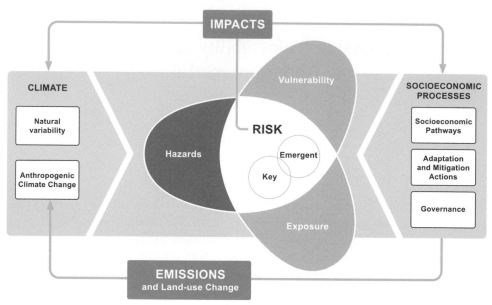

Figure 4.1. How risk relates to climate change vulnerability, hazards, and exposure.
Source: Figure 19-1 from Oppenheimer M, Campos M, Warren R, et al. Emergent risks and key vulnerabilities. In Field CB, Barros VR, Dokken DJ, et al., editors. *Climate Change 2014: Impacts, Adaptation, and Vulnerability Part A: Global and Sectoral Aspects. Contribution of Working Group II to the Fifth Assessment Report of the Intergovernmental Panel on Climate Change.* Cambridge, UK and New York: Cambridge University Press; 2014. pp. 1039–99. www.ipcc.ch/report/ar5/wg2/emergent-risks-and-key-vulnerabilities/wgii_ar5_fig19-1/.

For example, potential excess deaths from exposure to extreme heat will depend on the level and duration of the heat stress to which a population is exposed, the age distribution and social profile of the population (the elderly and the poor being at higher risk of heat-related death for a given exposure), the level of acclimatization and the population's access to protective technologies or interventions such as air conditioning. As a simple but useful short cut, risk is often expressed as the probability of occurrence of hazardous events or trends multiplied by the impacts if these events or trends occur (8).

The terminology can be confusing as 'vulnerability' is variously used as the starting point of an assessment, as a focal point, and as the end point (9). The starting point approach emphasizes social vulnerability and focuses on the need to address the effects of current climate variability. The end point perspective considers vulnerability to be the net expected impacts of a given level of global change (usually climate change), with a focus on the future (10). Key factors in determining social vulnerability are the availability of resources and the entitlement of communities and individuals to access those resources. Thus reduction in poverty, diversification of income sources, respect for common property management rights, and actions to promote collective security can all play useful roles in reducing vulnerability (9). The IPCC, in its fifth assessment report, notes that there has been an evolution in thinking about vulnerability, from framing it primarily as a

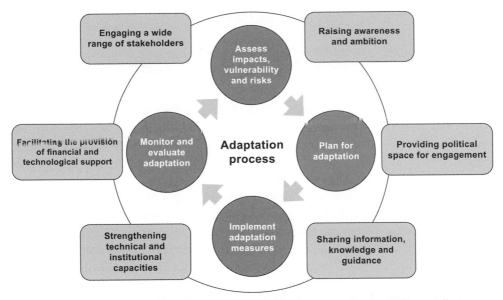

Figure 4.2. A summary of how the assessment of risks, impacts, and vulnerability contributes to the development of adaptation measures.

Source: UNFCCC. What do adaptation to climate change and climate resilience mean? 2020. Online: https://unfccc.int/topics/adaptation-and-resilience/the-big-picture/what-do-adaptation-to-climate-change-and-climate-resilience-mean.

biophysical concept, to emphasizing the social and economic drivers of vulnerability and how these affect the ability of individuals and communities to respond to climate change (11). It defines vulnerability as 'the propensity or predisposition to be adversely affected' but acknowledges that vulnerability encompasses a number of concepts, 'including sensitivity or susceptibility to harm and lack of capacity to cope and adapt'.

Four dimensions are relevant to describing vulnerability: (a) the system being analysed, which might be a population group, an ecosystem or an economic sector; (b) the valued attribute of the system that is threatened by exposure to a hazard, e.g. health or socio-economic development; (c) the nature of the hazard to which the system is exposed – usually external, such as climate change, but sometimes internal, such as corruption; and (d) the point in time or period of interest (10). Vulnerability should be seen as a dynamic and context-specific concept, influenced by human behaviours and by how society is organized and governed (8).

Climate Risk Assessments

Several approaches have been used to assess the level of risk of a given country to climate and other environmental changes. In 2014 the European Commission Joint Research Centre considered five of the open-access indices most commonly used to assess climate-change-related risks and to identify the most vulnerable countries (12):

- Global Climate Change Risk Index (CRI) (13)
- World Risk Index (WRI) (14)
- Notre Dame Global Adaptation Index (ND-GAIN) (15)
- Center for Global Development (CGDev) (16)
- Climate Vulnerability Monitor (17)

The comparison showed that the indices differ in their definitions of vulnerability and in some cases lack transparent methodologies, hampering assessment of their strengths and weaknesses (12). Some of the indices focus particularly on just one risk pathway, extreme events, but there is no consensus on which such events should be included. Although poor institutional capacity is used as an indicator of increased climate risk, no objective validation was presented. However, despite the differences in the studies, the three indices for which comparative data were available (CGDev, ND-GAIN, WRI) were generally consistent in their climate change risk ranking of 169 countries (18). They found European countries, the USA, and Australia to be the least vulnerable countries, and Sub-Saharan African countries to be highly vulnerable, while assessments of Asian and Latin American countries varied between indices.

ND-GAIN

A notable index is the University of Notre Dame's ND-GAIN (15) (gain.nd.edu), which originated at the Global Adaptation Institute in Washington, DC, and moved to the University of Notre Dame Climate Change Adaptation Program in 2013. The ND-GAIN Country Index uses 45 indicators to rank 181 countries annually based upon their vulnerability and their readiness to adapt successfully, and the ND-GAIN Urban Adaptation Assessment applies similar methods to 278 US cities. The ND-GAIN vulnerability indicators reflect exposure, sensitivity, and adaptive capacity across six sectors (food, water, health, ecosystem service, human habitat, and infrastructure), and thus extend beyond a narrow assessment of vulnerability. Some assess health directly and others, such as those for ecosystem services, food, human habitats, infrastructure, and water, are relevant to health. ND-GAIN also assesses a county's readiness to leverage private and public sector investment for adaptation, with economic, governance, and social indicators (**Table 4.1**). The economic dimension of readiness is based on the World Bank 'Ease of Doing Business' rankings (19). The ND-GAIN values vary substantially across the world, with highest vulnerability in Africa (**Figure 4.3**).

The ND-GAIN index, in turn, finds influential use by the World Bank. The Bank plays an important role in assisting countries to adapt to and mitigate climate change through low-carbon development. This includes support for cross-sectoral efforts to reduce the impacts of climate change and its drivers, which integrate energy, transport, economic development, agriculture, environmental management, and health care. To help prioritize its efforts the Bank developed 'Disaster and Risk Screening' tools across multiple sectors (climatescreeningtools.worldbank.org). In addition, using a variant of the ND-GAIN approach, the Bank identified 'hotspots', defined as countries that are experiencing or

Table 4.1. *ND-GAIN country-level readiness indicators*

Component	Indicators			
Economic readiness	Ease of doing business rankings			
Governance readiness	Political stability and non-violence	Control of corruption	Rule of law	Regulatory quality
Social readiness	Social inequality	Information and communications technology infrastructure	Education	Innovation

Source: https://gain.nd.edu/our-work/country-index/methodology/.

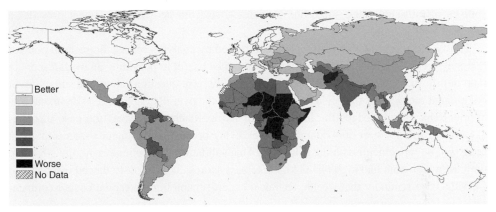

Figure 4.3. The ND-GAIN Country Index, summarizing countries' vulnerability to climate change and other global challenges in combination with readiness to improve resilience. Source: Data from ND-GAIN. https://gain.nd.edu/our-work/country-index/. Image credit: Rochelle Schneider dos Santos.

likely to experience an increased burden of disease, due either to climate change itself (an 'impact hotspot') or to air pollution emissions associated with the drivers of climate change (an 'emissions hotspot') (20).

Identifying impact hotspots draws on the ND-GAIN sub-indices for health and human habitat (which are fairly highly correlated). The most vulnerable countries are defined as those with high scores on both. This information is combined with the ND-GAIN readiness index to identify impact hotspots – a designation that helps identify where improvements in country capacity could reduce risks (20). Specific locations are associated with high risks of different health outcomes from climate change. For example, people living in deltaic regions and on floodplains are likely to confront floods, those living in low-latitude mountain regions may experience transmission of malaria at higher altitudes, those living in tropical and sub-tropical regions are likely to suffer extreme heat stress leading to declines in labour productivity, and disruptions to food availability are likely in Sub-Saharan Africa, South Asia, and the Pacific islands.

Emissions hotspots are identified based on the Global Burden of Disease estimates of fine particulate ($PM_{2.5}$) air pollution health burden – both the burden itself and its relative importance compared to other major causes of death or disease in each country (21). The World Bank approach identifies countries in which household air pollution ranked among the top five national contributors to disease burden ($n = 39$) and countries in which ambient air pollution ranked among the top nine national contributors ($n = 28$). Countries with high levels of both household and ambient air pollution according to this analysis include Afghanistan, Chad, Central African Republic, Guinea-Bissau, Mali, Sierra Leone, South Sudan, Democratic Republic of Congo, Niger, North Korea, Guinea, and Laos. While emissions hotspots are meant to align air pollution emissions with GHG and SLCP emissions, this approach oversimplifies the links between the two. This is in part because PM air pollution includes a mixture of warming and cooling aerosols depending on the sources of emissions, and some fine particles arise from aeolian dust, but it provides a practical way forward.

The World Bank hotspot analysis is accompanied by analyses of specific sectoral policies that may yield large benefits. Such approaches can be useful in providing an overall score to help in prioritizing actions. However, they inevitably raise questions about the suitability of component metrics and their relative contribution to the overall quantitative assessment. The World Bank acknowledges the limitations of its approach, but makes the case that it is essential to make a start on a complex and evolving process of this nature – accepting that there will be refinements as more is learned. This validation will be important; the USA and the UK, which both had high ND-GAIN readiness scores, had seriously flawed responses to COVID-19, a reminder that wealth, technical expertise, and human capital cannot compensate for maladroit governance.

The Climate Vulnerability Monitor

Another notable approach is the Climate Vulnerability Monitor, first published in 2010 to serve as a resource to Climate Vulnerable Forum (CVF) officials undertaking climate change negotiations and policy development, and updated in 2012 (17). The CVF (thecvf .org/) is a partnership of 48 climate vulnerable countries, initiated by the Maldives in 2009. It advocates for more effective policies and greater resources for adaptation to and mitigation of climate change internationally, and it supports national actions for adaptation and mitigation. The Monitor addresses four areas of climate change impact, with a focus on their economic impact: environmental disasters, habitat change, health impact, and industry stress. It considers both exposure and vulnerability, taking into account socioeconomic factors, including safety nets, education and access to information, as well as biophysical factors such as geography, topography, vegetation, and natural resources. In addition to the focus on climate change impacts, the Monitor provides indicators of the impacts of the carbon economy, i.e. the emissions of all climate-active pollutants and their effects on health and the wider economy. Examples include the effects of acid rain on corrosion and water resources, the loss of biodiversity and primary productivity of plants from acid rain and tropospheric ozone, the damaging effects of oil spills and oil sands, the adverse effects of ambient air pollution on agriculture, and both ambient and household air pollution on health. It classifies vulnerability for 2010 and 2030 in one of five categories (acute, severe, high, moderate, and low), and for each indicator it includes an assessment of uncertainty ranging from robust (highest confidence) to speculative (lowest confidence). For climate, it also indicates uncertainties at the regional level reflecting the extent of disagreement between different climate models, e.g. for temperature and rainfall.

A couple of caveats about interpreting Climate Vulnerability Monitor findings should be mentioned. First, classifying a country as having low vulnerability does not imply that impacts will be trivial – the assessments relate to the near future and are not comprehensive. They do not, for example, take into account indirect effects such as human migration from vulnerable to less vulnerable locations. Second, the carbon impacts used in the Monitor are very heterogeneous and not directly proportionate to emissions of GHGs which, in high-income nations, are much higher but accompanied by tighter emission controls that reduce air pollution levels (albeit still dangerous, as discussed in Chapter 3).

Assessing Future Risk and Vulnerability

The evidence available, so far, suggests that more work needs to be done to ensure consistent use of terms in various rating scales, to evaluate their effectiveness in correctly identifying which countries are particularly vulnerable to climate and other environmental changes, and to assess their ability to benefit from investment and specific interventions for adaptation and resilience.

While some risks are well established, others are emergent, implying complex phenomena with the potential to interact and to evolve over time. Many risk assessments do not address emergent risks adequately because of their inherent dependence on complex systems. Examples of these emergent risks include:

- how biodiversity loss increases the risks to agriculture and freshwater supply through the loss of ecosystem services such as pollination, water purification, and storm protection (see Chapter 3);
- how depletion of aquifers reduces freshwater reserves and thus increases vulnerability to climate change through related reductions in precipitation;
- how competition for land between food and bioenergy crops might undermine food security;
- how a web of factors – sea level rise, ocean acidification, increased storm surges, loss of protective wetlands and mangroves, salination of freshwater, and subsidence from excessive groundwater extraction – together with urbanization and amplification of heat stress as a result of the urban heat island, could interact to threaten urban populations;
- the effects of population displacement on health and on societies more generally.

Risks increase progressively as temperature increases exceed 2 °C above pre-industrial levels with accompanying increases in the probability of large-scale singular events, such as the collapse of the Greenland ice cap with potential global mean sea level rise of up to 7 metres, particularly after 3 °C heating (8, 22, 23). The IPCC (11) identified a range of key risks, defined as the interaction between exposure to hazards and vulnerability of exposed populations, leading to potentially severe adverse consequences (see **Table 4.2**). Whilst risk assessments will inevitably tend to focus on the most pressing – and thus near-term – risks, it will be increasingly important to consider long-term and potentially far-reaching risks, which if unaddressed could destabilize whole societies. This is particularly important when constructing infrastructure that can 'lock in' countries and settlements to patterns of development that will render them unable to respond effectively to future large-scale challenges (see Chapter 5 for discussion).

Given the growing urgency of ensuring effective expenditure of limited resources to address climate and other environmental impacts, it is regrettable that so little attention has been given to developing the methods that enable the reliable assessment of risks.

So far we have discussed indices using widely available data that can be used across different countries and settings, but there is also a need to develop the capacity to undertake national and sub-national climate impact assessments based on local data collection and analysis.

National Risk Assessments

An increasing number of countries have undertaken their own national climate change impact, vulnerability, and risk assessments using a range of approaches and methods. A recent survey of European Environment Agency (EEA) member countries elicited responses from 24 of 33 countries (24). The main reported motivation for undertaking national assessments was to develop nationally relevant adaptation policies and plans (see Chapter 5). All national assessments were targeted at national governments but some also addressed sub-national governments, politicians in general, non-governmental organizations, the academic community, and, in a minority of cases, the general public.

Of the 24 countries responding, only 7 used existing guidelines such as those outlined above and another 5 used some combination of them, suggesting that current guidelines are not universally perceived as useful or not widely known. However, it is possible that countries adopt some of the recommendations in the guidelines when developing their own strategies.

Most countries attempted to encompass multiple sectors in national assessments. The responding countries had engaged an average of 12 sectors and thematic areas per assessment, with a maximum of 19. The most frequent sectors involved were water and agriculture, followed by biodiversity, energy, forestry, and human health. All of the national assessments used climate projections and most used socioeconomic and/or demographic projections. Nineteen of the 24 made a systematic assessment of non-climate factors and adaptive capacity. Fourteen made sub-national assessments, reflecting the diversity of risks

Table 4.2. *Examples of global key risks for different sectors, including the potential for risk reduction through adaptation and mitigation, as well as limits to adaptation. Each key risk is assessed as very low, low, medium, high, or very high. Risk levels are presented for three time frames: present, near term (2030–2040) and long term (2080–2100). In the near term, projected global mean temperature increases do not diverge substantially across different emission scenarios. For the long term, risk levels are presented for two possible futures (2°C and 4°C global mean temperature increase above pre-industrial levels). For each time frame, risk levels are indicated for continued current adaptation and assuming high levels of current or future adaptation. Risk levels are not necessarily comparable, especially across regions. Relevant climate variables are indicated by icons. The WG notations in the table refer to Working Group sources in the IPCC 5th Assessment Report.*

Climate-related drivers of impacts									Level of risk & potential for adaptation	
Warming trend	Extreme temperature	Drying trend	Extreme precipitation	Damaging cyclone	Flooding	Storm surge	Ocean acidification	Carbon dioxide fertilisation	Risk level with **high** adaptation	Risk level with **current** adaptation

Global Risks

Key risk	Adaptation issues & prospects	Climatic drivers	Timeframe	Risk & potential for adaptation
Reduction in terrestrial carbon sink: Carbon stored in terrestrial ecosystems is vulnerable to loss back into the atmosphere, resulting from increased fire frequency due to climate change and the sensitivity of ecosystem respiration to rising temperatures (*medium confidence*) [WGII 4.2, 4.3]	• Adaptation options include managing land use (including deforestation), fire and other disturbances, and non-climatic stressors.	(Warming trend, Extreme temperature, Drying trend)	Present; Near term (2030–2040); Long term (2080–2100) 2°C / 4°C	Very low — Medium — Very high
Boreal tipping point: Arctic ecosystems are vulnerable to abrupt change related to the thawing of permafrost, spread of shrubs in tundra and increase in pests and fires in boreal forests (*medium confidence*) [WGII 4.3, Box 4-4]	• There are few adaptation options in the Arctic.	(Warming trend, Extreme temperature)	Present; Near term (2030–2040); Long term (2080–2100) 2°C / 4°C	Very low — Medium — Very high
Amazon tipping point: Moist Amazon forests could change abruptly to less-carbon-dense, drought- and fire-adapted ecosystems (*low confidence*) [WGII 4.3, Box 4-3]	• Policy and market measures can reduce deforestation and fire.	(Warming trend, Extreme temperature, Drying trend)	Present; Near term (2030–2040); Long term (2080–2100) 2°C / 4°C	Very low — Medium — Very high
Increased risk of species extinction: A large fraction of the species assessed is vulnerable to extinction due to climate change, often in interaction with other threats. Species with an intrinsically low dispersal rate, especially when occupying flat landscapes where the projected climate velocity is high, and species in isolated habitats such as mountaintops, islands or small protected areas are especially at risk. Cascading effects through organism interactions, especially those vulnerable to phenological changes, amplify risk (*high confidence*) [WGII 4.3, 4.4]	• Adaptation options include reduction of habitat modification and fragmentation, pollution, over-exploitation and invasive species; protected area expansion; assisted dispersal; and *ex situ* conservation.	(Warming trend, Extreme temperature, Drying trend)	Present; Near term (2030–2040); Long term (2080–2100) 2°C / 4°C	Very low — Medium — Very high
Global redistribution and decrease of low-latitude fisheries yields, paralleled by a global trend to catches having smaller fishes (*medium confidence*) [WGII 6.3 to 6.5, 30.5, 30.6]	• Increasing coastal poverty at low latitudes as fisheries become smaller – partially compensated by the growth of aquaculture and marine spatial planning, as well as enhanced industrialized fishing efforts	(Warming trend)	Present; Near term (2030–2040); Long term (2080–2100) 2°C / 4°C	Very low — Medium — Very high
Reduced growth and survival of commercially valuable shellfish and other calcifiers (e.g., reef building corals, calcareous red algae) due to ocean acidification (*high confidence*) [WGII 5.3, 6.1, 6.3, 6.4, 30.3, Box CC-OA]	• Evidence for differential resistance and evolutionary adaptation of some species exists, but they are *likely* to be limited at higher CO₂ concentrations and temperatures. • Adaptation options include exploiting more resilient species or protecting habitats with low natural CO₂ levels, as well as reducing other stresses, mainly pollution, and limiting pressures from tourism and fishing.	(Ocean acidification)	Present; Near term (2030–2040); Long term (2080–2100) 2°C / 4°C	Very low — Medium — Very high
Marine biodiversity loss with high rate of climate change (*medium confidence*) [WGII 6.3, 6.4, Table 30-4, Box CC-MB]	• Adaptation options are limited to reducing other stresses, mainly pollution, and limiting pressures from coastal human activities such as tourism and fishing.	(Warming trend, Extreme temperature, Ocean acidification)	Present; Near term (2030–2040); Long term (2080–2100) 2°C / 4°C	Very low — Medium — Very high

Table 4.2. (*cont.*)

Global Risks				
Key risk	**Adaptation issues & prospects**	**Climatic drivers**	**Timeframe**	**Risk & potential for adaptation**
Negative impacts on average crop yields and increases in yield variability due to climate change (*high confidence*) [WGII 7.2 to 7.5, Figure 7-5, Box 7-1]	• Projected impacts vary across crops and regions and adaptation scenarios, with about 10% of projections for the period 2030–2049 showing yield gains of more than 10%, and about 10% of projections showing yield losses of more than 25%, compared to the late 20th century. After 2050 the risk of more severe yield impacts increases and depends on the level of warming.		Present / Near term (2030–2040) / Long term 2°C 4°C (2080–2100)	Very low — Medium — Very high
Urban risks associated with water supply systems (*high confidence*) [WGII 8.2, 8.3]	• Adaptation options include changes to network infrastructure as well as demand-side management to ensure sufficient water supplies and quality, increased capacities to manage reduced freshwater availability, and flood risk reduction.		Present / Near term (2030–2040) / Long term 2°C 4°C (2080–2100)	Very low — Medium — Very high
Urban risks associated with energy systems (*high confidence*) [WGII 8.2, 8.4]	• Most urban centers are energy intensive, with energy-related climate policies focused only on mitigation measures. A few cities have adaptation initiatives underway for critical energy systems. There is potential for non-adapted, centralised energy systems to magnify impacts, leading to national and transboundary consequences from localised extreme events.		Present / Near term (2030–2040) / Long term 2°C 4°C (2080–2100)	Very low — Medium — Very high
Urban risks associated with housing (*high confidence*) [WGII 8.3]	• Poor quality, inappropriately located housing is often most vulnerable to extreme events. Adaptation options include enforcement of building regulations and upgrading. Some city studies show the potential to adapt housing and promote mitigation, adaptation and development goals simultaneously. Rapidly growing cities, or those rebuilding after a disaster, especially have opportunities to increase resilience, but this is rarely realised. Without adaptation, risks of economic losses from extreme events are substantial in cities with high-value infrastructure and housing assets, with broader economic effects possible.		Present / Near term (2030–2040) / Long term 2°C 4°C (2080–2100)	Very low — Medium — Very high
Displacement associated with extreme events (*high confidence*) [WGII 12.4]	• Adaptation to extreme events is well understood, but poorly implemented even under present climate conditions. Displacement and involuntary migration are often temporary. With increasing climate risks, displacement is more likely to involve permanent migration.		Present / Near term (2030–2040) / Long term 2°C 4°C (2080–2100)	Very low — Medium — Very high
Violent conflict arising from deterioration in resource-dependent livelihoods such as agriculture and pastoralism (*high confidence*) [WGII 12.5]	Adaptation options: • Buffering rural incomes against climate shocks, for example through livelihood diversification, income transfers and social safety net provision • Early warning mechanisms to promote effective risk reduction • Well-established strategies for managing violent conflict that are effective but require significant resources, investment and political will		Present / Near term (2030–2040) / Long term 2°C 4°C (2080–2100)	Very low — Medium — Very high
Declining work productivity, increasing morbidity (e.g., dehydration, heat stroke and heat exhaustion), and mortality from exposure to heat waves. Particularly at risk are agricultural and construction workers as well as children, homeless people, the elderly, and women who have to walk long hours to collect water (*high confidence*) [WGII 13.2, Box 13-1]	• Adaptation options are limited for people who are dependent on agriculture and cannot afford agricultural machinery. • Adaptation options are limited in the construction sector where many poor people work under insecure arrangements. • Adaptation limits may be exceeded in certain areas in a +4°C world.		Present / Near term (2030–2040) / Long term 2°C 4°C (2080–2100)	Very low — Medium — Very high
Reduced access to water for rural and urban poor people due to water scarcity and increasing competition for water (*high confidence*) [WGII 13.2, Box 13-1]	• Adaptation through reducing water use is not an option for the many people already lacking adequate access to safe water. Access to water is subject to various forms of discrimination, for instance due to gender and location. Poor and marginalised water users are unable to compete with water extraction by industries, large-scale agriculture and other powerful users.		Present / Near term (2030–2040) / Long term 2°C 4°C (2080–2100)	Very low — Medium — Very high

Source: IPCC. Climate Change 2014: Synthesis Report. Contribution of Working Groups I, II and III to the *Fifth Assessment Report of the Intergovernmental Panel on Climate Change* [Core Writing Team, RK Pachauri and LA Meyer (editors)]. Geneva: Intergovernmental Panel on Climate Change; 2014.

within countries. The importance of stakeholder engagement at all stages, including of non-governmental organizations, was emphasized in the conclusions.

Whilst a survey of this kind can provide useful descriptive data about both positive experiences (e.g. improved ability to prioritize across sectors) and challenges (e.g. data gaps), it cannot tell us how effective such activities have been – only ongoing monitoring and evaluation can provide this information.

An increasing array of resources is available to those wishing to undertake climate change impacts, vulnerability and risk assessments (CCIV). One example is the comprehensive guidance provided by UN's Global Programme of Research on Climate Change Vulnerability, Impacts and Adaptation (25) with a strong emphasis on active stakeholder involvement, which aims to foster the co-production of knowledge.

The WHO in conjunction with the UN Framework Convention on Climate Change (UNFCCC) have developed country profiles on health and climate change for a growing range of countries. These summarize the current and projected climate hazards and the risks arising from exposure to those hazards; the current exposures and risks due to exposure to air pollution (because air pollution is at least in part related to the emissions of GHGs and short-lived climate pollutants); the co-benefits to health from mitigation policies that reduce these emissions; current and committed future greenhouse gas emissions; and policy responses. These profiles make clear that different countries face different risks, e.g. diarrhoea and malaria in Ethiopia as opposed to Lyme disease in the UK.

Many national climate change assessments are becoming more detailed over time as knowledge evolves and the range of potential impacts expands. For example, the fourth US National Climate Assessment, published in 2018, is the most comprehensive yet, encompassing 10 regional chapters and 16 national-level topic chapters, including a chapter on health (26). That chapter summarizes evidence of how different groups are particularly vulnerable to the effects of climate change on health, including indigenous communities, minorities, people living alone, and people with disabilities. **Figure 4.4**, from that chapter, illustrates factors that influence vulnerability and exposure to climate change.

National risks are also affected by other environmental changes. For example, a growing number of countries are unable to meet their domestic demands for freshwater because of depletion of finite aquifers, with serious implications for their food security. These countries depend on food trade to meet their growing demands, which results in a transfer of virtual freshwater from one (production) location to another (consumption) location. In this way, some countries have been able to sustain their population above carrying capacity, but this has rendered them vulnerable to cuts in food exports from their source countries. North and South America, Australia, and the former Soviet Union are 'water-rich' regions overall (although some parts are not) and the source of food exports, whereas Europe, Mexico, and western South America are dependent on the trade in virtual water. Both water-rich and water-poor countries rely on the same finite 'pool' of freshwater. This makes water-poor countries increasingly vulnerable to decreases in exports of food from water-rich nations, which at some point in the future may need to curtail their exports to address their own increasing demands. In the absence of cooperation, one study suggests

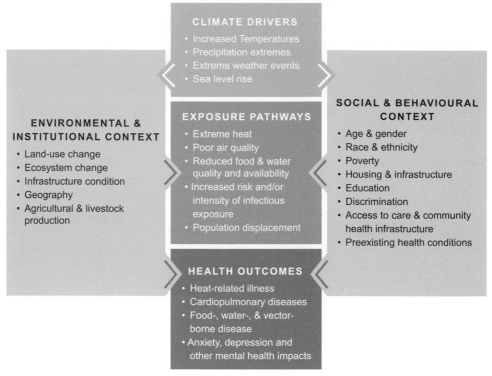

Figure 4.4. Key factors that influence vulnerability and exposure to climate change for individuals, including social determinants and behaviours (right box) and large-scale influences such as governance systems and institutions, natural and built environments (left box). Source: Adapted from Crimmins A, Balbus J, Gamble JL, et al. *The Impacts of Climate Change on Human Health in the United States: A Scientific Assessment.* Washington, DC: US Global Climate Research Program; 2016.

that this could happen as early as 2030 (27). A cooperative regime, whereby food exporting nations maintain some of their exports, might make it possible to sustain larger populations in food importing countries, in which case population declines in water-poor nations could be delayed to 2040–2060. Sudden changes in exports may occur. For example, when drought caused poor harvests in 2007–8 and 2010–11, Russia, Ukraine, and Kazakhstan temporarily restricted grain export (28). Strategies to increase efficiency, decrease waste, and decrease global per capita food consumption can reduce, but not abolish, vulnerability (see Chapter 5). Overall, it seems that the susceptibility to instability and perturbations to the global food system have increased over the past 25 years (29).

Another study, using a high GHG emission (RCP8.5) scenario, assessed the effects of climate change and constraints on freshwater availability for irrigation. It found that climate change could cause maize, soybean, wheat, and rice losses of between 400 and 2600 petacalories (Pcal, or 10^{15} calories) (8–43% of present-day total). Limited freshwater availability in regions that depend heavily on irrigation for agriculture could cause the reversion of 20–60 megahectares (Mha, or 10^6 hectares) of cropland from irrigated to rain-fed management, and a further loss of 600–2900 Pcal. Abundant freshwater availability in other regions could help counterbalance these losses by enabling more food to be grown in these regions, but that would be dependent on substantial investment in infrastructure (30). In many cases, those regions with the largest potential for yield increases from increased irrigation are also those with the greatest limits on water availability.

Local Vulnerability and Risks

Many countries feature a diversity of ecological and climatic zones, making downscaled, place-based risk assessment imperative. There is an array of resources to support screening at the local or project level for climate-related risks; some are particularly focused on large-scale investments in projects, and others at the community level (31). They have different target audiences, including project planners, national policymakers, donor agencies, international NGOs, communities, and local NGOs. A key common objective is strengthening the capacity to make structured risk assessments and to implement policies and programmes to reduce risks.

Many local approaches emphasize the importance of participatory community engagement, which aims to harness local knowledge of current risks in order to assess future risks. One example is the Community-based Risk Screening Tool – Adaptation and Livelihoods (CRiSTAL), developed by the International Institute for Sustainable Development in Canada (32). CRiSTAL arose from an earlier initiative that demonstrated how projects to promote sustainable livelihoods and/or ecosystem management and restoration can help reduce climate risks and strengthen adaptation to climate change, in response to demands from project planners and managers. It aims to create an international community of users to share their experiences and is regularly updated in response to changing knowledge and feedback from users, ensuring that it is a dynamic resource.

A study in Ethiopia showed how this tool can be used at the community level by helping users collect, synthesize, and organize information about (a) the locally relevant development context, (b) the current and future climate, (c) climate-related impacts and risks, and (d) the design of appropriate adaptation activities. In this study the focus was on two informal settlements in the city of Shashemene, located in the Ethiopian highlands, which were only 1 km apart and yet showed significant differences in several characteristics affecting climate risks (33). This part of Ethiopia has already experienced about 0.3 °C heating per decade and increasing rainfall variability since 1960. Using interviews of households and focus groups in the communities, the investigators identified locally important climate-sensitive health outcomes as well as vulnerabilities related to local socioeconomic and environmental conditions. Both communities perceived climate-sensitive health conditions (e.g. typhoid, malaria, diarrhoeal diseases, and trachoma) as serious threats. Whilst both communities relied extensively on water vendors for freshwater, Community A reported more frequent periods without access to clean water than did Community B, and was also significantly more likely to have experienced inadequate income, food, medicines, and fuel during the preceding year. Both communities perceived flooding as a serious problem. However, Community B was more likely to report drought as a serious problem than was Community A because there were more upstream users of water. Community B also had a larger number of children, and a higher incidence of doctor-diagnosed malaria and typhoid amongst children. This study illustrates the importance of local knowledge in understanding and addressing vulnerability to climate and other environmental exposures.

Research carried out on coastal communities in Mozambique and Tanzania shows that many poor communities perceive themselves as powerless in the face of environmental change. Multiple stressors, such as drought, rising food prices, and upstream river basin management in Mozambique combine to drive smallholder farmers and fishing communities deeper into poverty, forcing them to skip meals and increasing the risks of ill-health (34). Sometimes policies intended to increase sustainability of natural resources, such as by creating protected areas, can increase the vulnerability of communities that depend on those areas for food or other essential ecosystem services. For example, in Mnazi Bay, Tanzania, the establishment of a marine protected area led to feelings of disempowerment and exclusion by reducing the area in which people were permitted to fish, and forcing them further from the shore, thus increasing risks to fishing communities. The potential adverse effects of protected areas on local people were recognized at least as far back as 1992, in the Convention on Biological Diversity (CBD) arising from the Rio Summit. A systematic review of the effects of protected areas on human well-being found only nine studies addressing health, with a dearth of quantitative data on the effects, although qualitative data suggested that perceived impacts were negative (35). Several studies showed that use by local communities for health- and nutrition-related activities, such as hunting and consumption of wild foods, did little damage to ecosystems compared with large-scale commercial exploitation such as logging, which can have devastating effects.

Local risks can also arise from pollution sources, for example from point sources of air pollution and the disposal of hazardous waste (see Chapter 3).

The Influence of Social and Demographic Factors on Risk

Population numbers and demographic profiles affect risk through a number of pathways. Population growth is high in many regions that are projected to be increasingly exposed to hazards as a result of environmental change. Projected population size and age structure are influenced by assumptions about future development. As described in Chapter 2 and shown in Figure 2.3, Shared Socioeconomic Pathways (SSPs) portray diverse future development trajectories: they comprise a narrative storyline and a set of quantified measures of development (36). The SSPs are 'reference' pathways in that they assume no climate change or climate impacts, and no new climate policies, but they can be combined with different GHG emission trajectories – the RCPs. The different SSPs imply very different global population size and educational attainment by the end of the century, as shown in **Figures 4.5** and **4.6**. Population size varies between roughly 7 and 12 billion and educational attainment differs markedly, with SSP4 (Inequality) having over 1 billion people

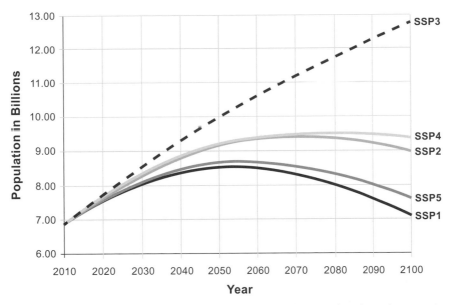

Figure 4.5. Future population projections under different shared socioeconomic pathways (SSPs).

Source: KC S, Lutz W. The human core of the shared socioeconomic pathways: population scenarios by age, sex and level of education for all countries to 2100. *Global Environmental Change.* 2017;42:181–92.

without education, whereas SSPs 1 (Sustainability), 2 (Middle of the road), and 5 (Fossil-fuelled development) have few uneducated people. These projections illustrate how different development futures will substantially affect exposures and risks (37). Because educational attainment is an important predictor of health, fertility, and vulnerability, these differences could have far-reaching effects on future prospects. SSP1 offers the best prospects for the future of humanity.

For example, a study of the effects of climate change and population increase on exposure to extreme levels of heat over a large region of Asia, the Middle East, and East Africa showed that, while the magnitude of temperature increases had the largest effect, projected population changes by the end of the century under SSP1 and SSP2 scenarios can exacerbate exposures by a factor of 1.2 (1.0–1.3) and 1.5 (1.3–1.7), respectively, compared with keeping population constant. Furthermore, the effect of population growth is projected to be greater when global mean temperature increase exceeds 2 °C. The effects of population growth are particularly pronounced in East Africa (38).

A study assessed global exposure to multi-sector climate risks using indicators from the land, water, and energy sectors. The number of people exposed to multi-sector risks approximately doubles with an increase in global mean temperature from 1.5 °C to 2 °C above pre-industrial levels, and doubles again with 3 °C change. The number exposed is approximately six-fold greater in the worst-case scenario (SSP3, Regional rivalry, and 3 °C) compared to the best-case scenario (SSP1, Sustainability, and 1.5 °C) – an increase

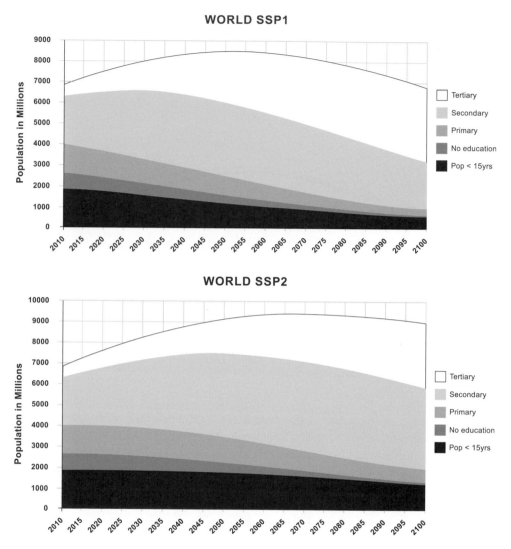

Figure 4.6. Population growth and educational attainment under different SSPs.
Source: KC S, Lutz W. The human core of the shared socioeconomic pathways: population scenarios by age, sex, and level of education for all countries to 2100. *Global Environmental Change*. 2017;42:181–92.

from about 0.8 billion people exposed to about 4.7 billion people exposed. For populations vulnerable to poverty, exposure to multi-sector climate risks is an order of magnitude (8–32 times) greater under SSP3 than under SSP1. In Africa and Asia, 91–98% of the populations may be exposed to multi-sector climate risks depending on the SSP–temperature combination (39).

Population ageing can also amplify the effects of climate change and air pollution on deaths. Older people are more likely to die as a result of exposure to heat and pollution than

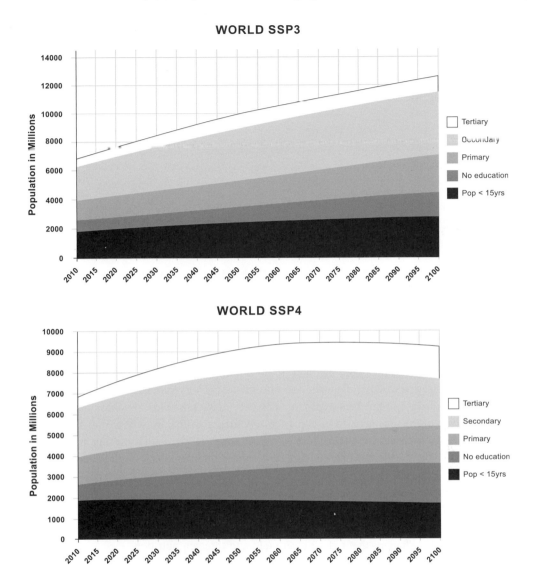

Figure 4.6. (*cont.*)

younger adults. **Figure 4.7** shows the numbers of deaths from different sources of pollution worldwide and age of death (40). Death rates are much higher at older ages although there is a second, smaller peak in childhood. However, taking into account the lost (disability adjusted) life-years, which are clearly much larger for deaths in childhood, the picture changes and the childhood peak becomes much larger, whilst the effects in the elderly become smaller. Vulnerability to the non-fatal effects of pollutants can also vary by age. For example, young children are more sensitive than adults to endocrine-disrupting chemicals or neurotoxins (41).

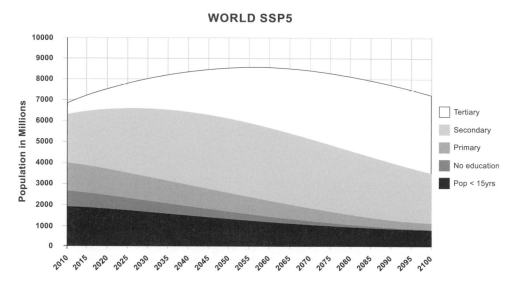

Figure 4.6. (*cont.*)

In the case of climate change, vulnerability varies by age differently depending on the pathway. The majority of deaths from heat exposure occur in the elderly because the ability to thermoregulate declines with age and because elderly people are more likely to suffer from cardiovascular disease, diabetes, and chronic respiratory disease, which increase the risk of deaths from extreme heat exposure (42). People with pre-existing chronic disease or mental health problems may also be at particularly increased risk from exposure to extreme climate events such as floods. In the case of stunting due to undernutrition in the first 1000 days of life, the disease burden increases in early childhood, although stunting leaves an enduring legacy (43, 44). Therefore, in low-income settings where infectious diseases and undernutrition are rife, deaths in childhood are likely to comprise a higher proportion of climate-change-related deaths than in high-income settings. HIV/AIDS increases the case fatality rate in severe acute malnutrition (45) and thus is likely to increase vulnerability to environmental changes that reduce crop yield and threaten nutrition.

The Paradox of Intensifying Environmental Changes and Improving Health

Although many Earth systems are increasingly degraded, health overall is still improving in many parts of the world. (COVID-19 may of course change the picture, but at the time of writing the situation is still evolving.) A key question then is why health damage from environmental change is not readily apparent from current health trends (46). One likely explanation is that the transformation of natural systems to grow more food has improved the health of the world population, which (so far) has exceeded the negative effects of declining biodiversity and other environmental consequences. Advances in health care have contributed greatly too, particularly through the scaling up of access to effective and affordable medicines, vaccines, and surgical interventions. Progress in access to safe water

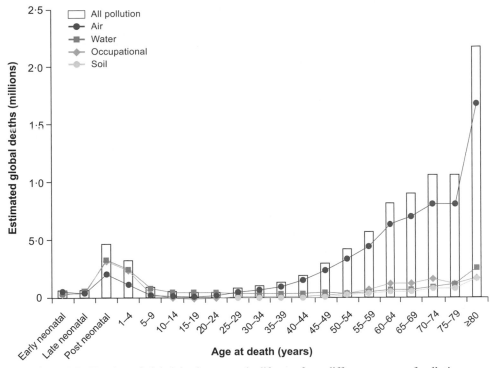

Figure 4.7. Number of global deaths across the lifespan from different sources of pollution.
Source: Landrigan PJ, Fuller R, Acosta NJR, et al. The *Lancet* Commission on Pollution and Health. *The Lancet*. 2018;391:462–512.

and sanitation is another key factor in improving health, together with improvements in housing.

There are likely to be time lags between environmental change and the negative consequences for health (47). Health impacts will become particularly apparent when certain tipping points are exceeded, for example the melting of ice caps or collapse of natural systems. It is also likely that, as a result of decreasing poverty, vulnerability has declined in many regions and populations are adapting to environmental change through planned and unplanned activities and the uptake of technologies such as air conditioning (48). Following the COVID-19 pandemic poverty is likely to increase, at least temporarily, and consequently vulnerability to environmental change will increase commensurately (see also Chapter 12).

Assessing the Adaptation Gap

There is a pressing need to reduce the risks of environmental change through the development and implementation of effective adaptation policies and technologies. Globally, spending for climate change adaptation remains well below the commitments made under the Paris Agreement. In 2018, global spending on adaptation to climate change that focused specifically on health was estimated to be £13 billion (5% of all adaptation spending), and health-related spending was estimated at £35 billion (13.5%). Compared with 2016–17, spending on adaptation to climate change for health and health care increased by 11.2% in 2017–18, suggesting that the importance of adaptation to protect health is being belatedly recognized (49).

The UN Adaptation Gap Report (50) outlined a number of barriers to making a comprehensive assessment of progress in adaptation and the gap between the current level of adaptation and the level required to achieve a societal goal of minimizing the impacts of climate change. First, the burgeoning and complex landscape of adaptation legislation and policy, with blurred boundaries between many development policies on the one hand and those policies designed specifically with adaptation as the primary objective on the other, makes it difficult to identify and assess all the relevant policy measures. Second, detailed analysis of national laws and regulations to foster adaptation is planned but not yet available. Third, although much critical action takes place at the sub-national level (e.g. at the level of local government or city administration), there is no comprehensive database of such actions. Fourth, there is no agreed set of metrics by which to assess the effectiveness of implementation of measures, and the evidence base for the cost-effectiveness of various policies and interventions is fragmented and weak because many of them have not been subject to rigorous evaluation.

The next chapter summarizes the evidence about effective approaches to adaptation and outlines challenges and limits to adaptation. Although much of the effort to date has focused on climate change, the chapter also addresses the importance of the other changes that characterize the Anthropocene Epoch. This broad approach to safeguarding humanity and vital natural systems is needed if current and future generations are to have reasonable prospects of achieving and sustaining healthy and fulfilling lives. At the same time, the potential for adaptation cannot be used as an excuse to delay action to address the driving forces propelling us towards an uncertain and potentially hazardous future. The imperative for transformational change to tackle the root causes of our current unsustainable pathways of development is addressed from Chapter 6 onwards.

References

1. National Research Council. *Risk Assessment in the Federal Government: Managing the Process*. Washington, DC: The National Academies Press; 1983.
2. National Research Council. *Science and Decisions: Advancing Risk Assessment*. Washington, DC: The National Academies Press; 2009.
3. EPA. *Guidelines for Ecological Risk Assessment*. Washington, DC: US Environmental Protection Agency; 1998.
4. Harwell MA, Gentile JH, McKinney LD, et al. Conceptual framework for assessing ecosystem health. *Integrated Environmental Assessment and Management*. 2019; 15(4):544–64. doi: 10.1002/ieam.4152.
5. Forbes VE, Galic N. Next-generation ecological risk assessment: predicting risk from molecular initiation to ecosystem service delivery. *Environment International*. 2016;91:215–19. doi: 10.1016/j.envint.2016.03.002.

6. Suter GW, Vermeire T, Munns WR, Sekizawa J. An integrated framework for health and ecological risk assessment. *Toxicology and Applied Pharmacology*. 2005;207(2, Supplement):611–16. doi: 10.1016/j.taap.2005.01.051.

7. Forbes VE, Calow P. Use of the ecosystem services concept in ecological risk assessment of chemicals. *Integrated Environmental Assessment and Management*. 2013;9(2):269–75. doi: 10.1002/ieam.1368.

8. Oppenheimer M, Campos M, Warren R, et al. Emergent risks and key vulnerabilities. In Field CB, Barros VR, Dokken DJ, et al., editors. *Climate Change 2014: Impacts, Adaptation, and Vulnerability Part A: Global and Sectoral Aspects Contribution of Working Group II to the Fifth Assessment Report of the Intergovernmental Panel on Climate Change*. Cambridge, UK and New York: Cambridge University Press; 2014. pp. 1039–99.

9. Kelly PM, Adger WN. Theory and practice in assessing vulnerability to climate change and facilitating adaptation. *Climatic Change*. 2000;47(4):325–52. doi: 10.1023/A:1005627828199.

10. Füssel H-M. Vulnerability: a generally applicable conceptual framework for climate change research. *Global Environmental Change*. 2007;17(2):155–67.

11. IPCC. *Climate Change 2014: Synthesis Report. Contribution of Working Groups I, II and III to the Fifth Assessment Report of the Intergovernmental Panel on Climate Change* [Core Writing Team, RK Pachauri and LA Meyer (editors)]. Geneva: Intergovernmental Panel on Climate Change; 2014.

12. Miola A, Simonet C. Concepts and Metrics for Climate Change Risk and Development. Towards an Index for Climate Resilient Development. Ispra, Italy: European Commission Joint Research Centre; 2014. Report EUR 26587EN.

13. Eckstein D, Künzel V, Schäfer L. Global Climate Risk Index 2018: Who Suffers Most from Extreme Weather Events? Weather-Related Loss Events in 2016 and 1997 to 2016. Bonn: GermanWatch.

14. Welle T, Birkmann J. The World Risk Index: an approach to assess risk and vulnerability on a global scale. *Journal of Extreme Events*. 2015;02(01):1550003. doi: 10.1142/S2345737615500037.

15. Chen C, Noble I, Hellmann J, et al. University of Notre Dame Global Adaptation Index. Country Index Technical Report. Notre Dame, IN: University of Notre Dame; 2015.

16. Wheeler D. Quantifying Vulnerability to Climate Change: Implications for Adaptation Assistance. Working Paper 240. Washington, DC: Center for Global Development; 2011.

17. DARA, Climate Vulnerable Forum. *Climate Vulnerability Monitor 2nd Edition: A Guide to the Cold Calculus of a Hot Planet*. Madrid: Fundación DARA Internacional; 2012.

18. Miola A, Simonet C. Indicators and maps comparing countries' vulnerability to climate change: fit for purpose metrics to guide global action for climate resilient development within the context of the new Sustainable Development Goals. Ispra, Italy: European Commission, Joint Research Centre; 2014.

19. World Bank. Ease of Doing Business rankings 2020. Available from www.doingbusiness.org/en/rankings.

20. Kleiman G, Bouley T, Meiro-Lorenzo M, Wang H. *Geographic Hotspots for World Bank Action on Climate Change and Health*. Washington, DC: World Bank; 2017. Contract No. 113571.

21. Forouzanfar MH, Alexander L, Anderson HR, et al. Global, regional, and national comparative risk assessment of 79 behavioural, environmental and occupational, and metabolic risks or clusters of risks in 188 countries, 1990–2013: a systematic analysis

for the Global Burden of Disease Study 2013. *The Lancet*. 2015;386(10010):2287–323. doi: 10.1016/S0140-6736(15)00128-2.

22. Lenton TM, Rockström J, Gaffney O, et al. Climate tipping points – too risky to bet against. *Nature*. 2019;575:592–5. doi: 10.1038/d41586-019-03595-0.

23. Xu C, Kohler TA, Lenton TM, Svenning J-C, Scheffer M. Future of the human climate niche. *Proceedings of the National Academy of Sciences*. 2020:201910114. doi: 10.1073/pnas.1910114117.

24. EEA. National Climate Change Vulnerability and Risk Assessments in Europe, 2018. Copenhagen: European Environment Agency; 2018. EEA Report 1/2018.

25. Hinkel J, Bharwani S, Bisaro A, et al. PROVIA Guidance on Assessing Vulnerability, Impacts and Adaptation to Climate Change: Consultation Document. Nairobi: United Nations Environment Programme, Global Programme of Research on Climate Change Vulnerability, Impacts and Adaptation (PROVIA); 2013.

26. Reidmiller DR, Avery CW, Easterling DR, et al. *Impacts, Risks, and Adaptation in the United States: Fourth National Climate Assessment, Volume II*. Washington, DC: US Global Change Research Program (USGCRP); 2018.

27. Suweis S, Rinaldo A, Maritan A, D'Odorico P. Water-controlled wealth of nations. *Proceedings of the National Academy of Sciences*. 2013;110(11):4230–3. doi: 10.1073/pnas.1222452110.

28. Fellmann T, Hélaine S, Nekhay O. Harvest failures, temporary export restrictions and global food security: the example of limited grain exports from Russia, Ukraine and Kazakhstan. *Food Security*. 2014;6(5):727–42. doi: 10.1007/s12571-014-0372-2.

29. Suweis S, Carr JA, Maritan A, Rinaldo A, D'Odorico P. Resilience and reactivity of global food security. *Proceedings of the National Academy of Sciences*. 2015;112(22): 6902–7. doi: 10.1073/pnas.1507366112.

30. Elliott J, Deryng D, Müller C, et al. Constraints and potentials of future irrigation water availability on agricultural production under climate change. *Proceedings of the National Academy of Sciences*. 2014;111(9):3239–44. doi: 10.1073/pnas.1222474110.

31. Trærup S, Olhoff A. *Climate Risk Screening Tools and Their Application: A Guide to the Guidance*. Nairobi: United Nations Environment Programme; 2011.

32. IISD. CRiSTAL User's Manual Version 5. Community-Based Risk Screening Tool – Adaptation and Livelihoods. Winnipeg: International Institute for Sustainable Development; 2012.

33. Bambrick H, Moncada S, Briguglio M. Climate change and health vulnerability in informal urban settlements in the Ethiopian Rift Valley. *Environmental Research Letters*. 2015;10(5):054014. doi: 10.1088/1748-9326/10/5/054014.

34. Bunce M, Rosendo S, Brown K. Perceptions of climate change, multiple stressors and livelihoods on marginal African coasts. *Environment, Development and Sustainability*. 2010;12(3):407–40. doi: 10.1007/s10668-009-9203-6.

35. Pullin A, Bangpan M, Dalrymple S, et al. Human well-being impacts of terrestrial protected areas. *Environmental Evidence*. 2013;2(1):19. doi: 10.1186/2047-2382-2-19.

36. O'Neill BC, Kriegler E, Riahi K, et al. A new scenario framework for climate change research: the concept of shared socioeconomic pathways. *Climatic Change*. 2014;122(3):387–400. doi: 10.1007/s10584-013-0905-2.

37. KC S, Lutz W. The human core of the shared socioeconomic pathways: population scenarios by age, sex and level of education for all countries to 2100. *Global Environmental Change*. 2017;42:181–92. doi: 10.1016/j.gloenvcha.2014.06.004.

38. Harrington LJ, Otto FEL. Changing population dynamics and uneven temperature emergence combine to exacerbate regional exposure to heat extremes under 1.5 °C and 2 °C of warming. *Environmental Research Letters*. 2018;13(3):034011.

39. Byers E, Gidden M, Leclère D, et al. Global exposure and vulnerability to multi-sector development and climate change hotspots. *Environmental Research Letters*. 2018; 13(5):055012. doi: 10.1088/1748-9326/aabf45.
40. Landrigan PJ, Fuller R, Acosta NJR, et al. The *Lancet* Commission on Pollution and Health. *The Lancet*. 2018;391:462–512. doi: 10.1016/S0140-6736(17)32345-0.
41. Landrigan PJ, Fuller R, Fisher S, et al. Pollution and children's health. *Science of The Total Environment*. 2019;650(Pt 2):2389–94. doi: 10.1016/j.scitotenv.2018.09.375.
42. Kenny GP, Yardley J, Brown C, Sigal RJ, Jay O. Heat stress in older individuals and patients with common chronic diseases. *Canadian Medical Association Journal/ Journal de l'Association medicale canadienne*. 2010;182(10):1053–60. doi: 10.1503/cmaj.081050.
43. Dewey KG, Begum K. Long-term consequences of stunting in early life. *Maternal and Child Nutrition*. 2011;7(Suppl. 3):5–18. doi: 10.1111/j.1740-8709.2011.00349.x.
44. Berti C, Agostoni C, Davanzo R, et al. Early-life nutritional exposures and lifelong health: immediate and long-lasting impacts of probiotics, vitamin D, and breastfeeding. *Nutrition Reviews*. 2017;75(2):83–97. doi: 10.1093/nutrit/nuw056.
45. Wagnew F, Worku W, Dejenu G, Alebel A, Eshetie S. An overview of the case fatality of inpatient severe acute malnutrition in Ethiopia and its association with human immunodeficiency virus/tuberculosis comorbidity: a systematic review and meta-analysis. *International Health*. 2018;10(6):405–11. doi: 10.1093/inthealth/ihy043.
46. Raudsepp-Hearne C, Peterson GD, Tengö M, et al. Untangling the environmentalist's paradox: why is human well-being increasing as ecosystem services degrade? *BioScience*. 2010;60(8):576–89. doi: 10.1525/bio.2010.60.8.4.
47. Filippelli GM. Exploring the paradox of increased global health and degraded global environment: how much borrowed time is humanity living on? *GeoHealth*. 2018;2(8): 226–8. doi: 10.1029/2018GH000155.
48. Whitmee S, Haines A, Beyrer C, et al. Safeguarding human health in the Anthropocene Epoch: report of The Rockefeller Foundation–*Lancet* Commission on Planetary Health. *The Lancet*. 2015;386(10007):1973–2028. doi: 10.1016/S0140-6736(15) 60901-1.
49. Watts N, Amann M, Arnell N, et al. The 2019 report of The *Lancet* Countdown on Health and Climate Change: ensuring that the health of a child born today is not defined by a changing climate. *The Lancet*. 2019;394(10211):1836–78. doi: 10.1016/S0140-6736(19)32596-6.
50. UNEP. Adaptation Gap Report 2018. Nairobi: United Nations Environment Programme; 2018.

5

Adaptation and Resilience to Planetary Change

Adaptation and Resilience: Companion Concepts

Many of the dramatic changes across the planet during the Anthropocene Epoch cannot be reversed within our lifespans, so it becomes imperative to adapt to change as far as possible. According to the IPCC, adaptation is 'the process of adjustment to actual or expected climate and its effects. In human systems, adaptation seeks to moderate or avoid harm or exploit beneficial opportunities. In some natural systems, human intervention may facilitate adjustment to expected climate and its effects' (1, p. SPM 5). While this definition refers only to climate, the context in which adaptation has been most thoroughly considered, the concept of adaptation is applicable to the full range of planetary changes. As implied by the IPCC definition, an adaptation action might be taken *proactively*, to reduce harm in advance of an impact, or *reactively*, in response to a perceived or real health risk.

Resilience is closely allied to adaptation. The word derives from the Latin *resilire*, meaning to 'leap back' or 'recoil'. As a physical concept, resilience is the ability of an object to recover its original form after being bent, compressed, or stretched. In psychological terms, resilience refers to a person's ability to 'bounce back' from adversity or trauma. And in the context of planetary change, resilience is the ability of a system to respond to stresses, shocks, or perturbations while preserving its identity and function (2). In general, resilient systems are able to absorb shocks and stresses, to self-organize, and to learn and adapt (2, 3). Examples of resilient systems might include a coastal city (that can recover following a severe storm), a forest (that can recover following a fire), or a more complex 'socio-ecological' system that includes social and natural elements (4, 5).

So adaptation and resilience are closely related – adaptation as a set of deliberate actions in response to a stimulus or stress, resilience as a system condition, one of whose features is the capacity to adapt. From this perspective, resilience becomes a useful framework to identify effective adaptation responses to planetary changes, and to assess how they fit in to a wider system that aims to maintain – and where possible enhance – essential functions under changing conditions (6, 7).

Spatial scale is relevant to understanding adaptation and resilience. While our focus in this book is on planetary changes, many of these, from climate change to pollution, play out on a local to regional scale, and many key responses are implemented at that smaller scale as well. Adaptation to climate change, for example, might involve developing a municipal heat action plan or a wetland management protocol.

Adaptation

Adaptation versus mitigation: Adaptation is often presented in counterpoise to mitigation actions that aim to reduce the magnitude of change by addressing the drivers of change at source. In the case of climate change this means reducing emissions of greenhouse gases (and short-lived climate pollutants). The aim is to achieve net-zero emissions as rapidly as possible to reduce the risks of dangerous climate change. Adaptation may be viewed as 'managing the unavoidable', while mitigation may be viewed as 'avoiding the unmanageable'. Adaptation and mitigation map readily to familiar public health concepts: mitigation to primary prevention, and adaptation to secondary and tertiary prevention as well as to disaster preparedness and response (8).

Mitigation and adaptation are often considered separately, with different communities of researchers and practitioners engaged in their development and implementation, but they should be seen as an integrated approach that aims to navigate humanity safely through the Anthropocene. There are several reasons for advocating closer integration. First, adaptation will be easier, less costly and more successful if environmental change can be kept within reasonably safe limits. Second, some adaptation strategies could increase the challenges to mitigation, for example the widespread uptake of air conditioning will reduce indoor heat-related risks but will increase energy demands and thus potentially fossil fuel emissions, as well as contributing to urban heat islands by displacing excess heat into the outdoor environment. Third, it would be more cost effective to consider the two approaches in a coherent way, for example in the design of the built environment both to reduce the impacts of extreme events and to reduce energy demands. It is a combination of effective adaptation and mitigation that offers the best prospects for safeguarding human health in the Anthropocene.

Adaptive capacity: Adaptation is a complex challenge, requiring expertise, knowledge of specific challenges, financing, effective partnerships, community engagement, and more. Adaptive capacity is defined as 'the combination of the strengths, attributes and resources available to an individual, community, society, or organization that can be used to prepare for and undertake actions to reduce adverse impacts, moderate harm, or exploit beneficial opportunities' (9, p. vii). While measurement of adaptive capacity is a complex challenge (10), research suggests that adaptive capacity is much higher in wealthy nations than in poor nations, and in wealthy communities than in poor communities. An important goal of development is building this capacity across all jurisdictions and spatial scales – including down to the neighbourhood and even household scale. For example, one study showed that in urban neighbourhoods in Latin America, the absence of formal, defined economic and political structures reduced neighbourhood adaptive capacity (11), emphasizing the need for such structures as a part of development processes. Even more aspirational is building transformative capacity, the 'capacity of individuals and organisations to be able to both transform themselves and their society in a deliberate, conscious way' (12, p. 2) – a scale of change that, as discussed below, is commensurate with the challenges of the Anthropocene.

Adaptation to what? The planet is changing in many ways, as surveyed in Chapter 1, but one of these changes – climate change – has dominated the adaptation literature. The concept of adaptation (and the concept of resilience) are less well developed in relation to other environmental threats. With pollution, for example, most focus, properly, is on reducing population exposure rather than adapting to pollutants, and with biodiversity the emphasis is on preventing species loss rather than on adapting to a biologically impoverished world. However, adaptation may sometimes be necessary where preventive actions are too slow, infeasible, or unsuccessful. For example, the use of domestic air purifiers in heavily polluted urban environments, or the use of artificial pollination of food crops, could be seen as adaptive responses. The suitability of adaptation strategies will vary across different planetary changes.

Those who plan adaptation actions must know to what they are adapting. Ideally, they would know this in advance, to give them time to act, and they would know the intensity, duration, and location of each particular challenge. Accordingly, there have been extensive efforts to characterize, model, and forecast such eventualities as malaria spread (13), rainfall patterns (14), and locust infestations (15). Climate information services require the participation of meteorological agencies, which generate much of the starting-point data, together with health, agriculture, urban planning, and other sectors (16). This work can be difficult. Because much adaptation is local, it requires the downscaling of data that may only be available at large scales. It requires consideration of local contextual issues (17), including political context (18). It requires dialogue across diverse professional disciplines, interoperability of sometimes incompatible databases, information sharing among agencies that may be as much rivals as partners, and effective communication that reaches the right users, based on a thorough understanding of those users' needs (16).

Adaptation as a system challenge: Adaptation involves the management of complex systems, and is best regarded not as an isolated intervention but as an ongoing iterative process (19). The commonly used, but potentially confusing, term for this approach to adaptation is adaptive management, derived from the environmental field (7, 20, 21). Adaptive management is a structured approach to decision making in the setting of uncertainty, complexity, and changing circumstances – precisely the characteristics of planetary systems that affect human health and well-being. Importantly, adaptive management can help avoid unintended and unwanted consequences of adaptation actions, which are discussed below.

Incremental versus transformational adaptation: It is useful to distinguish between incremental adaptation and transformational adaptation (22, 23). The former seeks to protect existing infrastructure and systems, typically at their current scale and in their current locations, from natural variations in climate and extreme events. The latter, in contrast, aims to ensure that infrastructure and systems are fundamentally realigned, generally through non-linear changes, to address future challenges, which may be of much greater magnitude than past risks. Transformational change generally has one or more of three characteristics: (a) it comprises much larger-scale measures; (b) it employs novel technologies and tools; and/or (c) it involves deep structural changes in the nature or location of economic activities (24).

Much current adaptation activity is essentially incremental and involves performing familiar tasks at somewhat higher levels of ambition. For example, of the 314 adaptations to climate impacts in seven sectors – agriculture and forestry, coastal areas, ecosystems, energy, health, transportation, and water resources – listed by the US National Research Council's Panel on Adapting to the Impacts of Climate Change, only 16 (5%) represented new approaches that had not previously been used somewhere in the USA (24). The distinction between incremental and transformational may not be an absolute one, if for example cumulative incremental change becomes large scale over time, nevertheless it is a useful distinction.

One example of large-scale adaptation is the development of farmer-managed greening of the Sahel in the face of drought. This capitalized on the restorative power of residual tree roots, which give rise to trees that provide fodder, fuel, and shelter when appropriately nurtured and pruned by farmers (25). In this way about 5 million hectares of green-belt land – detectable by satellite – have been created, and will contribute to climate resilience. Examples of new approaches include the development and dissemination of drought-resistant maize through innovative partnerships (comprising a mix of local, national, global partners) combining new breeding techniques, other improved agronomic practices, free distribution to farmers, and a long-term perspective that aims for sustained results (25). In this case the support of major donors such as the Bill and Melinda Gates and Buffett foundations helped catalyse large-scale change. A third example of transformative adaptation is the planned movement of people from low-lying islands that become uninhabitable as a result of sea level rise (26). Transformations become necessary to address highly vulnerable regions and populations whose very survival could be threatened in the absence of far-reaching responses, or accelerated environmental change, including where multiple environmental stressors interact.

Transformational adaptation is often difficult to implement in anticipation of a threat because it involves uprooting traditional patterns of activity and often substantial costs, sometimes in the face of uncertainty about the pace and extent of change. Established patterns of land use or ownership together with access to natural resources such as water may constitute barriers to change, as there is sometimes understandable reluctance to give up long-standing rights. At the same time, many low-income communities have insecure land tenure, which adversely affects their ability to implement changes that advance adaptation (and mitigation). When customary access to, and ownership of, land are unrecognized by national and local governments this can decrease adaptive capacity. A range of land use policies such as recognition of customary tenure, community mapping, redistribution, co-management, and regulation of rental markets, can provide both security and flexibility response to climate change (27). In Chapter 10, Box 10.1, we discuss the importance of expanding community-conserved areas – territories of life – in which indigenous communities can flourish, supporting biodiversity conservation, climate change adaptation and mitigation.

Adaptation with no regrets: Many policies can be implemented with no or few regrets because they both reduce vulnerability and yield other benefits – economic, social, environmental, and/or for health. Some examples of options for reducing the risks from extreme climate events are given in **Table 5.1**.

Resilience

Resilience has become an increasingly widely used term in fields ranging from engineering and architecture to security studies and international relations. Its use reflects the understanding that multiple stressors threaten the stability of globally interconnected economic and governance systems which are often ill-equipped to deal with them. Indeed, in 2013, *Time* magazine declared resilience to be the 'buzzword of the year' (29). This implies an evanescent concept, without clear focus, in danger of being blown by the winds of fashion in different directions. Resilience has been described as a loose 'boundary object' which through its relative vagueness can bring different communities and perspectives together as well as being a descriptive ecological concept (30). At the same time, it suggests a perception – perhaps even a yearning – for a concept that would help to provide stability and optimism for an uncertain future. This implies that we need more empirical evidence of what works in what circumstances to ensure that development pathways create resilient societies, able to anticipate and respond to threats in a timely and equitable fashion.

Resilience as development: Rather than considering climate change adaptation and development as independent processes, a more constructive approach is to integrate them as climate-resilient development, which aims to ensure that climate resilience is built into long-term planning, with the objective of identifying synergies and trade-offs (31). This integrates solutions to two legacies of the Anthropocene: the risk resulting from perturbed Earth systems, and the deep inequities resulting from unbalanced economic development – the 'unfinished agenda' of the Anthropocene.

The immiseration of large portions of the human population is not only morally wrong, it reduces their resilience and deepens their vulnerability to climate change and other planetary derangements. This unfinished agenda is exemplified by the 29% of the global population who lacked safely managed drinking water supplies, and the 55% who were without safely managed sanitation services in 2017 (32). In 2017, only 28% of the population in less-developed countries had basic handwashing facilities (32). Unsafe drinking water, inadequate sanitation, and lack of hygiene continue to be major global causes of premature death, accounting for some 870,000 deaths in 2016, mainly from diarrhoeal diseases, undernutrition and intestinal nematode infections (32). Although in the least developed countries the proportion of people with access to electricity doubled between 2000 and 2017, there were still nearly 1 billion people without access to electricity (32). While the proportion of people living in slums decreased from 28% in 2000 to 23.5% in 2018, the absolute numbers increased from 807 million to over 1 billion, and the downward trend reversed in the last few years of the 2010s decade, because of rapid population growth and urbanization in low-income countries (32). Worryingly, the

Table 5.1. *Illustrative examples of climate change risk management and adaptation options in the context of changing exposure, vulnerability, and occurrence of climate extremes. In each example, information is characterized at the scale directly relevant to decision making, because the direction and magnitude of changes, and/or degree of certainty, may differ across scales. See original source for definitions of confidence levels in observed and likelihood of projected trends of climate extremes.*

Example	Exposure and vulnerability at scale of risk management in the example	Information on climate extremes across spatial scales			Options for risk management and adaptation in the example
		GLOBAL Observed (since 1950) and projected (to 2100) global changes	REGIONAL Observed (since 1950) and projected (to 2100) changes in the example	SCALE OF RISK MANAGEMENT Available information for the example	
Inundation related to extreme sea levels in tropical small island developing states	Small island states in the Pacific, Indian, and Atlantic Oceans, often with low elevation, are particularly vulnerable to rising sea levels and impacts such as erosion, inundation, shoreline change, and saltwater intrusion into coastal aquifers. These impacts can result in ecosystem disruption, decreased agricultural productivity, changes in disease patterns, economic losses such as in tourism industries, and	**Observed**: Likely increase in extreme coastal high water level worldwide related to increases in mean sea level. **Projected**: Very likely that mean sea level rise will contribute to upward trends in extreme coastal high-water levels. High confidence that locations currently experiencing coastal erosion and inundation will continue to do so due to increasing sea level, in the absence	**Observed**: Tides and El Niño–Southern Oscillation have contributed to the more frequent occurrence of extreme coastal high-water levels and associated flooding experienced on some Pacific Islands in recent years. **Projected**: The very likely contribution of mean sea level rise to increased extreme coastal high-water levels, coupled with the likely increase in tropical cyclone maximum wind	Sparse regional and temporal coverage of terrestrial-based observation networks and limited in situ ocean observing network, but with improved satellite-based observations in recent decades. While changes in storminess may contribute to changes in extreme coastal high-water levels, the limited geographical coverage of studies to date and the uncertainties associated with storminess changes	Low-regrets options that reduce exposure and vulnerability across a range of hazard trends: • Maintenance of drainage systems • Well technologies to limit sal water contamination of groundwater • Improved early warning systems • Regional risk pooling • Mangrove conservation, restoration, and replanting Specific adaptation options include, for instance, rendering national economies more climate-independent and adaptive management involving

153

Table 5.1. (cont.)

		Information on climate extremes across spatial scales			
Example	Exposure and vulnerability at scale of risk management in the example	GLOBAL Observed (since 1950) and projected (to 2100) global changes	REGIONAL Observed (since 1950) and projected (to 2100) changes in the example	SCALE OF RISK MANAGEMENT Available information for the example	Options for risk management and adaptation in the example
	population displacement – all of which reinforce vulnerability to extreme weather events.	of changes in other contributing factors. Likely that the global frequency of tropical cyclones will either decrease or remain essentially unchanged. Likely increase in average tropical cyclone maximum wind speed, although increases may not occur in all ocean basins.	speed, is a specific issue for tropical small island states. See global changes column for information on global projections for tropical cyclones.	overall mean that a general assessment of the effects of storminess changes on storm surge is not possible at this time.	iterative learning. In some cases there may be a need to consider relocation, for example, for atolls where storm surges may completely inundate them.
Flash floods in informal settlements in Nairobi, Kenya	Rapid expansion of poor people living in informal settlements around Nairobi has led to houses of weak building materials being constructed immediately adjacent to rivers and to	**Observed**: Low confidence at global scale regarding (climate-driven) observed changes in the magnitude and frequency of floods. **Projected**: Low confidence in	**Observed**: Low confidence regarding trends in heavy precipitation in East Africa, because of insufficient evidence. **Projected**: Likely increase in heavy precipitation	Limited ability to provide local flash flood projections.	Low-regrets options that reduce exposure and vulnerability across a range of hazard trends: • Strengthening building design and regulation • Poverty reduction schemes • City-wide drainage and sewerage improvements

154

| Impacts of heat waves in urban areas in Europe | Factors affecting exposure and vulnerability include age, pre-existing health status, level of outdoor activity, socioeconomic factors including poverty and social isolation, access to and use of cooling, physiological and behavioural adaptation of the population, and urban infrastructure. | **Observed**: Medium confidence that the length or number of warm spells or heat waves has increased since the middle of the twentieth century, in many (but not all) regions over the globe. Very likely increase in number of warm days and nights at the global scale. **Projected**: Very likely increase in | **Observed**: Medium confidence in increase in heat waves or warm spells in Europe. Likely overall increase in warm days and nights over most of the continent. **Projected**: Likely more frequent, longer, and/or more intense heat waves or warm spells in Europe. Very likely increase in warm days and nights. | Observations and projections can provide information for specific urban areas in the region, with increased heat waves expected due to regional trends and urban heat island effects. | Low-regrets options that reduce exposure and vulnerability across a range of hazard trends:
• Early warning systems that reach particularly vulnerable groups (e.g. the elderly)
• Vulnerability mapping and corresponding measures
• Public information on what to do during heat waves, including behavioural advice |
| | blockage of natural drainage areas, increasing exposure and vulnerability. | projections of changes in floods because of limited evidence and because the causes of regional changes are complex. However, medium confidence (based on physical reasoning) that projected increases in heavy precipitation will contribute to rain-generated local flooding in some catchments or regions. | | indicators in East Africa. | The Nairobi Rivers Rehabilitation and Restoration Programme includes installation of riparian buffers, canals, and drainage channels and clearance of existing channels; attention to climate variability and change in the location and design of wastewater infrastructure; and environmental monitoring for flood early warning. |

Table 5.1. (*cont.*)

Example	Exposure and vulnerability at scale of risk management in the example	Information on climate extremes across spatial scales			Options for risk management and adaptation in the example
		GLOBAL Observed (since 1950) and projected (to 2100) global changes	REGIONAL Observed (since 1950) and projected (to 2100) changes in the example	SCALE OF RISK MANAGEMENT Available information for the example	
		length, frequency, and/or intensity of warm spells or heat waves over most land areas. Virtually certain increase in frequency and magnitude of warm days and nights at the global scale.			• Use of social care networks to reach vulnerable groups Specific adjustments in strategies, policies, and measures informed by trends in heat waves include awareness raising of heat waves as a public health concern; changes in urban infrastructure and land use planning, for example, increasing urban green space; changes in approaches to cooling for public facilities; and adjustments in energy generation and transmission infrastructure.
Increasing losses from hurricanes in the USA and the Caribbean	Exposure and vulnerability are increasing due to growth in population and increase in property values,	**Observed:** Low confidence in any observed long-term (i.e. 40 years or more) increases in tropical cyclone activity, after	See global changes column for regional projections	Limited model capability to project changes relevant to specific settlements or other locations, due to the inability of global	Low-regrets options that reduce exposure and vulnerability across a range of hazard trends:

	particularly along the Gulf and Atlantic coasts of the USA. Some of this increase has been offset by improved building codes.	accounting for past changes in observing capabilities. **Projected:** Likely that the global frequency of tropical cyclones will either decrease or remain essentially unchanged. Likely increase in average tropical cyclone maximum wind speed, although increases may not occur in all ocean basins. Heavy rainfalls associated with tropical cyclones are likely to increase. Projected sea level rise is expected to further compound tropical cyclone surge impacts.	models to accurately simulate factors relevant to tropical cyclone genesis, track, and intensity evolution.	• Adoption and enforcement of improved building codes • Improved forecasting capacity and implementation of improved early warning systems (including evacuation plans and infrastructures) • Regional risk pooling In the context of high underlying variability and uncertainty regarding trends, options can include emphasizing adaptive management involving learning and flexibility (e.g. Cayman Islands National Hurricane Committee).
Droughts in the context of food security in West Africa	Less advanced agricultural practices render region vulnerable to increasing variability in seasonal rainfall, drought, and weather extremes.	**Observed:** Medium confidence that some regions of the world have experienced more intense and longer droughts, but in some regions droughts have become	**Observed:** Medium confidence in an increase in dryness. Recent years characterized by greater interannual variability than previous 40 years,	Sub-seasonal, seasonal, and interannual forecasts with increasing uncertainty over longer timescales. Improved monitoring, instrumentation, and	Low-regrets options that reduce exposure and vulnerability across a range of hazard trends: • Traditional rain and groundwater harvesting and storage systems

Table 5.1. (*cont.*)

	Exposure and vulnerability at scale of risk management in the example	Information on climate extremes across spatial scales			Options for risk management and adaptation in the example
		GLOBAL Observed (since 1950) and projected (to 2100) global changes	REGIONAL Observed (since 1950) and projected (to 2100) changes in the example	SCALE OF RISK MANAGEMENT Available information for the example	
Example	Vulnerability is exacerbated by population growth, degradation of ecosystems, and overuse of natural resources, as well as poor standards for health, education, and governance.	less frequent, less intense, or shorter. **Projected**: Medium confidence in projected intensification of drought in some seasons and areas. Elsewhere there is overall low confidence because of inconsistent projections.	with the western Sahel remaining dry and the eastern Sahel returning to wetter conditions. **Projected**: Low confidence due to inconsistent signal in model projections.	data associated with early warning systems, but with limited participation and dissemination to at-risk populations.	• Water demand management and improved irrigation efficiency measures • Conservation agriculture, crop rotation, and livelihood diversification • Increasing use of drought-resistant crop varieties • Early warning systems integrating seasonal forecasts with drought projections, with improved communication involving extension services • Risk pooling at the regional or national level

Source: Field CB, Barros V, Stocker TF, et al. *Managing the Risks of Extreme Events and Disasters to Advance Climate Change Adaptation.* Cambridge, UK: IPCC; 2012.

estimated number of undernourished people, which had been falling, plateaued in 2015 and rose thereafter (see Chapter 10) (32). Only time will tell whether this is the beginning of an ominous trend; the COVID-19 pandemic heightens this concern. Both aid and government expenditure on agriculture have fallen, which seems short sighted in the face of adverse trends in nutrition and high food prices in many African countries. Overall, these trends indicate large numbers of people at high risk of adverse health and development outcomes which can be exacerbated by climate and other environmental changes. Resilience strategies should therefore target these populations. Recent declines in official development assistance (ODA) (by 2.7% between 2017 and 2018, the last year for which data are currently available) will make it more difficult for the poorest countries that cannot self-finance their development needs to develop rapidly enough to keep up with the pace of environmental change (32).

Adaptation and Resilience in Practice: Five Examples

Health sector: Increasing resilience to protect health necessitates both strengthening health systems and addressing the social determinants of health. A full discussion of how to achieve universal health coverage (see also Chapter 7 on the Sustainable Development Goals) is beyond the scope of this book, but in brief it will require addressing all the essential building blocks of health systems (**Figure 5.1**). In practice, the building blocks are not discrete independent entities but rather interdependent domains making up a more or less integrated system which differs in its performance and coherence depending on the national context. Global environmental changes add an extra dimension to the challenge of providing affordable and effective care to a growing world population, namely an increased ability to withstand sudden shocks and environmental stresses. Strategies to increase resilience seek to enable health systems to rebound from these stresses and shocks such that they perform at least as well as previously, and where possible at a higher level of performance (**Figure 5.2**) (33–35). In order to address environmental change, governance

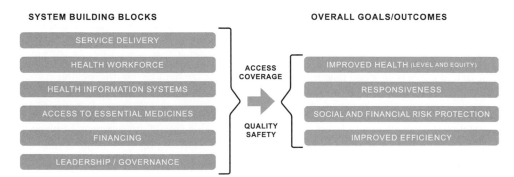

Figure 5.1. WHO health system framework.
Source: WHO. *Monitoring the Building Blocks of Health Systems: A Handbook of Indicators and Their Measurement Strategies.* Geneva: World Health Organization; 2010.

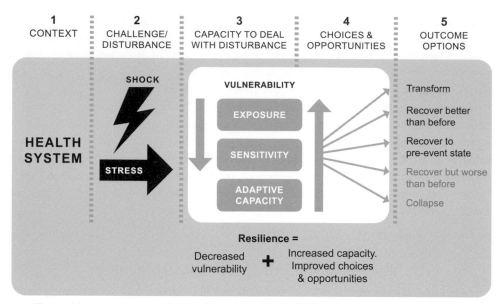

Figure 5.2. Features of resilient health systems. Under 'outcome options' to the right of the figure, the darker fonts indicate improved or neutral recovery, and the lighter fonts indicate residual negative impacts.

Source: WHO. *Operational Framework for Building Climate Resilient Health Systems.* Geneva: World Health Organization; 2015.

mechanisms need to consider specifically how shocks and environmental stress can affect the functioning of health systems and build links with sectors that influence resilience and adaptive capacity for health, such as water and sanitation, housing, food and agriculture, and energy (33). The health workforce needs to be trained to take on new roles in emergencies and to prioritize their activities in the face of changing threats. Information systems need to address: (a) provision of robust information on exposure and vulnerability to climate and other hazards as well as on the capacity of the system to respond, and identification of effective adaptation strategies; (b) the development of effective early warning systems for disease outbreaks or extreme events by integrating climate and other relevant information into disease surveillance and linking the warnings to effective responses; and (c) provision and implementation of the best available evidence from robust research to guide policy and practice (33). In this way, a resilient health information system can support effective adaptive responses.

Essential products and technologies must be harnessed to increase resilience and promote adaptation. Supply chains and stores of essential medication should, for example, be planned in the expectation that there may be fluctuations in demand and disruptions in availability after extreme events. Essential imaging and other equipment should not be sited in basements in flood-prone locations. Hospitals can be designed using passive ventilation principles, together with shading, which both reduce energy use and cool the fabric of the building, thus reducing exposure to heat stress. Training of health professionals should address the environmental determinants of ill-health, the importance of vulnerability assessment and adaptation planning, which are often neglected in both undergraduate and postgraduate education. By becoming increasingly aware of the changes in the distribution of diseases, the emergence of new and unfamiliar infections, and the vulnerability of their patients to environmental change, which has implications for both physical and mental health, health professionals can respond more effectively to the challenges of the Anthropocene. The wider role of health professionals in addressing these challenges is discussed in Chapter 11.

It is not uncommon for the electricity grid to fail during large-scale emergencies, and health facilities in particular need energy systems that allow them to function off-grid for periods of time. Many facilities in low-income settings lack reliable sources of power, which severely limits their ability to provide surgery, intensive care, imaging, and other functions that rely on uninterrupted power supply. In these conditions, renewable electricity through photovoltaics, together with storage batteries, can provide reliable electricity including for water pumps and vaccine chains; solar thermal panels can provide a reliable supply of hot water, and solar lamps can provide light for essential procedures. These technologies can enhance resilience by ensuring independent water and energy supplies even to remote rural facilities, and in emergencies, as well as reducing the emissions of climate-altering pollutants from traditional fossil-fuelled power generation.

Extreme heat: Adaptation to extreme heat occurs on many levels. Physiological adaptation occurs, but only to a point, beyond which people cannot continue to adapt (36). Community adaptation includes early warning systems that enable authorities to advise people of impending extreme heat, public education, mapping to identify vulnerable neighbourhoods, 'buddy' systems for looking in on vulnerable people such as elders living alone, cooling centres, provision of water, rescheduling or cancelling of outdoor activities during hot times of the day, extended hours in public pools, and support for utility bill payments so people are not inhibited from using their air conditioning (37–40). Longer-term adaptations include the replacement of dark urban surfaces (rooftops and pavements) with lighter colours to reduce the heat island effect and planting of trees for shade. As described below, many of these adaptive strategies have been evaluated.

Passive cooling avoids increased energy use and can help reduce urban heat exposure. A study of the potential impact of cool surfaces on heat-related death rates in the District of Columbia, USA, suggested that a 10% increase in urban surface reflectivity could reduce the number of deaths during heat events by an average of 6%. This could be achieved by increasing vegetation and by using reflective surfaces for roofs and pavements. Converting dark grey roofs to white roofs on approximately one-quarter of the District's buildings would result in an increase in District-wide roof reflectivity of 10% (41).

The use of external shutters to reduce incident sunlight may be a highly efficient and cost-effective means of reducing heat-related deaths, particularly when combined with other energy efficiency measures (42). Highly effective insulation together with reduced air flow to increase energy efficiency could potentially increase heat exposure, although in a temperate climate such as in the UK any modest increase in heat-related mortality would likely be outweighed by reduced cold-related mortality (42).

Green space in urban areas may reduce the risk of heat-related deaths by about 6% (43) by cooling the local environment, although the magnitude of the effect is likely to vary by location and site characteristics (see also Chapter 9) (44, 45).

Severe storms: Countries such as Bangladesh have shown remarkable progress in reducing death rates from tropical cyclones over recent decades as a result of improved early warning systems and storm shelters using integrated policies across ministries and effective community engagement, including by effective NGOs. Two severe cyclones struck Bangladesh in 1970 and 1991, causing approximately 500,000 and 140,000 deaths, respectively (46). In comparison, cyclone Sidr in 2007 was responsible for about 4200 deaths and cyclone Mora caused only six deaths in 2018 following the successful evacuation of 1 million people from high-risk areas. The key explanations for the successful adaptation include enshrining the legal framework for disaster management and preparedness in national legislation, combined with effective implementation at all levels, from national to village level, with the involvement of 2000 village disaster committees (9). The Prime Minister oversees the National Disaster Management Council, which has responsibility for training, management, coordination, and implementation of relevant activities. The system involves 50,000 volunteers and provides 2000 cyclone shelters. Community organizations empower women to participate in disaster-related activities. Another key feature is the restoration of 1200 km^2 of mangrove forests which provide protection against storm surges.

Floods: The Netherlands offers a powerful lesson in how development can reduce the risks of extreme events such as floods. In the thirteenth and fourteenth centuries, major storm surges breached sea dikes causing more than 200,000 deaths, which amounted to a lifetime risk of death from floods exceeding 5%. Growing wealth and improved technology, including the increasing use of windmills and the wind-powered Archimedes screw, made it increasingly possible to pump water and reclaim land. Since the last major flood in 1953, the lifetime risks have been very low, at 0.01%, 500 times lower than in the Middle Ages (28).

Examples such as these illustrate why it is not surprising that global death rates from climate-related disasters are falling. However, this is not grounds for complacency, as the absolute numbers affected and insurance losses are rising. In 2018, a total of 831 extreme weather events resulted in US$166 billion in economic losses, around half the value experienced in 2017, but still higher than any other year since 2005 (47).

The role of early warning systems: There is particular interest in the use of health early warning systems, not only for heat waves and other extreme events but also for infectious disease outbreaks. Early warning systems may be valuable for climate-sensitive diseases where they have epidemic potential or, in the case of heat and other hazardous exposures, where sudden changes in exposure greatly increase risk. They may also be valuable when climatic and other factors affect nutrition; for example, following widespread famine in Africa in the early 1980s, considerable research and development effort was focused on the development of famine early warning systems.

The framework outlined in **Figure 5.3** can provide a structure for the assessment of the suitability for and design of an early warning system (EWS) for a climate-sensitive health outcome (48). A range of diseases and health outcomes have been identified as potentially

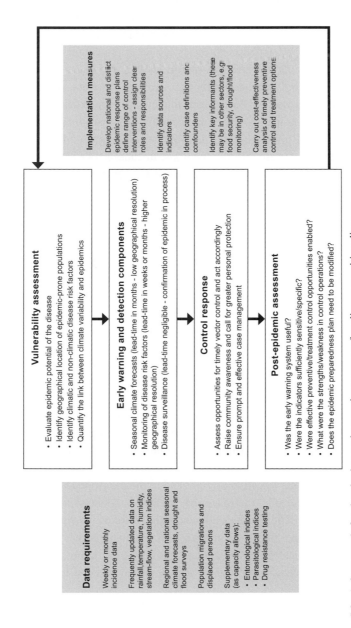

Figure 5.3. A framework for developing early warning systems for climate-sensitive diseases.

Source: Kuhn K, Campbell-Lendrum D, Haines A, Cox J. *Using Climate to Predict Infectious Disease Epidemics*. Geneva: World Health Organization; 2005.

suitable for an EWS, including dengue, malaria, meningococcal meningitis, cholera, and extreme heat-related deaths.

Overall, the evidence suggests that heat EWSs probably have beneficial effects (49, 50) although the methods used to evaluate their impact were relatively crude and the threshold for declaring an alert varies substantially. There is a need for iterative approaches because risks may vary over time, necessitating periodic review (51).

Using seasonal climate and El Niño forecasts allows a prediction to be made at the beginning of the year for the full dengue season in Ecuador (52). Combining active surveillance data with routine dengue reports improves the accuracy of estimates based on historical seasonal averages. Statistical models that quantify the extent to which environmental and socioeconomic indicators can explain variations in disease risk have been used, for example, to provide real-time, probabilistic estimates of dengue outbreak risk ahead of a mass gathering in Brazil, for all 550 microregions, in June 2014 (53). The same model was used successfully to predict dengue outbreak risk after a major El Niño event in southern coastal Ecuador (54).

Using seasonal forecasts from multi-model climate predictions, it was possible to provide an additional four months warning of impending malaria risk in Botswana, where the links between climate variability and malaria incidence are well established (13). In Ethiopia, predictions based on weather variables identified epidemics with reasonable accuracy and provided more time for public health responses than early detection systems (55). Early warning systems informed by such studies are likely to be particularly valuable in locations where malaria transmission is unstable and epidemics can occur.

Epidemics of meningococcal meningitis, which ravage the 'meningitis belt' of West Africa, including Burkina Faso, Mali, Togo, and Niger, are related to rainfall, wind, and dust in the pre-, post-, and epidemic season, with the strongest relationships in savannah areas (56, 57). These relationships can be used to provide early warning to increase the lead time for disease control strategies.

Sea surface temperature estimated by remote sensing has been used to assess the risks of cholera transmission in the Bay of Bengal (58). Current efforts are linking data from NASA satellites and other sources on sea surface temperature and height, river discharge, land use and vegetation cover, soil moisture, and water storage to predict cholera in Bangladesh, which has two peaks each year linked to drought and to the monsoon (58). Drought causes saltwater intrusion accompanied by increased risk of cholera transmission and the monsoon brings heavy rains which destroy water and sanitation facilities, spreading contaminated water across large areas. The plan is to inform local populations, who even in informal settlements have high mobile phone ownership, and who have expressed willingness to change their water procurement behaviours when supported by robust information about risks (59).

Encouragingly, by 2016, 61% of the population in the LDCs and 84% globally were covered by a third generation (3G) mobile broadband network (9), which could help adaptation efforts directly or indirectly by allowing warning of extreme events, transfer of funds within and between families, and giving market intelligence to subsistence farmers, allowing them to benefit from selling their crops at higher prices.

Whilst early warning systems show promise, they will only be effective where they support behaviour change through health services and direct communication with the public. Countries with the greatest needs are therefore likely to be those where implementation is most challenging.

Nature-Based Adaptation

There is growing interest in adaptation solutions that are based on protecting natural systems, building on their intrinsic resilience to environmental changes and avoiding some of the adverse effects that are sometimes associated with 'hard' engineered solutions. In some situations, such as landscape restoration to combat desertification and soil erosion, nature-based solutions (NBS) offer the only feasible way forward. In other cases, engineered approaches, such as the provision of clean piped water and sanitation systems, may be essential, but increasingly the two approaches must work together, achieving synergies where possible.

Globally up to 40% of rainfall results from evaporation over land, with the remainder from oceans (60). Thus, vegetation makes crucial contributions to rainfall, in many cases contributing to water recycling. At a local scale, land use management has a major effect on soil moisture, including crop variety, spacing, and rotation, as well as tillage practices. The NBS have particular promise in rain-fed agriculture, which is responsible for the bulk of food production including through family farms. Techniques such as no-till agriculture can protect soil integrity and reduce water loss from soils (60). Natural grasslands produce high-quality water, free from the overloading of nitrogen and phosphorus that often characterizes run-off from agricultural land in high-income settings (60) (**Figure 5.4**).

China is developing 16 pilot 'sponge' cities by 2020 across 450 km^2 with the intention of relieving waterlogging, reducing freshwater requirements, promoting sustainable industrial development, and increasing public satisfaction (61). Examples of measures include the use of green roofs, walls, and permeable pavements, as well as the restoration of degraded lakes and wetlands, which will then be better able to absorb excessive rainwater. Raingardens and bio-retention swales are used to collect run-off and remove selected pollutants. Some of this water can be stored in natural systems to provide a supply of freshwater for essential purposes in periods of drought. The cities aim to cope with extreme rainfall events and should be able to absorb rainwater from a 1 in 30-year heavy rainfall event – a great improvement on current urban standards in China (61).

The Royal Society reviewed the evidence for ecosystem-based, hard and hybrid approaches to natural hazard reduction (62). They noted how ecosystem approaches are often advocated because they are no- or low-regret options that provide a range of benefits, and they assessed different options designed to reduce the impact of four types of extreme event. It is clear that ecosystem or hybrid options should be given serious consideration in some instances and that these can support the achievement of other goals such as the Sustainable Development Goals (see Chapter 7) and the Aichi Biodiversity targets (see Box 7.1).

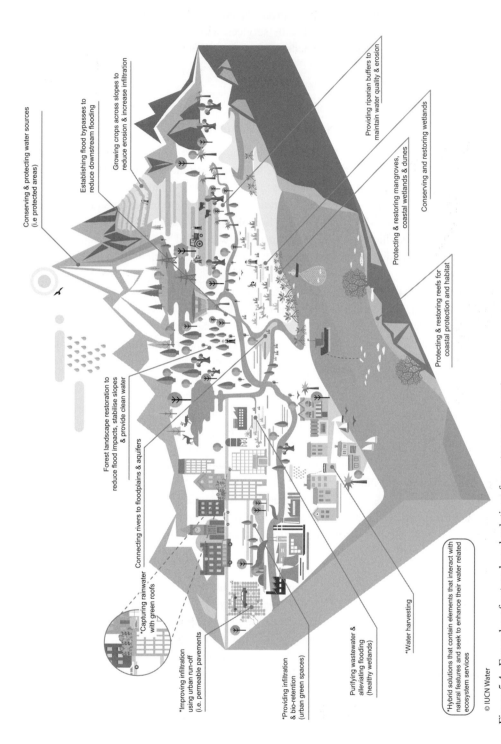

Conserving & protecting water sources
(i.e protected areas)

Establishing flood bypasses to
reduce downstream flooding

Growing crops across slopes to
reduce erosion & increase infiltration

Providing riparian buffers to
maintain water quality & erosion

Conserving and restoring wetlands

Protecting & restoring mangroves,
coastal wetlands & dunes

Protecting & restoring reefs for
coastal protection and habitat

Forest landscape restoration to
reduce flood impacts, stabilise slopes
& provide clean water

Connecting rivers to floodplains & aquifers

*Capturing rainwater
with green roofs

*Improving infiltration
using urban run-off
(i.e. permeable pavements)

*Providing infiltration
& bio-retention
(urban green spaces)

Purifying wastewater &
alleviating flooding
(healthy wetlands)

*Water harvesting

*Hybrid solutions that contain elements that interact with
natural features and seek to enhance their water related
ecosystem services

© IUCN Water

Figure 5.4. Examples of nature-based solutions for water management.

Source: Cohen-Shacham E, Janzen C, Maginnis S, Walters G. *Nature-Based Solutions to Address Global Societal Challenges*. Gland, Switzerland: International Union for Conservation of Nature and Natural Resources (IUCN); 2016.

166

Open Air Laboratories are an evolving concept for the co-design of potential NBS, integrating the shared knowledge and skills of a range of stakeholders, researchers, and end-users to develop, deploy, and assess the effectiveness of NBS to address evolving hydro-meteorological (floods, droughts) risks (63).

Such approaches can also be consistent with the use of 'traditional ecological knowledge' (TEK) – insights held by communities, particularly those from indigenous groups, based on knowledge communicated over generations, sometimes through oral traditions. Use of TEK may point to solutions that work well in both ecological and social terms (64, 65), but it is not always tapped in designing and implementing adaptation and resilience strategies, sometimes as a consequence of marginalization of indigenous communities. Three links have been identified between TEK and resilience (66). First, the knowledge itself can be resilient and dynamic, adapting to new realities, providing that communities have the capacity to use and manage that knowledge. Second, such knowledge may make important contributions to sustaining biodiversity and ecosystem services, as it has evolved over long periods of observation and co-existence with ecosystems. Third, traditional knowledge gives useful insights into the perceptions of key stressors affecting local communities, which may differ from the views of local elites or government institutions. In addition, the development of effective adaptation and resilience strategies may be an opportunity to strengthen community integrity and build on traditional knowledge to co-design solutions that minimize the potential to undermine existing cultures and livelihoods (see also Chapter 12).

Progress in Adaptation and Resilience

Like clinical interventions, adaptation measures need to be safe and effective. Safe, in this context, means free of serious adverse consequences. Effective means that the adaptation efforts result in greater resilience, and in reduced harm to people. In evaluating adaptation efforts against these criteria, several questions arise. Are adaptation efforts being pursued at sufficient scale and with sufficient funding? Do the adaptation efforts demonstrably build resilience? Do they demonstrably protect people's health? Are there unintended consequences and, if so, do the positive outcomes outweigh any negative outcomes? These questions can be answered through programme evaluation and effectiveness research.

General trends in resilience: Resilience increases with general improvements in health and security. There is no doubt that such improvements over recent decades have considerably boosted resilience to environmental change. In some cases, this is due to improvements in universal health coverage, housing, nutrition, and water and sanitation, which are being made independently of concerns about climate and other environmental changes. However, progress is uneven and insufficiently rapid to protect against increasing impacts in the future. For example, in low-income countries billions of people still lack access to safely managed sanitation and improvements are occurring at about 2.8% a year, so it will take many years for them to catch up with middle- and high-income countries. Although access to affordable health care is growing in low-income countries, the number of trained doctors and nurses is stagnating in many of them and progress is highly inequitable. Progress in some key health indicators has been impressive at a global level. For example, the under-5 mortality rate declined by 49%, and the neonatal mortality rate fell by 49%, from 2000 to 2017 (32). Official development assistance (ODA) from all donors for basic health care increased by 47% between 2008 and 2016, in real terms amounting to US$10.1 billion (67). More equitable and far-reaching development, with attention to the most marginalized populations, would do much to build resilience globally but may be under increasing threat from the post-COVID-19 recession.

More specific information about progress in adaptation comes from systematic efforts to assess adaptation over time and between and across jurisdictions, populations, and sectors, known as adaptation tracking (68). Article 7, paragraph 14 of the UNFCCC Paris Agreement commits signatories to a five-yearly assessment of observed adaptation to track progress and enable appropriate future commitments through the Nationally Determined Contributions (NDCs) and National Adaptation Plans. This is part of the larger monitoring, reporting, and evaluation (MRE) function of climate adaptation. Scholars have recommended systematic approaches to this tracking, including standardizing the objectives and categories of adaptation, the evidence sources to be used, and the data collection approach (69), and calling for consistency, comparability, comprehensiveness, and coherency in reporting (70).

Tracking of adaptation at the national level is reported by the United Nations Environment Programme in its bi-annual Adaptation Gap Report. The most recent UNEP report (9) shows mixed results. Only 40 developing countries had quantifiable adaptation targets in their current NDCs, while 49 included quantifiable targets in national laws and policies. The report focused specifically on the adaptation gap in health, defined as 'the difference between the climate-related health outcomes under actual adaptation efforts and the climate-related health outcomes that would occur under desirable levels of health adaptation efforts', and found that adaptation efforts fell far short of what would be required to avert substantial preventable morbidity and mortality from heat, severe weather, infectious diseases, and undernutrition. The report noted that international climate finance for health had been 'negligible'.

Adaptation can also be evaluated at the level of individual efforts, such as planting of novel crops or urban heat wave preparedness plans. Ideally such evaluation is comprehensive, combining the assessment of mitigation and adaptation initiatives (71), and considering a range of potential benefits including social and ecosystem benefits (72).

Adaptation to extreme heat is the most thoroughly evaluated of adaptations to global environmental change. This includes studies of changing vulnerability over time, and studies of the efficacy of specific interventions in reducing heat-related mortality.

There is limited evidence about changes in vulnerability to climate and other environmental changes. A systematic review of studies assessing changes in heat- and cold-related death rates over time found eleven papers that quantified changes in the relative risk of death, or the absolute heat-related mortality (73). Of these, ten found a decrease in susceptibility over time which was statistically significant in five studies. All of the studies were in high-income countries and the magnitude of the decrease varied by location. It was difficult to distinguish the effects of gradual acclimatization and specific adaptation measures. The changing use of air conditioning is one potential explanation for decreased susceptibility to heat, but in two studies that attempted to address the issue no significant association with air conditioning was found within or between cities over time. Another possible explanation is that heat-related deaths were relatively more likely to result from infectious causes such as diarrhoea in the early twentieth century, but improved sanitation and health care have reduced such deaths. However, this is unlikely to be the full explanation, as some studies found declines over more recent periods. One study examining trends for 105 US cities (covering 106 million people) during the summers between 1987 and 2005, using daily weather, air pollution, and age-stratified mortality rates, found a greater than 50% reduction in deaths attributable to heat (74). The decreases were greatest in those aged 75 and over, in cooler and more northern cities. The decline in heat-related deaths probably reflects the decline in cardiovascular disease death rates over the period of study (a result of hypertension treatment, cholesterol-lowering medications, reduced smoking, and advances in medical care), as cardiovascular disease may increase susceptibility to heat (74).

A lack of the protective effect of air conditioning in some studies may be because people in poverty lack air conditioning, and/or avoid utilizing it during periods of intense heat in an effort to hold down their electricity bills. Beneficial effects of air conditioning seem to depend on how it is provided, with central air conditioning probably providing greater benefit than air conditioning of individual rooms. In the USA, the increased risk of extreme heat exposure for black people compared with their white counterparts was partly related to reduced access to central air conditioning but not room air conditioning (75). Waste heat from air conditioning exacerbates the urban heat island effect, increasing outdoor temperature by more than 1 °C in some locations, particularly at night, thus increasing energy demands and greenhouse gas emissions (76).

Researchers have assessed the effects of interventions such as heat wave early warnings and cooling centres. Such assessment is hampered by weak study design and variation in the perception of what constitutes adequate evidence of impact (77). Simple comparisons of death rates in years before and after the introduction of heat warning systems do suggest benefits. For example, following the European heat wave of 2003, French authorities put in place a heat plan to reduce death rates, and following a period of high temperatures in 2006, 2065 excess deaths occurred, compared with about 6500 excess deaths predicted, suggesting that about 4400 deaths may have been prevented (78). A similar finding emerged from South Asia's first heat action plan, in Ahmedabad (79). Although it is difficult to account fully for other possible contributing factors in studies of this kind, the large effect size suggests that there is substantial impact. However, a cautionary note may be in order because there is much variability in heat-related deaths from year to year, with a high-mortality year sometimes followed by a low-mortality year (80), perhaps because many particularly susceptible people have already died. An additional challenge is that most deaths attributable to heat occur outside 'heat waves' which by definition reflect extreme but relatively rare events. The implication of this is that to reduce heat-related deaths strategies should aim to reduce risks throughout the summer season. This will require close involvement of primary care services because they are best placed to identify, monitor, and support those in the population who are at highest risk of heat-related death as a result of their age or co-morbidity (81).

Only 47 countries had heat-health action plans in place according to a survey in mid-2017, the majority of these in the WHO European Region and none in Africa (9). There are no reliable global-level data on preparedness for other extreme events, although the situation should improve under reporting provisions for the Sendai Framework and the Sustainable Development Goals. The number of countries providing climate services to the health sector increased from 55 in 2018 to 70 in 2019. These services, usually provided by national meteorological services, could be used to give early warning of heat waves and other extreme events; however, utilization of these services remains low. Only 4 of the 47 member states responding to a survey reported that the services were integrated into policymaking (47).

The health effects of other forms of adaptation have also been studied, but to a limited extent. For example, a survey of 300 households in rural Eswatini (formerly Swaziland) assessed a bundle of household-level adaptations to drought, including planting drought/heat-resistant crops, conservation farming methods to minimize soil erosion, keeping bees to sell honey, chicken husbandry, selling natural resources, selling handicrafts, pursuing off-farm work, and participating in training and capacity building provided by aid organizations. The outcome of interest was child nutrition. Adaptation was associated with better child nutrition, with the association especially strong among more vulnerable households (82). Rigorous assessment of the health impacts of adaptation will be important in guiding strategies from the national to the household level.

Trade-offs and Limits to Adaptation

Unintended adverse consequences can result from well-intentioned plans (83). Adaptation strategies need careful design to minimize the prospects of such outcomes. Maladaptation has been defined as adaptation efforts that actually increase vulnerability. Barnett and O'Neill identify five dimensions of maladaptation: actions that increase emissions of greenhouse gases or other adverse environmental impacts; disproportionately burden the most vulnerable; have high opportunity costs; reduce incentives to adapt; and/or set paths that limit future choice (84). The example of air conditioning given above shows how a technical solution may provide benefits but at a price, not just economic costs and greenhouse gas emissions, but also by displacing the excess heat to the external environment. With current technology it may be difficult to avoid trade-offs entirely but they can be minimized, e.g. by passive cooling.

Desalinated water production is an example of an adaptation technology that may have diverse adverse impacts. Its uptake is driven by declining freshwater supplies in many parts of the world, in some cases exacerbated by climate change. Currently about 16,000 operational desalination plants produce around 95 million cubic metres/day of desalinated water for human use, largely in high- or middle-income countries because of the high costs (85). The most commonly used technology is reverse osmosis, which produces about 70% of the total desalinated water (85). This is an energy-intensive process, and when fossil fuels are used as the energy source, the carbon footprint may be large (86). In addition, desalination leads to the formation of over 140 million m^3/day of brine, over half of which is produced by Saudi Arabia, UAE, Kuwait, and Qatar, which employ thermal desalination processes with low recovery ratios (85). The adverse impacts of desalination include the risk to marine life and ecosystems more generally from the discharge of untreated brine, including in some cases toxic chemicals used for descaling and anti-fouling purposes. Increasing the recovery rate of freshwater through desalination reduces the volume of brine for disposal but increases energy requirements. Costs may fall disproportionately on deprived communities and in some cases large sunk costs may preclude other options. Research is currently exploring the use of brine for growing halophytes (salt-tolerant plants) and fish, as well as the prospects of recovering metals and commercial salt.

Another example of maladaptation is provided by some wetland protection policies that capitalize on the potential to provide a buffer against flooding as well as protecting biodiversity, but can also provide a suitable habitat for insect vectors of disease. Dams and irrigation programmes provide water needed for agriculture, but can also result in increased habitats for malaria-transmitting mosquitoes if this potential hazard is not addressed. Potential solutions include building dams in non-malarial areas, optimizing water management by varying water levels to desiccate mosquito larvae, and situation of populations outside a buffer zone to prevent malaria transmission (87).

Potential adaptation policies and technologies should be screened to assess the potential for maladaptation in order to reduce the risks of unintended adverse consequences.

There are limits to adaptation. Although humanity has adapted to environmental change throughout its existence, the current pace of change and the potential for non-linear state shifts in critical natural systems will make adaptation increasingly difficult if current unsustainable trajectories continue. The limits to adaptation are likely to be defined by location (for example small low-lying island populations have little option to retreat and may be compelled to migrate). Countries in which physical labour becomes increasingly difficult because of heat stress will face difficult choices in attempting to provide livelihoods for large populations without the training or adequate infrastructure to do so. Technological limits may also be reached in some locations – for example heavily populated deltas in low-income countries may be unable to provide robust sea defences.

Financing Adaptation

Estimates of the costs of adapting to climate change are subject to wide uncertainties, ranging from US$140 billion to US$300 billion annually by 2030 and from US$280 billion to US$500 billion by 2050 according to the UN Adaptation Gap Report (9). These may be underestimates given that the added costs of making infrastructure resilient to future changes are estimated to be between 0.5% and 10% of the total project investment cost. Global investment in new infrastructure through 2030 could range from US$57 trillion to US$95 trillion (9).

Another consideration in trying to assess adaptation costs is that losses from extreme events were the highest on record in 2017, estimated at US$330 billion, of which US$136 billion were insured losses (9). The science of attribution of climate events to climate change is advancing rapidly, so future estimates may be able to provide more robust indications of the true cost.

The importance of adequate funding for health systems in preventing disaster-related deaths is supported by the clear relationship between deaths per capita from climate-related disasters and per capita health national spending (47). The advent of dedicated climate funds provides an additional source of funds for health systems, although it is currently underutilized. These funds include the Global Environment Facility (through the Least Developed Countries Fund and the Special Climate Change Fund), the Green Climate Fund, the Kyoto Protocol Adaptation Fund, and bilateral funds from donor countries. Nevertheless, currently a small proportion of successful bids address health directly or indirectly. According to a recent estimate, spending on adaptation for health represents only about 5% of all global adaptation spending (47). Although adaptation funding in health and health care rose by 11.2% in 2017–18 over the previous year, outpacing the overall rise in adaptation spending, per capita health adaptation spending remained below £1 in the African, Eastern Mediterranean, and Southeast Asian regions (47). Adaptation funding constitutes a potentially important source of support for low-income countries which lack internal resources from taxes and other income streams to finance universal health coverage in general, or adaptation specifically, from their own funds, but it is currently inadequate.

The projected sum of US$100 billion annually by 2020, agreed under the Paris Climate Agreement for mitigation and adaptation (88) is far from being reached. Assessing how much of the projected sum has actually been committed is complex and beset by data uncertainties and gaps. International public finance flows amounted to about US$58 billion in 2015–16 with between 21% and 29% of the total devoted to adaptation, depending on the source of funding (88). Most of the adaptation funding was in the form of grants and most of the mitigation funding was in the form of concessional loans, but there is an increasing trend to fund both adaptation and mitigation, which although a welcome development, from the perspective of integrated planning, makes it more difficult to track trends for the two categories separately.

Conclusions

Whilst climate change adaptation funds and policies drive much of the activity around adaptation and resilience, other considerations frequently influence the nature and effectiveness of such actions. There is a danger that the most marginalized and threatened communities will benefit least from adaptation and resilience activities in the absence of strong commitments, currently often lacking, to reduce inequities. Building general resilience through protecting livelihoods and strengthening water, sanitation, health, and food systems will have near-term benefits as well as improving the prospects for vulnerable populations. Nature-based solutions and hybrid approaches with conventional 'hard' solutions hold promise in many circumstances. Enhancing security of land tenure for low-income communities and protecting the rights of indigenous communities to manage natural resources can contribute to effective adaptation and reduce environmental degradation.

Failure to remain within planetary boundaries will make successful adaptation more difficult because limits to adaptation – physical, socioeconomic, technological, and behavioural – will be reached in an increasing number of locations. For this reason it is imperative to develop and rapidly implement policies that address the driving forces that are taking us beyond planetary boundaries and achieve equitable development at greatly reduced levels of environmental impact. In Chapter 6 we consider the barriers to developing and articulating this way forward in the face of formidable challenges posed by the magnitude of change, powerful vested interests and the relentless pursuit of weakly regulated consumption, using technologies rendered obsolete by the reality of the Anthropocene Epoch.

References

1. Field CB, Barros VR, Dokken DJ, et al., editors. *Climate Change 2014. Impacts, Adaptation, and Vulnerability. Part A: Global and Sectoral Aspects. Working Group II Contribution to the Fifth Assessment Report of the Intergovernmental Panel on Climate Change*. Cambridge, UK and New York: Cambridge University Press; 2014.
2. Berkes F. Environmental governance for the Anthropocene? Social-ecological systems, resilience, and collaborative learning. *Sustainability*. 2017;9. doi: 10.3390/su9071232.
3. Walker B, Holling CS, Carpenter SR, Kinzig A. Resilience, adaptability and transformability in social-ecological systems. *Ecology and Society*. 2004;9(2). doi: 10.5751/ES-00650-090205.
4. Folke C, Biggs R, Norström AV, Reyers B, Rockström J. Social-ecological resilience and biosphere-based sustainability science. *Ecology and Society*. 2016;21(3). doi: 10.5751/ES-08748-210341.
5. Levin S, Xepapadeas T, Crépin A-S, et al. Social-ecological systems as complex adaptive systems: modeling and policy implications. *Environment and Development Economics*. 2013;18(2):111–32. doi: 10.1017/S1355770X12000460.
6. Nelson DR, Adger WN, Brown K. Adaptation to environmental change: contributions of a resilience framework. *Annual Review of Environment and Resources*. 2007; 32(1):395–419. doi: 10.1146/annurev.energy.32.051807.090348.
7. Plummer R, Armitage D. A resilience-based framework for evaluating adaptive co-management: linking ecology, economics and society in a complex world. *Ecological Economics*. 2007;61(1):62–74. doi: 10.1016/j.ecolecon.2006.09.025.
8. Frumkin H, Hess J, Luber G, Malilay J, McGeehin M. Climate change: the public health response. *American Journal of Public Health*. 2008;98(3):435–45. doi: 10.2105/AJPH.2007.119362.
9. UNEP. Adaptation Gap Report 2018. Nairobi: United Nations Environment Programme; 2018.
10. Bossio CF, Ford J, Labbé D. Adaptive capacity in urban areas of developing countries. *Climatic Change*. 2019;157(2):279–97. doi: 10.1007/s10584-019-02534-2.
11. Romero-Lankao P, Hughes S, Qin H, et al. Scale, urban risk and adaptation capacity in neighborhoods of Latin American cities. *Habitat International*. 2014;42:224–35.
12. Ziervogel G, Cowen A, Ziniades J. Moving from adaptive to transformative capacity: building foundations for inclusive, thriving, and regenerative urban settlements. *Sustainability*. 2016;8(9). doi: 10.3390/su8090955.
13. Thomson MC, Doblas-Reyes FJ, Mason SJ, et al. Malaria early warnings based on seasonal climate forecasts from multi-model ensembles. *Nature*. 2006;439(7076):576–9. doi: 10.1038/nature04503.

14. Sillmann J, Thorarinsdottir T, Keenlyside N, et al. Understanding, modeling and predicting weather and climate extremes: challenges and opportunities. *Weather and Climate Extremes*. 2017;18:65–74. doi: 10.1016/j.wace.2017.10.003.
15. Gay P-E, Lecoq M, Piou C. Improving preventive locust management: insights from a multi-agent model. *Pest Management Science*. 2018;74(1):46–58. doi: 10.1002/ps.4648.
16. Thomson MC, Mason SJ, editors. *Climate Information for Public Health Action*. London: Routledge; 2018.
17. Guido Z, Knudson C, Campbell D, Tomlinson J. Climate information services for adaptation: what does it mean to know the context? *Climate and Development*. 2019: 1–13. doi: 10.1080/17565529.2019.1630352.
18. Eriksen SH, Nightingale AJ, Eakin H. Reframing adaptation: the political nature of climate change adaptation. *Global Environmental Change*. 2015;35:523–33. doi: 10.1016/j.gloenvcha.2015.09.014.
19. Hess JJ, McDowell JZ, Luber G. Integrating climate change adaptation into public health practice: using adaptive management to increase adaptive capacity and build resilience. *Environmental Health Perspectives*. 2012;120(2):171–9. doi: 10.1289/ehp.1103515.
20. Ebi K. Climate change and health risks: assessing and responding to them through 'adaptive management'. *Health Affairs*. 2011;30(5):924–30. doi: 10.1377/hlthaff.2011.0071.
21. Holling CC. *Adaptive Environmental Assessment and Management*. New York: Wiley; 1978.
22. Few R, Morchain D, Spear D, Mensah A, Bendapudi R. Transformation, adaptation and development: relating concepts to practice. *Palgrave Communications*. 2017;3(1): 17092. doi: 10.1057/palcomms.2017.92.
23. O'Brien K. Global environmental change II: from adaptation to deliberate transformation. *Progress in Human Geography*. 2011;36(5):667–76. doi: 10.1177/0309132511425767.
24. Kates RW, Travis WR, Wilbanks TJ. Transformational adaptation when incremental adaptations to climate change are insufficient. *Proceedings of the National Academy of Sciences*. 2012;109(19):7156–61. doi: 10.1073/pnas.1115521109.
25. Reij C, Tappan G, Smale M. *Agroenvironmental Transformation in the Sahel*. Washington, DC: IFPRI (International Food Policy Research Institute); 2009.
26. Dannenberg AL, Frumkin H, Hess JJ, Ebi KL. Managed retreat as a strategy for climate change adaptation in small communities: public health implications. *Climatic Change*. 2019;153(1):1–14. doi: 10.1007/s10584-019-02382-0.
27. IPCC. *Climate Change and Land: An IPCC Special Report on Climate Change, Desertification, Land Degradation, Sustainable Land Management, Food Security, and Greenhouse Gas Fluxes in Terrestrial Ecosystems*. 2019. Available from www.ipcc.ch/srccl/.
28. Field CB, Barros V, Stocker TF, et al. *Managing the Risks of Extreme Events and Disasters to Advance Climate Change Adaptation*. Cambridge, UK: IPCC; 2012.
29. Walsh B. Adapt or die: why the environmental buzzword of 2013 will be resilience. *Time*. 8 January 2013.
30. Brown K. *Resilience, Development and Global Change*. Abingdon, UK and New York: Routledge; 2015.
31. Fankhauser S, McDermott TKJ, editors. *The Economics of Climate-Resilient Development*. Cheltenham, UK and Northampton MA, USA: Edward Elgar; 2016.
32. UN DESA. *The Sustainable Development Goals Report 2019*. New York: United Nations; 2019.
33. WHO. *Operational Framework for Building Climate Resilient Health Systems*. Geneva: World Health Organization; 2015.
34. Bouley T, Roschnik S, Karliner J, et al. *Climate-Smart Healthcare: Low-Carbon and Resilience Strategies for the Health Sector*. Washington, DC: World Bank; 2017.

35. Paterson J, Berry P, Ebi K, Varangu L. Health care facilities resilient to climate change impacts. *International Journal of Environmental Research and Public Health*. 2014; 11(12):13097–116. doi: 10.3390/ijerph111213097.
36. Taylor NA. Human heat adaptation. *Comprehensive Physiology*. 2014;4(1):325–65.
37. Casanueva A, Burgstall A, Kotlarski S, et al. Overview of existing heat-health warning systems in Europe. *International Journal of Environmental Research and Public Health*. 2019;16(15):2657. doi: 10.3390/ijerph16152657.
38. Fernandez Milan B, Creutzig F. Reducing urban heat wave risk in the 21st century. *Current Opinion in Environmental Sustainability*. 2015;14:221–31. doi: 10.1016/j.cosust.2015.08.002.
39. Stone B, Jr, Vargo J, Liu P, et al. Avoided heat-related mortality through climate adaptation strategies in three US cities. *PLoS One*. 2014;9(6):e100852. doi: 10.1371/journal.pone.0100852.
40. Vu A, Rutherford S, Phung D. Heat health prevention measures and adaptation in older populations: a systematic review. *International Journal of Environmental Research and Public Health*. 2019;16(22):4370. doi: 10.3390/ijerph16224370.
41. Kalkstein LS, Sailor D, Shickman K, Sheridan S, Vanos J. Assessing the Health Impacts of Urban Heat Island Reduction Strategies in the District of Columbia. Washington, DC: Global Cool Cities Alliance; 2013.
42. Taylor J, Wilkinson P, Picetti R, et al. Comparison of built environment adaptations to heat exposure and mortality during hot weather, West Midlands region, UK. *Environment International*. 2018;111:287–94. doi: 10.1016/j.envint.2017.11.005.
43. Schinasi LH, Benmarhnia T, De Roos AJ. Modification of the association between high ambient temperature and health by urban microclimate indicators: a systematic review and meta-analysis. *Environmental Research*. 2018;161:168–80. doi: 10.1016/j.envres.2017.11.004.
44. Harlan SL, Declet-Barreto JH, Stefanov WL, Petitti DB. Neighborhood effects on heat deaths: social and environmental predictors of vulnerability in Maricopa County, Arizona. *Environmental Health Perspectives*. 2013;121(2):197–204. doi: 10.1289/ehp.1104625.
45. Dang TN, Van DQ, Kusaka H, Seposo XT, Honda Y. Green space and deaths attributable to the urban heat island effect in Ho Chi Minh City. *American Journal of Public Health*. 2018;108(S2):S137–43. doi: 10.2105/ajph.2017.304123.
46. Haque U, Hashizume M, Kolivras KN, et al. Reduced death rates from cyclones in Bangladesh: what more needs to be done? *Bulletin of the World Health Organization*. 2012;90(2):150–6. doi: 10.2471/blt.11.088302.
47. Watts N, Amann M, Arnell N, et al. The 2019 report of The *Lancet* Countdown on Health and Climate Change: ensuring that the health of a child born today is not defined by a changing climate. *The Lancet*. 2019;394(10211):1836–78. doi: 10.1016/S0140-6736(19)32596-6.
48. Kuhn K, Campbell-Lendrum D, Haines A, Conx J. *Using Climate to Predict Infectious Disease Epidemics*. Geneva: World Health Organization; 2005.
49. Toloo GS, Fitzgerald G, Aitken P, Verrall K, Tong S. Are heat warning systems effective? *Environmental Health*. 2013;12:27. doi: 10.1186/1476-069x-12-27.
50. Toloo G, FitzGerald G, Aitken P, Verrall K, Tong S. Evaluating the effectiveness of heat warning systems: systematic review of epidemiological evidence. *International Journal of Public Health*. 2013;58(5):667–81. doi: 10.1007/s00038-013-0465-2.
51. Hess JJ, Ebi KL. Iterative management of heat early warning systems in a changing climate. *Annals of the New York Academy of Sciences*. 2016;1382(1):21–30. doi: 10.1111/nyas.13258.

52. Lowe R, Stewart-Ibarra AM, Petrova D, et al. Climate services for health: predicting the evolution of the 2016 dengue season in Machala, Ecuador. *The Lancet Planetary Health*. 2017;1(4):e142–51. doi: 10.1016/S2542-5196(17)30064-5.

53. Lowe R, Coelho CA, Barcellos C, et al. Evaluating probabilistic dengue risk forecasts from a prototype early warning system for Brazil. *eLife*. 2016;5. doi: 10.7554/eLife.11285.

54. Lowe R. The impact of global environmental change on vector-borne disease risk: a modelling study. *The Lancet Planetary Health*. 2018;2:S1. doi: 10.1016/S2542-5196 (18)30086-X.

55. Teklehaimanot HD, Lipsitch M, Teklehaimanot A, Schwartz J. Weather-based prediction of *Plasmodium falciparum* malaria in epidemic-prone regions of Ethiopia I. Patterns of lagged weather effects reflect biological mechanisms. *Malaria Journal*. 2004;3:41.

56. Pérez García-Pando C, Stanton MC, Diggle PJ, et al. Soil dust aerosols and wind as predictors of seasonal meningitis incidence in Niger. *Environmental Health Perspectives*. 2014;122(7):679–86. doi: 10.1289/ehp.1306640.

57. Woringer M, Martiny N, Porgho S, et al. Atmospheric dust, early cases, and localized meningitis epidemics in the African meningitis belt: an analysis using high spatial resolution data. *Environmental Health Perspectives*. 2018;126(9):97002.

58. Lobitz B, Beck L, Huq A, et al. Climate and infectious disease: use of remote sensing for detection of *Vibrio cholerae* by indirect measurement. *Proceedings of the National Academy of Sciences*. 2000;97(4):1438–43. doi: 10.1073/pnas.97.4.1438.

59. Akanda AS, Aziz S, Jutla A, et al. Satellites and cell phones form a cholera early-warning system. *Eos*. 2018;99. Available from https://eos.org/science-updates/satellites-and-cell-phones-form-a-cholera-early-warning-system.

60. UNESCO WWAP. *Nature-Based Solutions for Water*. Paris: UNESCO World Water Assessment Programme; 2018.

61. Chan FKS, Griffiths JA, Higgitt D, et al. 'Sponge City' in China: a breakthrough of planning and flood risk management in the urban context. *Land Use Policy*. 2018; 76:772–8. doi: 10.1016/j.landusepol.2018.03.005.

62. Royal Society. *Resilience to Extreme Weather*. London: Royal Society; 2014.

63. Kumar P, Debele SE, Sahani J, et al. Towards an operationalisation of nature-based solutions for natural hazards. *Science of The Total Environment*. 2020;731:138855.

64. Berkes F, Colding J, Folke C. Rediscovery of traditional ecological knowledge as adaptive management. *Ecological Applications*. 2000;10(5):1251–62.

65. Menzies CR, editor. *Traditional Ecological Knowledge and Natural Resource Management*. Lincoln, NE: University of Nebraska Press; 2006.

66. Gomez-Baggethun E, Corbera E, Reyes-Garcia V. Traditional ecological knowledge and global environmental change: research findings and policy implications. *Ecology and Society*. 2013;18(4):72. doi: 10.5751/es-06288-180472.

67. OECD. Aid to Health 2020. Available from www.oecd.org/dac/stats/aidtohealth.htm.

68. Berrang-Ford L, Biesbroek R, Ford JD, et al. Tracking global climate change adaptation among governments. *Nature Climate Change*. 2019;9(6):440–9. doi: 10.1038/s41558-019-0490-0.

69. Tompkins EL, Vincent K, Nicholls RJ, Suckall N. Documenting the state of adaptation for the global stocktake of the Paris Agreement. *WIREs Climate Change*. 2018;9(5):e545. doi: 10.1002/wcc.545.

70. Ford JD, Berrang-Ford L. The 4Cs of adaptation tracking: consistency, comparability, comprehensiveness, coherency. *Mitigation and Adaptation Strategies for Global Change*. 2016;21(6):839–59. doi: 10.1007/s11027-014-9627-7.

71. Grafakos S, Trigg K, Landauer M, Chelleri L, Dhakal S. Analytical framework to evaluate the level of integration of climate adaptation and mitigation in cities. *Climatic Change*. 2019;154(1):87–106. doi: 10.1007/s10584-019-02394-w.

72. Richerzhagen C, Rodríguez de Francisco CJ, Weinsheimer F, et al. Ecosystem-based adaptation projects, more than just adaptation: analysis of social benefits and costs in Colombia. *International Journal of Environmental Research and Public Health*. 2019;16(21). doi: 10.3390/ijerph16214248.

73. Arbuthnott K, Hajat S, Heaviside C, Vardoulakis S. Changes in population susceptibility to heat and cold over time: assessing adaptation to climate change. *Environmental Health*. 2016;15(Suppl. 1):33. doi: 10.1186/s12940-016-0102-7.

74. Bobb JF, Peng RD, Bell ML, Dominici F. Heat-related mortality and adaptation to heat in the United States. *Environmental Health Perspectives*. 2014;122(8):811–16.

75. O'Neill MS, Zanobetti A, Schwartz J. Disparities by race in heat-related mortality in four US cities: the role of air conditioning prevalence. *Journal of Urban Health: Bulletin of the New York Academy of Medicine*. 2005;82(2):191–7.

76. Salamanca F, Georgescu M, Mahalov A, Moustaoui M, Wang M. Anthropogenic heating of the urban environment due to air conditioning. *Journal of Geophysical Research: Atmospheres*. 2014;119(10):5949–65. doi: 10.1002/2013JD021225.

77. Boeckmann M, Rohn I. Is planned adaptation to heat reducing heat-related mortality and illness? A systematic review. *BMC Public Health*. 2014;14:1112. doi: 10.1186/1471-2458-14-1112.

78. Fouillet A, Rey G, Wagner V, et al. Has the impact of heat waves on mortality changed in France since the European heat wave of summer 2003? A study of the 2006 heat wave. *International Journal of Epidemiology*. 2008;37(2):309–17. doi: 10.1093/ije/dym253.

79. Hess JJ, LM S, Knowlton K, et al. Building resilience to climate change: pilot evaluation of the impact of India's first heat action plan on all-cause mortality. *Journal of Environmental and Public Health*. 2018;2018:8. doi: 10.1155/2018/7973519.

80. Guo Y, Barnett AG, Tong S. High temperatures-related elderly mortality varied greatly from year to year: important information for heat-warning systems. *Scientific Reports*. 2012;2:830. doi: 10.1038/srep00830.

81. Williams L, Erens B, Ettelt S, et al. Evaluation of the Heatwave Plan for England: Final Report. London: London School of Hygiene and Tropical Medicine, Policy Innovation and Evaluation Research Unit; 2019. PIRU Publication 2019-24.

82. Bailey MK, McCleery AR, Barnes G, McKune LS. Climate-driven adaptation, household capital, and nutritional outcomes among farmers in Eswatini. *International Journal of Environmental Research and Public Health*. 2019;16(21).

83. Eriksen S, Aldunce P, Bahinipati CS, et al. When not every response to climate change is a good one: identifying principles for sustainable adaptation. *Climate and Development*. 2011;3(1):7–20. doi: 10.3763/cdev.2010.0060.

84. Barnett J, O'Neill S. Maladaptation. *Global Environmental Change*. 2010;20(2):211–13. doi: 10.1016/j.gloenvcha.2009.11.004.

85. Jones E, Qadir M, van Vliet MTH, Smakhtin V, Kang S-M. The state of desalination and brine production: a global outlook. *Science of The Total Environment*. 2019;657:1343–56. doi: 10.1016/j.scitotenv.2018.12.076.

86. Ameen F, Stagner JA, Ting DSK. The carbon footprint and environmental impact assessment of desalination. *International Journal of Environmental Studies*. 2018;75(1):45–58. doi: 10.1080/00207233.2017.1389567.

87. Kibret S. Time to revisit how dams are affecting malaria transmission. *The Lancet Planetary Health*. 2018;2(9):e378–9. doi: 10.1016/s2542-5196(18)30184-0.

88. UNFCCC. Summary and Recommendations by the Standing Committee on Finance on the 2018 Biennial Assessment and Overview of Climate Finance Flows. Bonn: UN Climate Change Secretariat (UNFCCC); 2018.

6

Addressing Conceptual, Knowledge, and Implementation Challenges

The Anthropocene Epoch offers both unprecedented challenges and opportunities for humanity. Which of those prevails will depend on decisions made at all levels, from global to individual, over coming years. There have been many warnings of the risks arising from our current myopic development pathways but the prospect of serious, and even catastrophic, effects may not motivate action on the scale and pace required; indeed it can result in feelings of impotence and resignation.

It is an urgent priority to develop and communicate a compelling vision of a society in which people thrive – achieving health, well-being, and happiness – in the context of a sustainable, just world economy. In this world, we would live within planetary boundaries, with minimal disruption of nature's essential life support systems. In order to do so the social, economic, and environmental dimensions of sustainable development must be addressed in an integrated way, avoiding unintended consequences and making trade-offs explicit.

Rapid decarbonization of the world economy will necessitate exponential increase in the deployment of clean renewable energy sources driven by reallocation of investments such that new investments in clean energy must significantly exceed fossil fuel investments by 2025. Transport systems based on active travel, public transit, and shared 'on demand' fleets of electric vehicles will be necessary. But moving to a post-carbon economy is only one, albeit arguably the most important, transformation required to set humanity on the path to sustainable prosperity. Conserving forests, green spaces (including in cities), oceans, and freshwater sources, whilst making urban and rural settlements more resilient and more liveable places, will be critical to human well-being and the protection of natural systems. Transformation of food systems to reduce waste and increase the supply of affordable healthy, nutritious food, together with implementation of ambitious policies to reduce the environmental footprint of food will also be essential. Finally, progress towards a circular economy (including minimizing waste through re-use, re-manufacturing, recycling and promoting the shared economy) will be indispensable for success. This is because industry is responsible for about one-third of the global total annual GHG emissions, and its contribution is rising rapidly. The circular economy also offers major prospects for reducing exposure to pollution, slowing depletion of finite resources, improving health, and providing employment, if these objectives are integrated from the outset.

The Rockefeller Foundation–*Lancet* Commission on Planetary Health (1) suggested that three broad challenges needed to be addressed in order to move towards a sustainable health-promoting economy within the 'safe operating space' provided by the planetary boundaries: imagination or conceptual challenges, knowledge challenges, and implementation challenges.

Conceptual Challenges

Some of the most pervasive imagination or conceptual challenges are economic. Widely held assumptions include: that perpetual economic growth of the sort that has characterized the Great Acceleration is possible; that consumption is a fitting driver of human behaviour; and that measures of economic growth are suitable measures of human progress. Each of these assumptions is highly questionable.

Perpetual Economic Growth?

The Great Acceleration has featured unprecedented expansion of the human enterprise, including growth of the global economy. Economic growth is in some circumstances necessary – it has lifted billions of people out of poverty – but in its current form, it brings with it disruption and even destruction of planetary systems, as described in Chapter 1. This process can be quantified; nations can be ranked on their environmental impact, both relative to the resources they have available and in absolute terms. One study considered 170 countries, using data on natural forest loss, habitat conversion, marine captures, fertilizer use, water pollution, carbon emissions, and species threat (2). The top ten countries with respect to relative environmental impact were Singapore, South Korea, Qatar, Kuwait, Japan, Thailand, Bahrain, Malaysia, Philippines, and the Netherlands. These countries all have economies that exploit far more resources than they have available within their borders. The countries with the highest absolute impact were Brazil, USA, China, Indonesia, Japan, Mexico, India, Russia, Australia, and Peru. Countries perform poorly for different reasons, with varying contributions from the range of environmental impacts assessed. Statistical analysis showed that, overall, wealthier countries (measured as total purchasing power parity-adjusted Gross National Income, GNI) tended to have worse environmental impacts than poorer countries. (GNI is equal to Gross Domestic Product, or GDP, plus wages, salaries, and property income the country's residents earned abroad together with net taxes and subsidies received from abroad.) Poor governance was also associated with slightly higher impacts. The environmental Kuznets curve hypothesis proposes that as countries grow they initially cause large environmental impacts, which plateau and decline when they reach a certain level of development (3). This analysis however found no evidence that this had occurred, implying that we cannot expect environmental impacts to decline automatically as conventional development advances.

One approach that has been increasingly advocated as a path to perpetual economic growth is progressively to decouple GDP growth from the accompanying environmental impacts of that growth. Decoupling is described as either relative ('weak') or absolute ('strong'). In China relative decoupling has occurred, with GDP growing more than 20-fold between 1990 and 2012, accompanied by a four-fold increase in energy use and five-fold increase in material use (4). In comparison Germany showed slower GDP growth than did China, but a 10% decline in energy use and a 40% decline in total material use. Globally only a relative decoupling has been experienced, with GDP rising between three- and four-fold over the same 22-year period and energy and material use increasing by 54% and 66%, respectively. When decoupling does occur it is important to understand the mechanism, as some progress can be illusory. Apparent decoupling could for example occur as a result of manufacturing, together with many of its adverse environmental impacts, being exported to other nations. Growing income and wealth inequalities could contribute to decoupling when only a small proportion of the population benefits from economic growth and most of the population does not increase material consumption. Further possible explanations could be when increasing flows of financial resources are disconnected from material or energy use or when one resource is substituted for another.

The terms 'green growth' and 'sustainable economic growth' are sometimes used to denote a way forward, featuring perpetual GDP growth without adverse environmental consequences. Aside from the understandable scepticism that any system can foster growth indefinitely, a key question is whether absolute decoupling of GDP and environmental impacts is possible. If it is not then the pursuit of indefinite global growth in GDP becomes a hazardous option and will undermine future prosperity by pushing essential natural systems to the point of collapse.

Whilst it is possible to conceive of the complete decarbonization of the world economy by the phaseout of fossil fuel sources of energy, this is only one of the challenges to be addressed in the pursuit of absolute decoupling. The $I = PAT$ equation, introduced earlier at page 10 (5), is fundamental to understanding the feasibility of absolute decoupling. Recall that in this equation, I is environmental impact, P is population, A is affluence (usually expressed as GDP per capita), and T represents technology. Technology in this context refers to how resource intensive a particular path to affluence is – to the environmental footprint of particular ways of using energy and resources. Advanced, efficient technologies such as electric vehicles will have lower T values than more rudimentary technologies such as polluting diesel vehicles. Both T and I will have different units depending on which resource or pollutant is considered. For example, in the case of climate change, where $I = PAT$ is known as the Kaya identity, T corresponds to both energy intensity and carbon intensity – that is, to energy used per unit of GDP, and GHG emissions per unit of energy used, respectively. As discussed earlier, one criticism of the $I = PAT$ equation is the failure to consider interdependencies, such as the potential for increased consumption to outpace efficiency savings from improved technologies. This implies that policies directed at the right side of the equation may not necessarily reduce the impact and argues for policies that directly target impact, such as taxes to reduce environmental impacts (see also Chapter 12). Despite such criticisms we find it a useful framework for understanding sustainability.

Subsequent chapters consider how to address these three determinants of environmental impact, but in considering the conceptual challenges confronting us it is clear that technology alone will be insufficient to decouple economic growth from impact, although better technologies are undoubtedly part of the solution. Using historical data and modelled projections it has been shown that, even with optimistic assumptions, GDP growth cannot be fully decoupled from material and energy use (4). Thus, perpetual growth in GDP, even if it could be achieved, is incompatible with climate goals and environmental sustainability more generally, because absolute decoupling is unattainable. The conclusion must surely be that whilst equitable GDP growth plays an important role in poverty alleviation in low-income nations, in affluent countries there is limited potential to advance human progress through continued GDP growth. We should therefore be agnostic about GDP growth as a political priority (6) – its salience will depend on the context and trade-offs, as well as how equitably the fruits of economic growth are distributed. In many high-income countries a key political challenge is how to advance human progress with a gradually declining rate (and perhaps reversal) of GDP growth. This will be essential to sustain health, provide worthwhile employment, and tackle growing populism which increasingly provides a vehicle for the expression of tensions related to inequalities. One implication is that increasingly the A in the I = PAT equation should be replaced with a more direct measure of human progress than affluence that also avoids undue dependence on the flawed metric of GDP. In Chapter 11 we discuss the relevance of population growth to sustainability (Box 11.1) and the potential to provide universal access to modern family planning, together with improved education and empowerment of women.

Consumption: To What End?

Another imagination challenge is embedded in consumption patterns. The term conspicuous consumption, coined by economist Thorsten Veblen (7), proved an enduring descriptor of the role of consumption as a status symbol. In a free market where competitive consumption prevails, each household tries to outcompete others in the consumption of goods. This ends up with everyone working harder in order to consume more than would be necessary if the norm were to work less and consume less (8). This may explain trends observed in classic studies of Roseto, Pennsylvania, an Italian–American immigrant community with low rates of cardiovascular disease at baseline (1935). Over 50 years of follow-up, as traditional norms of moderation and social cohesion gave way to more typically American consumption patterns and the associated competitive stress, cardiovascular disease rates rose dramatically (9). Scholars such as Partha Dasgupta and Paul Ehrlich have explored the implications of this dynamic, which they call 'competitive consumption' (10). It creates its own externalities, with implications both for the planet and for people (11). Less work and less consumption may be more conducive to a healthy and fulfilling life but, according to this analysis, that option is foreclosed by the relentless social pressures for increasing status.

One example is the increasing use of the private car in many countries which depends on use of an underpriced resource, namely oil, together with a vast and growing infrastructure of roads, filling stations, garages, and parking lots. The car-dominated transport infrastructure occupies much of the limited space in cities, exposes the public to road danger, air pollution, and noise, and contributes to sedentary lifestyles (12). Clearly the use of the private car also brings benefits, particularly in rural locations where public transport is sparse. Some adverse consequences can be moderated through electrification and autonomous vehicles (see Chapter 8 for discussion) but overall the ownership of the private car has been scaled up without objective assessment of the costs and benefits in comparison with investment in affordable public transport and promotion of active travel (walking and cycling), particularly in urban settings, where an increasing proportion of the world's population lives. Ownership of cars also becomes part of the competitive economy, with the choice of model and make signalling status and promoting the sale of large, inefficient vehicles which can also be justified as providing added driver protection (against other over-sized vehicles) and thus adds further impetus to growth in the sales of such vehicles.

Another driver of consumption – and perhaps a driver of failure of imagination – is conformity. People may consume to provide a sense of belonging, to establish or reinforce their identity in a group, a 'tribe', or a certain stratum of society (13–15). Like all social constructs, this one is in theory mutable; groups or entire societies might elect to achieve conformity at different levels of depletion of natural capital. If a low level of natural capital depletion were seen as socially desirable then such conformity might help transition towards sustainability (10, 16).

Measures of Economic Growth as Measures of Human Progress

The GDP is the standard measure of economic performance and is routinely used as a measure of societal progress more generally. The GDP is a measurement of the market value of all final goods and services produced in the economy during a defined time period (excluding the value of components that go into making final products, to avoid double counting – so if the GDP includes the market value of a bicycle, it does not separately include the market value of the wheels or seat that come with the bicycle).

But the use of GDP as a proxy for societal progress more generally is itself a conceptual failure. The limitations of the GDP as a measure are highly relevant to Planetary Health (17, 18). First, GDP does not account for the distribution of wealth, so a rising GDP can conceal deepening inequality in a society. Second, GDP fails to place a value on many activities of obvious value to people, such as a parent raising a child or the work of a community volunteer. Third, GDP values many economic transactions that represent failures and would be better avoided, such as the cleanup of toxic wastes, the repair of a damaged vehicle, or the treatment of preventable illness. Fourth, GDP fails to capture externalities. Externalities occur when an industrial or commercial activity affects other parties without these impacts being reflected in market prices. Polluting industries generally do not pay the costs of pollution that they generate and sometimes try unscrupulously to evade their responsibilities. For example, Volkswagen escaped paying for the illnesses and deaths that resulted from excess NO_x emissions from its diesel vehicles in 2015, which occurred when it deliberately concealed those emissions (19, 20). (The company finally faced the music when the subsequent scandal led to hefty fines and legal settlements.) Frequently, the entity harmed by externalities is a natural asset such as a forest or a river delta; a rational accounting system would reckon this loss of natural capital, but GDP does not.

Climate change is one of the consequences of this failure to price economic activities appropriately. Carbon pricing covers about 20% of emissions and the price is often too low to have a meaningful effect on carbon dioxide emissions (21) (see Chapters 8 and 12). This disadvantages companies that wish to base their business model on sustainable economic principles, because they cannot benefit from what are effectively subsidies from unpaid damages to human health and natural systems.

Importantly, GDP does not track with measures of human well-being. While increasing GDP is associated with increasing life expectancy, this relationship flattens out above a per capita GDP of about US$10,000, a relationship (depicted by the 'Preston curve') identified nearly a half-century ago (22, 23). Longitudinal studies in countries such as the UK, USA, and Japan show that well-being has remained stable for many decades despite GDP increasing between three- and eight-fold, a phenomenon known as the 'Easterlin paradox' (24). There are many reasons for this disjunction; the positive effects of economic growth are likely accompanied by substantial negative externalities with implications for health and well-being (25). According to this analysis, for example economic growth in these countries results in the positive and negative effects cancelling each other out. Meanwhile, economic activities become the driving forces taking humanity beyond planetary boundaries and outside a safe operating space, and undermining future prospects.

The deficiencies of GDP as a measure of human progress have been widely recognized by policymakers and economists but it has proved difficult to supplant, partly because it has the virtue of being represented by a single number which appears superficially attractive and gives a simple, but sometimes illusory, metric of progress. Discontent with the dominant role of GDP in political and economic discourse has resulted in many proposals for better measures of progress (**Table 6.1**). Below we consider several of these in some detail.

Economics for People: The Better Life Index

One prominent effort to rethink the GDP was the Commission on the Measurement of Economic Performance and Social Progress (26) (sometimes referred to as the Stiglitz–Sen–Fitoussi Commission after its three leaders). The Commission was launched by French President Sarkozy in 2008 to address perceived deficiencies in contemporary economic and social measures. The Commission's work coincided with the financial crash of 2008, which added to the sense of urgency (since sadly dissipated) surrounding its recommendations.

The Commission argued for better measures of economic performance in an increasingly complex economy, with a large proportion of activity devoted to the provision of services such as health, educational and information services, and research activities. Increasingly, it is not growth in volume but in quality of services that provides the major benefits, and failure to accurately reflect quality of services in summary measures of economic performance can misrepresent reality. The provision of 'collective services' such as health care and state education or public housing depends on local or national government action – often by direct provision, sometimes by regulation. Government output accounts for about 20% of economic activity (using the admittedly flawed GDP metric) in

Table 6.1. *GDP and alternative measures of progress*

	Domains measured	Organization	URL	Comments
Gross Domestic Product (GDP)	Economic activity	Most governments, multilateral agencies, and conventional economists		• Includes no measure of human welfare • Includes only market transactions; does not count uncompensated work • Does not consider distribution of wealth • Does not consider externalities such as pollution • Does not consider environmental sustainability • Treats depletion of natural capital as income • Credits economic activity reflecting harm e.g. illness treatment, pollution cleanup
Genuine Progress Indicator (GPI) (Originally Index of Sustainable Economic Welfare, ISEW)	*Economic* • Consumption • Income inequality • Underemployment • Consumer durables • Investment *Environmental* • Pollution • Loss of wetlands, soil, forests • CO_2 emissions • Ozone depletion	Redefining Progress (NGO)	https://gnhusa.org/genuine-progress-indicator/	• Embodies value judgements regarding aspects of social life • Combines diverse ('apples and oranges') elements • No measures of subjective life satisfaction and psychological well-being

	Indicators / Dimensions	Organization	Website	Notes
	• Depletion of non-renewables *Social* • Value of housework, parenting, volunteering • Commuting • Crime • Car crashes • Leisure time • Higher education • Family changes			
Better Life Index	*11 dimensions of well-being* • Housing • Income • Jobs • Community • Education • Environment • Civic engagement • Health • Life satisfaction • Safety • Work-life balance	OECD	www.oecdbetterlifeindex.org	• Grew out of the work of the International Commission on the Measurement of Economic Performance and Social Progress (Stiglitz–Sen–Fitoussi Commission, 2008–2009) and its successor, High Level Expert Group on the Measurement of Economic Performance and Social Progress (starting 2013) • Embodies value judgements regarding which aspects of social life have value • Combines diverse ('apples and oranges') elements • Does not measure economic activity
Social Progress Index (SPI)	51 indicators across three dimensions	Social Progress Imperative (NGO)	www.socialprogress.org/	

185

Table 6.1. (*cont.*)

Domains measured	Organization	URL	Comments
Basic human needs • Nutrition and basic medical care • Water and sanitation • Shelter • Personal safety *Foundations of well-being* • Access to basic knowledge • Health and wellness • Environmental quality *Opportunity* • Personal rights • Personal freedom and choice • Inclusiveness • Access to advanced education			• No measures of subjective life satisfaction and psychological well-being • Does not measure economic inequality
Human Development Index (HDI) *Three dimensions integrated into a composite score (using geometric means)* • Health (life expectancy at birth) • Knowledge (mean years of schooling for adults aged ≥ 25 and expected years of schooling for children of school-entering age) • Decent standard of living (per capita GNI using a log scale to reflect the lesser importance of	United Nations Development Programme	http://hdr.undp.org/en/content/human-development-index-hdi	Relatively simple index. It does not reflect poverty, inequalities, human security, etc., but the inequality-adjusted HDI does reflect the distribution of values within the population (http://hdr.undp.org/en/content/inequality-adjusted-human-development-index-ihdi)

Gallup-Sharecare Global Wellbeing Index	increased income at higher levels) *Five elements of self-assessed well-being* • Sense of purpose • Social relationships • Financial security • Relationship to community • Physical health	Gallup (private polling firm)	www.gallup.com/ 175196/gallup-healthways-index-methodology.aspx https://wellbeingindex .sharecare.com/	Based or responses to survey questions in Gallup World Poll
Gross National Happiness	*Social and Economic Development* • Living standards • Education • Health *Preservation and promotion of culture* • Cultural diversity and resilience • Community vitality • Psychological well-being • Time use *Conservation of environment* Ecological diversity *Good Governance*	Government of Bhutan	www .grossnationalhappiness .com/	Based on responses to survey questions
Thriving Places Index (TPI)	*Sustainability* • Energy use • Waste • Green infrastructure	Centre for Thriving Places (NGO)	www .thrivingplacesindex .org/	• Utilizec only in England and Wales • Uses data from established UK sources (e.g. Office for

Table 6.1. (*cont.*)

	Domains measured	Organization	URL	Comments
	Equality • Health • Income • Gender • Social • Ethnicity *Local Conditions* • Place and environment • Mental and physical health • Education and learning • Work and local economy • People and community			National Statistics, Public Health England)
Happy Planet Index	• Ecological footprint • Inequality of outcomes • Well-being • Life expectancy	New Economics Foundation (NGO)	http://happyplanetindex .org/	Assesses human well-being together with ecological footprint (see below)
Ecological Footprint	• *Demand side*: measures the ecological assets a population requires to produce the natural resources it consumes, e.g. land, forests, fisheries • *Supply side*: measures biocapacity, the productivity of a place's ecological assets • Expressed in global hectares	Global Footprint Network (NGO)	www.footprintnetwork .org/our-work/ ecological-footprint/	Not a comprehensive index; measures only resource consumption and waste generation relative to nature's ability to generate new resources and absorb waste

OECD countries, and total government expenditure for more than 40% (26). The key recommendations included giving closer attention to income and consumption rather than production. The latter provides an inadequate measure of well-being because of factors such as depreciation, flows of income in and out of a country, and differences between the price of outputs and the price of consumer goods. This implies that the focus should be on measures of household income and consumption, including services provided by government.

However, income and consumption are themselves limited in what they can portray; spending on consumption beyond what is necessary for health and well-being may undermine future well-being and erode wealth. Measures of the future stocks required to maintain well-being – including physical, natural, human, and social capital – are therefore required, a necessary but challenging exercise. The average values of statistics such as income or wealth do not tell us how equally distributed they are in society. More attention should be paid to trends in the median, which better reflects inequities than the average, together with the extremes of the wealth distribution – the very poor and the very rich.

Growing concern about pervasive inequities has led to many calls for policies to reduce persistent, and in some cases growing, gaps between rich and poor. According to economist Thomas Piketty (27), when the rate of return on capital exceeds the growth in output and in income, as it did in the nineteenth century and seems likely to do in the twenty-first century, it is inevitable that, absent corrective policies, within-country inequities widen, with adverse consequences for democracy and the cohesion of society. During much of the twentieth century there was an unprecedented combination of circumstances as a consequence of the shocks arising from two World Wars, resulting in progressive tax policies and exceptional economic growth for three decades following 1945. This resulted in the rate of return on capital being less than the growth rate. This state of affairs has now ended and according to Piketty is reverting to the situation prevailing in the nineteenth century. The upper decile own about 60% of wealth in Europe and more than 70% in the USA; whereas the poorer half of the population owns barely 5% of total wealth. Thus, if no effective intervention is made then wealth disparities will widen even further in the future.

The Commission on the Measurement of Economic Performance and Social Progress also recommended including both subjective and objective measures of well-being in assessments of societal progress, noting that these will need development and refinement. There is a consensus that health and education are core determinants of well-being, but people's everyday activities, including social relationships and interconnectedness, political voice and governance, the right to decent work and housing are clearly also germane.

The Commission issued its final report in 2009, but its work continued. In 2011, the Organisation for Economic Co-operation and Development issued its Better Life Index, in line with the Commission's recommendations. Two years later, the OECD convened a High Level Expert Group on the Measurement of Economic Performance and Social Progress, to extend the work of the Stiglitz–Sen–Fitoussi Commission. This work concluded in 2018 with the publication of additional useful insights (including recommendations for research) and models (28–30).

Happiness: The World Happiness Report

There have been considerable advances in knowledge about the determinants of and trends in happiness (or subjective well-being). This field blossomed in 2011, with a UN General Assembly resolution called 'Happiness: Towards a Holistic Definition of Development'. A UN high-level meeting the following year, chaired by Secretary General Ban Ki-moon and Prime Minister Jigme Thinley of Bhutan, was called 'Wellbeing and Happiness: Defining a New Economic Paradigm', reflecting the growing sense that economic thinking needed to be re-oriented towards human well-being. At that meeting, the first *World Happiness Report* was released. Successor annual reports have been published under the auspices of the UN Sustainable Development Solutions Network.

The *World Happiness Report* bases its findings on nationally representative surveys conducted by Gallup in 160 countries using as a major measure answers to the question: 'Please imagine a ladder, with steps numbered from 0 at the bottom to 10 at the top. The top of the ladder represents the best possible life for you and the bottom of the ladder represents the worst possible life for you. On which step of the ladder would you say you personally feel you stand at this time?' (also known as the 'Cantril Ladder') (31). This can be seen as a measure of life satisfaction, which is closely related to happiness but not necessarily identical to it. Another approach, the World Value Survey, has been operated independently by a network of social scientists since 1981. It asks directly about happiness 'taking all things together', with responses ranging from 'very happy' to 'not at all happy'. The WVS provides the most extensive longitudinal data series available, making it possible to study trends in happiness over time. For most countries where two or more measurements are available over time the average recent score is higher than the earlier score(s) suggesting that in many countries improvements are occurring.

Whilst life satisfaction and happiness are closely associated, the latter may reflect an experiential and emotional assessment and the former a more evaluative, cognitive view. In 2014, for example, people in Ghana, Rwanda, and Zimbabwe scored relatively low on life satisfaction but over 80% responded that they were either very or rather happy. Cultural factors are important, such as closeness with family and societal cohesion, which may explain why people in Latin America score higher on life satisfaction than expected at a given level of GDP; for example, Costa Rica has a higher mean score than the UK despite having a per capita GDP about one-third of the UK's (32).

National averages should be interpreted with care and comparisons between age groups may not account for cohort effects. For example whilst happiness generally increases with age, older people appear less than happy than younger age groups, because across all ages earlier generations are less happy than later generations (32). Overall a higher income, both between and within countries, is associated with higher happiness and life satisfaction. Increases in income generally have a stronger association with life satisfaction in poorer countries than in richer countries.

The five highest-ranking countries according to the 2020 *World Happiness Report* are Finland, Denmark, Switzerland, Iceland, and Norway (in that order), with relative stability over time in the top ranked countries (31). The 2018 report focused on the happiness of migrants showing that their happiness is largely determined by the level of happiness in destination countries, with a relatively small contribution of 10–25% from countries of origin (33). International comparisons of happiness suggest that, in addition to per capita income, five factors are relevant to understanding the differences between countries: population health (measured as health-adjusted life expectancy which adjusts life expectancy for the time spent in less than perfect health); strength of social networks; personal freedom (perceived ability to make important life decisions); social trust (public perception of corruption in business and government); and generosity.

A striking observation is how in the USA, despite steady economic growth that doubled per capita income since 1972, happiness has stagnated and even declined – the Easterlin paradox introduced above. This has been attributed to lack of change in relative income due to the large inequalities in income distribution and the consequent failure of economic growth to benefit many people, as illustrated by the much greater increase in the wealth of the top 10% compared with the median income (32). More recently, the lack of improvement in happiness in the USA has been attributed to declines in social capital, which have offset any modest increases in income for much of the population (34). In the USA, life expectancy had been steadily improving until 2010 when it plateaued, and it began to decline from 2014 (35, 36). The turnabout is attributed to obesity, substance (particularly opioid) abuse, and depression. The declines in life expectancy are occurring in the under 65s rather than the oldest age groups. There have also been declines in the strength of social networks and increases in perceptions of corruption in public life. It is difficult to be certain of the exact contribution of each of these to the decline in subjective well-being but it seems reasonable to propose that they are playing a role. This reinforces the need to consider how best to address non-income determinants of happiness and subjective well-being, as well as reducing income inequalities.

China also performs relatively poorly in terms of happiness in comparison to its relatively robust economic performance. By 2012 virtually every urban household had a range of electronic goods – colour TV, refrigerator, washing machine, and air conditioning, which would have been undreamt of a generation before. But well-being seems even lower than it was in 1990 (37). This is particularly true for older and low-income people; upper-income and younger people have enjoyed modest improvements in well-being. The strongest determinants of well-being were the labour market and social safety nets. When unemployment rises and the social safety net is slashed, well-being suffers markedly and socioeconomic differences in well-being, small at first, widen substantially.

There is a strong relationship between health and life satisfaction both within and between countries, but at low levels of life satisfaction there is considerable diversity in life expectancy (**Figure 6.1**). Mental health is clearly related to life satisfaction, but this relationship only holds within rather than between countries. Thus, countries with higher reported rates of depression do not have lower levels of life satisfaction, while within countries people diagnosed with anxiety or depression have lower levels of life satisfaction, even when physical health, income, and education are controlled for (32). This may reflect the likelihood that many cases of depression are undiagnosed and therefore unreported in

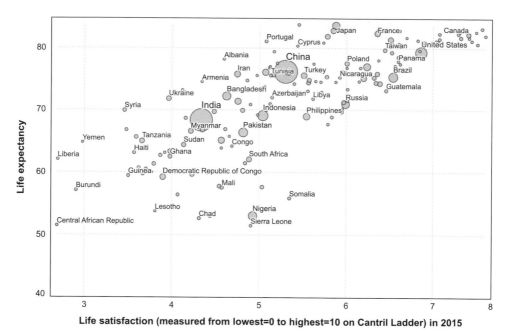

Figure 6.1. Life satisfaction versus life expectancy, by country, 2015. The vertical axis shows life expectancy at birth. The horizontal axis shows self-reported life satisfaction in the Cantril Ladder (0–10 point scale with higher values representing higher life satisfaction).

Source: Ortiz-Ospina E, Roser M. Happiness and life satisfaction. Our World in Data. 2017. https://ourworldindata.org/happiness-and-life-satisfaction.

official statistics, with the probability of diagnosis being lower in countries with weak health systems.

   While there is generally marked improvement in life expectancy over time in low- and middle-income countries the picture is more mixed in high-income nations. A study of trends in life expectancy in 18 high-income countries showed that the majority experienced declines in 2014–15, in some cases due to increases in respiratory disease mortality from influenza, but most bounced back in 2015–16 with robust gains that more than compensated for the declines (38). The two exceptions were the USA and the UK. In the UK, the declines in 2014–15 were mainly at older ages ($\geq$65) attributed largely to respiratory and circulatory diseases, Alzheimer's disease, nervous system diseases, and mental disorders, together with drug overdose for men. It is likely that declines in health and social service provision contributed to these increases (39, 40). Recent data on trends in death rates in the UK show that, after accounting for changes in age structure and population size, death rates have remained about the same since 2011 after decades of decline (41). It is clear therefore that we cannot assume continuing increases in life expectancy in high-income countries – these are dependent on government policies. At the time of writing the effects of the COVID-19 pandemic on life expectancy are becoming apparent but a comprehensive analysis is not yet possible.

Comprehensive Assessment of Human Well-Being: The Social Progress Index

One widely cited alternative to the GDP is the Social Progress Index. This index comprises 12 components divided into three broad groups: (a) Basic Human Needs (nutrition and basic health care, water and sanitation, shelter, personal safety); (b) Foundations of Well-being (access to basic knowledge, access to information and communications, health and wellness, environmental quality); and (c) Opportunity (personal rights, personal freedom and choice, inclusiveness, access to advanced education). Within these there are 51 individual indicators. By focusing on social and environmental indicators directly relevant to people and excluding economic indicators, it is possible to assess the independent relationship of the SPI with conventional economic measures such as GDP. **Figure 6.2** shows that the relationship flattens off above a per capita GDP of about US$40,000 per annum (adjusted for purchasing power). Such analyses are useful in showing how high levels of human progress are only partly dependent on GDP growth. However, the SPI is a complex indicator which depends on combining a range of disparate measures – a disadvantage when trying to understand and communicate results.

Parsimonious Measures of Well-Being: The Human Development Index

The Human Development Index (HDI), utilized by the UN Development Programme, is a simpler composite index focusing on three basic dimensions of human development: life expectancy at birth; mean years of schooling and expected years of schooling; and Gross National Income per capita. UNDP regularly updates the HDI. Although the global population increased from 5 to 7.5 billion between 1990 and 2017, the number of people in the low human development category fell from 3 billion (60% of the global population) to 926 million (12% of the world population), whilst the number of people experiencing high and very high human development rose from 1.2 (24%) to 3.8 billion (51%). (**Figure 6.3** shows how HDI increased by region.) As noted previously, persistent inequalities undermined this picture of progress and the discrepancy between the HDI and the inequality-adjusted HDI (IHDI) shows how inequality undermines progress (**Figure 6.4**).

High levels of human development are generally associated with high greenhouse gas emissions per capita (**Figure 6.5**) but there are large differences between countries. For example, Qatar emitted 45 tonnes of CO_2 per person in 2014 whereas Uruguay also achieved high human development but only emitted 2 tonnes of CO_2 per person (42). The relationship between HDI and deforestation is the converse (Figure 6.5), probably reflecting historical deforestation in many high-HDI countries which is now being reversed in some cases. In contrast, rapid deforestation is now occurring in many tropical countries as populations increase and forests are cleared for food production and economic activities, such as sale of timber or for palm oil plantations.

Incorporating Inequality and Ecological Footprint: The Happy Planet Index

One approach to assessing human progress in relation to ecological footprint is the Happy Planet Index (43) developed by the New Economics Foundation. The HPI aims

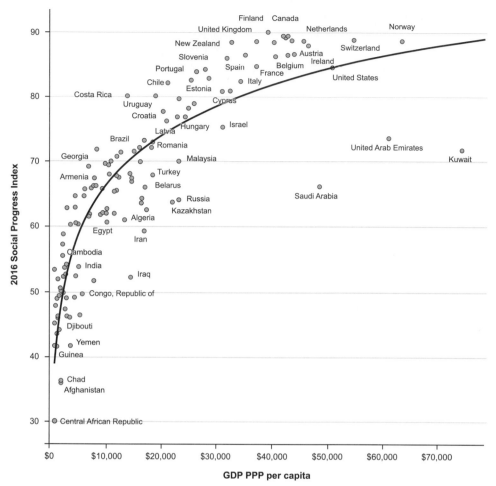

Figure 6.2. The relationship between GDP and the Social Progress Index. PPP indicates adjustment for purchasing power parity using a standardized basket of goods.
Source: Social Progress Index v GDP per capita. Econfix, 2017. https://econfix.wordpress.com/2017/01/14/social-progress-index-v-gdp-per-capita/.

to show how efficiently the residents of different countries use environmental resources to lead long and happy lives. It uses four measures: UN estimates of life expectancy; well-being as assessed by Gallup polling; inequality-adjusted life expectancy and well-being; and the ecological footprint estimated by the Global Footprint Network, expressed in standardized global hectares per person (**Table 6.1**) (44). Inequality-adjusted life expectancy (or well-being) is the average for the residents of a country, adjusted to reflect inequalities in the distribution across the national population, and expressed as a percentage, such that the value is adjusted downward more for countries with more unequal distributions.

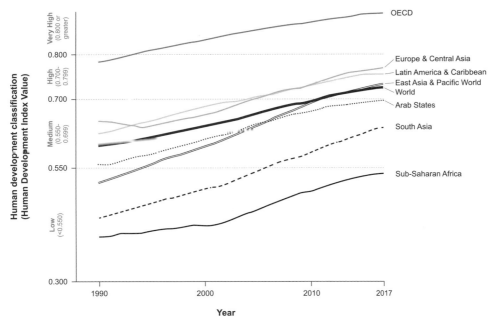

Figure 6.3. Human Development Index by country grouping, 1990–2017.
Source: UNDP. Human Development Indices and Indicators. 2018 Statistical Update. New York: UNDP, 2018. http://hdr.undp.org/en/2018-update.

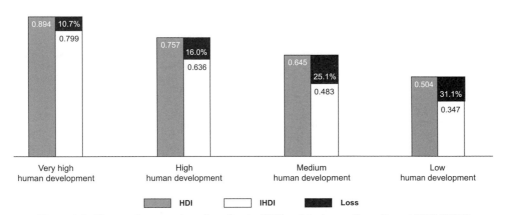

Figure 6.4. The gap between the values for the HDI and the inequality-adjusted HDI (IHDI) by human development group. Taking inequality into account reduces the overall levels of human development particularly in countries with lower levels of human development.
Source: Adapted from UNDP. Human Development Indices and Indicators. 2018 Statistical Update. New York: UNDP, 2018. http://hdr.undp.org/en/2018-update.

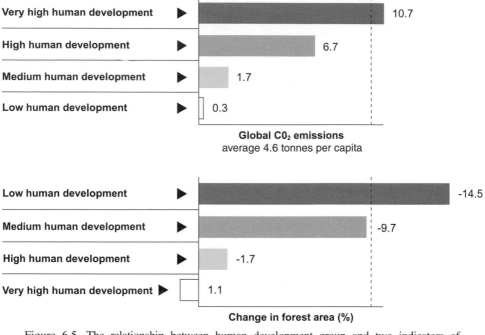

Figure 6.5. The relationship between human development group and two indicators of environmental performance: per capita CO_2 emissions (top) and change in forest area (bottom). Source: Adapted from UNDP. Human Development Indices and Indicators. 2018 Statistical Update. New York: UNDP, 2018. http://hdr.undp.org/en/2018-update.

In simplified terms,

$$\text{HPI} \approx \frac{\text{Life expectancy} \times \text{Well-being} \times \text{Inequality}}{\text{Ecological footprint}}$$

(This is approximate since it omits some statistical adjustments described in the HPI methods paper.) Costa Rica achieves the world's highest HPI, with higher levels of well-being than the UK and the USA and a higher life expectancy than the USA at one-third of its ecological footprint. This reflects deliberate national priorities: Costa Rica abolished its army in 1949 and redirected military spending to pensions, health care, and education; its constitution guarantees citizens the right to a healthy environment; and in 2008 it committed to carbon neutrality by 2021 (later shifted to 2050) (45, 46).

Recap: Measuring Human Progress

There is no single answer to the question 'which metric is best for measuring human progress and the environmental cost of that progress?' The answer will depend on the perspective taken and the value accorded to a particular outcome. Composite indices such as the Happy Planet Index attempt to summarize the overall picture in a single number, which is useful for communicating the broad picture but also raises important caveats.

These include whether an index such as the ecological footprint captures all the important environmental impacts of human societies and weights them appropriately in arriving at a single number. This is inevitably debatable because a single index can never capture the complexity of the multiple environmental impacts of human society. Judgements about inclusion and exclusion of specific metrics have to be made; for example, the ecological footprint excludes freshwater consumption and uses a standard global hectare to estimate the land required per capita to generate the resources and absorb pollutants. Providing the limitations of specific indices are understood, they can provide useful if not comprehensive perspectives on the extent to which progress is based on sustainable foundations.

Ultimately though the interpretation of composite metrics will always be contested and for this reason it is also wise to use simpler, more transparent indicators which assess human progress in relation to specific environmental impacts. So, for example, an assessment of the per capita greenhouse gas emissions in relation to, say, healthy life expectancy would indicate the extent to which a nation was contributing to climate change and thus undermining future prospects for health in achieving a given level of health of its current population. Similar analyses could be performed for other environmental factors such as freshwater use, air pollution, land use, fish catch, etc. Whilst in some cases the effects are predominantly local, such as freshwater depletion or fisheries decline, they have transboundary impacts, e.g. from the effects of increased demand for food imports on global trade flows. In fact, most environmental changes have both local and global effects but with a varying balance between them depending on the specific change, the geographical location, and the timescale over which impacts occur.

The choice of metrics to monitor trends in human progress ultimately depends on values and preferences, but life expectancy or healthy life expectancy are likely to be widely acceptable. Achieving population health has also been widely accepted as a governmental objective, including in the Sustainable Development Goals. SDG 3 specifically focuses on universal health coverage and other SDGs address many of the social and environmental determinants of health (see Chapter 7). It is clear that merely focusing on improvements in health and other indicators of human progress without taking into account the environmental costs of achieving these advances gives a partial view and fails to take into account how our actions today can prejudice the prospects for vulnerable populations now and in the future.

Tackling Externalities

A key driver of increasingly destructive environmental trends is the failure to account for the resulting damages in economic policies, such as internalizing the full economic costs of health damage due to climate change. These environmental externalities are difficult to eliminate, in part because some are difficult to define and on superficial analysis may appear to be small in magnitude. Many powerful interests base their business models on not paying the full economic costs of their activities and thus vigorously oppose attempts to internalize the costs. Furthermore, some externalities may be experienced in the distant future, increasing the uncertainties. The use of discount rates weakens the case for action since adverse outcomes experienced at some time in the future are less valued than those experienced now. This may be appropriate for some decisions but is conceptually flawed when it risks wide-ranging, irreversible damage to essential planetary life support systems, and condemns those who have contributed little to this state of affairs to suffer disproportionately. In this case the policy challenge can be reframed as 'What are the costs and benefits of action now and in the future?' and 'What is the most cost-effective approach to reducing unacceptable risks?' In developing answers to these questions we need to consider both the benefits (and any potential disbenefits) of action in the near term (coming months and years) as well in the mid-to longer term (decades to centuries).

If for example we can identify near-term benefits to action on environmental change, these should motivate decision makers (many of whom have a short-term perspective driven by imperatives of political office) to act with greater ambition than might otherwise be the case. This rationale underlies the work on the health co-benefits (ancillary benefits) of zero-carbon policies, which are addressed in Chapters 8 and 11. The presence of such (co)-benefits implies the presence of significant externalities in current policies and technologies and suggests that it could be in our self-interest to price them in a way that minimizes the adverse effects and capitalizes on the benefits. When externalities are taken into account through appropriate pricing mechanisms such as carbon taxes, they could yield major benefits to society, if these mechanisms are well designed. There is of course also the potential for adverse effects, as exemplified by the reaction of the French *gilets jaunes* to the 2018 fuel tax imposed by President Macron. This reaction arguably occurred because the tax burden fell on those who could not afford to pay and for whom there were no affordable alternatives to driving. This experience highlights the needs to integrate social and environmental outcomes when developing policies, as well as considering and minimizing the potential for unintended adverse consequences.

Global Health, Planetary Health, and Other Concepts

There is a range of partly overlapping concepts that can prove confusing for lay people and for professionals alike (47). An important distinction is between terms such as global health and international health, which only consider human health without reference to environmental impacts of achieving health and development, and those that place health to varying degrees within an environmental context – environmental health, Ecohealth, One Health and Planetary Health. The former implicitly assume that the challenges to health in the Holocene will continue broadly unchanged in the Anthropocene, and the latter to varying degrees conceptualize health as being influenced by or dependent on the environment on different scales, from local to planetary. Ecohealth has frequently focused on how ecosystems disruption has influenced disease patterns, particularly the transmission of vector-borne or zoonotic diseases, but may also encompass broader issues such as how biodiversity loss can affect health. One Health arose from collaboration between veterinary and human medicine, initially focusing on diseases transmitted from animals to humans, but has latterly broadened its perspective to encompass the relationship between human health and a range of living organisms. Planetary Health focuses on the potential for damage to the planet's natural (life support) systems to undermine and reverse progress in human health and emphasizes the importance of solutions that allow human societies to flourish within finite planetary boundaries at much lower levels of environmental change than today's. Planetary Health therefore subsumes major environmental trends such as climate change and can also be seen as an overarching concept within which other perspectives can be accommodated. That said, the commonalities among these various environmentally-based frameworks greatly outweigh the differences.

Ecosystem Services, Natural Capital, and Sustainable Prosperity

The economic and political systems that resulted in the transition to the Anthropocene Epoch are ill-suited to addressing the challenges we now face and are responsible for accelerating the risks in our current dangerous course. Since the pursuit of limitless economic growth cannot be sustained, the imperative is to reframe what constitutes a successful society in which all people have the potential to thrive within the constraints posed by the physical limits of our finite planet. Articulating this vision into a compelling discourse to replace the sterile consumerism of our age will challenge the conventional political landscape, including the traditional polarity of left and right, as neither of these ideological perspectives can claim to have addressed the need for new thinking in the Anthropocene Epoch. Indeed it could be argued that traditional political discourse has largely failed to realize that there has been a 'state shift' that profoundly threatens the straightjacket of conventional thinking. The environmental movement and its political allies such as Green parties have largely failed to make a decisive breakthrough, perhaps because they couch their message largely in terms of protecting the environment rather than enhancing the quality of human life. The myopia of conventional political discourse and the sterile polarization of views are perhaps amongst the central conceptual or imagination challenges we face.

Prosperity is about creating a society in which humanity can flourish in terms of health, education, satisfying employment, equity, and material sufficiency, rather than acquisition of material goods for its own sake. The implementation of that vision will depend on culture and context; a single blueprint is inappropriate and restricting. It will certainly require recognition not only of human rights but also of responsibilities, which are often overlooked as a necessary prerequisite for sustainability. These responsibilities are subject to debate but will have to address inequalities in consumption of material goods where these have significant externalities. Promoting acceptance of responsibilities cannot be achieved by top-down policy alone, although legislation and regulation have vital roles to play; it requires a change in values. This can only come from growing awareness of our predicament and the inter-generational and geographical injustice of our current development trajectory, which depends on encouraging narcissistic behaviour, fostering insecurity and anxiety through unrelenting competition, often weighted in favour of the privileged. It is notable for example that the children's strikes for climate change that blossomed beginning in 2018 used the slogan 'If you won't behave like adults – we will', a powerful condemnation of current 'adult' behaviour.

A major limitation of conventional economic approaches is their failure adequately to value the natural capital on which the sustainability of human societies ultimately depends. The natural capital approach integrates health and the natural environment into economic decision making (**Figure 6.6**). It recognizes and makes explicit the inevitable trade-offs between different objectives. It seeks to halt and reverse the substitution of natural capital with other forms of capital which has occurred over the last two centuries or so, bringing benefits from reduced poverty but now increasingly threatening future prospects (see also Chapter 12).

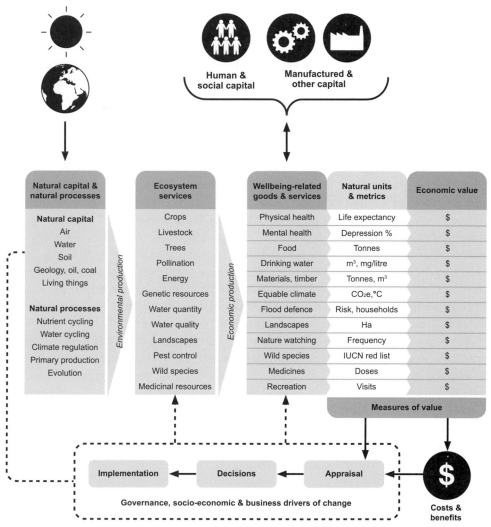

Figure 6.6. The natural capital approach, showing the integration of health and the natural environment into economic decision making.

Source: Bateman I, Wheeler B. Bringing Health and the Environment into Decision Making: The Natural Capital Approach. 2018. Rockefeller Foundation Economic Council on Planetary Health at the Oxford Martin School. www.planetaryhealth.ox.ac.uk/wp-content/uploads/sites/7/2018/06/Health-Env-in-Decision-Making.pdf.

Ecosystem services were introduced in Box 3.1 (pp. 101–2) Figure 6.6 summarizes how natural capital provides ecosystem services which in turn result in human well-being. The metrics that reflect different aspects of well-being can in turn be assigned a monetary value which allows them to be expressed in common units to inform decision making. There are several approaches to converting the metrics in the figure to economic valuations. This conversion can be justified because many ecosystem services provide valuable but unquantified benefits to humanity; for example, pollination by bees and other pollinators increases crop yield and contributes to nutrition. In this case the value of the food production attributable to natural pollinators and the potential effect of environmental factors, such as climate change, on pollination could be used to assess economic impacts.

Another approach looks at the revealed preferences which can be inferred for example from the time and travel costs people are prepared to pay to visit a specific natural park. Stated preferences can be elicited by presenting people with choices between different options, for example for improving health or the environment at different costs. Such approaches can be used to estimate the value of statistical life (VSL). The VSL represents the *ex ante* value of avoiding the death of an unidentified individual, by assessing a population's willingness to pay to prevent that death. So for example if people were prepared to pay on average US$30 to reduce the risk of dying from 3 in 100,000 to 2 in 100,000, the value of a statistical life would be US$3 million (30 × 100,000). Because it is time-consuming to measure VSL in different populations, in practice values are transferred from different settings with appropriate adjustments, e.g. for differences in per capita incomes (48). One can debate the utility of the results provided by these analyses, and it is important to be aware of potential biases and limitations, but in some cases they provide the only feasible approach and they are widely used to inform policy decisions.

Finally, cost-based approaches can be used for example to select among policy alternatives, such as deciding whether to pay to protect a water source from pollution or to pay for a water purification plant. However, such approaches do not reflect the true value of an intact and unpolluted watershed because the watershed provides many other services, such as recreation and biodiversity protection, which are often omitted in such costings. Such estimates should therefore be treated as a lower bound of the true value.

While valuation of natural capital has its challenges and is likely to result in underestimates because of the difficulty in accounting for all the natural capital used in socio-economic activities, failure to take its value into account will perpetuate and accentuate the destructive changes to natural systems that often characterize the Anthropocene.

Knowledge Challenges

Conceptual or imagination challenges, then, are substantial. So are knowledge challenges. Before considering policies and practices that can advance human progress within planetary boundaries, it is germane to consider what kinds of knowledge are needed for sustaining people and the planet in the Anthropocene, and how the knowledge gaps can be filled.

There has been extensive investment in biomedical research which has brought many advances ranging from improved vaccines to advanced surgical procedures, organ transplantation and, most recently, the promise of personalized medicine. All of us would want treatment tailored to our needs and personal circumstances if and when we become ill, but the term has become increasingly equated with genomic medicine, and generally overlooks the importance of the social and physical environments in causing disease. It is therefore likely to have limited impact and may contribute to the escalating costs which make many health systems increasingly unaffordable. It seems unlikely for example that personalized medicine in the biomedical sense could do much to address the plateauing and decline in life expectancy observed in the USA and the UK, and its ability to ameliorate the health effects of global environmental change is likely to be small. This is not to decry the very real advances from biomedicine, but rather to make the case for more balanced investment in addressing the urgent and complex challenges of the Anthropocene. This will require a transdisciplinary approach which transcends the narrow focus of much biomedical research and engages a range of sectors and disciplines that can contribute to better understanding.

Transcending Disciplinary Boundaries

While planetary boundaries flow from physical laws and are immutable, disciplinary boundaries are entirely human contrivances, and can (and often should) be breached. Interdisciplinary approaches seek to encourage the transfer of knowledge and methods from one discipline to another while maintaining disciplinary boundaries and identities. Transdisciplinary approaches, in contrast, aim to dissolve the methodological boundaries between disciplines (49, 50). Addressing the challenges of the Anthropocene Epoch will necessitate throwing off the shackles of siloed thinking and embracing the complexity of the systemic challenges that confront humanity. This does not mean that uni-disciplinary scientific knowledge will become irrelevant, but rather that advances in understanding of the challenges and their potential solutions will likely arise from the ferment of concepts and ideas that come from heterogeneous teams, where mutual learning is the norm and the methods employed are shaped by the nature of the problem to be addressed.

Synthesizing Evidence

Another dominant theme in the research landscape of the Anthropocene is the growing importance of research synthesis. A single study is rarely definitive and increasingly science advances by systematically reviewing all the available evidence on a given topic, seeking to minimize bias which occurs when a single high-profile study is given undue prominence. Because it is easier to publish studies with strongly positive results in high-impact journals and negative studies are either not published at all or in low-impact journals, depending on the results of the most widely quoted studies can be misleading. For this reason, systematic reviews of the literature are undertaken using published search criteria and rigorous assessment of the quality of included studies (51). This approach was pioneered in the health field, notably by the Cochrane Collaboration, which now involves 13,000 members and over 50,000 supporters who come from more than 130 countries. It focuses particularly on intervention studies of relevance to health but does not have a specific review group on interventions to reduce the effects of environmental change on health. Systematic reviews have increasingly been taken up in other fields. For example, the Campbell Collaboration (campbellcollaboration.org) produces systematic reviews to inform policy and practice and includes topics relevant to Planetary Health such as international development, food security, and climate solutions.

The Collaboration for Environmental Evidence (environmentalevidence.org) focuses on 'the effectiveness of environmental management interventions and the impact of human activities on the environment', which is relevant to the Planetary Health agenda but does not encompass it comprehensively.

The Rockefeller Foundation–*Lancet* Commission on Planetary Health (1) documented the limited evidence base for understanding the effects of large-scale environmental changes on health and potential solutions to the health challenges of the Anthropocene. The Commission identified 150 systematic reviews relevant to the topic following a search of 12 databases, but only 25 of these were considered to be of high quality. More kudos should accompany the publication of high-quality systematic reviews of the evidence on a specific topic, which can often illuminate the true state of knowledge in a way that a single study cannot do. Obviously primary research is still the driver of new knowledge, but no individual can keep up with every publication in a given field, nor can the potential distortion of knowledge by publication bias be addressed in the absence of systematic searching for and synthesizing research evidence. This is also a powerful method for finding gaps in knowledge and thus prioritizing research funding. The research output of the clinical research field vastly exceeds that for Planetary Health, with nearly 8000 quality assured systematic reviews in the Cochrane Library, over 95% of which concern interventions. Fortunately, innovative methods such as machine-searching of the literature based on artificial intelligence are making systematic reviews more feasible and less costly (52, 53). Clinical and public health research is of vital importance for underpinning progress in health and health care, but failing to address the challenges of Planetary Health will erode the progress made as a result of better health care and public health.

Research Needs

Several types of primary research are needed to address the challenges posed to health in the Anthropocene. First, there is a need to improve understanding of the complex, multi-scale mechanisms and pathways linking environmental change and health. Proposed research agendas on environment and health are increasingly addressing knowledge gaps on Planetary Health, especially in relation to climate change, but too often they remain siloed and dominated by traditional uni-disciplinary thinking (54, 55). Research should focus on vulnerable regions and populations because there is often an inverse relationship between the potential magnitude of effects and the availability of data, for example Sub-Saharan Africa is particularly under-represented in research on the effects of environmental change and potential solutions. Second, knowledge of effective adaptation strategies to environmental change becomes essential to minimize adverse effects of change that cannot be prevented. Third, the health co-benefits (and co-harms) of mitigation strategies to reduce the environmental footprint of societies should be quantified across different settings and sectors. Fourth, there is a need for research to develop and validate better metrics of human progress in relation to sustainability. These metrics should reflect the environmental costs of progress and the depletion of natural capital. Fifth, there is a need for implementation research to address on-the-ground realities of getting research findings into policy and practice, including the role of co-design and participatory engagement of key stakeholders.

Systems Thinking

Systems thinking is increasingly central to understanding complex inter-relationships in natural systems and human societies (56) and to achieving transformative change. Systems thinking is 'a holistic approach for understanding the dynamic interactions among complex economic, environmental, and social systems and for evaluating the potential consequences of interventions' (57). Systems thinking is necessary to identify and anticipate potential unintended consequences, trade-offs, and synergies between policies. These insights are essential to effectively inform policy and planning towards the achievement of long-term sustainability. Systems approaches can also illuminate the potential for sudden non-linear changes in natural systems which lead to potentially far-reaching consequences for humanity, for example abrupt shifts in marine ecosystems as a consequence of combinations of overfishing, increased nutrient inputs from fertilizer run-offs, climate change, urbanization and sewage inflow acting at different spatial and timescales (58). These combined effects can result in sudden declines in diversity and productivity of marine ecosystems with profound implications for communities dependent on them.

Resources to Support Research

An ambitious research agenda to address Planetary Health challenges will require considerable funding. Although definitive figures are unavailable because of the difficulties in classifying and monitoring financial commitments from diverse sources, it is likely that less than 0.1% of the global health research budget is devoted to research that aims to address issues related to Planetary Health. For example, the Wellcome Trust, the world's second largest charitable health research funder, during the first seven years of its Our Planet Our Health programme, allocated a fraction of 1% of its research funding to that area. Some research in the fields of agriculture, nutrition, renewable energy, urban planning and design, etc. address topics of relevance to Planetary Health but often fails to make the links with human health and sustainability. A combination of re-prioritizing research funds and making better use of existing funds by forging links between different discipline-specific or sectoral silos to address the systemic challenges of Planetary Health could accelerate the rate of progress.

Diverse Forms of Knowledge

Knowledge from different sources is needed to address the scale and speed of change required to address the burgeoning challenges that confront humanity. This has led to calls to abandon the view of knowledge that likens knowledge gaps to empty vessels into which 'knowledge' flows until they become full. The term 'knowledge systems' is used to denote the interconnectedness of different disciplines and perspectives, and the process of co-design with key stakeholders. This engagement aims to ensure that insights and potential solutions reflect stakeholder priorities and are compatible with their decision-making context (59, 60) (**Figure 6.7**).

The European Science Foundation conducted a Foresight exercise synthesizing the views of 100 experts from 30 countries to develop 'Responses to Environmental and Societal Challenges for Our Unstable Earth' (61). Their recommendations aimed to support the institutional development needed for the flourishing of open knowledge systems and include the need for greater use of participatory approaches to bring a range of societal actors into the research process from the outset to ensure that research questions and any potential policy options are shaped by the needs of those who can use the resulting knowledge to influence decisions. They also advocated greater recognition of transdisciplinary and interdisciplinary research in exercises that assess research quality, together with greater emphasis on research to address the barriers to and facilitators of the implementation of research findings.

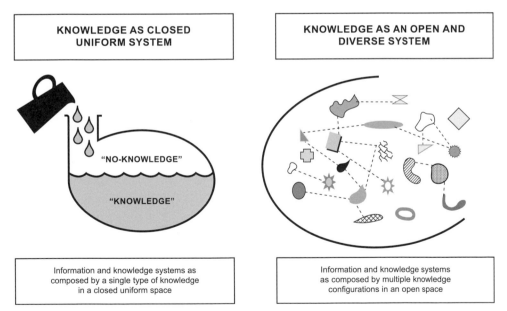

Figure 6.7. Images of closed versus open knowledge systems.

Source: Adapted from Tàbara JD. A new vision of open knowledge systems for sustainability: opportunities for social scientists. *World Social Science Report 2013: Changing Global Environment.* UNESCO; 2013. pp. 112–18.

There is also increasing recognition of the need to understand the human–environment relationship through lenses that supplement those of the biomedical and earth sciences. Qualitative research incorporating insights from the social and behavioural sciences can deepen our understanding of the human side of complex human–ecological systems (62, 63). Traditional ecological knowledge, gleaned from the perspectives of Indigenous communities and embodying time-tested beliefs and perceptions, can also provide invaluable insight (64–66).

Implementation Challenges

Even when we know what to do we frequently fail to make the necessary changes. It is well known in the health research field that there is frequently an 'implementation gap' between knowledge and action. This can result both in lack of access to effective interventions and the provision of ineffective or even harmful interventions such as unnecessary surgical procedures and medications (67). There are many reasons for failure to implement research findings including: the sheer volume of published research; questions about the generalizability of published results; and lack of demand for research evidence from clinicians, policymakers, and the wider public.

Systematic reviews of research evidence can help address the volume of research of varying quality as outlined above, but their findings need to be communicated in a timely fashion to those who can use the evidence. To be most useful to decision makers, reviews should be inclusive, rigorous, transparent, and accessible (68). The challenge is particularly pronounced for Planetary Health where a wide range of decision makers and professionals are involved. Not only is research synthesis needed, but so is effective communication of findings in various forms such as policy briefs, decision support tools, and targeted social media postings.

Confronting Disinformation

Confusing and misleading disinformation (including outright lies) is nothing new, but in the era of social media, it spreads with remarkable rapidity, and can effectively derail policies and practices that protect health and the environment. Such disinformation can be deadly (69). For example, disinformation about vaccine efficacy and safety has led to reductions in vaccination coverage and in some countries growing outbreaks of preventable diseases such as measles (70). With the outbreak of COVID-19, disinformation rapidly 'went viral' and undermined public health efforts such as social distancing and wearing masks in public (71). Such medical disinformation has parallels with climate change denial, in which vested interests, such as parts of the fossil fuel industry, have orchestrated disinformation campaigns to sow doubt, undermine public awareness of the scientific consensus on climate change, and polarize opinion (72).

In some cases, governments, politicians, or other actors weaponize disinformation, attempting to influence elections or destabilize adversary countries (73). Dangerous 'memes' (ideas, concepts, or behaviours) and conspiracy theories (74) can spread rapidly from person to person within a culture, particularly within subcultures whose members adhere strongly to ideological or other 'tribal' identities, and seek out sources that reinforce their views, ignoring contradictory evidence (75, 76). Some challenge the very notion that anything is true, giving rise to the so-called 'post-truth era' (77). These developments are inimical to the goal of basing Planetary Health policies and practices on the best science – indeed, to the scale of disruptive political change needed to live within planetary boundaries.

Disinformation is sometimes plainly ridiculous. A climate denial organization, the Oregon Institute of Science and Medicine, launched an online Global Warming Petition Project, claiming that 'over 31,000 American scientists have signed a petition stating that there is no scientific evidence that the human release of carbon dioxide will, in the foreseeable future, cause catastrophic heating of the Earth's atmosphere'. In fact fewer than 1% of signatories had a relevant scientific background, and among the signatories were Charles Darwin, characters from Star Wars, and a member of the Spice Girls (78)! With the outbreak of COVID-19, conspiracy theories abounded, suggesting that Bill Gates had created the SARS-CoV-2 virus in order to trick people into implanting microchips, that 5G mobile networks transmit the virus, that commercial disinfectants and ultraviolet light could be taken internally to cure the disease.

Emerging evidence has begun to identify effective strategies for counteracting online disinformation, including refutation by trusted people who are in the same networks where the disinformation is posted (79), citing sources rather than simply disagreeing (80), refutation by experts (81), crowd-sourcing the correction (82), and providing evidence of scientific consensus (78). It is helpful for social media platforms to provide warnings or direct readers to reliable sources of information (83). And not surprisingly, when corrective information is provided in appealing narrative form, people are more receptive (84). Based on these insights, efforts to counter disinformation will need to be systematically scaled up in order to increase public and political support for far-reaching policies to reduce the risks to humanity in the Anthropocene Epoch. The imperative of tackling disinformation and potential steps that could be taken are discussed further in Chapter 12.

Implementation Research

Implementation research is often underfunded and of low academic prestige because it is perceived as applied and perhaps mundane – 'just' programme evaluation. In reality it is a complex and challenging field requiring insights from many disciplines and performs a vital role linking knowledge generation to societal impact. To date, much of the implementation research in the health field has been in high-income countries; evidence from LMICs is scarce. The challenges for implementing policies to promote Planetary Health are greater in several respects than those for health-care interventions because they frequently involve more complex causal chains (85). Interventions often involve addressing institutional barriers, incentives, and behaviours on the ground rather than the uptake of a specific drug or diagnostic test and there may be trade-offs that differ according to the context. For example policies to incentivize biofuel production in order to meet climate goals could increase food prices, thus disadvantaging poor populations (86). For this reason, implementation research should aim systematically to assess the potential adverse consequences of implementing a technology, intervention, or policy as well as the desired effects.

Exposing these undesirable effects will help to improve accountability of decision makers and increase demands for adequate risk assessment before implementation begins.

Closing the 'Know-Do Gap'

Addressing the 'implementation gap' demands a range of actions to accelerate uptake of effective policies, technologies, and interventions to promote Planetary Health (85). First, more research should be co-designed with those who can make use of the findings to increase the likelihood that they will be implemented. Second, universities and research institutions should build more capacity to address implementation challenges supported by changes to incentives such as including measures of research impact in research assessment exercises. Third, researchers should embrace a range of methods to evaluate the implementation of evidence-based interventions as the traditional biomedical approaches such as randomized controlled intervention trials will rarely be feasible or appropriate. These approaches will need to capitalize on quasi-experimental approaches and comparative case studies of successes and failures using both quantitative and qualitative methods. Fourth, there is a need to evaluate approaches that garner political support for implementation efforts and are sufficiently compelling to overcome vociferous political opposition from those whose interests are threatened. For example introducing carbon taxes to accelerate implementation of zero-carbon technologies is more likely to be accepted if the income is used to reduce inequities and support essential services, such as health and social care, rather than merely generating additional government income. Finally, there is a need for better understanding of the actions of vested interests, including in the fossil fuel industry, in promoting disinformation (72) and to put in place more effective mechanisms to counter its pernicious effects.

Conclusions

The imperative of tackling the conceptual, knowledge, and implementation challenges of the Anthropocene demands fresh thinking and reassessment of traditional methods of generating, sharing, and using knowledge. Systems approaches are essential to address the challenges of the epoch and these will require transdisciplinary collaboration which is not constrained by the siloed thinking of conventional research. There is an important role for the science community to counter disinformation which can hinder the implementation of evidence-based solutions. Strategies to advance Planetary Health should be integrated as far as possible into wider development policies. The Sustainable Development Goals offer a major opportunity to do so, as discussed in the following chapter.

References

1. Whitmee S, Haines A, Beyrer C, et al. Safeguarding human health in the Anthropocene Epoch: report of The Rockefeller Foundation–*Lancet* Commission on Planetary Health. *The Lancet*. 2015;386(10007):1973–2028. doi: 10.1016/S0140-6736(15)60901-1.
2. Bradshaw CJA, Giam X, Sodhi NS. Evaluating the relative environmental impact of countries. *PLoS One*. 2010;5(5):e10440. doi: 10.1371/journal.pone.0010440.

3. Stern DI, Common MS, Barbier EB. Economic growth and environmental degradation: the environmental Kuznets curve and sustainable development. *World Development.* 1996;24(7):1151–60. doi: 10.1016/0305-750X(96)00032-0.

4. Ward JD, Sutton PC, Werner AD, et al. Is decoupling GDP growth from environmental impact possible? *PLoS One.* 2016;11(10):e0164733. doi: 10.1371/journal.pone.0164733.

5. Ehrlich PR, Holdren JP. Impact of population growth. *Science.* 1971;171:1212–17.

6. Raworth K. *Doughnut Economics: Seven Ways to Think Like a 21st-Century Economist.* London: Random House; 2017.

7. Veblen T. *The Theory of the Leisure Class: An Economic Study in the Evolution of Institutions.* New York: MacMillan; 1899.

8. de Graaf J, Wann D, Naylor TH. *Affluenza: The All-Consuming Epidemic.* San Francisco: Berrett-Koehler; 2002.

9. Egolf B, Lasker J, Wolf S, Potvin L. The Roseto effect: a 50-year comparison of mortality rates. *American Journal of Public Health.* 1992;82(8):1089–92.

10. Dasgupta PS, Ehrlich PR. Pervasive externalities at the population, consumption, and environment nexus. *Science.* 2013;340(6130):324–8. doi: 10.1126/science.1224664.

11. Miller D. *Consumption and Its Consequences.* Cambridge, UK and Malden, MA: Polity Press; 2012.

12. Nieuwenhuijsen M, Khries H, editors. *Integrating Human Health into Urban and Transport Planning.* Berlin/Heidelberg: Springer; 2018.

13. Ruvio AA, Belk RW, editors. *The Routledge Companion to Identity and Consumption.* Abingdon, UK and New York: Routledge; 2012.

14. Hogg MK, Michell PCN. Identity, self and consumption: a conceptual framework. *Journal of Marketing Management.* 1996;12(7):629–44. doi: 10.1080/0267257X.1996.9964441.

15. Douglas M. *The World of Goods: Towards an Anthropology of Consumption (Revised edition).* London and New York: Routledge; 1996.

16. Jackson T. *Motivating Sustainable Consumption: A Review of Evidence on Consumer Behaviour and Behavioural Change.* Guildford, UK: University of Surrey Centre for Environmental Strategy; 2004.

17. Kubiszewski I, Costanza R, Franco C, et al. Beyond GDP: measuring and achieving global genuine progress. *Ecological Economics.* 2013;93:57–68. doi: 10.1016/j.ecolecon.2013.04.019.

18. Stiglitz JE, Sen A, Fitoussi J-P. *Mismeasuring Our Lives: Why GDP Doesn't Add Up.* New York: The New Press; 2010.

19. Oldenkamp R, van Zelm R, Huijbregts MAJ. Valuing the human health damage caused by the fraud of Volkswagen. *Environmental Pollution.* 2016;212:121–7.

20. Hou L, Zhang K, Luthin M, Baccarelli A. Public health impact and economic costs of Volkswagen's lack of compliance with the United States' emission standards. *International Journal of Environmental Research and Public Health.* 2016;13(9):891.

21. Watts N, Amann M, Arnell N, et al. The 2018 report of the *Lancet* Countdown on Health and Climate Change: shaping the health of nations for centuries to come. *The Lancet.* 2018;392(10163):2479–514. doi: 10.1016/S0140-6736(18)32594-7.

22. Lutz W, Kebede E. Education and health: redrawing the Preston curve. *Population and Development Review.* 2018;44(2):343–61. doi: 10.1111/padr.12141.

23. Preston S. The changing relation between mortality and economic development. *Population Studies.* 1975;29:231–48. doi: 10.2307/2173509.

24. Easterlin R. Will raising the incomes of all improve the happiness of all? *Journal of Economic Behavior & Organization.* 1995;27(1):35–47. doi: 10.1016/0167-2681(95)00003-B.

25. Pretty J, Barton J, Bharucha ZP, et al. Improving health and well-being independently of GDP: dividends of greener and prosocial economies. *International Journal of Environmental Health Research*. 2016;26(1):11–36. doi: 10.1080/09603123.2015.1007841.
26. Stiglitz JE, Sen AK, Fitoussi J-P. Report by the Commission on the Measurement of Economic Performance and Social Progress. Paris: Commission on the Measurement of Economic Performance and Social Progress; 2009.
27. Piketty T. *Capital in the Twenty-First Century*. Cambridge, MA: The Belknap Press of Harvard University Press; 2014.
28. Stiglitz JE, Fitoussi J-P, Durand M. *Beyond GDP: Measuring What Counts for Economic and Social Performance*. Paris: OECD Publishing; 2018.
29. Stiglitz JE, Fitoussi J-P, Durand M. *For Good Measure: An Agenda for Moving Beyond GDP*. New York: New Press; 2019.
30. Stiglitz JE, Fitoussi J-P, Durand M. *For Good Measure: Advancing Research on Well-Being Metrics Beyond GDP*. Paris: OECD Publishing; 2018.
31. Helliwell JF, Layard R, Sachs JD, De Neve J-E. *World Happiness Report 2020*. New York: Sustainable Development Solutions Network; 2020.
32. Ortiz-Ospina E, Roser M. Happiness and life satisfaction. Our World in Data. 2017. Available from https://ourworldindata.org/happiness-and-life-satisfaction.
33. Helliwell JF, Layard R, Sachs JD. *World Happiness Report 2018*. New York: Sustainable Development Solutions Network; 2018.
34. Sachs JD. America's health crisis and the Easterlin paradox. In Helliwell JF, Layard R, Sachs JD, editors. *World Happiness Report 2018*. New York: Sustainable Development Solutions Network; 2018. pp. 146–59.
35. Woolf SH, Schoomaker H. Life expectancy and mortality rates in the United States, 1959–2017. *JAMA*. 2019;322(20):1996–2016. doi: 10.1001/jama.2019.16932.
36. Case A, Deaton A. Rising morbidity and mortality in midlife among white non-Hispanic Americans in the 21st century. *Proceedings of the National Academy of Sciences*. 2015;112(49):15078–83. 10.1073/pnas.1518393112.
37. Easterlin RA, Morgan R, Switek M, Wang F. China's life satisfaction, 1990–2010. *Proceedings of the National Academy of Sciences*. 2012;109(25):9775–80.
38. Ho JY, Hendi AS. Recent trends in life expectancy across high income countries: retrospective observational study. *BMJ*. 2018;362:k2562. doi: 10.1136/bmj.k2562.
39. Hiam L, Harrison D, McKee M, Dorling D. Why is life expectancy in England and Wales 'stalling'? *Journal of Epidemiology and Community Health*. 2018;72(5): 404–8. doi: 10.1136/jech-2017-210401.
40. Lewer D, Jayatunga W, Aldridge RW, et al. Premature mortality attributable to socioeconomic inequality in England between 2003 and 2018: an observational study. *The Lancet Public Health*. 2020;5(1):e33–41. doi: 10.1016/S2468-2667(19)30219-1.
41. Full Fact. UK death rates have stayed the same since 2011. 2020. Available from https://fullfact.org/health/deaths-in-the-UK-graph/.
42. UNDP. *Human Development Indices and Indicators: 2018 Statistical Update*. New York: United Nations Development Programme; 2018.
43. Abdallah S, Thompson S, Michaelson J, Nic M, Steuer N. *The Happy Planet Index 2.0*. London: New Economics Foundation; 2009.
44. Wackernagel M, Beyers B. *Ecological Footprint: Managing Our Biocapacity Budget*. Gabriola Island, BC: New Society Publishers; 2019.
45. Irfan U. Costa Rica has an ambitious new climate policy – but no, it's not banning fossil fuels. Vox. 17 July 2018.
46. Rodriguez S. Costa Rica launches 'unprecedented' push for zero emissions by 2050. Reuters. 25 February 2019.

47. Buse CG, Oestreicher JS, Ellis NR, et al. A public health guide to field developments linking ecosystems, environments and health in the Anthropocene. *Journal of Epidemiology and Community Health.* 2018;72(5):420–5. doi: 10.1136/jech-2017-210082.

48. OECD. *Mortality Risk Valuation in Environment, Health and Transport Policies.* Paris: Organisation for Economic Co-operation and Development; 2012.

49. Mauser W, Klepper G, Rice M, et al. Transdisciplinary global change research: the co-creation of knowledge for sustainability. *Current Opinion in Environmental Sustainability.* 2013;5(3–4):420–31.

50. Choi DC, Pak AW. Multidisciplinarity, interdisciplinarity and transdisciplinarity in health research, services, education and policy: 1. Definitions, objectives, and evidence of effectiveness. *Clinical and Investigative Medicine.* 2006;29(6):351–64.

51. Royal Society and Academy of Medical Sciences. Evidence Synthesis for Policy: A Statement of Principles. London: Royal Society and Academy of Medical Sciences; 2018.

52. Shemilt I, Simon A, Hollands GJ, et al. Pinpointing needles in giant haystacks: use of text mining to reduce impractical screening workload in extremely large scoping reviews. *Research Synthesis Methods.* 2014;5(1):31–49. doi: 10.1002/jrsm.1093.

53. Thomas J, Noel-Storr A, Marshall I, et al. Living systematic reviews: 2. Combining human and machine effort. *Journal of Clinical Epidemiology.* 2017;91:31–7.

54. WHO Regional Office for Europe. Setting Research Priorities in Environment and Health: Report of a Meeting in Cascais, Portugal, 27–28 April 2017. Copenhagen: WHO Regional Office for Europe; 2017.

55. NIEHS. NIEHS Strategic Plan 2018–2023: Advancing Environmental Health Sciences, Improving Health. Research Triangle Park, NC: National Institute for Environmental Health Sciences; 2018.

56. Pongsiri MJ, Gatzweiler FW, Bassi AM, Haines A, Demassieux F. The need for a systems approach to planetary health. *The Lancet Planetary Health.* 2017;1(7):e257–9.

57. Fiksel J, Bruins R, Gatchett A, Gilliland A, ten Brink M. The triple value model: a systems approach to sustainable solutions. *Clean Technologies and Environmental Policy.* 2014;16(4):691–702.

58. Rocha JC, Peterson GD, Biggs R. Regime shifts in the anthropocene: drivers, risks, and resilience. *PLoS One.* 2015;10(8):e0134639. doi: 10.1371/journal.pone.0134639.

59. O'Fallon LR, Wolfle GM, Brown D, Dearry A, Olden K. Strategies for setting a national research agenda that is responsive to community needs. *Environmental Health Perspectives.* 2003;111(16):1855–60. doi: 10.1289/ehp.6267.

60. Campbell H, Vanderhoven D. *Knowledge That Matters: Realising the Potential of Co-Production.* Manchester, UK: N8/ESRC Research Programme; 2016.

61. Jäger J, Pálsson G, Goodsite M, et al. *Responses to Environmental and Societal Challenges for Our Unstable Earth (RESCUE).* Strasbourg and Brussels: European Science Foundation Forward Look – ESF-COST 'Frontier of Science' joint initiative; 2012.

62. Scammell MK. Qualitative environmental health research: an analysis of the literature, 1991–2008. *Environmental Health Perspectives.* 2010;118(8):1146–54.

63. Brown P. Qualitative methods in environmental health research. *Environmental Health Perspectives.* 2003;111(14):1789–98. doi: 10.1289/ehp.6196.

64. Cámara-Leret R, Fortuna MA, Bascompte J. Indigenous knowledge networks in the face of global change. *Proceedings of the National Academy of Sciences.* 2019; 116(20):9913–18. doi: 10.1073/pnas.1821843116.

65. Finn S, Herne M, Castille D. The value of traditional ecological knowledge for the environmental health sciences and biomedical research. *Environmental Health Perspectives.* 2017;125(8):085006. doi: 10.1289/ehp858.

66. Gomez-Baggethun E, Corbera E, Reyes-Garcia V. Traditional ecological knowledge and global environmental change: research findings and policy implications. *Ecology and Society*. 2013;18(4):72. doi: 10.5751/es-06288-180472.

67. Haines A, Kuruvilla S, Borchert M. Bridging the implementation gap between knowledge and action for health. *Bulletin of the World Health Organization*. 2004;82(10): 724–31; discussion 32.

68. Donnelly CA, Boyd I, Campbell P, et al. Four principles to make evidence synthesis more useful for policy. *Nature*. 2018;558(7710):361–4. doi: 10.1038/d41586-018-05414-4.

69. Hopf H, Krief A, Mehta G, Matlin SA. Fake science and the knowledge crisis: ignorance can be fatal. *Royal Society Open Science*. 2019;6(5):190161.

70. Editorial. Vaccine hesitancy: a generation at risk. *The Lancet Child & Adolescent Health*. 2019;3(5):281.

71. Ball P, Maxmen A. The epic battle against coronavirus misinformation and conspiracy theories. *Nature*. 2020;581(7809):371–4. doi: 10.1038/d41586-020-01452-z.

72. Oreskes N, Conway EM. *Merchants of Doubt: How a Handful of Scientists Obscured the Truth on Issues from Tobacco Smoke to Global Warming*. London: Bloomsbury Press; 2010.

73. Bradshaw S, Howard PN. The global organization of social media disinformation campaigns. *Journal of International Affairs*. 2018;71(1.5):23–32.

74. Lewandowsky S, Gignac GE, Oberauer K. The role of conspiracist ideation and worldviews in predicting rejection of science. *PLoS One*. 2013;8(10):e75637.

75. Wang Y, McKee M, Torbica A, Stuckler D. Systematic literature review on the spread of health-related misinformation on social media. *Social Science & Medicine*. 2019; 240:112552. doi: 10.1016/j.socscimed.2019.112552.

76. Scheufele DA, Krause NM. Science audiences, misinformation, and fake news. *Proceedings of the National Academy of Sciences*. 2019;116(16):7662–9.

77. Lewandowsky S, Ecker UKH, Cook J. Beyond misinformation: understanding and coping with the 'post-truth' era. *Journal of Applied Research in Memory and Cognition*. 2017;6(4):353–69. doi: 10.1016/j.jarmac.2017.07.008.

78. van der Linden S, Leiserowitz A, Rosenthal S, Maibach E. Inoculating the public against misinformation about climate change. *Global Challenges*. 2017;1(2):1600008.

79. Margolin DB, Hannak A, Weber I. Political fact-checking on Twitter: when do corrections have an effect? *Political Communication*. 2018;35(2):196–219.

80. Vraga EK, Bode L. I do not believe you: how providing a source corrects health misperceptions across social media platforms. *Information, Communication & Society*. 2018;21(10):1337–53. doi: 10.1080/1369118X.2017.1313883.

81. Vraga EK, Bode L. Using expert sources to correct health misinformation in social media. *Science Communication*. 2017;39(5):621–45. doi: 10.1177/1075547017731776.

82. Pennycook G, Rand DG. Fighting misinformation on social media using crowdsourced judgments of news source quality. *Proceedings of the National Academy of Sciences*. 2019;116(7):2521–6. doi: 10.1073/pnas.1806781116.

83. Yaqub W, Kakhidze O, Brockman ML, Memon N, Patil S. Effects of credibility indicators on social media news sharing intent. In *Proceedings of the 2020 CHI Conference on Human Factors in Computing Systems*. Honolulu, HI: Association for Computing Machinery; 2020. pp. 1–14.

84. Jones MD, Crow DA. How can we use the 'science of stories' to produce persuasive scientific stories? *Palgrave Communications*. 2017;3(1):53. doi: 10.1057/s41599-017-0047-7.

85. Pattanayak SK, Haines A. Implementation of policies to protect planetary health. *The Lancet Planetary Health*. 2017;1(7):e255–6.

86. Hasegawa T, Fujimori S, Takahashi K, Masui T. Scenarios for the risk of hunger in the twenty-first century using Shared Socioeconomic Pathways. *Environmental Research Letters*. 2015;10(1):014010. doi: 10.1088/1748-9326/10/1/014010.

7

Health in the Sustainable Development Goals

The Sustainable Development Goals (SDGs) were adopted in 2015 by all UN member states and represent an ambitious, far-reaching programme of action. If implemented, they would set nations on a course that significantly enhances the prospects for sustaining human progress with a lower level of environmental impacts than today's development pathway. The SDGs are the latest manifestation of a process than can be traced at least as far back as the Earth Summit in Rio de Janeiro in 1992. There, over 170 countries adopted Agenda 21, a comprehensive action plan to catalyse a global partnership for sustainable development aimed at improving human lives and protecting the environment. The Earth Summit was a significant milestone; it established the fundamental principle that the attainment of healthy and productive lives in harmony with nature should be at the heart of the sustainable development agenda. It led to a number of crucial international conventions including the UN Framework Convention on Climate Change and the UN Convention on Biological Diversity. Nevertheless, subsequent progress was disappointing in several respects. The creation of a fair and just trading system which would foster the development of least developed countries proceeded slowly, with many wealthy nations maintaining subsidies favouring their own interests. Growth in development assistance proceeded at a slower rate than agreed and inequities between and within nations were pervasive, offsetting many of the potential benefits of economic growth. Environmental degradation continued apace, as we have seen, and the opportunity to capitalize on the momentum of the Earth Summit largely dissipated (1).

The Millennium Development Goals

The advent of UN Millennium Development Goals (MDGs), endorsed by world leaders in 2000, marked a period of renewed optimism. The MDGs focused on poverty reduction and human development in low- and middle-income countries (LMICs) with a target date of 2015 from a baseline of 1990. The MDGs and targets are summarized in **Table 7.1**. Progress towards the targets was assessed by changes in 60 indicators.

Achievement of the MDGs was variable, both between goals and regionally, but overall they can be considered a qualified success. According to the then UN Secretary General, Ban Ki-moon, they resulted in the 'most successful anti-poverty movement in history'

Table 7.1. *Millennium Development Goals: goals and targets*

Goal 1: Eradicate extreme poverty and hunger	
Target 1.A	Halve, between 1990 and 2015, the proportion of people whose income is less than one dollar a day
Target 1.B	Achieve full and productive employment and decent work for all, including women and young people
Target 1.C	Halve, between 1990 and 2015, the proportion of people who suffer from hunger
Goal 2: Achieve universal primary education	
Target 2.A	Ensure that, by 2015, children everywhere, boys and girls alike, will be able to complete a full course of primary schooling
Goal 3: Promote gender equality and empower women	
Target 3.A	Eliminate gender disparity in primary and secondary education, preferably by 2005, and in all levels of education no later than 2015
Goal 4: Reduce child mortality	
Target 4.A	Reduce by two-thirds, between 1990 and 2015, the under-5 mortality rate
Goal 5: Improve maternal health	
Target 5.A	Reduce by three-quarters, between 1990 and 2015, the maternal mortality ratio
Target 5.B	Achieve, by 2015, universal access to reproductive health
Goal 6: Combat HIV/AIDS, malaria and other diseases	
Target 6.A	Have halted by 2015 and begun to reverse the spread of HIV/AIDS
Target 6.B	Achieve, by 2010, universal access to treatment for HIV/AIDS for all those who need it
Target 6.C	Have halted by 2015 and begun to reverse the incidence of malaria and other major diseases
Goal 7: Ensure environmental sustainability	
Target 7.A	Integrate the principles of sustainable development into country policies and programmes and reverse the loss of environmental resources
Target 7.B	Reduce biodiversity loss, achieving, by 2010, a significant reduction in the rate of loss
Target 7.C	Halve, by 2015, the proportion of people without sustainable access to safe drinking water and basic sanitation
Target 7.D	By 2020, to have achieved a significant improvement in the lives of at least 100 million slum dwellers
Goal 8: Develop a global partnership for development	
Target 8.A	Develop further an open, rule-based, predictable, non-discriminatory trading and financial system. Includes a commitment to good governance, development and poverty reduction – both nationally and internationally
Target 8.B	Address the special needs of the least developed countries. Includes: tariff and quota free access for the least developed countries' exports; enhanced programme of debt relief for heavily indebted poor countries (HIPC) and cancellation of official bilateral debt; and more generous official development assistance (ODA) for countries committed to poverty reduction
Target 8.C	Address the special needs of landlocked developing countries and small island developing states (through the Programme of Action for the Sustainable Development of Small Island Developing States and the outcome of the twenty-second special session of the General Assembly)
Target 8.D	Deal comprehensively with the debt problems of developing countries through national and international measures in order to make debt sustainable in the long term
Target 8.E	In cooperation with pharmaceutical companies, provide access to affordable essential drugs in developing countries
Target 8.F	In cooperation with the private sector, make available the benefits of new technologies, especially information and communications

credited with helping to pull about 1 billion people out of extreme poverty (2). In the 1990s half of the extremely poor lived in East Asia and the Pacific; by 2015 only 6% did so (3). This shows how rapid poverty reduction is possible with appropriate policies and political will. However, profound inequities persist. Women experience a higher risk of living in extreme poverty than men, particularly in households headed by widows, single mothers, and women who are separated or divorced. The remaining people living in extreme poverty are relatively concentrated, in 2015, of the approximately 736 million people living on less than US$1.90 per day (equivalent to US$1.00 per day in 1996 prices), over half lived in Sub-Saharan Africa (SSA) and one-third in South Asia.

Substantial improvements also occurred in other domains. The proportion of undernourished people in developing regions fell by almost half, youth literacy increased (with a decline in gender disparity), the under-5 mortality rate declined by more than half, new HIV infections fell by 40% and HIV treatment rose dramatically, access to improved drinking water rose, and overseas development assistance from high-income nations increased.

Amongst the criticisms directed at the MDGs were that progress would have happened anyway, that progress was driven by middle-income nations, particularly China and India, from their own resources, with little contribution from multinational efforts, and that progress was driven by economic growth rather than by specific policies to prioritize the MDGs. Assessing the impact of the MDGs is challenging in the absence of a counterfactual (no MDGs) for comparison. But studies of the MDGs have suggested that they helped accelerate progress, particularly when there was capacity to monitor indicators (4, 5).

Nevertheless, the ambitious targets enshrined in a number of the MDGs were not met and there were major regional differences in performance. For example, while China made great strides in reducing undernourishment, the reduction in the Caribbean, Oceania, Southern Asia, and Sub-Saharan Africa failed to achieve the target. Climate-related disasters, conflict, economic recession, reduced employment, and higher food and energy prices have all contributed to undernourishment in Africa and parts of Western Asia. In the case of under-5 mortality the target of two-thirds reduction was also unmet. Mother's education is the most powerful determinant of inequalities in survival and those children living in rural areas have higher risks of death than those in urban centres. Although the primary school enrolment rate reached 91% by 2015, major socioeconomic differences in the proportion completing education persisted, and overall about one in six adolescents in LMICs failed to complete primary education. The target of a three-quarters reduction in maternal mortality was missed by a wide margin, reflecting the lack of skilled care at one in four deliveries worldwide.

Despite the real successes of the MDGs this unfinished agenda means that many people are entering an increasingly hazardous period afflicted by a burden of disadvantage which makes it more difficult for them to adapt to environmental change and puts them at particular risk of adverse effects. In addition, the lack of attention to environmental sustainability in those countries most responsible for undermining the planet's life support systems meant that the MDGs could only play a secondary role in tackling the burgeoning challenges of the Anthropocene Epoch.

The Sustainable Development Goals

Twenty years after the Earth Summit, many world leaders met again for the UN Conference for Sustainable Development – Rio+20, in Rio de Janeiro in 2012. Notable by their absence were President Obama, Prime Minister Cameron, and Chancellor Merkel, ostensibly distracted by the ongoing sovereign debt crisis. Perhaps they were also signalling ambivalence about putting the sustainable development agenda centre stage, particularly in light of the significant challenges it poses to the current 'Western' concept of economic progress. Nevertheless, the Conference resulted in a number of substantial achievements, including launching the process that delivered the Sustainable Development Goals (SDGs), successors to the MDGs, in 2015. Rio+20 participants also agreed to form the High-Level Political Forum on Sustainable Development, which was tasked to review progress on sustainable development annually, although these reviews are based on voluntary, state-led reports of progress, which clearly have limitations.

The SDGs differ in important respects from the MDGs. While the MDGs focused on development in the Global South, the SDGs apply to countries at all levels of development. And while the MDGs focused on economic and social goals, the SDGs explicitly address environmental sustainability. The SDGs comprise 17 goals, 169 targets, and 232 indicators, compared with 8 goals and 21 targets in the MDGs, and are thus much more ambitious and comprehensive in scope than the MDGs.

One way to categorize the SDGs is shown in **Figure 7.1**. The central circle contains six people-centred goals that focus on the enhancement of well-being. The ring outside that contains seven infrastructure goals that relate to the provision of clean water, energy, food, and other essential requirements for health and well-being. The outer ring contains three environmental goals that focus on natural systems including climate, land, and water, and by implication biodiversity and air (6). SDG 17 is an overarching and enabling goal which aims to enhance the prospects for implementation and reinforce development cooperation. The SDGs aim to achieve 'the triple bottom line' of economic prosperity, social inclusion, and environmental sustainability for all nations, underpinned by good governance and high levels of global cooperation (7).

The SDGs should be seen as a coherent and indivisible whole, which implies that the goals are interdependent and mutually reinforcing. Poorly designed policies that address a goal in isolation could backfire; for example, accelerating tropical deforestation to grow more food might help address SDG 2 (no hunger) in the short term but would also undermine goals 13 (climate action) and 15 (life on land), contributing to climate change and biodiversity loss. Similarly, the overuse of fertilizers to increase crop yield (SDG 2) can lead to excess nutrient run-off resulting in water pollution, eutrophication, and damage to marine and freshwater ecosystems (SDG 14). Conversely, well-crafted policies and practices can simultaneously advance multiple SDGs. For example, safely managed composting of human waste can both deliver sanitation services (SDG 6) and reduce greenhouse gas emissions (SDG 13) (8).

Such interactions between the SDGs have been systematically mapped and evaluated. One study, by the International Science Council (ICSU), considered four of the

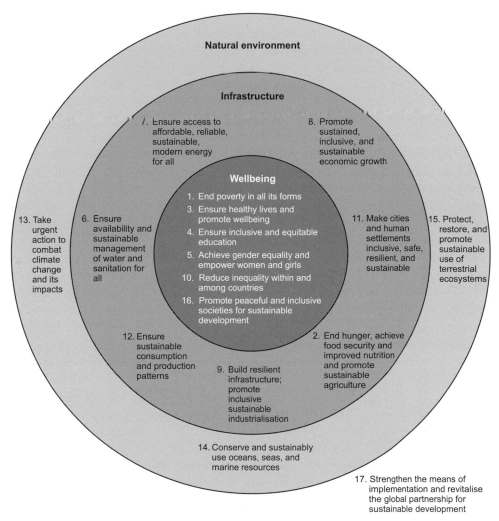

Figure 7.1. Framework for examining interactions between the SDGs. This approach situates human well-being in larger environmental and planetary contexts.

Source: Waage J, Yap C, Bell S, et al. Governing the UN Sustainable Development Goals: interactions, infrastructures, and institutions. *The Lancet Global Health*. 2015;3(5):e251–2.

SDGs – 2 (hunger), 3 (health), 7 (energy), and 14 (marine life) – and rated the possible interactions on a scale from highly positive to highly negative. (The terms used, from the most positive to the most negative, were indivisible, reinforcing, enabling, consistent, constraining, counteracting, and cancelling, with the first three classified as positive, the middle term neutral and the last three as negative) (9, 10).) The investigators identified a total of 316 interactions, of which 238 were positive, 66 were negative, and 12 were neutral. In no case was an interaction 'cancelling' (meaning that achieving one goal would preclude achieving another) but they did identify many constraints and

conditionalities in jointly achieving goals. For example, although increased food production may improve incomes for the rural poor and reduce food insecurity, this does
not invariably happen. In some regions of Brazil, the focus on export-led food production, which is heavily mechanized, has brought a decline in rural employment (9). As a
result of such interactions, the authors called for 'coordinated policy interventions to
shelter the most vulnerable groups, promote equitable access to services and development opportunities, and manage competing demands over natural resources to support
economic and social development within environmental limits' (9, p. 8).

Another study, by investigators at the Potsdam Institute for Climate Impact Research
(PIK) in Germany, was more comprehensive; it mapped interaction among all the
SDGs (11). The PIK investigators designated a positive correlation between two SDG
indicators as a 'synergy' and a negative correlation as a 'trade-off', and considered
interactions on both global and national scales. Like the ICSU study, this study found
substantially more synergies than trade-offs. SDG 1 (no poverty) showed the greatest
number of synergies with the other goals, whereas SDG 12 (responsible consumption
and production) was the goal most commonly associated with trade-offs.

The very complexity of the SDGs may be a barrier to the transformative actions needed
to implement them. A useful approach is to bundle the SDGs according to the transformations they require. Six such transformative actions have been identified, which comprise
discrete yet interacting components engaging different business and civil society actors,
with clear communication pathways aiming to develop and disseminate problem-focused
solutions (12). These transformations comprise (1) education, gender, and inequality;
(2) health, well-being, and demography; (3) energy decarbonization and sustainable industry; (4) sustainable food, land, water, and oceans; (5) sustainable cities and communities;
and (6) digital revolution for sustainable development. The proposed transformations are
mutually exclusive and collectively exhaustive; systems-based; aligned with government
organization; easily communicable; and few in number although consistent with the
17 SDGs, together with their targets and indicators. They address potential trade-offs in
three ways. First, they explicitly consider potentially antagonistic actions (such as increased
food production, biodiversity protection, and land for biofuels) in a systems framework that
identifies feedback loops and trade-offs. Second, they embody a strong commitment to
equity and leaving no-one behind. Third, they move towards a circular economy that aims
to decouple economic activity from environmental impacts.

Health in the SDGs

The SDG explicitly dedicated to human health is SDG 3, but health is inherent in most
if not all of the SDGs. Both the deliberations that shaped SDG 3 and the specific targets
that form it (**Table 7.2**) focused heavily on the health sector, especially on attainment
of universal health coverage (UHC), with two of the SDG 3 targets – reduced burdens of
road traffic injuries (3.6) and hazardous chemicals and pollution (3.9) – being exceptions
to that rule.

Table 7.2. *Targets for SDG 3: Good health and well-being*

3.1	By 2030, reduce the global maternal mortality ratio to less than 70 per 100,000 live births.
3.2	By 2030, end preventable deaths of newborns and children under 5 years of age, with all countries aiming to reduce neonatal mortality to at least as low as 12 per 1000 live births and under-5 mortality to at least as low as 25 per 1000 live births.
3.3	By 2030, end the epidemics of AIDS, tuberculosis, malaria, and neglected tropical diseases and combat hepatitis, water-borne diseases, and other communicable diseases.
3.4	By 2030, reduce by one-third premature mortality from non-communicable diseases through prevention and treatment and promote mental health and well-being.
3.5	Strengthen the prevention and treatment of substance abuse, including narcotic drug abuse and harmful use of alcohol.
3.6	By 2020, halve the number of global deaths and injuries from road traffic accidents.
3.7	By 2030, ensure universal access to sexual and reproductive health-care services, including for family planning, information and education, and the integration of reproductive health into national strategies and programmes.
3.8	Achieve universal health coverage, including financial risk protection, access to quality essential health-care services and access to safe, effective, quality and affordable essential medicines and vaccines for all.
3.9	By 2030, substantially reduce the number of deaths and illnesses from hazardous chemicals and air, water and soil pollution and contamination.
3.A	Strengthen the implementation of the WHO Framework Convention on Tobacco Control in all countries, as appropriate.
3.B	Support the research and development of vaccines and medicines for the communicable and non-communicable diseases that primarily affect developing countries, provide access to affordable essential medicines and vaccines, in accordance with the Doha Declaration on the TRIPS Agreement and Public Health, which affirms the right of developing countries to use to the full the provisions in the Agreement on Trade-Related Aspects of Intellectual Property Rights regarding flexibilities to protect public health, and, in particular, provide access to medicines for all.
3.C	Substantially increase health financing and the recruitment, development, training and retention of the health workforce in developing countries, especially in least developed countries and small island developing States.
3.D	Strengthen the capacity of all countries, in particular developing countries, for early warning, risk reduction and management of national and global health risks.

High-quality clinical care for all is without doubt a desirable goal. However, excessive focus on curative clinical services often comes at the expense of health promotion, disease prevention, and action on social determinants of health (13). It is not hard to imagine this imbalance yielding more health-care services but worse population health outcomes with more pronounced inequities, for several reasons (13). First, expanding health-care access, without deliberate action to target the underserved, often increases inequity by disproportionately benefitting the wealthy (14, 15). Second, clinical facilities and providers, particularly those based on hospital care, tend to emphasize specialist curative interventions over preventive and more basic curative interventions in primary care. Third, political pressures can skew health budgets towards alluring but costly clinical services at the expense of public health. Fourth, many LMICs lack a sufficient health-care workforce, limiting the potential benefits of health-care delivery. Fifth, many of the SDG 3 targets cannot be achieved without population-level interventions. Relative neglect of 'upstream' health risks such as severe weather events, food shortages, and hazardous chemical exposures can substantially worsen health. While achievement of affordable and effective UHC can profoundly reduce mortality and morbidity from many diseases, it is not sufficient; the social, economic, and environmental determinants of health are largely outside the control of the health-care sector and addressing these will play a central role in achieving progress on health.

These considerations highlight the need for an integrated approach to health in the SDGs – a need that many commentators have recognized (16–19). Just as achieving SDG targets other than those in SDG 3 can promote health, strategies to achieve SDG 3 targets could advance other SDGs – which would, in turn, protect people. For example, reducing the burden of water- and vector-borne diseases that are sensitive to climate and other environmental changes (Target 3.3) will entail progress towards SDGs 6 (clean water and sanitation), 13 (climate action), 14 (life below water), and 15 (life on land). And progress towards decarbonizing the energy and transport sectors (SDGs 7, clean energy, and 11, sustainable cities and communities, respectively) will yield improved cardiovascular and respiratory health.

Indeed, many of the SDGs other than SDG 3 directly or indirectly reflect or influence the social, economic, and environmental determinants of health. Overall, 12 SDGs, 33 targets, and 57 indicators have been identified as health-related (17). Global environmental threats that stall progress in reaching any of these goals impede the achievement of health for all. Several of the most pertinent links between other SDGs, the environment, and health are described in **Table 7.3**, and discussed further in the paragraphs that follow.

SDG 1 aspires to end poverty in all its forms everywhere. Achieving this goal would dramatically improve prospects for health. However, SDG 1 is threatened by climate change, which could drive 100 million people into extreme poverty by 2030 (20) as a result of decreased crop yield and declines in labour productivity from extreme heat exposure. The extreme poverty rate is projected to be 6% in 2030, thus failing to reach the global target to eradicate extreme poverty (21). Children are over-represented amongst the poor, with one in five children living in extreme poverty. Unsustainable levels of household and commercial debt in many countries do not bode well for future prospects, nor does the global economic downturn triggered by the COVID-19 pandemic. Direct economic losses from disasters have increased by more than 150% in the last 20 years, with a disproportionate burden falling on vulnerable low-income countries.

SDG 2 aims to end hunger, achieve food security and improved nutrition, and promote sustainable agriculture. This goal is clearly threatened by environmental trends including climate change, loss of pollinators, fishery decline, and freshwater depletion. After years of decline, the number of undernourished people worldwide began to rise in 2015 (see Chapter 10) (22). One-fifth of the African population (over 275 million people), and nearly one in three people in East Africa, suffer from undernutrition, a problem compounded by drought, conflict, and most recently locust swarms. In order to address this challenge successfully the productivity of small-scale farmers, currently lower than that of large-scale farms, must be improved in the face of growing environmental threats. But at a time when international aid to agriculture and government expenditure on agriculture as a proportion of the total budget are declining, this improvement seems elusive.

Table 7.3. *Examples of linkages between the SDGs, health outcomes, and environmental change (N.B. the list is not exhaustive)*

SDG	Relevant targets	Link to health outcomes and environmental change
SDG 1: By 2030, to eradicate extreme poverty for all people everywhere.	1.1 To 'eradicate extreme poverty . . ., currently measured as people living on less than $1.25 a day' (In 2015, the World Bank reset the poverty line to $1.90 a day.) 1.2 To 'reduce at least by half the proportion of men, women and children of all ages living in poverty in all its dimensions according to national definitions'	Poor populations lack access to clean energy and are likely to be heavily exposed to household air pollution (resulting in increased risks of lower respiratory infections and NCDs) as well as more vulnerable to the effects of climate and other environmental changes.
SDG 2: End hunger, achieve food security and improved nutrition, and promote sustainable agriculture.	2.2 To 'end all forms of malnutrition, including achieving, by 2025, the internationally agreed targets on stunting and wasting in children under 5 years of age . . .' 2.4 By 2030, 'ensure sustainable food production systems and implement resilient agricultural practices that increase productivity and production, that help maintain ecosystems, that strengthen capacity for adaptation to climate change . . .'	Climate and other environmental changes increase risks of stunting. Stunting causes cognitive impairment and may increase risks of obesity and related NCDs in later life. Undernutrition increases risks of infectious diseases. More sustainable diets including increased fruit and vegetable consumption can reduce NCD risks and the environmental footprint of the food system.
SDG 3: Ensure healthy lives and promote well-being for all at all ages.	3.4 By 2030, 'reduce by one-third premature mortality from non-communicable diseases through prevention and treatment and promote mental health and well-being . . .' 3.9 By 2030, 'substantially reduce the number of deaths and illnesses from hazardous chemicals and air, water and soil pollution and contamination'	Policies to promote clean, zero-carbon energy, increase active travel and use of public transport and healthy, sustainable diets can reduce GHG emissions, provide other environmental benefits and reduce NCD risks. Air pollution is a major risk factor for NCDs and many pollutants are co-emitted with carbon dioxide and short-lived climate pollutants (SLCPs).
SDG 7: Ensure access to affordable, reliable, sustainable and modern energy for all	7.1 By 2030, 'ensure universal access to affordable, reliable and modern energy services' 7.2 By 2030, 'increase substantially the share of renewable energy in the global energy mix'	Reduces exposure to household and ambient air pollution and GHG/SLCP emissions with consequent health co-benefits.

Table 7.3. (*cont.*)

SDG	Relevant targets	Link to health outcomes and environmental change
SDG 11: Make cities inclusive, safe, resilient and sustainable	11.2 By 2030, 'provide access to safe, affordable, accessible, and sustainable transport systems for all, improving road safety, notably by expanding public transport'	Potential to reduce NCD risks through increased physical activity and reduced urban air pollution as well as contributing to climate change mitigation.
	11.5 By 2030, 'significantly reduce the numbers of deaths and people affected and substantially decrease the direct economic losses relative to global gross domestic product caused by disasters … with a focus on protecting the poor and people in vulnerable situations'	Reduce risk of people with NCDs exposed to disasters and consequent breakdown of health care. Nature-based solutions, such as protection of wetlands, mangroves, and coral reefs can reduce risks to health from natural disasters and protect biodiversity. Increased green space can reduce urban heat island effect and improve physical and mental health (links also to SDGs 14 and 15).
	11.6.2 Level of ambient particulate matter (PM 10 and PM 2.5)	Air pollution exposure increases risks of IHD, stroke, LRTIs, etc. Decarbonizing the energy and transport systems reduces air pollution and GHG emissions.
SDG 13: Take urgent action to combat climate change and its impacts	13.1 'Strengthen resilience and adaptive capacity to climate-related hazards and natural disasters in all countries'	See above.
	13.2 'Integrate climate change measures into national policies, strategies and planning' 13a 'Implement the commitment undertaken by developed-country parties to the United Nations Framework Convention on Climate Change to a goal of mobilizing jointly $100 billion annually by 2020'	Potential funding for policies to reduce GHG/SLCP emissions which also result in health co-benefits through reduced air pollution exposure, etc.

Source: Adapted from Haines A, Amann M, Borgford-Parnell N, Leonard S, Kuylenstierna J, Shindell D. Short-lived climate pollutant mitigation and the Sustainable Development Goals. *Nature Climate Change.* 2017;7(12):863–9.

SDG 4 aims to ensure inclusive and equitable quality education and promote lifelong learning opportunities for all. More than 50% (617 million) of children and adolescents of primary and lower-secondary school age worldwide are not achieving minimum proficiency in mathematics and literacy (21). This has major implications for their life prospects, and of course for health, because of the close relationship between educational attainment and health throughout the life course. In fact, mean years of schooling are more closely related to improvements in life expectancy worldwide than are increases in income (23).

SDG 6 aims to ensure that water and sanitation be available to all people, and sustainably managed. Achieving this goal would help reduce vulnerability to climate change, the loss of ecosystem services from watershed degradation, and pollution of water sources. According to the United Nations Economic and Social Council, about 785 million people still lack access to even basic drinking water services and 701 million still practise open defecation. Currently, achieving universal access to basic sanitation services by 2030 would require doubling the current annual rate of progress (21). Rural–urban differences are pervasive, with access to education, health care, and safe water being lower amongst rural populations in many low-income countries.

SDG 7 aims to ensure access to affordable, reliable, sustainable, and modern energy for all. This goal, if achieved, will reduce household and ambient air pollution as well as reducing the emissions of greenhouse gases and other climate active pollutants. This SDG is particularly relevant to the approximately 3 billion people, mainly living in Asia and Sub-Saharan Africa, who are still cooking without the benefit of clean fuels and technologies. Current progress is slow, amounting to an annual increase of 0.46%, whereas a 2.66% annual increase would be required to achieve universal access by 2030 (21).

SDG 8 promotes sustained, inclusive and sustainable economic growth, full and productive employment, and decent work for all. The question of whether sustained economic growth is even possible is addressed in Chapter 6, and as discussed later in this chapter, critics have claimed that the SDGs emphasize economic growth excessively relative to environmental sustainability and human well-being. But clearly, decent work for all is a laudable goal. Many countries are failing to create employment opportunities for young people, who are three times more likely to be unemployed than older adults. Globally, only 40% of employed people are women and they occupy only 25% of managerial positions. In those countries where data are available there is a gender pay gap of 12% (21).

SDG 11 focuses on making cities and human settlements inclusive, safe, resilient, and sustainable. Between 1990 and 2016, the proportion of the global urban population living in slums declined from 46% to 23%, but this improvement was largely offset by population growth and rural-to-urban migration. Currently over 800 million people still live in slums. About 2 billion people lack access to waste collection services and 3 billion to controlled waste disposal facilities. Only 18% of urban dwellers in SSA have convenient access to public transport (21). These conditions significantly undermine the health and well-being of those who live in slums (24).

SDG 12 addresses sustainable consumption and production but has had little impact thus far. Material consumption has increased by over 250% since 1970 and the rate of extraction

---

Box 7.1. **Global Biodiversity Commitments in the Convention on Biological Diversity (CBD) and the Sustainable Development Goals (SDGs)**

The CBD vision: 'By 2050, biodiversity is valued, conserved, restored and wisely used, maintaining ecosystem services, sustaining a healthy planet and delivering benefits essential for all people.'

CBD Aichi Target 12 states: 'By 2020 the extinction of known threatened species has been prevented and their conservation status, particularly of those most in decline, has been improved and sustained.'

SDGs 14 and 15 aim, by 2030, to 'Conserve and sustainably use the oceans, seas and marine resources' (SDG 14) and 'Sustainably manage forests, combat desertification, halt and reverse land degradation, halt biodiversity loss' (SDG 15). Target 15.5: 'Take urgent and significant action to reduce the degradation of natural habitats, halt the loss of biodiversity and protect and prevent the extinction of threatened species.'

---

of material resources has increased every year since 2000. Although SDG 13 is intended to support urgent action on climate change, its scope is limited as the Paris Agreement is the main vehicle for cutting GHG emissions and the financial flows to support adaptation and mitigation have so far been inadequate.

It is perhaps not surprising that countries overall perform worst on SDG 13 (climate action), SDG 14 (life below water), and SDG 15 (life on land) because human progress has been so dependent on the unsustainable exploitation of nature's bounty, which is rapidly becoming exhausted. Addressing biodiversity loss will also be essential for the achievement of the SDGs. This is a looming challenge, underlined by the failure to meet the Aichi Biodiversity targets, part of the Convention on Biological Diversity (CBD) strategic plan for 2011–2020 (25), by 2020 (26). Biodiversity plays major roles in adapting to and mitigating climate change, maintaining freshwater quality, supporting the productivity of the oceans, and providing resilient supplies of food, fuel, and fibre for today's and future populations (27). **Box 7.1** summarizes global biodiversity commitments in the CBD and SDGs.

Several generalizations apply to the health-related SDGs. First, they are numerous, reflecting the many pathways through which human choices, and resulting social and environmental conditions, determine health. Second, many of the SDGs are more fully met in urban areas than in rural areas, emphasizing a persistent urban–rural health gap. Third, Africa and parts of Asia lag the rest of the world across many SDGs, emphasizing the need for particular attention to those regions. Overall, the UN concludes that the prospects for achieving the SDGs have deteriorated since 2015. The weakening of multilateral cooperation diminishes the ability to mobilize resources and to implement commitments under global treaties such as the Paris Climate Agreement.

Challenges in Monitoring Progress

One of the major conceptual and practical challenges of the SDGs is monitoring progress. Relevant data are collected and reported at the country level, as well as by various international organizations (World Bank, UNDP, the Sustainable Development Solutions Network, Our World in Data's SDG Dashboard, OECD, WHO). The different reports do not always correspond closely, a function of different data sources and data processing methods, different proxy measures chosen when preferred data are unavailable, and other factors (17).

A principal mechanism for tracking progress on the SDGs is the High-Level Political Forum on Sustainable Development (HLPF), a successor to the UN Commission on Sustainable Development, which functioned for about two decades until 2013. The HLPF has convened annually since 2013 under the auspices of the UN General Assembly and the UN Economic and Social Council. As part of SDG process, member states are encouraged to conduct regular reviews of progress at the national and sub-national levels, which are in turn reviewed by the HLPF. Of the 193 UN member states, 46 conducted these voluntary national reviews in 2018, 47 in 2019, and 47 in 2020 – roughly a quarter of the total, each year, reflecting the shortfalls in tracking SDG progress.

Another tracking process is carried out by the Sustainable Development Solutions Network, which issues an annual report, including an SDG Index and Dashboard, tracking country performances on the 17 SDGs (weighted equally). The 2020 report (28) provided country scores for 166 of 193 UN member states; the remaining 27 countries were unable to meet standards of data availability and quality to be included in the ranking. The number of indicators included in this report increased from 60 in 2016 to 88 in 2018 (29), and to 115 in 2020 (including 85 assessed globally, and 30 more only for OECD countries) (28). These represent a subset of the full range of 232 indicators for which data are available.

Health-related SDGs are the focus of several tracking efforts. The World Health Organization publishes an annual report, *World Health Statistics*, which monitors health for the SDGs (30). This report considers 43 health-related indicators, mostly corresponding to SDG 3 targets, but also drawing from SDGs 1, 6, 7, 11, 16, and 17. According to the 2019 report, whilst progress had been made on 24 of 43 health-related indicators, progress had stalled or reversed for five indicators (child overweight, malaria incidence, alcohol consumption, water sector ODA, and road traffic mortality). (Trends had not yet been reported for 14.) At least half of the world's population was not receiving essential health services when needed. The WHO estimated a global deficit of nearly 18 million health workers by 2030, largely in LMICs. An analysis of health facilities in 16 countries, largely in Africa and the Americas, showed that only 15% of health facilities surveyed in 2016 provided affordable medicines (30). Declining trends in HIV and tuberculosis incidence are insufficient to meet the SDG targets, with drug-resistant tuberculosis constituting a growing threat in some countries. After years of progress malaria incidence was on the rise again. In ten of the highest burden SSA countries there were an estimated 3.5 million more cases in 2017 than in 2016 (21).

The Global Burden of Diseases, Injuries, and Risk Factors Study (GBD) (31) identified 52 health-related SDG indicators (related to SDGs 1, 2, 3, 5, 6, 7, 8, 11, 13, 16, and 17) and assessed progress on 41 of them, in 2017. This assessment yielded a health-related SDG index for 195 countries and territories for the period 1990–2017, and projected trends to 2030. There were dramatic differences in trends between countries, and for many countries there were substantial internal inequities. The indicators for which the most countries had at least 95% probability of target attainment by 2030 were under-5 mortality, neonatal mortality, maternal mortality ratio, and malaria indicators. No countries were projected to meet the targets for NCD mortality and suicide mortality. For some indicators including child undernutrition, several infectious diseases, and most violence measures, no country was achieving rapid enough progress to meet the SDG targets. The GBD has since expanded its set of health-related SDG indicators to 57 (17). The COVID-19 pandemic has had pervasive but as yet unquantified impacts on progress towards the achievement of the SDGs.

Progress on a sustainable environment and intact Earth systems has been equally difficult to monitor. Across the SDGs (with the exception of Goal 10) there are 93 indicators directly related to the environment. However, only 34 of these have existing agreed methodology and data available from most countries (32). Biodiversity, a highly complex outcome, is an example of the difficulty of measuring environmental progress. Aggregated indices are needed. Changes in terrestrial biotic integrity (the 'health' of the biota) can be estimated and mapped globally using existing indices such as the Biodiversity Intactness Index (BII), based on estimates of the average abundance of originally present species for any defined area relative to their abundance in undisturbed habitat. Near-future global losses of species (extinctions) may be estimated using the Red List Index (RLI). Trends in the abundance of wild species are reflected by population-level indicators such as the Living Planet Index (LPI). These latter two indices are measured across the Earth as a whole and reflect the diversity and abundance of species globally (see (27) for discussion). At present widely accepted indicators that directly link biodiversity loss and human health are proving elusive, but this should not undermine attempts to preserve biodiversity, loss of which is irreversible.

The SDGs highlight the challenges, but also emerging opportunities, in identifying and tracking appropriate indicators of the state of the planet and its people. Increasingly countries will need to integrate data from Earth observations, official sources and national surveys, and ground-based monitoring (including by citizen scientists), as well as making use of the potential of big data in order to have a fuller picture of progress.

Opportunities for Multiple Benefits across the SDGs

Despite the sobering assessment of progress so far there is much that can be done to galvanize political will, foster more rapid progress, and achieve the SDGs. Addressing single targets across the SDGs in isolation is inefficient and risks unintended adverse consequences. Identifying policies and interventions that can yield multiple benefits offers the prospect of more rapid progress and optimal resource use. An example of such synergy is the multiple benefits of reducing short-lived climate pollutants (particularly black carbon and methane) (33). Because methane contributes to (tropospheric) ozone formation which in turn damages crops and human health, measures to reduce methane emissions support the achievement of SDG 2 (zero hunger) and SDG 3 (good health and well-being) by reducing ozone air pollution, and help to mitigate climate change (SDG 13). Similarly, the progress towards Goals 2 and 3 would benefit from reduced black carbon pollution from a variety of measures such as clean cooking technologies, replacing kerosene wick lamps with clean efficient lighting technologies, eliminating highly polluting on- and off-road diesel vehicles, reducing open-field burning of agricultural waste, and replacing traditional brick kilns and coke ovens with high-efficiency technologies (see Chapter 8 for wider discussion of the health co-benefits of clean energy). Some interventions for reducing these emissions also have beneficial effects on the achievement of other goals and targets.

At the same time, potential trade-offs must be identified to enable the risks and benefits to be assessed. Examples include the promotion of costly zero-carbon energy which disadvantages the poor, the potential effects of biofuel policies which result in increased food prices, and increases in household air pollution exposure following improved insulation and sealing air leaks in houses without providing adequate ventilation (see Chapters 8 and 9 for further discussion).

Financing the SDGs

Can the SDGs be feasibly achieved or are they a utopian dream that creates unrealistic expectations? The answer depends on many factors, including governance mechanisms, garnering sufficient public support, and vanquishing vested interests. However, a key enabler is financial resources. An initial attempt to assess the incremental investment needed for SDG achievement arrived at an estimate of US$343–360 billion per annum for low-income countries and US$900–944 billion per annum for LMICs (34). In another estimate, the IMF assessed annual spending needs for 155 developing countries in five areas – education, health, roads, electricity, and water and sanitation – and estimated total needs of US$2.6 trillion by 2030 of which US$2.1 trillion was accounted for by emerging economies (35). They estimated an additional expenditure of about US$500 billion annually would be needed by 49 low-income countries, amounting to about 15% of their GDP by 2030. While these are substantial sums they are only a small proportion of the global economy (which currently stands at US$80 trillion in nominal terms and over US$100 trillion if adjusted for purchasing power parity) and could in theory be met by a combination of development aid, private investment, and domestic government sources from the countries concerned. The almost instantaneous redeployment of large amounts of funding in response to the COVID-19 pandemic in 2020 suggests that the world economic system has more flexibility than is typically supposed.

Currently though, the resources are inadequate and the 2020 UN report on financing the SDGs sounded the alarm about negative trends (36). These included capital outflows of above US$200 billion from low-income countries in 2018; growing inequalities in many countries; 30 least developed and other vulnerable countries facing or at a high risk of debt distress; increasing trade restrictions; and faltering growth in real wages. Only five of the 30 OECD Development Assistance Committee (DAC) countries had achieved the UN target of spending 0.7% of their GDP on development. The average was 0.31% and aid fell in real terms in 2017. One-quarter of bilateral aid was going to humanitarian spending on refugees, up from one-sixth in 2010 (35). By 2020, with the COVID-19 pandemic raging and its full impacts not yet predictable, these trends had only worsened (36).

Faced with these challenges, the UN calls for reform of the multilateral trading system, more effective mechanisms to deal with sovereign debt which are contingent on countries' ability to pay, greater mobilization of private sector finance and harnessing technology, for example to provide banking services to those on low incomes. Some of these potential solutions are however of uncertain effectiveness. For example, there is doubt about whether, and in which circumstances, blended finance including public–private partner-ships provides good value for money, and less than 10% goes to low-income countries.

Short-term decision making may distort development priorities and divert limited resources from strategic plans that could propel lasting progress. In response, the UN advocates integrated national financing frameworks. Co-financing across sectors could be used as a more efficient approach to raising and deploying resources than single sector approaches (37). One example is the UNDP's Solar for Health initiative (www.undp-capacitydevelopment-health.org/en/capacities/focus/solar-for-health/), which aims to pro-vide solar energy for health facilities. In doing so it helps to save lives (SDG 3), ensure access to clean sources of electricity (SDG 7), eliminate unreliable and environmentally damaging energy sources (SDGs 12 and 13), and provide opportunities for private sector investors to contribute to sustainability (SDG 17). Currently the resources required for SDG financing are spread across a range of stakeholders and the costs may be too great for a single funder to bear. However, if spillover effects (i.e. effects on multiple outcomes relevant to the SDGs) are considered, then a compelling case can be made for investments from multiple sources. For co-financing to be worthwhile, by enhancing the efficiency of expend-iture of a particular sector, it would have to reduce the cost of achieving a unit of outcome for the sector concerned compared with its current least efficient intervention or programme.

In order to make co-financing work it would be necessary to quantify the multiple outcomes across sectors, together with the costs of the intervention or programme, and to understand the cost of achieving the outcome of interest for a given sector using alternative approaches. The latter should represent the willingness to pay by the sector in question. Such approaches should be attractive to rational decision makers and could be used both when central budget holders (such as the Ministry of Finance) control expenditure and when responsibilities are shared among budget holders. Negotiations could be more complex in the latter case, and other considerations such as competition for total resources may come into play, but nevertheless such approaches offer potential for bringing limited resources together more efficiently and accelerating progress towards achieving the SDGs.

Wider issues of resource mobilization to address the challenges of the Anthropocene are discussed in subsequent chapters.

The Limitations of the SDGs

Despite the multiplicity of goals, targets, and indicators, the SDGs feature several limita-tions. One was discussed above: the difficulty of monitoring progress. Observers have also argued that many of the SDGs are aspirational or vague (38). Critics have noted that the SDGs do not directly confront the biophysical limits of the planetary boundaries; for instance they do not require an absolute reduction of resource use, which amounts to implicitly prioritizing economic growth over sustainability (39). This in turn may constrain the ability of the SDGs to drive transformative change at the scale, and with the rapidity, required. Still another criticism holds that the SDGs are implemented mostly by governments and multilateral agencies, whereas lasting change requires extensive engagement by cities, businesses, and civil society (40), and downscaling from the global to the local (41).

A major conceptual defect is the inability to assess whether, and if so to what extent, a given country is achieving progress by unsustainably exploiting natural systems and thus diminishing the prospects of today's or future populations. Chapter 6 discussed a number of indices that attempt to assess human progress in relation to environmental impact. Each of these has its strengths and weaknesses, but inclusion of both composite and single issue indicators that encompass human progress and environmental sustainability would greatly facilitate assessment of progress towards sustainable living within planetary boundaries. One example is the relationship between carbon emissions and life expectancy (**Figure 7.2**) (42). As shown in the chart inset, some countries have achieved life expectancies well above 70 years at annual average carbon emissions less than one tonne per capita. These countries are quite diverse in terms of climate, location, and culture, although they include no high-income countries; they do include colder countries that generally have higher per capita emissions because of the need to heat buildings. This analysis suggests that achieving a high

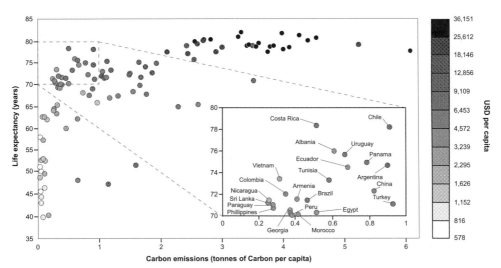

Figure 7.2. Life expectancy (vertical axis), consumption-based emissions (horizontal axis), and income (vertical grey-scale). The inset shows countries with life expectancy over 70 years and less than one tonne of carbon emissions per capita. The highest life-expectancy levels are attained at a wide range of carbon emissions and incomes.

Source: Steinberger JK, Roberts JT, Peters GP, Baiocchi G. Pathways of human development and carbon emissions embodied in trade. *Nature Climate Change.* 2012;2(2):81–5.

level of health is compatible with low GHG emissions but that currently a high per capita GDP is not.

Another limitation of the SDGs, perhaps related to their complexity, is the lack of broad-based popular awareness and support in many countries. While some NGOs are campaigning for the SDGs it is difficult to mobilize mass support behind such a complex and heterogeneous collection of goals and targets. The pursuit of economic growth in contrast is an apparently simple and often seductive priority, despite its flawed basis. It is essential in low-income countries to lift people out of poverty. But economic expansion continues to dominate the political agenda in high-income countries as well, where its benefits are often equivocal as it often disproportionately benefits the rich, erodes natural capital, and undermines sustainable development. Advancing the SDGs will require new forms of political discourse, built on widespread popular support for quality of life, equity, and sustainability. Support for 'Green New Deal' proposals in Europe and North America provide a basis for cautious optimism (see also Chapter 12).

The Way Forward

The SDGs will dominate the global landscape for sustainable development until 2030, after which they will be superseded by, hopefully, an even more comprehensive and far-reaching framework to support human progress within environmental limits. This need not mean a more complex set of goals and indicators but rather a focus on key transformative policies that can deliver this ambitious vision. Meanwhile, capitalizing on the opportunities of the SDGs can yield major benefits and insights that will shape the post-SDG agenda. Each of the SDGs potentially contributes to reaching four aspirations: prosperity, social inclusion, environmental sustainability, and good governance (local to global). Judicious selection of policies that can achieve multiple benefits can accelerate progress and could help lever the necessary resources by encouraging co-financing across sectors.

An approach to reduce the complexity of the SDGs and make action more feasible is to focus on a number of transformative policies that address multiple goals (see also Chapter 12). The World in 2050 (43), for example, focuses on six transformations across global, regional, and local scales that address major drivers of future changes: (a) human capacity and demography; (b) consumption and production; (c) decarbonization and energy; (d) food, biosphere, and water; (e) smart cities; and (f) digital revolution. (A similar framework has been proposed by the Sustainable Development Solutions Network (12) (see p. 217).) The TWI 2050 framework is designed to allow modellers to explore the effects of different policy options on the achievement of all of the SDGs, taking into account synergies and trade-offs and using common assumptions in some cases, in order to facilitate comparisons.

Subsequent chapters address opportunities to improve health through sustainable development policies that aim to safeguard the future of humanity within planetary boundaries, including the need for transformative change to adapt to the environmental change which cannot be prevented and to reduce profoundly the environmental footprint of humanity in coming decades.

References

1. Haines A, Alleyne G, Kickbusch I, Dora C. From the Earth Summit to Rio+20: integration of health and sustainable development. *The Lancet*. 2012;379(9832): 2189–97. doi: 10.1016/s0140-6736(12)60779-x.
2. UN. The Millennium Development Goals Report. New York: United Nations; 2015. Available from www.undp.org/content/dam/undp/library/MDG/english/UNDP_MDG_ Report_2015.pdf.
3. World Bank, Poverty & equity data portal. 2020. Available from http://povertydata .worldbank.org/poverty/home/.
4. McArthur JW, Rasmussen K. Change of pace: accelerations and advances during the Millennium Development Goal era. *World Development*. 2018;105:132–43. https://doi .org/10.1016/j.worlddev.2017.12.030.
5. Jacob A. Mind the gap: analyzing the impact of data gap in Millennium Development Goals' (MDGs) indicators on the progress toward MDGs. *World Development*. 2017;93:260–78. https://doi.org/10.1016/j.worlddev.2016.12.016.
6. Waage J, Yap C, Bell S, et al. Governing the UN Sustainable Development Goals: interactions, infrastructures, and institutions. *The Lancet Global Health*. 2015;3(5): e251–2. doi: 10.1016/s2214-109x(15)70112-9.
7. Sachs JD. From Millennium Development Goals to Sustainable Development Goals. *The Lancet*. 2012;379(9832):2206–11. http://dx.doi.org/10.1016/S0140-6736(12)60685-0.
8. McNicol G, Jeliazovski J, François JJ, Kramer S, Ryals R. Climate change mitigation potential in sanitation via off-site composting of human waste. *Nature Climate Change*. 2020;10(6):545–9. doi: 10.1038/s41558-020-0782-4.
9. ICSU. *A Guide to SDG Interactions: From Science to Implementation*. Paris: International Science Council; 2016. Available from https://council.science/publica tions/a-guide-to-sdg-interactions-from-science-to-implementation/.
10. Nilsson M, Griggs D, Visbeck M. Map the interactions between Sustainable Development Goals. *Nature*. 2016;534:320–2.
11. Pradhan P, Costa L, Rybski D, Lucht W, Kropp JP. A systematic study of Sustainable Development Goal (SDG) interactions. *Earth's Future*. 2017;5(11):1169–79. doi: 10.1002/2017ef000632.
12. Sachs JD, Schmidt-Traub G, Mazzucato M, et al. Six transformations to achieve the Sustainable Development Goals. *Nature Sustainability*. 2019;2(9):805–14. doi: 10.1038/s41893-019-0352-9.
13. Schmidt H, Gostin LO, Emanuel EJ. Public health, universal health coverage, and Sustainable Development Goals: can they coexist? *The Lancet*. 2015;386(9996): 928–30. doi: 10.1016/S0140-6736(15)60244-6.
14. Gwatkin DR, Ergo A. Universal health coverage: friend or foe of health equity? *The Lancet*. 2011;377(9784):2160–1. doi: 10.1016/s0140-6736(10)62058-2.
15. Sanogo NdA, Fantaye AW, Yaya S. Universal health coverage and facilitation of equitable access to care in Africa. *Frontiers in Public Health*. 2019;7:102. doi: 10.3389/fpubh.2019.00102.
16. Hussain S, Javadi D, Andrey J, Ghaffar A, Labonté R. Health intersectoralism in the Sustainable Development Goal era: from theory to practice. *Globalization and Health*. 2020;16(1):15. doi: 10.1186/s12992-020-0543-1.
17. Asma S, Lozano R, Chatterji S, et al. Monitoring the health-related Sustainable Development Goals: lessons learned and recommendations for improved measure-ment. *The Lancet*. 2020;395(10219):240–6. doi: 10.1016/S0140-6736(19)32523-1.
18. Lucas PL, Hilderink HBM, Janssen PHM, et al. Future impacts of environmental factors on achieving the SDG target on child mortality: a synergistic assessment.

Global Environmental Change. 2019;57:101925. https://doi.org/10.1016/j.gloenvcha .2019.05.009.

19. Stafford-Smith M, Griggs D, Gaffney O, et al. Integration: the key to implementing the Sustainable Development Goals. *Sustainability Science*. 2017;12(6):911–19. doi: 10.1007/s11625-016-0383-3.

20. Hallegate S, Bangalore M, Bonzanigo L, et al., editors. *Shock Waves: Managing the Impacts of Climate Change on Poverty*. Washington, DC: World Bank; 2016.

21. Secretary-General. Progress Towards the Sustainable Development Goals. Report of the Secretary-General. New York: United Nations Economic and Social Council; 2019. Contract No. E/2019/68. Available from https://unstats.un.org/sdgs#.

22. FAO, IFAD, UNICEF, WFP, WHO. The State of Food Security and Nutrition in the World 2019: Safeguarding Against Economic Slowdowns and Downturns. Rome: FAO; 2019. Available from www.fao.org/publications/sofi/en/.

23. Lutz W, Kebede E. Education and health: redrawing the Preston curve. *Population and Development Review*. 2018;44(2):343–61. doi: 10.1111/padr.12141.

24. Ezeh A, Oyebode O, Satterthwaite D, et al. The history, geography, and sociology of slums and the health problems of people who live in slums. *The Lancet*. 2017;389(10068):547–58. doi: 10.1016/S0140-6736(16)31650-6.

25. Marques A, Pereira HM, Krug C, et al. A framework to identify enabling and urgent actions for the 2020 Aichi Targets. *Basic and Applied Ecology*. 2014;15(8):633–8. https://doi.org/10.1016/j.baae.2014.09.004.

26. IPBES. Global Assessment Report on Biodiversity and Ecosystem Services. Bonn, Germany: Intergovernmental Science-Policy Platform on Biodiversity and Ecosystem Services; 2019. Available from www.ipbes.net/assessment-reports.

27. Mace GM, Barrett M, Burgess ND, et al. Aiming higher to bend the curve of biodiversity loss. *Nature Sustainability*. 2018;1(9):448–51. doi: 10.1038/s41893-018-0130-0.

28. Sachs J, Schmidt-Traub G, Kroll CN, LaFortune G, Fuller G, Woelm F. Sustainable Development Report 2020: The Sustainable Development Goals and COVID-19. Cambridge, UK: Cambridge University Press, 2020. Available from www.sdgindex.org/.

29. Lafortune G, Fuller G, Moreno J, Schmidt-Traub G, Kroll C. SDG Index and Dashboards: Detailed Methodological Paper. New York: Sustainable Development Solutions Network; 2018.

30. WHO. World Health Statistics 2019: Monitoring Health for the SDGs. Geneva: World Health Organization; 2019. Available from www.who.int/gho/publications/world_ health_statistics/2019/en/.

31. GBD 2017 SDG Collaborators. Measuring progress from 1990 to 2017 and projecting attainment to 2030 of the health-related Sustainable Development Goals for 195 countries and territories: a systematic analysis for the Global Burden of Disease Study 2017. *The Lancet*. 2018;392(10159):2091–138. doi: 10.1016/s0140-6736(18) 32281-5.

32. UNEP. GEO-6: Healthy Planet, Healthy People. United Nations Environment Programme; 2019. Available from https://wedocs.unep.org/handle/20.500.11822/27539.

33. Haines A, Amann M, Borgford-Parnell N, et al. Short-lived climate pollutant mitigation and the Sustainable Development Goals. *Nature Climate Change*. 2017;7(12): 863–9. doi: 10.1038/s41558-017-0012-x.

34. Schmidt-Traub G, Sachs JD. Financing for Sustainable Development: Implementing the SDGs through Effective Investment Strategies and Partnerships. New York: Sustainable Development Solutions Network; 2015. Available from https://resources.unsdsn.org/

financing-for-sustainable-development-implementing-the-sdgs-through-effective-invest
ment-strategies-and-partnerships.

35. UN Inter-agency Task Force on Financing for Development. Financing for Sustainable Development Report 2019. New York: United Nations; 2019. Available from https://developmentfinance.un.org/fsdr2019.

36. UN Inter-agency Task Force on Financing for Development. Financing for Sustainable Development Report 2020. New York: United Nations; 2020. Available from https://developmentfinance.un.org/fsdr2020.

37. UNDP. Financing Across Sectors for Sustainable Development: Guidance Note. New York: UN Development Programme; 2019. Available from www.undp.org/content/undp/en/home/librarypage/hiv-aids/financing-across-sectors-for-sustainable-development.html.

38. Murray CJL. Shifting to Sustainable Development Goals: implications for global health. *New England Journal of Medicine*. 2015;373(15):1390–3. doi: 10.1056/NEJMp1510082.

39. Eisenmenger N, Pichler M, Krenmayr N, et al. The Sustainable Development Goals prioritize economic growth over sustainable resource use: a critical reflection on the SDGs from a socio-ecological perspective. *Sustainability Science*. 2020;15(4):1101–10. doi: 10.1007/s11625-020-00813-x.

40. Hajer M, Nilsson M, Raworth K, et al. Beyond cockpit-ism: four insights to enhance the transformative potential of the Sustainable Development Goals. *Sustainability*. 2015;7(2). doi: 10.3390/su7021651.

41. Sterling EJ, Pascua P, Sigouin A, et al. Creating a space for place and multidimensional well-being: lessons learned from localizing the SDGs. *Sustainability Science*. 2020;15(4):1129–47. doi: 10.1007/s11625-020-00822-w.

42. Steinberger JK, Roberts JT, Peters GP, Baiocchi G. Pathways of human development and carbon emissions embodied in trade. *Nature Climate Change*. 2012;2(2):81–5. doi: 10.1038/nclimate1371.

43. TWI2050. The World in 2050: Transformations to Achieve the Sustainable Development Goals. Laxenburg, Austria: International Institute for Applied Systems Analysis (IIASA); 2018. Available from www.twi2050.org.

8

Transforming Energy and Industry: Towards a Net-Zero Circular Economy for Health

The industrial revolution, the rise of nation states, and the emergence of market societies represented a turning point in the history of human civilization – a Great Transformation, as memorably characterized by economic historian Karl Polanyi (1). Indeed, there are echoes of Polanyi's phrase in the Great Acceleration, the vast upscaling of the human enterprise that has brought us up against planetary boundaries (2). We can assert, without hyperbole, that another civilizational transformation is now needed – a transformation in how energy and materials are used, in how humans co-exist with the natural world, and in the accompanying social and economic underpinnings of modern societies (3, 4).

This chapter begins with an account of decarbonization – the process of phasing out fossil fuel combustion as a primary source of energy. But decarbonization is only one part of a larger transformation, one that will require reduced consumption by people with high-consumption lifestyles, balanced by increased consumption by those now enduring lives of deprivation. It will also require a circular economy that replaces the 'take–make–waste' status quo – systematically designing out waste and pollution, keeping products and materials in use, and regenerating natural systems. These changes, if implemented and managed well, will deliver substantial short-term benefits for human health and well-being. Throughout the chapter, our vision of a clean, healthy, equitable, and sustainable economy is animated by a spirit of 'constructive hope' that helps motivate action (5) and by the conviction that the needed changes, while difficult, are within reach.

The Challenge of Decarbonization

Decarbonization is fundamental to stabilizing the Earth's climate and to managing the impacts of climate change. It is a pressing but elusive goal. The scale of the challenge is made clear by consideration of carbon budgets – calculations of how much carbon-based fuel can be burned, and the amount of GHGs that can be emitted, within a defined time period, to keep within specific temperature limits.

Carbon Budgets

Carbon budgets are based on Earth system models (ESMs) which simulate the behaviour of the carbon cycle and the responses of the oceans, atmosphere, soil, and other Earth systems, to estimate the Earth's temperature over multi-decadal timescales (6–8). Some uncertainty is inherent in these calculations; they require assumptions about complex feedback loops, GHGs other than carbon dioxide (CO_2), and technologies and policy choices that are at this point unpredictable. One benchmark is the 2014 IPCC estimate, that to have at least a 50% chance of keeping average global temperature increases to $2\,°C$ above pre-industrial levels during the twenty-first century, the cumulative emissions of CO_2 between 2011 and 2050 must be restricted to about 1100 gigatonnes (870–1240 Gt CO_2) (9). In the decade that began in 2011, the starting point for IPCC calculations, global CO_2 emissions from fossil fuels averaged about 35 Gt/year (8, 10), which would reduce the IPCC's emissions ceiling to about 750 Gt. By comparison, the carbon emissions contained in known fossil fuel reserves are estimated to be well over 3000 Gt, and total fossil fuel resources potentially recoverable with future technologies, irrespective of economic conditions, could exceed 10,000 Gt. The mathematics is clear: to keep within the $2\,°C$ limit, most known fossil fuels (over 80% of coal, 33% of oil, and 50% of gas) must remain in the ground. In particular, the USA and former Soviet bloc countries would all have to use less than 10% of their known reserves of both hard coal and lignite. All Arctic resources are considered unburnable, and Canada would only be able to burn about 25% of its oil reserves as its natural bitumen deposits should remain largely unexploited (11).

Could more fossil fuels be burned while complying with carbon budget limits if the CO_2 released were captured or if CO_2 were removed from the atmosphere? Both carbon capture and storage (CCS) and carbon dioxide removal (CDR) have been promoted as ways of continuing to exploit fossil fuel reserves whilst limiting CO_2 levels in the atmosphere. However, these technologies remain speculative at present, and are very far from being deployed at any meaningful scale. They are prohibitively expensive, especially in relation to increasingly affordable renewable energy. Even if CCS can be scaled up cost effectively in the future (which seems doubtful at present), this would have only a modest impact on the reserves that can be exploited. For coal, CCS would reduce the proportion of coal needing to remain unburned globally from 88% only to 82% (11). The conclusion is clear: rapid, aggressive decarbonization of the global economy is required, while ongoing research efforts aim to improve the prospects for CDR.

The Current Status of Decarbonization

The world is not now on track to decarbonize sufficiently to reach the $2\,°C$ target. GHG emissions from fossil fuel combustion rose each year between 2000 and 2019 (at 3%/year in the 2000s, declining to 0.9%/year after 2010), stabilizing only briefly between 2014 and 2016, and reaching 36.8 Gt in 2019. (2020 data are not available at the time of writing, but will likely reflect a modest temporary downturn related to COVID-19.) Total GHG emissions, including from land use change, reached a record high of 43.1 $GtCO_2eq$ in 2019 (8). If countries implement only the unconditional Nationally Determined Contributions (NDCs) agreed in Paris, the cumulative gap between emissions and needed reductions is about 15 $GtCO_2eq$ (for the $2\,°C$ target) and 32 $GtCO_2eq$ (for the $1.5\,°C$ target) by 2030. If conditional NDCs (e.g. conditional on international support) are also implemented, the gap falls by about 3 GtCO$_2$eq. Implementing current unconditional commitments consistently throughout the twenty-first century would lead to a global mean temperature rise of about 3.2 °C (range 2.9–3.4 °C) by 2100. Current policies that only reduce global emissions by about 6 GtCO$_2$eq by 2030 are therefore grossly inadequate in their scale and level of ambition (12).

There are some encouraging signs, particularly with regard to coal. While coal exploitation is not over, coal plant construction has fallen in recent years. However, the decline has not been consistent; in 2019 the global coal fleet grew by 34.2 gigawatts (GW), the first increase since 2015, with nearly two-thirds of newly commissioned capacity in China. State-owned financial entities, particularly in China, continue to back coal plants in countries such as Bangladesh, South Africa, and Egypt. Of the OECD countries, Japan is the largest driver of new coal plant construction, with 11.9 GW under development domestically and 24.7 GW financed by Japan globally (13). But 33 nations now have coal phaseout plans. Over 100 financing institutions are restricting funding for new coal plants. Coal plant construction indicators in India declined by half from 2018 to 2019, and declines reached 22% in Southeast Asia, 40% in Africa, and 60% in Latin America. In India increasingly low tariffs for solar and wind energy are competing successfully with coal, and coal plant development permits in 2018 dropped to 10% of the annual 2008–2012 level. Coal plant costs will increase in the future to comply with increasingly strict pollution standards and become even less competitive with renewables.

In Europe the picture is mixed. While GDP increased by 14% from 2010 to 2019, electricity consumption fell by 4%, with the greatest declines in the UK, Germany, France, and Sweden (and with increases in Poland, Hungary, Lithuania, Romania, Malta, and Ireland) (14). Between 2010 and 2019, wind generation tripled, solar generation rose six-fold, and coal generation fell by 43%. In 2019, for the first time, wind and solar provided more electricity in Europe than did coal (18% versus 15%) (14). Six European countries were coal-free as of 2019, and 14 had committed to phasing out coal power by 2030, along with Greece by 2028 and Germany by 2038 (13). Poland has not committed to phasing out coal, and in fact commissioned 1.8 GW of coal in 2019. To stay within the Paris Agreement commitments about 70% of coal power stations should go off-line by 2025 (15).

Concern about inadequate progress has led to greater interest in the role of non-state actors (NSAs) including private firms and cities. There has been growing momentum, with over 7000 cities from 133 countries and over 6000 companies with at least US$36 trillion in revenue pledging mitigation action. Their potential impact is substantial – up to 19 GtCO$_2$eq emission reductions in excess of current policies by 2030, according to one estimate – but there are major uncertainties about how much private firms can mobilize change (16), particularly in countries whose national governments are uncommitted.

The above estimates may even understate the magnitude of the challenge. Some of the most recent climate models in the Coupled Model Intercomparison Project, CMIP 6, show much higher climate sensitivities than earlier models, suggesting long-term global average temperature increases of about 5 °C with a doubling of CO$_2$, compared with pre-industrial levels (17). This has been validated by short-term forecasts (18) and is due to changes in the understanding of cloud microphysics, a crucial determinant of climate system responsiveness. This emerging evidence suggests the need for even greater levels of ambition and adds to our perception of a climate emergency.

Economic and Social Policies to Support Decarbonization

Decarbonizing society requires both economic and social policies, and development and implementation of a range of technologies.

Reducing Fossil Fuel Subsidies

Fossil fuels are heavily subsidized throughout the world, with the result that the full economic costs of fossil fuels go unpaid. Two conceptual frameworks are useful in considering fossil fuel subsidies. The narrower concept, pre-tax subsidies, refers to the difference between the amount consumers pay for fuel and the true (opportunity) cost of supplying the fuel. For example, many oil-producing governments in the Middle East and North Africa have kept domestic oil prices below international prices. The global debate on fossil fuel subsidies typically refers to pre-tax subsidies. The broader approach, post-tax subsidies, refers to the difference between consumer fuel prices and true supply costs *plus* the taxes that would ordinarily be imposed on the full-price transactions, and the costs of environmental and health damage resulting from the fuel use (19). The concept of post-tax subsidies incorporates health and environmental costs that are generally externalized. According to an International Monetary Fund estimate, post-tax subsidies (US$5.2 trillion in 2017, or 6.5% of global GDP) were about 17 times higher than pre-tax subsidies (19–21). About 22% of the post-tax subsidy is accounted for by climate change costs (the magnitude of which, as we have seen, is open to debate) and 46% by air pollution costs, with China (US$1.8 trillion in 2013) and the USA (US$0.6 trillion in 2013) accounting for largest post-tax subsidies. Eliminating these subsidies would bring major benefits, including reducing carbon emissions by 21% and cutting fossil fuel air pollution deaths by about 55%, as well as increasing revenue by 4% of global GDP, and social welfare by 2.2% (using 2013 data).

All nations are committed, under Article 2.1.c of the Paris Agreement, to 'making finance flows consistent with a pathway towards low greenhouse gas emissions and climate-resilient development'. In 42 countries responsible for the majority of fossil fuel subsidies, subsidies declined between 2012 and 2016, but the subsidies increased in both 2017 and 2018, reaching US$319 billion and US$427 billion respectively – almost back to 2014 levels (22). Fossil fuel subsidies are substantial in some oil-producing nations such as Saudi Arabia and some coal-using countries such as Indonesia. The money devoted to fossil fuel subsidies globally is over three-fold greater than that received by renewables and exceeds the financing required to meet the SDGs for basic social protection and universal coverage with health, and education services by a similar amount (23).

In September 2016, the G20 countries reaffirmed their 2009 commitment to phase out (pre-tax) fossil fuel subsidies. However, removing only pre-tax subsidies would have limited impact; it would reduce the carbon price required to hold CO_2 levels in the atmosphere to 550 ppm by only 2–12%, if oil prices were low (24). Abolition of oil and gas subsidies might even lead to the unintended consequence of increased emissions, if these fuels were substituted by coal. Removing subsidies would have the largest benefits in high-income oil- and gas-producing countries where there are few poor people who would be adversely affected by higher prices and where the reductions in GHG emissions would exceed those from their national emission pledges under the Paris Agreement.

Reduction of (pre-tax) fossil fuel subsidies has sometimes been met by public opposition, including violent protests in some countries such as Nigeria, Indonesia, and Sudan. But there are opportunities to earn political support for subsidy removal by redirecting subsidies to popular social programmes including universal health coverage (25) (see Chapter 12).

Putting a Price on Carbon

The economically most efficient way of decarbonizing the economy is to put a price on carbon. Carbon pricing aims to internalize the cost of carbon emissions, thereby increasing the costs of fossil fuels compared with zero-carbon sources of energy and thus reducing the risks of dangerous climate change. Carbon pricing also helps ensure that emissions are cut where it costs least to do so. There are two main types of carbon pricing, emissions trading systems and carbon taxes. Emissions trading systems (ETSs) set a cap on the amount of GHG emissions permitted by covered sources. Companies or other entities receive a limited number of emission allowances which they can trade with each other, depending on their requirements for more or less emissions than their allotted allowance. The cap is reduced over time to drive emissions reductions. In the European Trading System, which covers about 45% of EU emissions, companies can also buy a limited number of international credits from emission-reducing projects around the world. Each year companies must surrender sufficient emission certificates to cover their emissions, or face heavy fines. Initial implementation of the European ETS was rocky (26). Over-issuing of allocation allowances beginning in 2005 led to price volatility and collapse of the carbon price, which in turn delayed investments in zero-carbon technologies. Subsequently, between 2008 and 2012, the economic crisis reduced demand and led to further price volatility. Weak governance mechanisms because of lobbying by high-emitting industries led to lax controls over national emission caps, thus compounding the oversupply of allowances, and to illegal activity including Value Added Tax (VAT) fraud. Reforms have been instituted to address these challenges, including the introduction of a market stability reserve to regulate prices of allowances, but the negative experiences have tarnished the reputation of the EU ETS specifically and ETSs more generally.

A 2019 World Bank report (27) documented 57 functioning carbon pricing schemes, including 28 ETSs in regional, national, and sub-national jurisdictions, and 29 carbon taxes, primarily at the national level. Examples of sub-national schemes include those in a number of US states, including California, and in the Canadian provinces of British

Columbia and Quebec. In total, carbon pricing initiatives cover 11 GtCO$_2$eq, or about 20% of global GHG emissions. The use of such schemes is likely to increase in the future as 96 of the 185 Parties to the UNFCCC (responsible for 55% of global GHG emissions) that have submitted their NDCs have stated that they are planning or considering the use of carbon pricing. Governments raised US$44 billion in revenues from carbon pricing in 2018, up from US$11 billion the previous year. However, the effectiveness of current pricing schemes is undermined by the wide range of carbon prices, which range from US$1 to US$127 per tonne of CO$_2$eq (the latter in Sweden), with about half the emissions being priced below US$10. Most current carbon prices are far below the levels required to meet the temperature targets enshrined in the Paris Agreement in a cost-effective manner. According to the High-Level Commission on Carbon Prices, prices should be at least US$40–80/tCO$_2$eq by 2020 and US$50–100/tCO$_2$eq by 2030 (28).

In order to keep to the much more ambitious (and probably unachievable) target of a 1.5 °C temperature increase (with 50–66% probability), the IPCC estimates that carbon prices should be in the range of US$135–6050/tCO$_2$eq in 2030, US$245–14,300/tCO$_2$eq in 2050, US$420–19,300/tCO$_2$eq in 2070, and US$690–30,100/tCO$_2$eq in 2100 (undiscounted values) (29). This would require far-reaching reforms in fiscal and tax policies, posing serious challenges to any government wishing to introduce them.

While carbon pricing is an efficient mechanism for decarbonizing the economy, it faces many of the same political hurdles as does the withdrawal of subsidies, because of opposition from powerful fossil fuel interests and public concerns about the impacts on the cost of living and employment. In December 2018, for example, the government of French President Macron was forced to backtrack on plans for an increased national fuel tax after sometimes violent protests by the *gilets jaunes* (yellow vests) movement. This experience demonstrates that any policy that aims to address environmental sustainability must also consider the impacts on the social and economic determinants of health and well-being (30, 31). However, well-designed policies, which aim to reduce inequalities as well as addressing public concerns about the environment, can motivate change and overcome resistance (32).

Carbon taxes exemplify a more general principle, namely that taxes should be targeted at activities that undermine sustainability and/or damage health and well-being, rather than at social goods (such as income). VAT could be reformed to disincentivize extractive industries such as mining or primary production and exempt activities that encourage re-use, repair, and re-manufacture. Carbon credits could be awarded to activities that prevent carbon emissions as well to those that reduce emissions. Any increases in government income can be used to support pro-poor services and reduce the burden from other taxes, particularly those that are regressive. In Chapter 12 we outline some lessons from the implementation of carbon pricing in different contexts.

The Economics of Decarbonization

Carbon prices cannot be equated with the aggregate economic costs of cutting carbon for two major reasons. First, emission prices reflect the marginal cost, i.e. the cost of an additional unit of emission reductions, whereas total economic costs reflect all mitigation activities that have been implemented. Second, emission prices interact with other measures to promote decarbonization, such as funding of research and development or subsidies for zero-carbon technologies, and will therefore reflect a lower marginal cost if these other measures have made significant contributions to mitigation efforts. Mitigation costs tend to rise over time and are of course on average higher for more stringent mitigation scenarios than for weaker ambitions, although with considerable variation depending on assumptions, for example about the technological options available and their costs.

The highly influential 2007 Stern Review concluded that 'stabilisation of greenhouse gases at levels of 500–550 ppm CO_2eq will cost, on average, around 1% of annual global GDP by 2050' (33). According to the IPCC (9), the decline in global consumption to reach levels of 430–480 ppm CO_2eq by 2100 will range from 1% to 4% in 2030, 2–6% in 2050, and 3–11% in 2100 relative to the baseline case without explicit mitigation. These equate to an annual loss of between 0.06 and 0.20 percentage points from 2010 to 2030. In contrast, studies project annual average consumption growth without mitigation of between 1.9% and 3.8% per year until 2050. Most studies reviewed by the IPCC report 1.5- to 3-fold higher losses of global consumption and GDP for scenarios reaching 430–530 ppm CO_2eq by 2100 compared to the 530–650 ppm CO_2eq range. An atmospheric CO_2 concentration of no more than 450 ppm is required to keep to a 2 °C temperature increase or 430 ppm for 1.5 °C. Overall these are modest costs when considered against the background rate of economic growth, or to the future costs of not taking action now. Furthermore, the resulting government income can be used to improve public services and/or reduce the tax burden on the poor. While equitable economic growth is important to alleviate poverty and improve health and well-being in LMIC economies, as discussed earlier, other factors such as income distribution and social policies are likely to be more important than economic growth for health and well-being in high-income countries.

Strategies for Keeping within 2 °C above Pre-Industrial Levels

Deep and rapid cuts in CO_2 and short-lived climate pollutant (SLCP) emissions are needed to give a reasonable probability of remaining within the 2 °C increase in mean global temperature agreed in Paris (**Figure 8.1**). Net emissions need to decline to near zero by mid-century in order to keep within the target temperature. A number of initiatives have suggested how this could be achieved. The Committee to Prevent Extreme Climate Change developed a suite of ten policies grouped as four 'building blocks' to address both GHGs and SLCPs (**Figure 8.2**) (34). In addition to implementing the emissions reductions agreed under the Paris Treaty (first building block), several sister agreements must be implemented (second building block). These include the Kigali Amendment to the Montreal Protocol to phase down hydrofluorocarbons (HFCs, which are GHGs used as refrigerants), efforts to address aviation emissions and black carbon (BC) emissions from shipping, and the commitment to reduce BC emissions by up to 33% by the eight countries

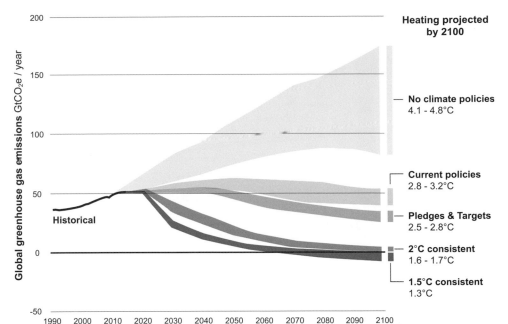

Figure 8.1. Global heating projections to 2100. Emissions and expected heating based on pledges and current policies. The 1.5 °C and 2 °C consistent pathways limit the amount of carbon dioxide removal, based on sustainability and economic constraints.
Source: Climate Action Tracker. Warming Projections Global Update. 2018. https://climateactiontracker.org/publications/warming-projections-global-update-dec-2018/.

making up the Arctic Council. Numerous sub-national and city-scale climate action plans must also be scaled up. One prominent example is the Under2 Coalition, signed by over 220 jurisdictions from 37 countries covering 43% of the world economy. The third building block comprises drastic reduction of SLCP emissions beginning now and completed by 2030, together with decarbonizing the global economy by 2050. Finally, the fourth building block was added as an insurance policy against delays and surprises and would require scaling up of carbon dioxide removal strategies. The amounts of CO_2 that would be necessary to remove will depend on the speed with which SLCP and GHG reductions occur and could range from zero if they decrease by 2020, to one trillion tonnes, if CO_2 emissions continue to increase until 2030.

Technical Approaches to Decarbonizing the Economy

In addition to the economic and policy approaches discussed above, decarbonization depends on technology. Some needed technologies are now readily available, and require only the appropriate incentives for widespread adoption. Other technologies are in development.

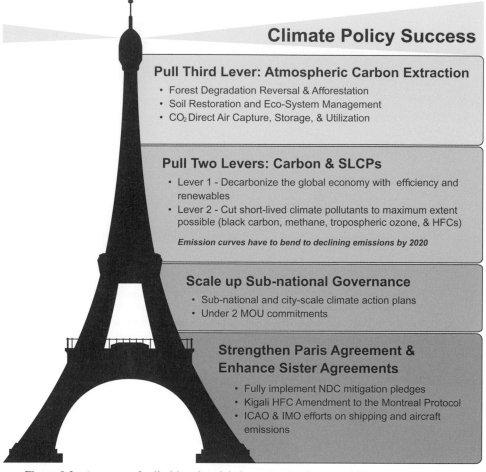

Figure 8.2. A strategy for limiting the global temperature increase to less than 2 °C above pre-industrial levels: four building blocks, three levers. SLCP: short-lived climate pollutants, HFCs: hydrofluorocarbons, MOU commitments: the Under2 Memorandum of Understanding, a climate agreement for sub-national governments (www.under2coalition .org/), NDC: Nationally Determined Contributions under the Paris Agreement, ICAO: International Civil Aviation Organization, IMO: International Maritime Organization.

Source: Ramanathan R, Molina ML, Zaelke D. *Well Under 2 Degrees Celsius: Fast Action Policies to Protect People and the Planet from Extreme Climate Change.* Paris: Climate & Clean Air Coalition; 2017. www.ccacoalition.org/en/resources/well-under-2-degrees-celsius-fast-action-policies-protect-people-and-planet-extreme.

Currently Feasible Approaches

A great deal of decarbonization can be accomplished with currently available technologies. This view underlay early analyses such as the 'stabilization wedges' model of Princeton professors Stephen Pacala and Robert Socolow (35), and with rapid technological development over recent decades, from storage batteries to electric vehicles, it has only become

Table 8.1. *The top ten solutions to climate change from the 2020 Drawdown analysis*

		Scenario 1: Drawdown by mid-2060s		Scenario 2: Drawdown by mid-2040s
Rank	Solution	Minimum CO_2eq (Gt) reduced/ sequestered (2020–2050)	Solution	Minimum CO_2eq (Gt) reduced/ sequestered (2020–2050)
1	Reduced food waste	86.7	Onshore wind turbines	147.7
2	Health and education	85.4	Utility-scale solar PV	119.1
3	Plant-rich diets	64.8	Reduced food waste	94.6
4	Refrigerant management	57.7	Plant-rich diets	91.7
5	Tropical forest restoration	54.5	Health & education	85.4
6	Onshore wind turbines	47.2	Tropical forest restoration	85.1
7	Alternative refrigerants	43.5	Improved clean cookstoves	72.8
8	Utility-scale solar PV	42.3	Distributed solar PV	68.6
9	Improved clean cookstoves	31.3	Refrigerant management	57.7
10	Distributed solar PV	28.0	Alternative refrigerants	50.5

Source: Project Drawdown. *The Drawdown Review 2020: Climate Solutions for a New Decade.* San Francisco: Project Drawdown; 2020. www.drawdown.org/drawdown-framework/drawdown-review-2020.

more valid. In 2017, Project Drawdown identified 100 solutions to climate change, ranked them in terms of the amount of CO_2eq reduced or sequestered, and estimated the associated costs; these were updated in 2020 (36, 37). (**Table 8.1** shows the top ten of the 100 solutions, updated in 2020, under each of two scenarios considered, one gradual and one rapid.) The estimated cost of purchasing and installing the 100 proposed solutions is between US\$22.5 and 28.4 trillion. Over their lifetimes, the solutions would collectively generate between US\$95.1 and 145.5 trillion in operating revenues. According to the Drawdown analysis, when compared to business as usual, the net lifetime profit would be between US\$15.6 and 28.7 trillion.

Outstanding Technical Challenges

Negative emission technologies are a staple of many scenarios for reaching a stable climate. Such scenarios assume GHG emissions at levels that would in and of themselves raise global temperatures unacceptably (called 'overshoot'), but compensate with carbon dioxide removal (CDR) and carbon capture and storage (CCS). A widely discussed example is bioenergy with carbon capture and storage (BECCS). BECCS refers to the use of biomass – energy crops and waste material such as agricultural and forestry residues – as a renewable energy source, coupled with capture and storage of the carbon released in its combustion. In the IPCC AR5, of the 116 scenarios that stabilized CO_2 concentrations between 430–480 ppm, over 100 relied on BECCS to deliver global net negative emissions. These assumed up to 22 Gt/year of CO_2 removal, with a median value of about 616 Gt CO_2 cumulatively removed by 2100 (38, 39). Such assumptions are necessary because we have already passed the point at which emissions reductions alone can achieve climate targets. However, they are heroic assumptions in view of current realities. As of 2019, just five facilities in the world were using BECCS technologies, collectively capturing approximately 1.5 million tonnes of CO_2 annually (40) – a tiny fraction of 1% of the level assumed in IPCC calculations. Moreover, the upscaling of BECCS would place enormous burdens on land and water supplies, thereby threatening biodiversity and food availability, and would pose the challenge of fine particulate air pollution from biomass burning (41). While negative emissions technologies will be essential, they are far from being available, and they pose major challenges if implemented (39, 41) – emphasizing the urgent need to decarbonize.

Even renewable energy technologies can release GHGs, as exemplified by the use of sulfur hexafluoride (SF_6) in electrical switchgear, including wind turbines, for its insulating properties. This compound has a global warming potential (GWP) 22,800 times that of CO_2 over a 100-year period with an atmospheric lifetime of 3200 years. The estimated atmospheric release of SF_6 from electrical equipment in England, Scotland, and Wales during 2015–16 amounted to about 11,320 kg, the equivalent of 258,110 tonnes of CO_2. Thus on a global scale the contributions could be concerning and the use of alternatives should be mandatory, but faces technical and implementation barriers (42).

Another set of technical challenges, unmet to date, relates to sectors in which GHG emissions are devilishly difficult and expensive to achieve, including certain heavy industries (cement, steel, and chemicals) and certain transport sectors (road vehicles weighing over 4.5 tonnes, shipping, and aviation). These are often overlooked in the NDCs submitted under the Paris Agreement. They are currently responsible for about 30% of the emissions from the energy and industrial sectors, and under business as usual development they are projected to increase in coming decades (43). The Energy Transitions Commission concluded that it was technically possible to decarbonize these sectors at a cost of about 0.5% of global GDP. If the challenges posed by these sectors can be successfully addressed, then decarbonization of the whole economy is feasible.

A final set of outstanding technical challenges relates to geoengineering. Radiative forcing geoengineering aims to reduce the amount of heating of the atmosphere by solar radiation and depends on mobilizing technologies such as injecting aerosols into the stratosphere or using space mirrors to reflect solar radiation (**Figure 8.3**). These techniques are highly controversial for a variety of reasons including regional variability of impacts, failure to address the direct effects of carbon dioxide excess such as ocean acidification and nutrient decline in food crops, and the significant risks accompanying large-scale experiments with Earth systems (44–46). There are major challenges for governance and informed consent of populations affected that currently appear intractable. Although some technologies have potential to contribute to reducing climate change, they are all in the early stages of development and carry great uncertainties and risks.

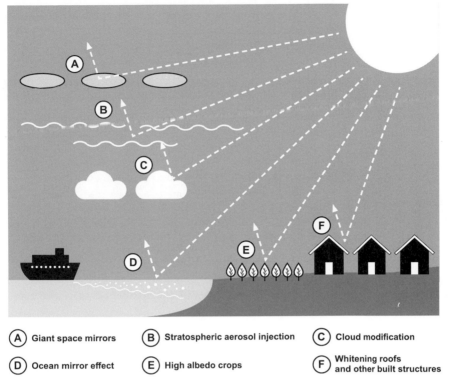

(A)	Giant space mirrors	(B)	Stratospheric aerosol injection	(C)	Cloud modification
(D)	Ocean mirror effect	(E)	High albedo crops	(F)	Whitening roofs and other built structures

Figure 8.3. Solar geoengineering. The six most commonly proposed solar geoengineering technologies are aerosol injection, marine cloud brightening, high-albedo crops and buildings, 'ocean mirrors' (using ocean foam to reflect sunlight), cloud thinning, and space sunshades or mirrors.

Source: Adapted from Carbon Brief, 2018. www.carbonbrief.org/explainer-six-ideas-to-limit-global-warming-with-solar-geoengineering. Image by Emanuel Santos.

Deep Decarbonization Pathways

The three main routes to decarbonization are reducing demand, greater efficiency, and decarbonizing the energy supply. Reducing demand can be achieved partly through transition to a circular economy based on re-manufacturing, re-use, and recycling, together with shared use of some products, as discussed below. Decarbonizing the energy supply relies on four principal technologies: hydrogen for heat storage or as a reduction agent; massive electrification from renewables; tightly regulated use of biomass for heat or as a feedstock for plastics; and finally a modest but important role for carbon capture and storage (see **Figure 8.4** for summary).

The most appropriate combination of technologies and policies to achieve deep cuts in carbon emissions will differ from country to country depending on energy sources, political priorities, economic system, and infrastructure. The Deep Decarbonization Pathways Project has developed methods for policy-relevant national analyses of decarbonization options and has assessed 16 developed and emerging economies representing 74% of the

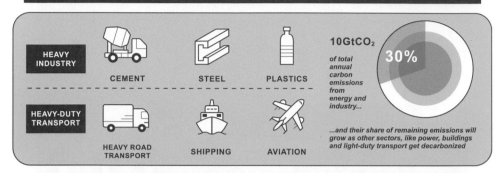

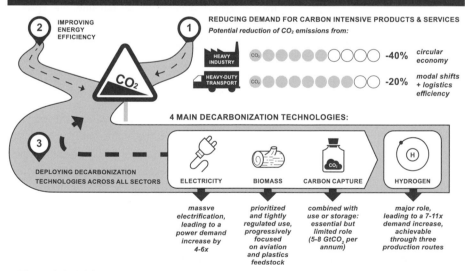

Figure 8.4. Major routes to decarbonization by mid-century.

Source: Energy Transitions Commission. Mission Possible: Reaching Net-Zero Carbon Emissions from Harder-To-Abate Sectors by Mid-Century. London: Energy Transitions Commission; 2018. www .energy-transitions.org/mission-possible.

energy-related CO_2 emissions which influenced the climate-policy debate of these countries (47). Their approach addresses four methodological challenges. First, faced with country-specific uncertainties there is a need to explore multiple plausible future options. Second, to be useful for policymaking, national scenarios should describe not just future emissions trajectories but also give transparent detail about the sectors, technologies, and policies involved. Third, to allow comparability between countries a consistent 'dashboard' of key metrics for assessing progress is needed. Fourth, analysis of the different pathways should start from the present, showing how realistic future targets for decarbonization can be achieved. This approach has great potential to develop credible policy options but will need engagement of policymakers at an early stage to get buy-in from those who must implement solutions. There is also an opportunity to capitalize more effectively on the near-term (co-)benefits of many policies and technologies, including for health.

Health Benefits of Decarbonization

Decarbonization need not be a story of deprivation and sacrifice; it can be a story of opportunity and benefit. Some of the benefits flow directly from eliminating fossil fuel combustion, while others flow from the expanded use of renewable energy. Many of these benefits relate to human health (48).

Health Benefits of Reducing Fossil Fuel Use

The most direct pathway from zero-carbon policies to improved health is through reduced air pollution, both household and ambient exposures. Many studies have shown that fine particulate and ozone air pollution will be reduced by policies to reduce CO_2 and SLCP emissions through various combinations of energy efficiency, fuel substitution, and replacing fossil fuels with clean renewable energy from sources such as solar and wind. A systematic review of studies up to 2015 identified eight published papers showing health co-benefits from policies in the energy sector (49). Several subsequent studies also demonstrated the potential for large health co-benefits. The International Energy Agency, for example, suggested that a 7% increase in investment would be needed to achieve a 'Clean Air Scenario' with a peak in CO_2 emissions in 2020, which would provide universal energy access and avert about 3 million premature deaths annually worldwide by 2040 compared with business as usual (50).

A study using a well-validated atmospheric chemistry model showed that phase-out of fossil fuels could reduce premature deaths by about 3.6 million annually from reduced ambient air pollution. A broader transformation, controlling all anthropogenic sources (including from agriculture and household sources contributing to ambient pollution), could reduce premature mortality by about 5.5 million annually at 2015 population (51). **Figure 8.5** shows where these reductions would occur. China (~1.5 million avoided premature deaths) and high-income economies such as the USA (~190,000) or the EU (~430,000) would reap major health dividends particularly from a fossil fuel phase-out.

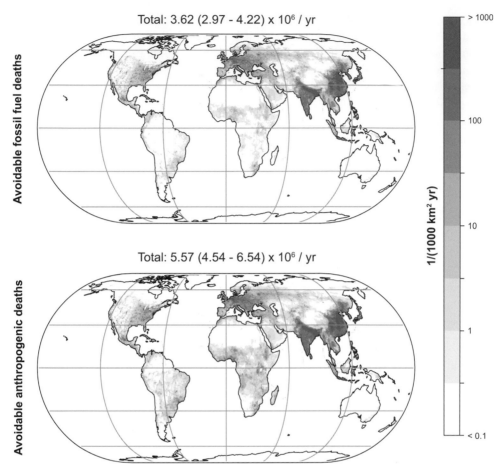

Figure 8.5. Avoidable excess mortality rate from air pollution (in deaths per 1000 km²/year). The upper panel shows excess deaths that may be avoided by the phasing out of fossil fuels, and the lower panel shows excess deaths that may be avoided by eliminating all anthropogenic emissions (e.g. ambient pollution from household and agricultural sources). Source: Lelieveld J, Klingmüller K, Pozzer A, Burnett RT, Haines A, Ramanathan V. Effects of fossil fuel and total anthropogenic emission removal on public health and climate. *PNAS*. 2019;116(15): 7192–7.

By contrast, Sub-Saharan Africa would experience smaller benefits from a fossil fuel phase-out and proportionally larger benefits from addressing other anthropogenic sources because fewer fossil fuels are consumed. This study also shows that burning fossil fuels produces cooling aerosols as well as GHGs. The reduction in these cooling aerosols impedes achievement of the 1.5 °C temperature increase goal. However, an additional benefit of reducing aerosols is substantial increases in precipitation in many regions, which could help reduce the risk of drought and food insecurity.

Another study estimated the health co-benefits of policies to reduce GHG emissions sufficiently to achieve three different ambitions – the Nationally Determined Contributions

(NDCs) to GHG emission reductions under the Paris Agreement and the more stringent 2 °C and 1.5 °C targets – and compared the costs and economic benefits of each scenario to a baseline case without climate policies (52). In the NDC scenario, cumulative premature deaths between 2020 and 2050 decrease by around 5% relative to the baseline (over 120 million deaths). The more stringent mitigation scenarios achieve substantially greater reductions in premature deaths: 21–27% for the 2 °C target and 28–32% for the 1.5 °C target. This study performed economic valuation using a standard approach, the value of a statistical life (VSL) – the monetary value of a reduced risk of dying, as determined by people's willingness to pay. (Because empirical studies to estimate VSL have not been conducted in all countries, the values are adjusted according to the national GDP and GDP growth rate, a widely used approach.) The analysis showed that globally the value of the health co-benefits substantially exceeded the cost of achieving emissions reductions for all scenarios studied. The ratio of the value of the health benefits to the costs of mitigation ranged from 1.4 to 2.45 across different scenarios, with major variation between countries. In LMICs where the health burden of air pollution is large (China and India), health benefits fully covered mitigation costs, while in wealthier regions (Europe and the USA), health benefits covered 7–84% and 10–41% of mitigation costs, respectively, depending on the scenario.

Another analysis (53) suggested that in the USA monetized health benefits from improvements in air quality can offset 26–1050% of the cost of carbon policies. Flexible policies that aim to minimize costs, such as cap-and-trade, yield higher net co-benefits than policies that target specific sectors. Inevitably, as effective policies lower future air pollution levels, the incremental health benefits from additional emissions reductions will decline. Nevertheless, the immediate health benefits of clean air provide impetus for deep cuts in emissions now.

The contributions of sectoral policies required to reduce emissions of GHGs and air pollutants vary from country to country, as illustrated by Figure 3.2, in Chapter 3, which shows the proportional contributions of different sources to air pollution deaths from ambient $PM_{2.5}$ and ozone in the USA and India. Traffic, power, and agriculture are proportionally much more important contributors to deaths in the USA than in India, where residential sources are the largest single source of ambient air pollution deaths (54).

Other studies have focused particularly on the reduction of SLCPs, including black carbon, methane, and tropospheric ozone. In one study, investigators identified a bundle of 14 technically feasible measures targeting black carbon and methane (**Table 8.2**). They calculated that, if implemented, these measures would reduce projected mean global warming by about 0.5 °C by 2050 (55). In addition to the climate benefits, the investigators projected that these measures would avoid between 0.7 and 4.7 million premature deaths from outdoor air pollution each year, and increase annual crop yields by 30 to 135 million tonnes due to ozone reductions (methane being an ozone precursor). Economic analyses using the VSL approach, the social cost of carbon and world market prices for crops, valued reductions in methane emissions at between US$700 and US$5000 per tonne, much larger than typical marginal abatement costs (less than US$250). Black carbon reductions yield major health and regional benefits as outlined above, but because they are often co-emitted with other pollutants, some of which are cooling, the net effects on global

Table 8.2. *A bundle of 14 technically feasible measures to reduce black carbon and methane (CH₄) emissions, which would avert about 0.5 °C global warming by 2050, save millions of lives, and increase crop yields*

Measures	Sector
Measures to reduce methane emissions	
Extended pre-mine degasification and recovery and oxidation of CH_4 from ventilation air from coal mines	Extraction and transport of fossil fuels
Reduced fugitive emissions from oil and natural gas production	
Reduced gas leakage from long-distance transmission pipelines	
Separation and treatment of biodegradable municipal waste through recycling, composting and anaerobic digestion as well as landfill gas collection with combustion/utilization	Waste management
Upgrading primary wastewater treatment to secondary/tertiary treatment with gas recovery and overflow control	
Control of CH_4 emissions from livestock, mainly through farm-scale anaerobic digestion of manure from cattle and pigs	Agriculture
Intermittent aeration of continuously flooded rice paddies	
Measures to reduce black carbon emissions	
Diesel particle filters for road and off-road vehicles	Transport
Elimination of high-emitting vehicles in road and off-road transport	
Clean-burning biomass cooking and heating stoves in developing countries	Residential
Substitution of clean-burning cookstoves using modern fuels (LPG or biogas) for traditional biomass cookstoves in developing countries	
Replacing traditional brick kilns with vertical shaft and Hoffman kilns	Industry
Replacing traditional coke ovens with modern recovery ovens, including the improvement of end-of-pipe abatement measures in developing countries	
Ban on open burning of agricultural waste	Agriculture

Source: Shindell D, Kuylenstierna JCI, Vignati E, et al. Simultaneously mitigating near-term climate change and improving human health and food security. *Science.* 2012;335(6065):183–9.

climate are more uncertain compared with the effects of methane reduction. Aggressive reduction of SLCP emissions can (a) reduce damages over coming decades from the rapid pace of climate change, e.g. those arising from biodiversity loss; (b) slow amplifying feedbacks, such as changes in albedo that are highly sensitive to deposition of black carbon on snow and ice in regions such as the Arctic and the Himalayas; (c) reduce the risk of potential non-linear changes, such as release of carbon from melting permafrost or ice sheet collapse; and (d) yield major benefits for health and crops. Because remaining within the 2 °C guardrail necessitates cutting both SLCP and CO_2 emissions (56), countries should track and report these emissions.

Health Benefits of Renewable Energy

Even when compared with natural gas, renewable energy sources show substantial benefits. For example, a comparison of wind energy at two locations in the USA with natural gas showed that wind energy will avoid costs of 1.8–11.8 cents/kWh (Altamont Pass) and 1.5–8.2 cents/kWh (Sawtooth), respectively, when human health and climate-related externalities are taken into account (57).

A range of symptoms have been attributed to wind generators including environmental sensitivity to sub-audible levels of infrasound. However, there is no empirical evidence of an effect and it is most likely due to the nocebo response from transmission of negative stories about windfarms propagated by social and conventional media (58). A systematic review concluded that there was reasonable evidence of a dose–response association between wind turbine noise, sleep disturbance, and annoyance but only above a threshold that can be addressed by regulation, as occurs in countries such as Denmark (59). The systematic review revealed no evidence of any other adverse health effects. It is widely accepted that noise limits for wind turbines should be set at a level at which less than 10% of exposed people are annoyed. A field study from the USA showed that only 4% of a population living within about 600 metres from wind turbines experienced significant annoyance (60). An open and inclusive planning process, taking potential noise effects into account, could reduce levels of annoyance, which may be lower than those associated with traffic noise (61), and in any event compared with the pervasive effects of fossil fuel combustion any impacts are minor.

Compared with fossil fuels, photovoltaics also have a favourable health and environmental safety profile. The advantage could be further improved by better life-cycle management of all components, the use of 'green chemistry' approaches, including bio-degradable compounds, and better training and regulation (62). Unsurprisingly, onshore wind and solar photovoltaics occupy three of the top ten 'solutions' to climate change outlined above in Drawdown (36).

Nuclear power has a more contentious balance of benefits and risks arising from the potential for accidents, deliberate attacks on reactors, and diversion of nuclear fuel to weapons, and the long-term hazards posed by nuclear waste storage. Nuclear energy as a proportion of global electricity generation peaked at 17.5% in 1996, and in absolute terms at 2660 TWh in 2006. By 2018 it represented about 10.2% of global electricity production, with the USA, France, Japan, Russia, and South Korea accounting for about 70% of global nuclear power generating capacity (63). Nuclear energy is driven, in part, by its favourable pollution and climate change profiles, but obstacles include public reluctance (especially after the 2011 Fukushima disaster) and very high up-front costs. New fuel cycles resulting in less waste, and safer reactor technologies, may alter the balance in the future but since solar, wind, and other technologies present few downsides they are likely to be dominant, particularly as storage technologies become cheaper and more reliable (64).

Hydropower is the major source of renewable energy at present. Although the rate of growth is far lower than for solar PV and wind energy, a current wave of dam construction is projected to double the number of large hydroelectric dams by 2030 (65), leaving almost no free-flowing major river systems. Dams affect not just water flow, but also fluxes of nitrogen, phosphorus, iron, silica, and sulfur (66) – sometimes with global-scale implications (67–69). Dammed reservoirs can be significant sources of greenhouse gases (70). River ecosystems are altered, species composition changes, and fish populations can be seriously compromised (66). Major dam projects can displace large numbers of people, with attendant risks ranging from infectious disease to depression and anxiety (71). Changing conditions along rivers can raise the local risk of malaria, schistosomiasis, and other infectious diseases (72–76). River alterations can even trigger the formation and release of toxic materials such as methylmercury (77). Many of these risks can be managed. For example, the direction of prevailing winds, the presence of marginal pools (small water bodies that form adjacent to reservoirs, in which mosquitoes can breed), and other local factors affect malaria risk near dams (78), so the locations of human settlements around dams need careful consideration to reduce malaria risk. Hydroelectric potential may change substantially over coming decades as rainfall, river flows, glacial melting, and other hydrological changes alter local conditions (79).

Burning solid biomass can have serious adverse effects on health (see household energy section below). Other potential biomass technologies may be less polluting, although presenting additional challenges such as competition with food crops for land (see Chapter 10).

Household Energy

Replacing current highly polluting domestic sources of energy with electricity (increasingly generated from clean renewable sources or, until then, from clean burning fuels) can also dramatically cut deaths from household air pollution which, as discussed in Chapter 3, is a major killer, particularly in low-income countries. While estimates vary, recent analyses suggest that the burden of disease from this source declined from about 1.9 million premature deaths in 2007 to about 1.6 million in 2017, reflecting the decline in reliance on domestic solid-fuel burning (80). A number of intervention studies have assessed the effects of 'clean cookstoves' on health outcomes. Improved efficiency biomass stoves have moderately reduced household air pollution exposures but clear evidence of improved respiratory and other health outcomes has been elusive (81, 82). This apparent lack of effect may reflect incomplete adoption of the new stoves by families, continued exposure from other sources of particulate pollution such as rubbish burning, and/or the failure of such stoves sufficiently to reduce fine particle pollution exposure. Biogas, ethanol, liquefied petroleum gas (LPG), and natural gas burn cleaner than solid fuels.

Household air pollution is a major source of ambient air pollution in some countries. A study in India (83) estimated that phasing out biomass for cooking and space- and water-heating, and kerosene for lighting, would reduce the population-weighted average annual ambient $PM_{2.5}$ exposure by about 17, 12, and 1%, respectively. If all these household sources of air pollution were mitigated, the Indian population breathing air meeting that country's ambient air quality standard (40 $\mu g/m^3$) would increase from 398 million to 585 million. While the Indian standard is far higher than the World Health Organization (WHO) guideline

(10 $\mu g/m^3$), widespread compliance with the Indian standard would nevertheless represent a major public health advance. The Government of India (GoI) in 2016 launched a major effort, *Pradhan Mantri Ujjwala Yojana* (PMUY), to provide 80 million households living 'below poverty level' with liquefied petroleum gas (LPG) by 2019. That goal was achieved, although many of the beneficiary homes have not fully embraced LPG use (84, 85). Although LPG is a fossil fuel there are nevertheless some modest climate benefits from using it in lieu of firewood because of reduced pressure on forests, although the magnitude depends on assumptions about the proportion of firewood from non-renewable sources and the range of climate-active emissions (86). The GoI also initiated a programme in 2014 to electrify all villages in India, thus eradicating emissions from kerosene lighting, an important source of pure black carbon which contributes significantly to climate change and air pollution (87).

An integrated assessment of combined ambient and household air pollution exposure in China showed that about 90% of the reduction in exposure from 2005 to 2015, resulting in about 0.40 (0.25–0.57) million fewer premature deaths annually, could be attributed to declines in household solid-fuel use, largely due to rapid urbanization and increased incomes rather than specific air pollution control policies (88). If all remaining household solid fuels were replaced by clean fuels, an additional 0.51 (0.40–0.64) million premature deaths annually could be prevented.

Even in high-income countries, household burning of wood can contribute significantly to air pollution, particularly in winter months. For example in London the particulate air pollution due to wood burning, the majority of which is in open fires, exceeds the modest reduction in air pollution from the Low Emission Zone (89), which aims to reduce transport-related emissions, suggesting the need to tighten the implementation of existing clean air legislation.

These data show that integrated approaches to address both household and ambient air pollution through providing access to clean fuels and electricity, increasingly generated from clean renewable sources, could both mitigate climate change and prevent millions of premature deaths annually. The benefits resulting from such policies are experienced in the near term in addition to the medium- to long-term benefits of reducing climate change, thus reinforcing the motivation for change.

Reducing Consumption

Achieving the goals of the Paris Agreement and other global sustainability initiatives will require technological innovation. However, efficient technology alone will not suffice to achieve climate targets. One important reason is that greater efficiency, as important as it is, may backfire. In many situations gains in efficiency stimulate demand through reduced prices, as described by William Stanley Jevons in 1905 (90). Jevons noted that more 'economical' use of coal in engines paradoxically increased the overall consumption of coal, iron, and other resources, instead of reducing their consumption as expected. This rebound effect of improved technological efficiency can reduce, and in some cases negate, the purported benefits of the efficiency. The degree to which this happens and the circumstances in which it occurs have been the subject of considerable debate (91, 92). The Jevons paradox, however, implies that while efficiency measures are useful in reducing energy and resource use, they will need to be accompanied by other policies to abate emissions or reduce consumption (93). Among them are strategies to change behaviour and address the value systems underlying consumption.

In addition to the limits on what efficiency can deliver, reductions in consumption and production are also needed for at least three reasons (94). First, private consumption with its accompanying production of material goods is a major driver of GHG emissions and other environmental changes (95, 96). Second, major investments in more sustainable infrastructure, including renewable energy, are required over coming decades. This will result in substantial GHG emissions which will use up much of the available carbon budget. Third, the world's poor require additional services and goods to achieve healthy and productive lives; addressing these unmet needs will take up an additional share of the carbon budget and move humanity closer to other planetary boundaries. The current inequities in carbon emissions are profound: the top 10% of the world's population according to income is responsible for about one-third of global GHG emissions, while the bottom 50% account for only 15%. Increasing the income of the global poor to US$3–8 per day will consume about 66% of the available 2 °C global carbon budget (97). Reduced consumption by the wealthy, therefore, needs to be accompanied by increased consumption by the poor – a version of the 'contract and converge' framework that has helped inform thinking about carbon emissions (98). The growing recognition that profligate economic growth and consumption are unsustainable, and that a rebalancing of goods and services to benefit the world's poor is needed, has inspired alternative ways of thinking about what constitutes a successful society and how prosperity can be sustained within planetary boundaries (see also Chapter 12). The opportunities to improve health and sustainability by addressing population growth are addressed in Chapter 11, which outlines the potential to provide universal access to modern family planning, together with the education and empowerment of women.

The Circular Economy

Increasing concern about the profligate use of resources and overexploitation of natural systems has led to political support in some regions, notably the EU, for initiatives that aim to develop a circular economy (CE) – a transformation from the current linear flow of materials from primary sources to products that are then largely disposed of as 'waste' after a variable, but often short, period of use. Formative influences on the CE concept include the Hannover Principles, developed by architect William McDonough and chemist Michael Braungart in advance of the Expo 2000 (**Table 8.3**). There is no single definition of the CE but two broad approaches have been described (99). The first focuses on the need for a 'closed-loop' economy in which waste is minimized and material flows are circular rather than linear, resulting in much lower consumption of primary resources. The second goes beyond the management of material resources to advocate new economic and social

Table 8.3. *The Hannover Principles*

- Insist on the right of humanity and nature to co-exist in a healthy, supportive, diverse and sustainable condition.
- Recognize interdependence.
- Respect relationships between spirit and matter.
- Accept responsibility for the consequences of design decisions upon human well-being, the viability of natural systems and their right to co-exist.
- Create safe objects of long-term value.
- Eliminate the concept of waste.
- Rely on natural energy flows.
- Understand the limitations of design.
- Seek constant improvement by sharing knowledge.

Source: McDonough W. The Hannover Principles: Design for Sustainability. 1992. https://mcdonough.com/writings/the-hannover-principles/.

relationships that reduce consumption and promote well-being. A widely used definition, from the Ellen MacArthur Foundation (EMF), describes a CE as 'one that is restorative and regenerative by design and aims to keep products, components, and materials at their highest utility and value at all times' (100).

The need for a CE approach is illustrated by the extraordinarily wasteful nature of the current European economy. The average European used 16 tonnes of materials in 2012, while 60% of discarded materials were landfilled or incinerated. About 95% of the material and energy value were lost to the economy with only 5% captured by recycling of waste and waste to energy generation (101). About 30% of food is wasted along the value chain and the average European car is parked 92% of the time. The average manufactured asset (excluding buildings) lasts only 9 years.

Consistent themes in descriptions of the CE include the use of zero-carbon renewable energy; re-use, recycling, and re-manufacturing of products; treating products as services; promoting extended product life and designing out waste; phasing out waste incineration and landfill (**Figure 8.6**). In reality CE approaches cannot guarantee 100% recycling of materials and full circularity is unachievable, not least because the second law of thermodynamics prevents it. Even a CE cannot therefore sustain indefinite economic growth and will be constrained by the finite limits of natural systems. It will need to be embedded within the broader concept of a 'steady state economy', which aims to achieve an equilibrium, balancing the resource requirements for human development with environmental sustainability. A steady state economy would be a dynamic rather than a static system because temporary adjustments to changing values, technological developments, or environmental conditions would still occur and these developments would drive the economy rather than vice versa (102). The long-term tendency would be towards a stable equilibrium consistent with the ecological concept of 'carrying capacity', the finite limit at which a population can be sustained.

Nevertheless, the CE concept is useful because it is increasingly accepted by policy-makers and it embodies many of the changes needed to increase the prospects for sustainability. It is increasingly clear for example that decarbonization cannot be achieved without

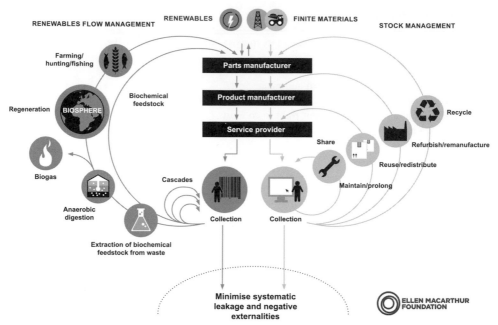

Figure 8.6. The circular economy concept: an industrial economy that is restorative by design.

Source: Adapted from original by Ellen MacArthur Foundation. www.ellenmacarthurfoundation.org/explore/the-circular-economy-in-detail.

a transition towards circularity. About 40% of emissions are from industry and demand is set to increase two- to four-fold as the world economy grows. Scale-up of zero-carbon energy is essential but not sufficient to achieve decarbonization. Of the remaining carbon budget, no more than 300 Gt can be allocated to the production of materials, whereas if current trends continue, material production by itself would require in excess of 900 Gt of emissions (103). Energy efficiency and decarbonizing energy supply can contribute to reducing these emissions but the majority are intrinsic to the processes involved, either because they arise from the chemical processes involved (e.g. cement and steel production), or because they are built into the products and released at the end of their life cycle (plastics). For these reasons, reducing demand for products will play an important role in mitigating climate change. There are three main routes by which this can happen: material recirculation (recycling), increased efficiency, and circular business models.

Recycling requires far less energy than new production. In the case of the EU, for example, recycling could supply about 75% of steel, 50% of aluminium, and 56% of plastics required by 2050 (103). In the EU about 60% of steel is currently made from iron ore, but in decades to come the majority of steel demand could be met from recycled material. Using electricity from renewable sources, this would reduce GHG emissions by about 90% per tonne of steel produced. This will require improving the quality of recycled steel; most currently recycled steel is contaminated with copper, and even at levels of 0.2–0.3% this makes it suitable for basic construction but not more exacting applications.

Only 10% of plastics are currently recycled, but there are opportunities to increase the proportion to over 50% by mechanical approaches and an additional 11% by chemical approaches. Although cement cannot be recycled, structural elements can be re-used and new methods are being developed to extract unreacted cement from concrete, which can then be used to make new concrete. Overall, analysis of the EU economy suggests that, of the policies to encourage circularity, material recirculation makes the largest contribution to GHG emissions reduction, amounting to some 178 Mt annually. The EU 2030 targets include recycling 65% of municipal waste and 75% of packaging waste, and reducing landfill to a maximum of 10% of municipal waste. The municipal waste targets are mandatory, but other targets depend on implementation of national laws.

Increased efficiency of use could make a smaller but still substantial contribution to EU emissions reductions of about 56 Mt annually. Efficiency-enhancing approaches use less material to achieve the same ends through using high-strength, lightweight materials and by optimizing design. Materials used in industry are often made to a standard shape and specification which may not represent their most efficient use. For example, beams could be made up to 30% lighter without sacrificing structural integrity and similar techniques could be used for car components, the 'rebar' used to reinforce concrete, and steel cans used for food (104). During current production processes substantial amounts of raw materials are discarded as scrap, but there is potential to greatly reduce this wasted resource. Old components such as steel beams from demolished buildings and car parts from crashed or scrap vehicles can often be re-used rather than thrown away. New markets will be needed to foster trade in items that are considered 'waste' today, such as old electric motors, bearings, and computer chips (105).

Circular business models aim to prolong product life and increase the utilization of material-intensive products. For example, a system of fleet-managed, shared vehicles, which could be selected according to need, could substantially reduce the number of cars in use and therefore material inputs compared with the current system of individually owned private cars. As well as reducing GHG emissions this also means much lower requirements for steel and these could therefore much more easily be met by recycled steel. Business models that boost the value realized from each product can greatly improve the economic returns from strategies to improve the material efficiency of products and can therefore lead to transformative change. Finally, fundamental changes to industrial processes would also be necessary, for example hydrogen could replace coal for primary steel production, and bioplastics could replace plastics from fossil fuel feedstocks.

The end-of-life care of refrigeration and air conditioning equipment poses particular challenges because of potential leakage of HFC refrigerants. These HFCs, unlike CFCs, do not degrade the ozone layer, but their capacity to heat the atmosphere is 1000–9000 times greater than the equivalent amount of carbon dioxide, depending on their chemical composition. The phase-out of HFCs has been agreed under the Kigali Amendment of the Montreal Protocol (which aimed to phase out substances that deplete the ozone layer), starting in high-income countries in 2019, in some low-income countries in 2024, and in others by 2028. Over the next 30 years or so large amounts of HFCs already in circulation could be released as air conditioning units and refrigerators are scrapped. If it were possible to collect 87% of these, over a period of 30 years the equivalent of nearly 90 Gt of CO_2 emissions could be avoided. Refrigerant management is among the top ten potential solutions to climate change in the Drawdown analysis (36) (Table 8.1).

The CE approach links to a number of SDGs (see Chapter 7), including SDG 8 (decent work and economic growth) and SDG 12 (responsible production and consumption). Within SDG 12, two especially relevant targets are sound management of chemicals and all wastes throughout their life cycles (Target 12.4), and the reduction of food waste (Target 12.3). Two other conceptual approaches, the Bioeconomy (BE) and Green Economy (GE), also aim to integrate social, economic, and environmental development, with the former focusing on innovations in land use and biological resources (see Chapter 10) and the latter having a broader and more inclusive vision of balancing social progress and environmental sustainability (106). The three (CE, BE, and GE) conceptual approaches vary considerably in how they are interpreted and applied. They can be seen as reflecting evolving thinking about how to reconcile social progress with environmental sustainability.

Circular economy and allied approaches have great promise in decoupling human development from the environmental footprints of societies, but there are a number of unresolved challenges that must be addressed to achieve their full potential (107). These include defining the systems boundaries, potential rebound effects, and the potential for harmful production to be transferred to low-income countries. The implications for health and well-being need to be considered from the outset to reduce the risks of unintended adverse consequences. For example, recycling may pose occupational exposures to toxic chemicals and dismantling of products may create injury risks in poorly regulated working conditions. **Table 8.4** summarizes the conclusions of a rapid assessment by the WHO European Region of how implementing circular economy models could affect human health and welfare (99).

The potential negative effects of the CE are largely unintended consequences of exposures during recycling and waste management. For example, poorly regulated battery recycling in low-income countries is estimated to expose 16.8 million people to hazardous levels of lead (108). As discussed in Chapter 3, large numbers of chemicals are released into the environment as a result of a range of industrial and other activities. Chemicals of concern include bisphenol A (BPA) and brominated flame retardants (BFRs) which are present in a wide range of products. Exposure to endocrine-disrupting chemicals, including those found in plastics, has been estimated to cause health and economic costs in the EU of over €150 billion annually (109), but further research is needed to identify the potential impacts of CE approaches on these costs. In six EU countries children's toys have been found to contain two polybrominated diphenyl ether (PBDE) flame retardants, which are used in plastics for electronics (110), and flame retardants have also been found in plastic products such as thermo-cups and kitchen utensils (109). The Stockholm Convention on Persistent Organic Pollutants prohibits chemicals such as these from being used in consumer products, products for children, and those coming into contact with food because of concerns about adverse health and environmental effects. These findings suggest the potential for widespread exposure to chemical hazards from poorly regulated recycling and waste management activities.

Table 8.4. *Impacts of the circular economy on health. Potential positive effects in plain text, potential negative effects in italics*

Reduced use of primary resources (production)

Process/Action	Source of potential health implications positive or negative	Health impact (direct or indirect)	Nature of potential health endpoint
Recycling	Food waste: redistribution of edible food	Improved food availability (direct), particularly for the poor	Improved nutrition
	Food waste: composting	*Health risks from inhalation of bioaerosols (direct)*	*Asthma or extrinsic allergic alveolitis*
	Food waste: risk if food safety is compromised	*Microbiological contamination of food (direct)*	*Food poisoning including diarrhoeal diseases*
	Chemicals in food packaging (BPA, phthalates, PFCs)	*Exposure to chemicals (direct)*	*Epigenetic effects*
	Use of brominated flame retardants (BFR) in manufacturing	*Exposure to chemicals (direct)*	*Endocrine, reproductive, and behavioural effects*
	E-waste recycling components	*Toxic by-products in soil, water, and food (direct and indirect)*	*Contact with hazardous waste, increased risk of injury in recycling process*
	Informal recycling sites	*Occupational health risks at poorly regulated sites (direct)*	*Injuries and exposure to hazardous materials (e.g. lead from battery recycling)*
	Waste reduction and recycling in health sector	Reduced health care costs (direct)	Reduced costs allow improved health service delivery
	Use of recycled materials in manufacturing processes	Indirect impact via reduced manufacturing air/water emissions	Reduced exposure to pollution and, in the long term, climate change
Efficient use of resources	*Use of sewage sludge in agriculture with contaminants (e.g. persistent industrial chemicals, pharmaceuticals, pesticides)*	*Change of soil/water quality causing direct and indirect health effects*	*Wide range of infectious diseases: e.g. typhoid, dysentery, diarrhoeal diseases* / *Health effects of exposures to a wide range of chemicals*

Table 8.4. (cont.)

Process/Action	Source of potential health implications positive or *negative*	Health impact (direct or indirect)	Nature of potential health endpoint
	Resource-efficient agricultural practices (including reduction in fertilizer and pesticide use), regenerative farming practices, closed loops of nutrients and other materials (see Chapter 10)	Direct and indirect effects	More efficient use of resources could reduce poverty in low-income/subsistence farmers Improved nutrition and reduction in excessive nitrate or pesticide exposure
Use of renewable energy sources	General move to renewable energy and energy efficiency across sectors in the circular economy	Lower air pollutants (direct) and GHGs (indirect)	Reduced cardiovascular and respiratory effects of air pollution (see text for discussion) and climate change effects
	Reduced energy recovery (incineration)	Reduced generation of pollutants during energy recovery process	Possible reduced cancers, respiratory and negative birth outcomes. The evidence is not conclusive and research continues
Maintain the highest value of materials and products (production)			
Re-manufacturing, refurbishment, and re-use of products and components	'Circular buildings'	Improved indoor air quality and use of non-toxic materials	Various, including occupational health and safety issues, mental health and respiratory effects
Product life extension	Reduced waste generation and production emissions	Reduced indirect impacts from waste management (landfill, incineration, recycling, etc.) and from manufacturing air/water emissions	Various, including reduced cancer, negative birth outcomes, and respiratory risks
	Resource savings through extension of product life in hospitals	Direct impact on health sector via reduced costs	Reduced costs allow improved health services benefitting a range of health outcomes
Change utilization patterns (consumption)			
Product as service	Performance models in health-care sector and other sectors	Direct impact on health sector via reduced costs	Reduced costs allow improved health services
		Indirect impact for various sectors (e.g. transport) via reduced manufacturing	Conditions related to emissions from manufacturing are reduced

Sharing models	Product- and service-sharing platforms (business to business, business to consumer, and consumer to consumer), e.g. car sharing	Indirect impact via reduced manufacturing emissions Direct impacts on air quality and noise from car sharing (COVID-19 permitting)	Reduced respiratory and cardiovascular conditions due to lower emissions
Shift in consumption patterns	Increasing consumption of healthier plant-based diets (Chapter 10)	Direct impacts on health	Reduction in conditions related to poor diet including cardiovascular diseases and some cancers
	Shift from material to virtual products or services	Indirect impact for various sectors via reduced manufacturing	Reduced harmful conditions related to manufacturing emissions
Combinations of actions			
Efficient use of resources, shift in consumption, new approaches	Healthier food production (Chapter 10)	Potential for indirect health benefits from reduced GHG and other emissions from changes in food production	Reduction in conditions related to poor diets, obesity, cancers Reduct on in health outcomes (esp. respiratory and cardiovascular) related to harmful emissions
Efficient use of resources, sharing models; shift in consumption	Mobility (see Chapter 9)	Indirect impacts from reduced air emissions. Possible improvements in road safety	Respiratory, road accident deaths and injuries
Efficient use of resources, eco design, use of renewable energy	Built environment	Improved indoor air quality and use of non-toxic materials	Various, including occupational health and safety issues, mental ill-health and respiratory conditions
Recycling, efficient use of resources, shift in consumption	Reduced landfill and incineration	Reduced direct impacts from air, water, and soil pollution and GHG emissions	Reduced cancer, negative birth outcomes, and respiratory diseases
Recycling, efficient use of resources, shift in consumption	Substitution and reduced use of hazardous materials resulting in reduced need for disposal of hazardous materials in long term	Reduced direct impacts from water and soil pollution	Multiple potential impacts including on cancers, birth outcomes, and diseases of the cardiovascular and nervous systems

Source: Adapted from WHO. *Circular Economy and Health: Opportunities and Risks.* World Health Organization, Regional Office for Europe; 2018. www.euro.who.int/en/publications/abstracts/circular-economy-and-health-opportunities-and-risks-2018.

Climate change will reduce the capacity for physical labour and could increase the mobilization of some chemicals in the environment, so additional precautions will be needed to scale up circular economy approaches in future. Vulnerable populations include those living close to or working on poorly regulated waste dumps, particularly children. Appropriate regulation could greatly reduce the potential adverse effects on health and well-being. Strengthening capacity for monitoring and regulation where it is currently weak, e.g. in some LMICs, will be essential for the safe global implementation of CE approaches.

Positive effects include those related to reduced waste, for example reduced waste burning (e.g. fine particulate air pollution) together with the benefits of reduced waste in food systems and the health sector. Reduced private car use could lead to improved road safety, physical activity, and air quality. Beneficial effects on health could also accrue from well-designed economic policies to support progress towards a CE. For example, a shift from labour-based to resource-based taxes could benefit the economy and employment as well as reducing the health and ecosystem costs from exposure to air and other forms of pollution. Waste disposal only generates 0.1 jobs per 1000 tonnes of material compared to 2 per 1000 tonnes for recycling (111). In Europe about 6% of tax income currently comes from environmental taxes and 50% from labour taxes and social contributions. An analysis of reforming the tax system to shift the balance from taxes on labour towards environmental taxes suggested that it would result in an increase of 2.9% in employment (equivalent to about 6.6 million additional jobs) as well as an average increase of 2% in (the admittedly flawed metric) GDP, compared with a business-as-usual scenario (101). The same study, focusing on the EU 27 countries, showed that, across three major sectors (mobility, food, and the built environment), CO_2 emissions would drop by up to 48% by 2030 using a CE approach (compared with 31% on the current development pathway) and 83% by 2050 (compared to 61% on the current development pathway), relative to 2012 levels.

Tracking Progress Towards Decarbonization and the Circular Economy

The International Energy Agency tracks progress in the global energy system with respect to three aims: keeping well below 2 °C temperature increase, the Paris Agreement climate goal, providing universal energy access, and substantially reducing air pollution. Its tracking report makes sobering reading as only six (solar PV, bioenergy, electric vehicles, rail, building lighting, data centres and networks) of 40 monitored technologies across six sectors are on track to achieve the aims (112). Even for those technologies with adequate progress, uptake is often patchy; for example, advances in energy storage (grid-scale and residential batteries) are dominated by a few countries (South Korea, China, the USA, and Germany) and in 2019 regressed from 'on track' to 'more efforts needed' when annual installations of energy storage technologies fell.

The 2020 Renewable Energy report from REN21, a Paris-based think tank, shows that progress to date in the uptake of renewables has been confined to the power sector, driven by consistent policies and increased cost-competitiveness of renewable energy compared with fossil fuels (113). There was far less progress in heating, transport, and industry. By the end of 2019, renewables provided over 27% of global electricity generation (17% of

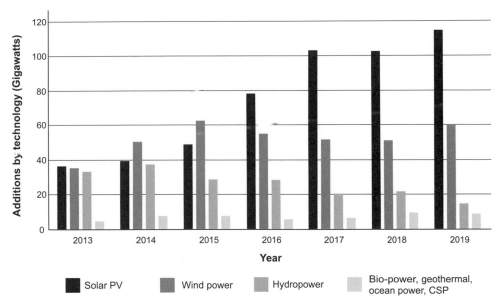

Figure 8.7. Annual additions of renewable power capacity, by technology, 2013–2019.
Source: Adapted from REN21. *Renewables Global Status Report.* Paris: REN21; 2020. www.ren21.net/reports/global-status-report/.

total final energy consumption) (**Figure 8.7**). Nevertheless, even in the power sector there are significant setbacks. Global energy-related CO_2 emissions increased by an estimated 1.7% in 2018 due to higher fossil fuel combustion, and remained level in 2019. This unwelcome increase was no doubt partly due to the hundreds of millions of dollars spent by the fossil fuel industry in lobbying activities to impede climate change policies and advertisements to influence public opinion against such policies (114).

Progress in the heating, cooling, and transport sectors was slow, with a lack of coherent policies to promote change and development of new technologies. In 2018 only 47 countries had policies for promoting renewable heating and cooling. Renewable energy provides only 9.8% of total heating and cooling energy (of which most is from modern bioenergy, solar thermal, and geothermal heat) and 3.3% of transport energy (of which most is from biofuels).

Conclusions

Many benefits will result from rapid transition to a zero-carbon and increasingly circular economy, particularly if policies are designed to minimize unintended adverse consequences and implemented in a way that addresses inequities in access to clean energy. These benefits include both the medium- to long-term benefits of reducing the risks of dangerous climate change and the near-term benefits, including of reduced air pollution, for health, crop yields, and the wider economy. Currently progress is inadequate because societies do not pay the full economic costs of polluting fossil fuels and in many cases subsidies favour powerful interests. When the value of health and other (co-)benefits of action are taken into account they partly or wholly offset the costs and should make action more attractive to policy-makers and the public. Subsequent chapters explore potential policies for sustainable and healthy cities, food systems, and the role of education and health care. The impacts of the COVID-19 pandemic on the prospects for decarbonization are discussed in Chapter 12.

References

1. Polanyi K. *The Great Transformation: The Political and Economic Origins of Our Time*. Boston: Beacon Press; 1944.
2. Steffen W, Broadgate WJ, Deutsch L, Gaffney O, Ludwig C. The trajectory of the Anthropocene: the Great Acceleration. *The Anthropocene Review*. 2015;2:81–98. doi: 10.1177/2053019614564785.
3. Görg C, Plank C, Wiedenhofer D, et al. Scrutinizing the Great Acceleration: the Anthropocene and its analytic challenges for social-ecological transformations. *The Anthropocene Review*. 2019;7(1):42–61. doi: 10.1177/2053019619895034.
4. Haberl H, Fischer-Kowalski M, Krausmann F, Martinez-Alier J, Winiwarter V. A socio-metabolic transition towards sustainability? Challenges for another Great Transformation. *Sustainable Development*. 2011;19(1):1–14. doi: 10.1002/sd.410.
5. Marlon JR, Bloodhart B, Ballew MT, et al. How hope and doubt affect climate change mobilization. *Frontiers in Communication*. 2019;4(20). doi: 10.3389/fcomm.2019.00020.
6. Rogelj J, Forster PM, Kriegler E, Smith CJ, Séférian R. Estimating and tracking the remaining carbon budget for stringent climate targets. *Nature*. 2019;571(7765): 335–42. doi: 10.1038/s41586-019-1368-z.
7. Millar RJ, Fuglestvedt JS, Friedlingstein P, et al. Emission budgets and pathways consistent with limiting warming to 1.5 °C. *Nature Geoscience*. 2017;10:741. doi: 10.1038/ngeo3031.
8. Friedlingstein P, Jones MW, O'Sullivan M, et al. Global Carbon Budget 2019. *Earth System Science Data*. 2019;11(4):1783–838. doi: 10.5194/essd-11-1783-2019.
9. Clarke L, Jiang K, Akimoto K, et al. Assessing transformation pathways. In Edenhofer O, Pichs-Madruga R, Sokona Y, et al., editors. *Climate Change 2014: Mitigation of Climate Change Contribution of Working Group III to the Fifth Assessment Report of the Intergovernmental Panel on Climate Change*. Cambridge, UK and New York: Cambridge University Press; 2014. pp. 413–510.
10. Peters GP, Andrew RM, Canadell JG, et al. Carbon dioxide emissions continue to grow amidst slowly emerging climate policies. *Nature Climate Change*. 2020;10(1): 3–6. doi: 10.1038/s41558-019-0659-6.
11. McGlade C, Ekins P. The geographical distribution of fossil fuels unused when limiting global warming to 2 °C. *Nature*. 2015;517(7533):187–90. doi: 10.1038/nature14016.
12. UNEP. Emissions Gap Report 2019. Nairobi: United Nations Environment Programme; 2019. Available from www.unenvironment.org/emissionsgap.
13. Shearer C, Myllyvirta L, Yu A, et al. Boom and Bust 2020: Tracking the Global Coal Plant Pipeline. San Francisco: Global Energy Monitor, Sierra Club, Greenpeace, and Centre for Research on Energy and Clean Air; 2020. Available from https://endcoal .org/2020/03/new-report-global-coal-power-under-development-declined-for-fourth-year-in-a-row/.
14. Redl C, Hein F, Buck M, Graichen P, Jones D. The European Power Sector in 2019: Up-to-Date Analysis on the Electricity Transition. Berlin and Brussels: Agora

Energiewende and Sandbag; 2020. Available from https://ember-climate.org/project/power-2019/.

15. Rocha M, Yanguas Parra P, Sferra F, et al. A Stress Test for Coal in Europe under the Paris Agreement: Scientific Goalposts for a Coordinated Phaseout and Divestment. Berlin: Climate Analytics; 2017. Available from https://climateanalytics.org/publications/2017/stress-test-for-coal-in-the-eu/.

16. UNEP. Emissions Gap Report 2018. Nairobi: United Nations Environment Programme; 2018.

17. Palmer T. Short-term tests validate long-term estimates of climate change. *Nature*. 2020;582(7811):185–6. doi: 10.1038/d41586-020-01484-5.

18. Williams KD, Hewitt AJ, Bodas-Salcedo A. Use of short-range forecasts to evaluate fast physics processes relevant for climate sensitivity. *Journal of Advances in Modeling Earth Systems*. 2020;12(4):e2019MS001986. doi: 10.1029/2019MS001986.

19. Coady D, Parry I, Le N-P, Shang B. Global Fossil Fuel Subsidies Remain Large: An Update Based on Country-Level Estimates. Washington, DC: International Monetary Fund; 2019. Working Paper No. 19/89. Available from www.imf.org/en/Publications/WP/Issues/2019/05/02/Global-Fossil-Fuel-Subsidies-Remain-Large-An-Update-Based-on-Country-Level-Estimates-46509.

20. Coady D, Parry IWH, Sears L, Shang B. How Large Are Global Energy Subsidies? Washington, DC: International Monetary Fund; 2015. Available from www.imf.org/en/Publications/WP/Issues/2016/12/31/How-Large-Are-Global-Energy-Subsidies-42940.

21. Coady D, Parry I, Sears L, Shang B. How large are global fossil fuel subsidies? *World Development*. 2017;91:11–27. https://doi.org/10.1016/j.worlddev.2016.10.004.

22. Watts N, Amann M, Arnell N, et al. The 2019 report of the *Lancet* Countdown on Health and Climate Change: ensuring that the health of a child born today is not defined by a changing climate. *The Lancet*. 2019;394(10211):1836–78. doi: 10.1016/S0140-6736(19)32596-6.

23. UN DESA. Accelerating SDG7: Policy Briefs in Support of the First SDG7 Review at the UN High-Level Political Forum 2018. New York: United Nations Department of Economic and Social Affairs; 2018. Available from https://sustainabledevelopment.un.org/content/documents/18041SDG7_Policy_Brief.pdf.

24. Jewell J, McCollum D, Emmerling J, et al. Limited emission reductions from fuel subsidy removal except in energy-exporting regions. *Nature*. 2018;554(7691):229–33. doi: 10.1038/nature25467.

25. Gupta V, Dhillon R, Yates R. Financing universal health coverage by cutting fossil fuel subsidies. *The Lancet Global Health*. 2015;3(6):e306–7. doi: 10.1016/S2214-109X(15)00007-8.

26. Borghesi S, Montini M. The best (and worst) of GHG emission trading systems: comparing the EU ETS with its followers. *Frontiers in Energy Research*. 2016;4(27). doi: 10.3389/fenrg.2016.00027.

27. World Bank. State and Trends of Carbon Pricing 2019. Washington, DC: World Bank; 2019. Available from https://openknowledge.worldbank.org/handle/10986/31755.

28. Stiglitz JE, Stern NH, Duan M, et al. Report of the High-Level Commission on Carbon Prices. Washington, DC: Carbon Pricing Leadership Coalition; 2017. Available from www.carbonpricingleadership.org/report-of-the-highlevel-commission-on-carbon-prices.

29. IPCC. *Global Warming of 1.5°C. An IPCC Special Report on the impacts of global warming of 1.5°C above pre-industrial levels and related global greenhouse gas emission pathways, in the context of strengthening the global response to the threat of climate change, sustainable development, and efforts to eradicate poverty*. Masson-Delmotte V, Zhai P, Pörtner H-O, et al., editors. Geneva: IPCC and WMO; 2018.

30. Le Bras H. Cars, gilets jaunes, and the Rassemblement national. *Études* 2019;4:31–44. www.cairn-int.info/journal-etudes-2019-4-page-31.htm.

31. Salies E. Gilets Jaunes: Is the Energy Transition Possible While Still Reducing Inequality? Paris: Observatoire français des conjonctures économiques (French Economic Observatory, OFCE); 2019. Available from www.ofce.sciences-po.fr/blog/gilets-jaunes-is-the-energy-transition-possible-while-still-reducing-inequality/.

32. Cuevas S, Haines A. Health benefits of a carbon tax. *The Lancet*. 2016;387(10013): 7–9. doi: 10.1016/S0140-6736(15)00994-0.

33. Stern NH. *The Economics of Climate Change: The Stern Review*. Cambridge, UK and New York: Cambridge University Press; 2007.

34. Ramanathan R, Molina ML, Zaelke D. Well Under 2 Degrees Celsius: Fast Action Policies to Protect People and the Planet from Extreme Climate Change. Paris: Climate & Clean Air Coalition; 2017. Available from www.ccacoalition.org/en/resources/well-under-2-degrees-celsius-fast-action-policies-protect-people-and-planet-extreme.

35. Pacala S, Socolow R. Stabilization wedges: solving the climate problem for the next 50 years with current technologies. *Science*. 2004;305(5686):968–72. doi: 10.1126/science.1100103.

36. Project Drawdown. *The Drawdown Review 2020: Climate Solutions for a New Decade*. San Francisco: Project Drawdown; 2020. Available from www.drawdown.org/drawdown-framework/drawdown-review-2020.

37. Hawken P, editor. *Drawdown: The Most Comprehensive Plan Ever Proposed to Reverse Global Warming*. London and New York: Penguin; 2017.

38. Fuss S, Canadell JG, Peters GP, et al. Betting on negative emissions. *Nature Climate Change*. 2014;4(10):850–3. doi: 10.1038/nclimate2392.

39. Mander S, Anderson K, Larkin A, Gough C, Vaughan N. The role of bio-energy with carbon capture and storage in meeting the climate mitigation challenge: a whole system perspective. *Energy Procedia*. 2017;114:6036–43. https://doi.org/10.1016/j.egypro.2017.03.1739.

40. Consoli C. Bioenergy and Carbon Capture and Storage: 2019 Perspective. Melbourne: Global CCS Institute; 2019. Available from www.globalccsinstitute.com/resources/global-status-report/.

41. Galik CS. A continuing need to revisit BECCS and its potential. *Nature Climate Change*. 2020;10(1):2–3. doi: 10.1038/s41558-019-0650-2.

42. Widger P, Haddad A. Evaluation of SF6 leakage from gas insulated equipment on electricity networks in Great Britain. *Energies*. 2018;11(8). doi: 10.3390/en11082037.

43. Energy Transitions Commission. Mission Possible: Reaching Net-Zero Carbon Emissions from Harder-to-Abate Sectors by Mid-Century. London: Energy Transitions Commission; 2018. Available from www.energy-transitions.org/mission-possible.

44. Lawrence MG, Schäfer S, Muri H, et al. Evaluating climate geoengineering proposals in the context of the Paris Agreement temperature goals. *Nature Communications*. 2018;9(1):3734. doi: 10.1038/s41467-018-05938-3.

45. Frumhoff PC, Stephens JC. Towards legitimacy of the solar geoengineering research enterprise. *Philosophical Transactions of the Royal Society A: Mathematical, Physical and Engineering Sciences*. 2018;376(2119):20160459. doi: 10.1098/rsta.2016.0459.

46. UCS. UCS Position on Solar Geoengineering. Boston: Union of Concerned Scientists; 2019. Available from www.ucsusa.org/resources/what-climate-engineering.

47. Waisman H, Bataille C, Winkler H, et al. A pathway design framework for national low greenhouse gas emission development strategies. *Nature Climate Change*. 2019;9(4):261–8. doi: 10.1038/s41558-019-0442-8.

48. Haines A, McMichael AJ, Smith KR, et al. Public health benefits of strategies to reduce greenhouse-gas emissions: overview and implications for policy makers. *The Lancet*. 2009;374(9707):2104–14. doi: 10.1016/s0140-6736(09)61759-1.

49. Gao J, Kovats S, Vardoulakis S, et al. Public health co-benefits of greenhouse gas emissions reduction: a systematic review. *Science of The Total Environment*. 2018;627:388–402. doi: 10.1016/j.scitotenv.2018.01.193.

50. International Energy Agency. Energy and Air Pollution: World Energy Outlook Special Report. Paris: International Energy Agency; 2016. Available from www.iea .org/reports/energy-and-air-pollution.

51. Lelieveld J, Klingmüller K, Pozzer A, et al. Effects of fossil fuel and total anthropogenic emission removal on public health and climate. *Proceedings of the National Academy of Sciences*. 2019;116(15):7192–7. doi: 10.1073/pnas.1819989116.

52. Markandya A, Sampedro J, Smith SJ, et al. Health co-benefits from air pollution and mitigation costs of the Paris Agreement: a modelling study. *The Lancet Planetary Health*. 2018;2(3):e126–33. doi: 10.1016/S2542-5196(18)30029-9.

53. Thompson TM, Rausch S, Saari RK, Selin NE. A systems approach to evaluating the air quality co-benefits of US carbon policies. *Nature Climate Change*. 2014;4(10): 917–23. doi: 10.1038/nclimate2342.

54. Lelieveld J, Haines A, Pozzer A. Age-dependent health risk from ambient air pollution: a modelling and data analysis of childhood mortality in middle-income and low-income countries. *The Lancet Planetary Health*. 2018;2(7):e292–300. doi: 10.1016/s2542-5196(18)30147-5.

55. Shindell D, Kuylenstierna JCI, Vignati E, et al. Simultaneously mitigating near-term climate change and improving human health and food security. *Science*. 2012; 335(6065):183–9. doi: 10.1126/science.1210026.

56. Shindell D, Borgford-Parnell N, Brauer M, et al. A climate policy pathway for near- and long-term benefits. *Science*. 2017;356(6337):493–4. doi: 10.1126/science.aak9521.

57. McCubbin D, Sovacool BK. Quantifying the health and environmental benefits of wind power to natural gas. *Energy Policy*. 2013;53:429–41. https://doi.org/10.1016/j .enpol.2012.11.004.

58. Crichton F, Petrie KJ. Health complaints and wind turbines: the efficacy of explaining the nocebo response to reduce symptom reporting. *Environmental Research*. 2015;140:449–55. doi: 10.1016/j.envres.2015.04.016.

59. Schmidt JH, Klokker M. Health effects related to wind turbine noise exposure: a systematic review. *PLoS One*. 2014;9(12):e114183. doi: 10.1371/journal.pone.0114183.

60. Hessler DM, Hessler GF. Recommended noise level design goals and limits at residential receptors for wind turbine developments in the United States. *Noise Control Engineering Journal*. 2011;59(1):94–104. https://doi.org/10.3397/1.3531795.

61. Pohl J, Gabriel J, Hübner G. Understanding stress effects of wind turbine noise: the integrated approach. *Energy Policy*. 2018;112:119–28. https://doi.org/10.1016/j .enpol.2017.10.007.

62. Bakhiyi B, Labrèche F, Zayed J. The photovoltaic industry on the path to a sustainable future: environmental and occupational health issues. *Environment International*. 2014;73:224–34. doi: 10.1016/j.envint.2014.07.023.

63. Schneider M, Froggatt A. The World Nuclear Industry: Status Report 2019. Mycle Schneider Consulting; 2019. Available from www.worldnuclearreport.org/-World-Nuclear-Industry-Status-Report-2019.

64. Smith KR, Frumkin H, Balakrishnan K, et al. Energy and human health. *Annual Review of Public Health*. 2013;34:159–88. doi: 10.1146/annurev-publhealth-031912-114404.

65. Zarfl C, Lumsdon AE, Berlekamp J, Tydecks L, Tockner K. A global boom in hydropower dam construction. *Aquatic Sciences*. 2015;77(1):161–70. doi: 10.1007/s00027-014-0377-0.

66. Van Cappellen P, Maavara T. Rivers in the Anthropocene: global scale modifications of riverine nutrient fluxes by damming. *Ecohydrology & Hydrobiology*. 2016;16(2):106–11. doi: 10.1016/j.ecohyd.2016.04.001.

67. Maavara T, Dürr HH, Van Cappellen P. Worldwide retention of nutrient silicon by river damming: from sparse data set to global estimate. *Global Biogeochemical Cycles*. 2014;28(8):842–55. doi:10.1002/2014GB004875.

68. Maavara T, Parsons CT, Ridenour C, et al. Global phosphorus retention by river damming. *Proceedings of the National Academy of Sciences*. 2015;112(51):15603–8. doi: 10.1073/pnas.1511797112.

69. Maavara T, Lauerwald R, Regnier P, Van Cappellen P. Global perturbation of organic carbon cycling by river damming. *Nature Communications*. 2017;8:15347. doi: 10.1038/ncomms15347.

70. Wehrli B. Climate science: renewable but not carbon-free. *Nature Geoscience*. 2011;4:585–6.

71. McDonald-Wilmsen B, Webber M. Dams and displacement: raising the standards and broadening the research agenda. *Water Alternatives*. 2010;3(2):142–61.

72. Zhang X, Peng L, Liu W, et al. Response of primary vectors and related diseases to impoundment by the Three Gorges Dam. *Tropical Medicine & International Health*. 2014;19(4):440–9. doi: 10.1111/tmi.12272.

73. Sanchez-Ribas J, Parra-Henao G, Guimaraes AE. Impact of dams and irrigation schemes in Anopheline (Diptera: Culicidae) bionomics and malaria epidemiology. *Revista do Instituto de Medicina Tropical de Sao Paulo*. 2012;54(4):179–91. www.scielo.br/pdf/rimtsp/v54n4/a01v54n4.pdf.

74. Keiser J, De Castro MC, Maltese MF, et al. Effect of irrigation and large dams on the burden of malaria on a global and regional scale. *American Journal of Tropical Medicine and Hygiene*. 2005;72(4):392–406. www.ajtmh.org/content/72/4/392.abstract.

75. Steinmann P, Keiser J, Bos R, Tanner M, Utzinger J. Schistosomiasis and water resources development: systematic review, meta-analysis, and estimates of people at risk. *The Lancet Infectious Diseases*. 2006;6(7):411–25. doi: 10.1016/s1473-3099(06)70521-7.

76. Kibret S, Wilson GG, Ryder D, Tekie H, Petros B. The influence of dams on malaria transmission in Sub-Saharan Africa. *Ecohealth*. 2017;14(2):408–19. doi: 10.1007/s10393-015-1029-0.

77. Calder RS, Schartup AT, Li M, et al. Future impacts of hydroelectric power development on methylmercury exposures of Canadian indigenous communities. *Environmental Science and Technology*. 2016;50(23):13115–22. doi: 10.1021/acs.est.6b04447.

78. Endo N, Eltahir EAB. Prevention of malaria transmission around reservoirs: an observational and modelling study on the effect of wind direction and village location. *The Lancet Planetary Health*. 2018;2(9):e406–13. doi: 10.1016/S2542-5196(18)30175-X.

79. Hamududu B, Killingtveit A. Assessing climate change impacts on global hydropower. *Energies*. 2012;5(2). doi: 10.3390/en5020305.

80. GBD Risk Factor Collaborators. Global, regional, and national comparative risk assessment of 84 behavioural, environmental and occupational, and metabolic risks or clusters of risks for 195 countries and territories, 1990–2017: a systematic analysis for the Global Burden of Disease Study 2017. *The Lancet*. 2018;392(10159):1923–94. doi: 10.1016/s0140-6736(18)32225-6.

81. Mortimer K, Ndamala CB, Naunje AW, et al. A cleaner burning biomass-fuelled cookstove intervention to prevent pneumonia in children under 5 years old in rural Malawi (the Cooking and Pneumonia Study): a cluster randomised controlled trial. *The Lancet*. 2017;389(10065):167–75. doi: 10.1016/S0140-6736(16)32507-7.

82. Smith KR, McCracken JP, Weber MW, et al. Effect of reduction in household air pollution on childhood pneumonia in Guatemala (RESPIRE): a randomised controlled trial. *The Lancet*. 2011;378(9804):1717–26. doi: 10.1016/s0140-6736(11)60921-5.

83. Chowdhury S, Dey S, Guttikunda S, et al. Indian annual ambient air quality standard is achievable by completely mitigating emissions from household sources. *Proceedings of the National Academy of Sciences*. 2019;116(22):10711–16. doi: 10.1073/pnas.1900888116.

84. Dabadge A, Sreenivas A, Josey A. What has the Pradhan Mantri Ujjwala Yojana achieved so far? *Economic & Political Weekly*. 2018;53(20). www.epw.in/journal/2018/20/notes/what-has-pradhan-mantri-ujjwala-yojana-achieved-so-far.html.

85. Mani S, Jain A, Tripathi S, Gould CF. The drivers of sustained use of liquified petroleum gas in India. *Nature Energy*. 2020;5(6):450–7. doi: 10.1038/s41560-020-0596-7.

86. Singh D, Pachauri S, Zerriffi H. Environmental payoffs of LPG cooking in India. *Environmental Research Letters*. 2017;12(11):115003. doi: 10.1088/1748-9326/aa909d.

87. Lam NL, Smith KR, Gauthier A, Bates MN. Kerosene: a review of household uses and their hazards in low- and middle-income countries. *Journal of Toxicology and Environmental Health Part B, Critical Reviews*. 2012;15(6):396–432. doi: 10.1080/10937404.2012.710134.

88. Zhao B, Zheng H, Wang S, et al. Change in household fuels dominates the decrease in $PM_{2.5}$ exposure and premature mortality in China in 2005–2015. *Proceedings of the National Academy of Sciences*. 2018,115(49):12401–6. doi: 10.1073/pnas.1812955115.

89. Fuller GW, Tremper AH, Baker TD, Yttri KE, Butterfield D. Contribution of wood burning to PM_{10} in London. *Atmospheric Environment*. 2014;87:87–94. https://doi.org/10.1016/j.atmosenv.2013.12.037.

90. Alcott B. Jevons' paradox. *Ecological Economics*. 2005;54(1):9–21. https://doi.org/10.1016/j.ecolecon.2005.03.020.

91. Giampietro M, Mayumi K. Unraveling the complexity of the Jevons paradox: the link between innovation, efficiency, and sustainability. *Frontiers in Energy Research*. 2018;6(26). doi: 10.3389/fenrg.2018.00026.

92. Sorrell S. Jevons' paradox revisited: the evidence for backfire from improved energy efficiency. *Energy Policy*. 2009;37(4):1456–69. https://doi.org/10.1016/j.enpol.2008.12.003.

93. Garrett TJ. No way out? The double-bind in seeking global prosperity alongside mitigated climate change. *Earth System Dynamics*. 2012;3(1):1–17. doi: 10.5194/esd-3-1-2012.

94. Alfredsson E, Bengtsson M, Brown HS, et al. Why achieving the Paris Agreement requires reduced overall consumption and production. *Sustainability: Science, Practice and Policy*. 2018;14(1):1–5. doi: 10.1080/15487733.2018.1458815.

95. Ivanova D, Stadler K, Steen-Olsen K, et al. Environmental impact assessment of household consumption. *Journal of Industrial Ecology*. 2016;20(3):526–36. doi: 10.1111/jiec.12371.

96. C40 Cities. The Future of Urban Consumption in a 1.5 °C World. London: C40 Cities, Arup and the University of Leeds; 2019. Available from www.c40.org/press_releases/new-research-shows-how-urban-consumption-drives-global-emissions.

97. Hubacek K, Baiocchi G, Feng K, et al. Global carbon inequality. *Energy, Ecology and Environment.* 2017;2(6):361–9. doi: 10.1007/s40974-017-0072-9.

98. Meyer A. *Contraction and Convergence: The Global Solution to Climate Change.* Cambridge, UK: UIT Cambridge; 2000.

99. WHO. Circular Economy and Health: Opportunities and Risks. World Health Organization, Regional Office for Europe; 2018. Available from www.euro.who.int/en/publications/abstracts/circular-economy-and-health-opportunities-and-risks-2018.

100. Ellen MacArthur Foundation. Towards a Circular Economy: Business Rationale for an Accelerated Transition. Cowes, UK: Ellen MacArthur Foundation; 2015. Available from www.ellenmacarthurfoundation.org/publications/towards-a-circular-economy-business-rationale-for-an-accelerated-transition.

101. Ellen MacArthur Foundation. Growth Within: A Circular Economy Vision for a Competitive Europe. Cowes, UK: Ellen MacArthur Foundation; 2015. Available from www.ellenmacarthurfoundation.org/publications/growth-within-a-circular-economy-vision-for-a-competitive-europe.

102. Czech B, Daly HE. In my opinion: the steady state economy – what it is, entails, and connotes. *Wildlife Society Bulletin.* 2004;32(2):598–605. doi: 10.2193/0091-7648 (2004)32[598:IMOTSS]2.0.CO;2.

103. Material Economics. The Circular Economy: A Powerful Force for Climate Mitigation. Stockholm: Material Economics Sverige AB; 2018. Available from https://materialeconomics.com/publications/the-circular-economy-a-powerful-force-for-climate-mitigation-1.

104. Allwood JM, Cullen JM. *Sustainable Materials Without the Hot Air: Making Buildings, Vehicles and Products Efficiently and with Less New Material,* 2nd ed. Cambridge, UK: UIT; 2015.

105. Stahel WR. The circular economy. *Nature.* 2016;531(7595):435–8. doi: 10.1038/531435a.

106. D'Amato D, Droste N, Allen B, et al. Green, circular, bio economy: a comparative analysis of sustainability avenues. *Journal of Cleaner Production.* 2017;168:716–34. https://doi.org/10.1016/j.jclepro.2017.09.053.

107. Korhonen J, Honkasalo A, Seppälä J. Circular economy: the concept and its limitations. *Ecological Economics.* 2018;143:37–46. https://doi.org/10.1016/j.ecolecon.2017.06.041.

108. Landrigan PJ, Fuller R, Acosta NJR, et al. The *Lancet* Commission on Pollution and Health. *The Lancet.* 2018;391:462–512. doi: 10.1016/S0140-6736(17)32345-0.

109. Trasande L, Zoeller RT, Hass U, et al. Estimating burden and disease costs of exposure to endocrine-disrupting chemicals in the European Union. *The Journal of Clinical Endocrinology and Metabolism.* 2015;100(4):1245–55. doi: 10.1210/jc.2014-4324.

110. DiGangi J, Strakova J. Toxic Toy or Toxic Waste: Recycling POPs into New Products. Gothenburg, Sweden: IPEN; 2015. Available from https://ipen.org/documents/toxic-toy-or-toxic-waste-recycling-pops-new-products.

111. Goldstein J, Electris C, Morris J. More Jobs, Less Pollution: Growing the Recycling Economy in the U.S. Boston: Tellus Institute with Sound Resource Management; 2011.

112. IEA. Tracking Clean Energy Progress: Assessing Critical Energy Technologies for Global Clean Energy Transitions. 2020. Available from www.iea.org/topics/tracking-clean-energy-progress.

113. REN21. Renewables 2020 Global Status Report. Paris: REN21; 2020. Available from www.ren21.net/gsr-2020/.

114. REN21. Renewables 2019 Global Status Report. Paris: REN21; 2019. Available from www.ren21.net/reports/global-status-report/.

9

Sustaining Urban Health in the Anthropocene Epoch

Urban living is becoming increasingly predominant, with 55% of the world's population currently living in cities and 68% projected to do so by 2050 (1). While megacities with more than 10 million inhabitants command much attention, they account for less than 10% of the world's urban population. In contrast, nearly half of all urban residents – over 2 billion people – live in cities with populations under 500,000 (**Table 9.1**). It is in these smaller cities where population growth rates tend to be highest. The USA, Latin America, and Japan are very highly urbanized but both Africa as a whole and India remain well below 50% urbanized (**Figure 9.1**). Many low- and lower-middle-income countries in particular are projected to urbanize rapidly in coming decades, with a projected increase of 2.5 billion people in urban areas by 2050 – accounting for essentially all projected human population growth through mid-century.

Urban population growth results not only from an excess of births over deaths in urban areas but also from rural–urban migration, and in some cases from immigration from other countries. Urban economic development is a dominant driver of environmental change, with cities currently responsible for about 70–75% of GHG emissions and 80–85% of economic activity – contributions that are likely to increase in future (2). Cities are centres of prosperity and material consumption, which can drive disproportionately high carbon emissions (3); at the same time they offer great potential for achieving low-carbon and health goals, with compact living spaces, reduced travel demand, shared infrastructure, and other advantages (4–6). The path to healthy, sustainable cities is complex and nuanced; a systems approach is indispensable to minimize risks of unintended adverse effects.

In most world regions, urban population growth has been accompanied by declining death rates and fertility. In general, urban development has been a positive force for economic development, poverty reduction, employment, and health, but there is much variation. Urban dwellers tend to be better educated than those in rural areas, and enjoy greater access to essential services such as health, clean water, sanitation, and education. Delivery of these services and provision of infrastructure is helped by economies of scale and technological innovation. Nevertheless, building decent housing and essential infrastructure has failed to keep pace with urban population growth on a global scale, such that the number of people living in slums increased from an estimated 807 in 2000 to 883 million in 2014, largely in East and Southeast Asia (332 million), Central and Southern Asia (197 million) and Sub-Saharan Africa (189 million) (7).

Table 9.1. *The global urban population according to city size*

City size	<300K	300K – <500K	500K – <1M	1M – <5M	5M – <10M	>10M	Total
Population in millions	1750	275	415	926	325	529	4220
% of urban total	22.9	3.6	5.4	12.1	4.3	6.9	100

Source: UN DESA. *World Urbanization Prospects: The 2018 Revision.* New York: United Nations Department of Economic and social Affairs, Population Division; 2018. Contract No. ST/ESA/ SER.A/420.

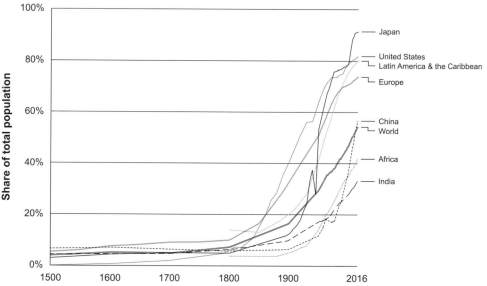

Figure 9.1. Urbanization over the past 500 years. Urban areas are based on national definitions and may vary by country.
Source: Ritchie H, Roser M. Urbanisation. Our World in Data. 2019. https://ourworldindata.org/ urbanization.

The term 'slum' first emerged in London in the 1820s where it was used to describe the urban areas with poorest quality housing and the most unsanitary conditions. Slums were often perceived as the origins of epidemics which commonly ravaged cities at the time, as well as the focus of crime, prostitution, and drug abuse. Nowadays 'slum' is often used loosely to describe urban settlements where basic services are absent or poorly functioning. The term can be stigmatizing, which is why other terms such as 'informal settlements' are increasingly used (8). For LMICs, therefore, a key challenge is how to accelerate progress to ensure that informal settlements are provided with the necessary services to meet the needs of today's residents, to anticipate and accommodate future growth, and to enhance resilience against increasing environmental threats. Nearly 60% of cities with populations of 300,000 or more are at high risk of at least one of six types of natural disasters (cyclones, droughts, floods, earthquakes, landslides, and volcanic eruptions) (1). Much of the growth in urban poverty is occurring in locations vulnerable to natural disasters, such as low-elevation coastal zones and arid regions, and the frequency and intensity of intense cyclones, droughts, floods, and landslides are projected to increase under climate change.

Cities are projected to experience a range of environmental challenges in future, which, although intensified by global processes such as climate change, will manifest themselves differently according to the location of the city. One example is the level of temperature increase, which depends not only on future emissions trajectories but also on the latitude. A study of 246 globally representative cities projected that, under a low GHG emission scenario (RCP2.6), no city would exceed a mean temperature rise of 2 °C from 2017 values by 2100. In comparison, under a high-emission scenario (RCP8.5), all cities are expected to do so, with some increasing by as much as 7 °C (9). Mid- to high-latitude cities will experience larger temperature increases than those at low latitudes. Increases are predicted to be larger in humid temperate and in dry regions compared with humid tropical regions, although the latter are already experiencing high levels of heat exposure and even modest increases could have severe effects. In the hottest months increases in temperature under high emission scenarios will pose serious challenges to adapting to climate change, in some locations exceeding human physiological capacity to adapt (10). Higher temperatures will require large amounts of power for air conditioning unless full use is made of strategies to reduce the urbal heat island (11) and passive cooling approaches (12), using for example natural ventilation, green roofs, and low-albedo materials are widely adopted.

A City's Carbon Emissions: Territorial or Consumption-Based?

Most cities engaged in climate change mitigation measures target their GHG reduction efforts on sources within their boundaries for both pragmatic and political reasons (13). This approach, usually called a territorial or production perspective, attributes the emissions to 'actors' within the city's geographical confines or under its administrative control. Cities are increasingly recognized as complex systems with regional and global linkages because of their demands for energy, material resources, and food. Understanding and addressing the environmental footprint of cities also requires accounting for flows of material resources and embedded energy from other locations. In this perspective the emissions are attributed to the final consumer of the good or service (14). City-level decision makers may overlook consumption-related emissions because they are unaware of their important contributions to overall emissions, or perhaps more importantly because they believe they can have little influence on them. The various city networks often have different guidelines regarding the collection of emissions data, making comparisons between cities in different networks difficult. The difference between production and consumption perspectives becomes larger as supply chains become longer, and nowadays the 'upstream' emissions from trade and resource flows linked to consumption patterns often exceed territorial emissions.

Increasingly it is possible to estimate the carbon emissions of cities. The 'top-down' approach, as exemplified by the Gridded Global Model of City Footprints (GGMCF), downscales national consumption-based emissions (also called carbon footprints, CFs) into a 250 m gridded model. In contrast, the 'bottom-up' approach considers energy sources, urban form, building infrastructure, and other determinants of carbon emissions at the local level – often varying from study to study. While the top-down approach inevitably yields approximations, its great advantage is the potential to be comprehensive and consistent. In order to estimate emissions, GGMCF uses national data on population and purchasing power, and sub-national data from CF studies from the USA, China, EU, and Japan. Even when uncertainties are taken into account the results show that emissions are strongly concentrated in a small proportion of world cities (3). For example the highest emitting 100 urban areas are responsible for 18% of the global CF. In general, high-emitting cities reflect the national carbon emissions from their country but there are exceptions. Of the top emitting 200 cities, 41 are in countries with relatively low carbon emissions but are showing characteristics of high-income high-CF cities. However, the most rapid population growth is occurring in cities with low per capita carbon emissions. This implies that if these cities develop along a low-carbon trajectory, major increases in emissions can be prevented.

A study of the UK's upstream emissions estimated the energy and process-related CO_2 emission component of the CF (i.e. excluding CO_2 emissions from land use change and short-lived climate pollutants). The study linked the production of goods and services across global supply chains to local consumption activities for 434 municipalities in the UK, using data from national surveys and other sources (15). The study showed that population density had little impact on emissions and there was no obvious regional pattern, with London boroughs having both the highest per capita CO_2 emissions nationally (15.51 tonnes per year, in City of London) and the lowest (10.21, Newham), compared to the UK average of 12.5. Per capita emissions increased with income, level of education, and car ownership, as well as decreasing household size. The consumption-based emissions were higher than territorial emissions, with 90% of the settlements in the UK being net carbon importers. Although the data are from 2004, consumption patterns are unlikely to have changed much subsequently; in fact, given the progress in decarbonizing the UK energy supply they are likely to be even more dominant now.

The international variation in urban GHG footprints is large, ranging from 2.4 tonnes CO_2eq per capita per year in Delhi to 60 tCO_2eq/cap/year in Luxembourg, but comparisons are hampered by lack of standardized methods and limited coverage of cities (for discussion see (13)). The GHG footprint combines direct emissions from local consumption sectors in the city (space and water heating, cooking, fuel use from combustion engines) with upstream emissions attributable to local consumption but arising from global supply chains. A study using a standardized approach to estimate these GHG footprints avoided the double counting that can occur when, for example, part of the supply chain for products and services is within the territorial limits of the city. It applied this approach to compare four cities (Berlin, Delhi, Mexico City, and New York metropolitan area), showing that upstream emissions from urban household consumption were comparable in magnitude to cities' overall territorial emissions, being higher in the case of Berlin (130%) and New York and lower for Delhi and Mexico City (80%) (13). Housing, transport, and food contribute more than three-quarters of the GHG footprints of all cities, with remaining contributions from health services and consumer goods. Although the small number of cities limits the strength of the conclusions, as expected, the wealthier cities, Berlin and New York, have longer supply chains than Delhi and Mexico City. The shares of non-domestic upstream emissions (i.e. not arising from the country in which the city is located) range between 16% (Delhi) and 52% (Berlin). Berlin relies on gas from Russia and trade within the EU whereas Delhi sources most of its fuel, food, and other goods and services from within India. The total emissions from such approaches are underestimates because they do not include government consumption and investments in fixed infrastructure. Nevertheless, an important conclusion is that, since housing and transport are major contributors to both territorial and upstream emissions, urban policymakers can influence a substantial proportion of their cities' carbon footprints.

Reducing the Environmental Footprint of Cities

For many cities, addressing socioeconomic inequities is a priority, as is increasing resilience. An additional challenge, particularly in high-income countries, is how to 'retrofit' existing cities to reduce their environmental footprint drastically – while concurrently protecting and enhancing the health of urban dwellers. Despite slow and inadequate progress in decarbonization by most national governments, it is heartening that there is real appetite for change amongst the leadership of many cities. This is exemplified by the Global Covenant of Mayors for Climate and Energy, which at the time of writing has over 10,000 member cities from 138 countries representing over 800 million people (16) and has ambitious GHG reduction targets. It aims to foster innovation in national governments, the private sector, and academia as well as among cities and local governments themselves. The Global Covenant aims to support the academic sector to recruit 10 million new students worldwide to study climate change-related topics by 2025. The C40 Cities include 96 cities, accounting for 25% of global GDP; this network promotes structured activities in energy and buildings, transport and urban planning, food, waste, water, air quality, and climate adaptation (www.c40.org). The 20 major cities in the Carbon Neutral Cities Alliance (https://carbonneutralcities.org/) have pledged to become climate neutral, aiming to cut GHG emissions by 80–100% by 2050 at the latest. They have identified seven 'game changing' strategies to achieve this goal: adopt a zero-emissions standard for new buildings; build a ubiquitous electric-vehicle charging infrastructure; mandate the recovery of organic material; electrify buildings' heating and cooling systems; designate car-free and low-emissions vehicle zones; empower local producers and buyers of renewable electricity; and set a city climate budget to drive decarbonization (17).

There is early evidence that cities participating in such urban networks are reducing their GHG emissions. For example, the Covenant of Mayors initiative, which aimed to implement sustainable energy policies, was launched by the EC in 2008 and, according to results from over 300 monitoring inventories (covering 25.5 million EU inhabitants largely between 2012 and 2014), achieved a 23% overall reduction in emissions. This was driven by a reduction of over one-third in building-related emissions from heating and cooling together with a 17% decrease in emissions from electricity due to a less carbon-intensive fuel mix and more efficient generation. However, there was less progress on transport-related emissions, with only a 7% emission reduction (18).

There are opportunities to develop urban settlements in low-income countries, where population growth is greatest, providing increasing quality of life but a much lower environmental footprint than cities in high-income settings. However, this will require forward planning and appropriate infrastructure development, including for accessible public transport, energy-efficient buildings, clean water, effective sanitation, and green space. Urban infrastructure often has a lifespan of 30–50 years, so decisions made in the near future will have implications for decades and could 'lock in' growing cities to obsolete patterns of development, with enduring implications for environmental sustainability and health. While well-planned urbanization offers the prospects of minimizing environmental impacts of human societies and capitalizing on the benefits of concentration of populations to provide essential services and efficient mobility, urban sprawl, weak governance, and dysfunctional public institutions frequently pose major barriers to progress (1). Sustainable Development Goal 11 aims to foster sustainable urban development; Table 7.3 in Chapter 7 summarizes how the targets in SDG 11 link to health and sustainability.

Contemporary cities are growing twice as fast in terms of land area as population, such that urban expansion could nearly triple the urban land area between 2000 and 2030, eroding biodiversity, accelerating deforestation and contributing to carbon emissions (1). Urban sprawl is associated with low-density land use, heavy reliance on private cars for transportation, segregation of land uses, and constrained economic opportunity for some groups, especially those in inner cities who are far from where jobs are created, and have no easy way to get to those jobs. Sprawl affects exposure to air pollution, heat, physical activity patterns, the risks of motor vehicle crashes, pedestrian injuries and fatalities, water quality and quantity, mental health, and social capital (19). It may be associated with some benefits to higher-income groups and comparatively more adverse effects on lower-income populations. Sprawling development makes the achievement of urban sustainability more difficult because of the associated high levels of transport emissions and the need for extensive infrastructure. Emerging transportation technologies, especially autonomous vehicles, could have paradoxical effects of reducing the 'pain' of travel and encouraging further urban sprawl (20, 21).

Not all cities are expanding; some cities, particularly in high-income countries, are experiencing declines in population. Between 1990 and 2018, populations fell in 95 cities, which were home to 300,000 people or more in 1990. In most of these, population is projected to decline further between 2018 and 2030 (1). These shrinking cities are mainly located in the Russian Federation, Ukraine, and other European countries. The population declines resulted from decreases in fertility, in some cases compounded by out-migration due to economic and political factors. These cities face different challenges from the majority of expanding cities, including sustaining employment and retaining young people, but also have the potential to expand green space and improve the quality of life for those who remain.

Transforming Cities for Health and Sustainability

Seven urban infrastructure sectors needed for livelihoods, health and well-being are linked to natural systems regionally and globally through transboundary flows (**Figure 9.2**). Understanding the nature and size of these flows will allow cities to reduce their environmental footprint more effectively without prejudicing their development prospects.

Basic principles have been proposed for transforming cities to support improved health and well-being, together with environmental sustainability (22). These can be summarized as:

1. Providing basic infrastructure, including energy, clean water, food, shelter, sanitation, and transport, particularly to address the needs of burgeoning populations living in informal settlements. This infrastructure needs to be designed and implemented in anticipation of changing climatic and other environmental conditions, with resilience to floods, droughts, storms, and heat (Chapter 5). While much contemporary discussion of infrastructure focuses on high technology, more basic, equitable innovations, such as point-of-use water treatment and non-motorized transportation infrastructure, are promising and feasible.
2. Pursuing multi-sectoral health strategies that, for example, aim to prioritize equitable access to affordable health care and education, adequate nutrition and green space.
3. Implementing policies which aim for four- to ten-fold increases in resource efficiency rather than incremental gains. These need a focus on urban form and achieving synergies between sectors. For example, co-location of essential services can minimize travel demands. Advanced district energy systems allow potentially wasted energy from one sector to be harnessed by other users, together with local power generation, e.g. through micro-grids, which reduces transmission losses.
4. Recognizing the need for diverse and culturally sensitive approaches, avoiding an undue emphasis on purely technological solutions. For example tightly insulated centrally heated or cooled buildings may be appropriate in some North American and European contexts but vernacular approaches to building design using passive cooling and heating may be more appropriate in other contexts.
5. Harnessing local knowledge to provide powerful insights for rapid change. For example solid waste in low-income cities provides less energy than in high-income settings because waste pickers have sifted through and extracted any potentially recyclable items. Integrating the insights of the waste pickers with modern waste-to-energy solutions can increase efficiency.
6. Increasing understanding of transboundary flows to estimate the environmental footprint of cities and to decide on the appropriate balance between local infrastructures, such as rooftop solar energy, water treatment, rainwater harvesting, and urban and peri-urban farms, and more distant infrastructure.

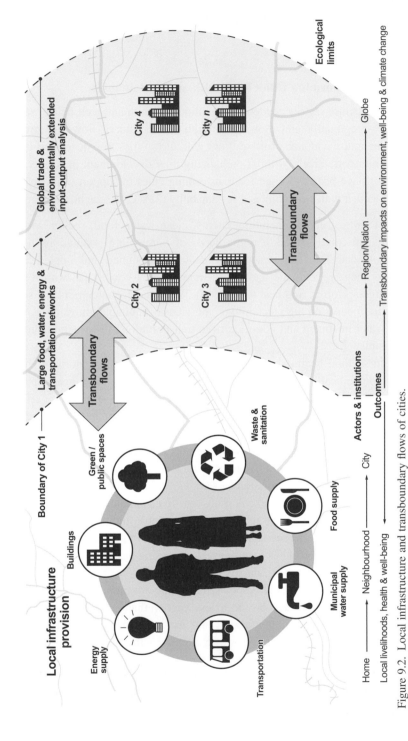

Figure 9.2. Local infrastructure and transboundary flows of cities.

Source: Ramaswami A, Russell AG, Culligan PJ, Sharma KR, Kumar E. Meta-principles for developing smart, sustainable, and healthy cities. *Science*. 2016;352 (6288):940–3. doi:10.1126/science.aaf7160.

7. Ensuring appropriate integration of local and distant sources of essential goods and services, including integrated monitoring of their impacts. Examples include how sensor technology, increasingly available at low cost, can enable more effective monitoring of a diverse range of water sources and of local air pollution (23).

8. Creating city-level capacity to implement high-impact, cross-sector, cross-scale solutions – including analytic, administrative, and political capacity. With increasing private sector involvement in high-technology solutions, questions of data ownership, equity, accountability, and public good arise (24, 25). These require robust governance capable of 'orchestration' (26), open to both public engagement and scientific study.

Data Revolution

Urban form (morphology) reflects the dynamic relationship between physical space and human activities in cities and strongly affects infrastructure. Increasingly, data repositories that combine remotely collected (satellite) data with locally collected data, for example, on the built environment, natural resources (water, biodiversity, etc.), population growth, health, and socioeconomic development, will revolutionize efforts to monitor progress in sustainable development and support decision makers with better evidence of urban trends and policy impacts. Currently there are major gaps in data because many of the datasets are biased towards larger cities in high/middle-income countries (27). Membership-based organizations of cities require cities to apply to join, and to submit information to track progress, thus disadvantaging LMIC cities that lack trained staff and functioning data collection systems. The most geographically representative database of cities, called Geonames, includes over 23,000 urban settlements with populations over 15,000. The UN Sample of Cities collects open source information from a sample of 200 globally representative cities, using satellite imagery, government, and local observers. Increasingly it will be possible to integrate remote sensing data from satellites with more representative locally collected data.

Energy

The previous chapter described the health and other co-benefits of zero-carbon energy, particularly as a result of reduced air pollution. Over 100 cities now get at least 70% of their energy from renewable sources such as hydro, geothermal, wind, and solar, including Auckland, Nairobi, Oslo, Seattle, and Vancouver. Over 40 cities, including Burlington, Basel, and Reykjavík, report getting 100% of their energy from renewables (28). Burlington, the largest city in Vermont, USA, has its own utility and grid, using a mix of solar, wind, hydro, and biomass to generate electricity. Basel gets 90% of its renewable energy from hydropower and 10% from wind, whereas Reykjavík uses hydropower and geothermal energy. Thus each city will have different energy mixes depending on locally available sources. Micro-grids powered by photovoltaics may be particularly suitable for rural areas but they can also provide resilience in urban areas where disruptions to supply may be common and help to ensure that vital facilities such as hospitals can function when mains supply is interrupted. As demands for air conditioning increase and surges in power demand occur due to heat waves, micro-grids could provide a sustainable alternative to diesel generators.

Greenhouse gas emissions must be cut by a half by 2030 to have a reasonable chance of constraining temperature rises to 2 °C above pre-industrial levels. Many cities could achieve an electrical mix of 50–70% from zero-emission renewables by 2030, representing 20–45% of the required GHG reductions. Depending on the location, energy-efficient buildings can address 20–55% of the gap and more efficient waste management could contribute another 10% towards the target (29).

Transport

Changes in urban transport present an opportunity to achieve multiple goals: reducing GHG emissions; promoting physical activity; reducing air pollution; reducing urban noise; reducing injuries; and liberating urban space for uses such as parks and cycleways.

The dominance in many cities of motor vehicles, especially the private car, contributes to a grossly inefficient and unhealthy transport system, and to the expropriation of large amounts of land. The average car is parked more than 90% of the time (30, 31), and when in operation carries on average between one and two people; if replaced by ride-sharing a small fraction of existing vehicles would be needed (32, 33). Motor vehicle infrastructure consumes a substantial share of urban space – in some cities, such as Houston or Little Rock, as much as 50 or 60% of surface area (34). In Britain, where about 80% of dwellings were built with a front garden, almost a third – 7 million plots – have been converted to car parking, an area roughly equivalent to 100 Hyde Parks (31). Reallocation of urban space from parking and road surface could yield land for housing, trees, parks, and other health-promoting amenities (35, 36). Up to one-third of particulate matter pollution in cities derives from traffic (with variation by region) (37); shifts from private cars to other modes (walking, cycling, and public transport) can yield substantial improvements in air quality (38) and health (39). Cities with predominantly public transport systems have much lower transport-related GHG emissions compared with those in which private car use is dominant. If mass transit systems can be scaled up to cover 40% of urban travel by 2050, compared with the projected decline from 35% to 21% as a result of the increasing dominance of private car use in low-income countries, about 6.6 Gt CO_2 emissions can be prevented between 2020 and 2050 (40).

One of the key benefits of a well-designed transportation system is physical activity (PA). PA lowers risks of cardiovascular disease, hypertension, diabetes, and breast and colon cancer. It also benefits mental health, delays the onset of dementia, and can help to maintain healthy weight. Insufficient PA is defined as less than 150 minutes of moderate-intensity, or 75 minutes of vigorous-intensity PA per week, or any equivalent combination of the two. Data from 358 surveys in 168 countries including 1.9 million participants showed that, globally, physical inactivity was commoner amongst women (32%) than men (23%) and in high-income countries (37%) compared with low-income countries (16%). Inactivity has increased since 2001 in high-income countries by about 6% (41). A global assessment estimated that inactivity caused 9% (range 5.1–12.5%) of premature deaths in 2008, amounting to more than 5.3 million of the 57 million deaths that occurred worldwide (42), probably a conservative estimate. A recent systematic review of prospective studies has shown that, using objective assessments of physical activity measured by accelerometers, there is a stronger relationship between reduced death rates and physical activity than when self-reported measures are used. Higher levels of physical activity and lower times spent sedentary were strongly protective. Maximal risk reductions are seen at an average of 24 minutes/day moderate-to-vigorous physical activity or 375 minutes of light physical activity (43). Options for increasing PA in whole populations are very limited, particularly as occupational PA is declining in many countries with increasing automation. Travel that entails walking or cycling – 'active travel' – therefore offers an attractive way of increasing PA while advancing environmental goals.

There is of course a potential for unintended consequences; while shifting to active travel increases PA, it could increase the risk of injuries, and it could increase exposure to air pollution as pedestrians and cyclists travel along busy streets. Studies estimating the health co-benefits of active travel, taking into account both positive and negative impacts, have used a variety of methods and modelling assumptions, making it difficult simply to combine results (44, 45). However, studies consistently find reductions in mortality, disability-adjusted life-years (DALYs), and years of life lost (YLL) with active travel, even when potential disbenefits are taken into account.

A modelling study in San Francisco (46) addressed the question of injuries. It showed that increasing walking and bicycling from a median daily duration of 4 to 22 minutes would reduce the burden of cardiovascular disease and diabetes by 14%, and decrease GHG emissions by 14%. The traffic injury burden would increase by 39%, but in absolute terms the reductions in cardiovascular disease and diabetes burden outweigh the additional injury burden by more than five-fold, with the relative benefits greatest in middle-aged and older populations. The risk of injuries can be reduced by policies to reduce traffic speed and separate motorized and non-motorized traffic.

Pedestrians and cyclists on traffic-filled urban streets breathe polluted air – during strenuous cycling, several times more air than they breathe at rest. Could the deleterious effects of this exposure eclipse the physical activity benefits of active travel? The answer depends, in part, on the levels of ambient pollution. At the global average urban background $PM_{2.5}$ concentration (22 $\mu g/m^3$), a modelling study found that the health benefits of PA greatly exceed increased risks from air pollution, even with extreme assumptions regarding duration of exposure and high respiratory rates (47). Another modelling study considered higher $PM_{2.5}$ levels (50 $\mu g/m^3$), and weighed the benefits of greater PA among cyclists against the risk from air pollution exposure while cycling (48). It found that, up to 5 hours/day of cycling, the PA benefits outweigh the air pollution-associated risk, beyond which more cycling poses more risk from air pollution than benefit from additional PA. Even in a highly polluted city such as São Paulo, Brazil, a counterfactual scenario in which active travel was encouraged in combination with reduced private car and motorcycle use projected considerable health gains (49). A range of other studies on low-carbon transport and active travel in London, Barcelona, the Netherlands, and Midwestern US cities also showed large health co-benefits from increased physical activity, outweighing the risks of pollution exposure and injuries (50, 51).

Walking and cycling are not the only modes of urban travel that increase PA; the use of public transport does so as well. Built into nearly every bus, trolley, or train journey is a walk from the trip origin to the boarding point, and from disembarkation to the trip destination – repeated on the return trip. For example, in England about one-third of public transport users achieved 30 minutes of PA daily during their journeys, with an average duration of physical activity of about 20 minutes daily (52). A substantial literature has corroborated this finding in the USA, Brazil, and elsewhere (53, 54). While there is some evidence that transit-related PA displaces other PA, in general the PA associated with transit use is additive to baseline PA (55), increasing the total and thereby yielding health benefits.

Another benefit of walking, cycling, and transit use is reduction of urban noise, much of which derives from vehicular traffic. Road traffic noise has been associated with increased risk of sleep disturbance, behavioural problems in children, cardiometabolic disorders, anxiety and depression, and other outcomes (56, 57), with the risk falling disproportionately on ethnic minorities (58).

Urban form is an important determinant of transportation patterns, with more compact cities being particularly conducive to walking, cycling, and public transport use, and thus lower per capita transport-related GHG emissions. A compelling example is the comparison of Atlanta with Barcelona, both of which have populations of about 5 million but with dramatically different population density and transport emissions (**Figure 9.3**). Research by Newman, Kenworthy, and colleagues showed a strong association between urban density and transport-related energy use across a range of cities (59–61) (**Figure 9.4**). However,

Figure 9.3. The built-up areas of Barcelona and Atlanta, shown on the same spatial scale. The per capita transport emissions are much lower in compact Barcelona than in sprawling Atlanta, despite their similar population sizes.

Source: LSE Cities 2014, from research for NCE Cities, March 2015. Available from https://lsecities.net/wp-content/uploads/2014/11/LSE-Cities-2014-Transport-and-Urban-Form-NCE-Cities-Paper-03.pdf.

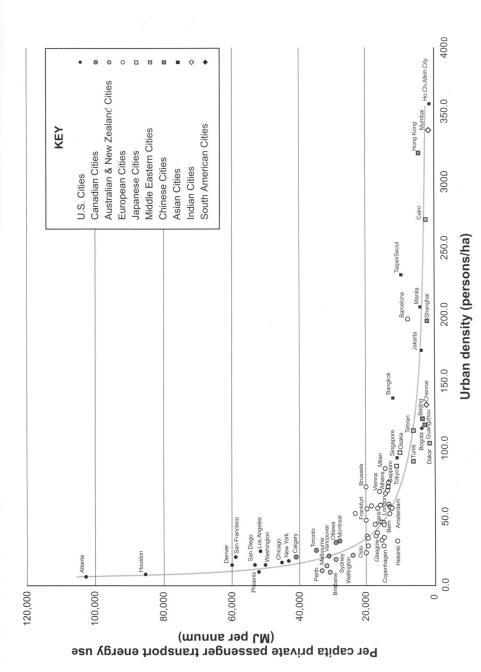

Figure 9.4. Urban density and per capita private transport energy use.

Source: Newman P. Density, the sustainability multiplier: some myths and truths with application to Perth, Australia. *Sustainability*. 2014;6(9):6467–87. www.mdpi .com/2071-1050/6/9/6467.

an analysis of 157 large US urbanized areas shows that population density alone explains only a small fraction of the variation in vehicle miles travelled (VMT). The relative accessibility of local neighbourhoods to the rest of the region is a significant influence (62). Meta-analysis of over 200 studies found that measures of accessibility to destinations and secondarily street network design were most strongly related to VMT (63). Measures of land use diversity, intersection density, and the number of destinations within walking distance were most closely related to walking. When these factors were taken into account, population and job density were only weakly related to VMT. Perhaps these results should not surprise us because accessibility to destinations, such as proximity to the city core, is likely to be greater in more compact cities with higher population density.

A study in Ontario, Canada, based on 3 million participants, assessed the relationship between neighbourhood walkability (measured using a validated index) and obesity and diabetes in adults aged 30–64. Between 2001 and 2012 obesity and overweight increased in the least walkable three quintiles of neighbourhoods and remained unchanged in the most walkable two quintiles of neighbourhoods, taking into account potential confounding factors such as poverty and leisure time activity. The diabetes incidence was lower in the most walkable quintile in 2001 and declined between 2001 and 2012 in the most walkable two quintiles, while remaining unchanged in the least walkable three quintiles (64). Rates of walking and cycling, together with public transport use, were significantly higher and car use lower in the most walkable neighbourhood quintile compared with the least walkable. Although a study design of this type cannot prove definitively that more walkable cities would reduce diabetes and obesity, the results are strongly suggestive.

Currently urban spending on supporting private car use often far exceeds spending on cycling and walking infrastructure. In low-income countries about 70% of urban transport budgets support car-oriented infrastructure, while typically 70% of journeys are on foot or by mass transit (40). Promotion of walking and cycling requires a systemic change in many urban transport policies to develop wider pavements, dedicated cycleways, and improved zoning, so that essential services are accessible within short distances of people's homes. According to the most recent Drawdown report, scaling up bicycle infrastructure could reduce global GHG emissions by 2.6–6.6 $GtCO_2eq$ between 2020 and 2050, scaling up electric bicycle use could save 1.3–4.1 $GtCO_2eq$, and more walkable cities could avert 1.4–5.5 $GtCO_2eq$ over the same period (65). Electric bikes make it possible for older and relatively sedentary people to cycle longer distances and may yield substantial physical activity benefits as a result (66). They may be more attractive to some users than conventional cycles (67) and could potentially accelerate the transition from car dependence in urban areas. Scaling up public transit systems in comparison could contribute 7.5–23.4 $GtCO_2eq$ in GHG emission reductions between 2020 and 2050.

Three other large-scale trends in urban transportation, operating individually and together, have transformative potential, including potential health and environmental benefits: electrification of vehicles, autonomous vehicles, and shared vehicles (32). For example, electrification of vehicles, combined with renewable electricity generation, can improve air quality and reduce noise in urban areas. Autonomous vehicles – immune from fatigue, distraction, and inebriation – can reduce mortality from motor vehicle crashes.

However, some benefits may be less robust than expected. For example, electric vehicles eliminate emissions from internal combustion engine exhaust, but brake pads, tyre wear, road surface abrasion, and road dust resuspension still pollute the air (68, 69). Moreover, the details matter greatly. For example, even if the electricity that propels electric vehicles is generated from biomass, its carbon, land, and water footprints are dramatically larger than if the electricity comes from solar-based hydrogen, which in turn has slightly larger land and water footprints than electricity from photovoltaic panels (70).

As in any complex system, unanticipated negative consequences may arise from well-intended transportation innovations. For example, contrary to expectation, the rise of car-sharing in some cities has *increased* traffic congestion, as large numbers of vehicles circulate while awaiting passengers (71), and a substantial part of car-sharing ridership has shifted not from private cars but from public transport (72, 73). Similarly, the ease of commuting in autonomous vehicles could propel further suburban sprawl (74, 75). A full understanding of the impacts of emerging transportation technologies will come with time, and careful management to optimize health and environmental outcomes will be needed.

A final factor relevant to the future of urban transportation is the impact of COVID-19. As the pandemic unfolded in 2020, crowded buses and trains emerged as potential sites of disease transmission. Some studies (using ecological designs that analyse data at the population or group level, rather than individual level) suggested associations between public transport use and disease transmission (76); in the USA, one study attributed racial disparities in COVID-19 incidence to racial disparities in transit use (77). Transit ridership plummeted in cities around the world; the falling revenues threatened the solvency of many systems. As it happened, there were few if any documented instances of transmission on trains or buses, and fears may have been overblown (78), but concerns remained. Strategies for restoring transit use in pandemic conditions include: limiting ridership to enable physical distancing; staggering use throughout the day (which entails staggering school and work hours); installing protective equipment; conducting frequent cleaning of buses and trains; running trains and buses more frequently and configuring longer trains; deploying contactless payment systems; requiring passengers to wear masks; providing hand sanitizer for passengers; and screening boarding passengers for symptom and/or fever (although such screening will fail to pick up asymptomatic infections) (79, 80).

On the other hand, demand for pedestrian infrastructure soared during the pandemic, as people craved opportunities to get outside and walk or bicycle. Many cities converted portions of roadways from vehicular use to pedestrian and bicycle use. Some cities took the opportunity to make these changes permanent. It is possible that COVID-19 will mark an inflection point in urban infrastructure for active travel (81). At this writing, the story of both active travel and transit resilience in the aftermath of COVID-19 remains to be written.

Housing

Housing is much more than a place of shelter; it also forms the foundation of home and community life and the opportunity for social interactions with friends and family which are pillars of human well-being. Safe, secure housing is a fundamental human need – and when such housing is environmentally sustainable, it not only advances public health, it also helps protect the planet. Across all climate zones, wealth levels, and cultural patterns, housing should offer stability and security from displacement; affordability; ample services such as water, sanitation, and energy; and freedom from dangerous exposures. Housing should also be resilient to changing environmental conditions and to disasters, an increasingly important necessity (82–84). Substandard housing may threaten mental and physical health through a range of pathways, including exposure to damp and mould, extreme temperatures, household air pollution (including from solid fuels and second-hand tobacco smoke), lead and other toxic substances, crowding (which increases transmission of infectious diseases such as tuberculosis and COVID-19), and risk of physical injury (85).

Some housing challenges are particularly associated with low-income settings, from informal settlements in cities of the Global South (86) to deprived neighbourhoods in wealthy countries. Population growth has outstripped housing construction around the world, creating a growing need for affordable, comfortable, and sustainable housing. By one definition, unimproved housing in low-income settings has at least one of four characteristics: unimproved water supply; unimproved sanitation; more than three people per bedroom; and construction using natural or unfinished materials (e.g. earth, sand, dung, or palm flooring) (87). Considerable progress is being made; the prevalence of unimproved housing in cities across Sub-Saharan Africa fell from 68% (65–71%) in 2000 to 47% (44–50%) in 2015 (87). However, major inequalities persist, with about 50% of urban dwellers and 80% of rural dwellers living in unimproved housing. The odds of living in improved housing were over double in the upper wealth quartile compared with all other households.

The COVID-19 pandemic affected housing and related health issues in at least two ways. First, households at lower income levels were more crowded, impeding physical distancing and likely contributing to disease spread. Second, the economic fallout of the pandemic disproportionately affected low-resource households; those unable to make rent or mortgage payments experienced housing insecurity (88). COVID-19 highlighted how housing vulnerability can interact with other vulnerabilities.

Many housing features can improve health, reduce vulnerability to environmental change, and limit environmental impacts in low-income settings, as shown in **Table 9.2** (89).

About 80–100% of malaria transmission in Sub-Saharan Africa occurs indoors; modern housing is associated with reduced malaria transmission compared with thatched huts and other traditional houses. However, there may be adverse consequences of improvements that fail to take into account environmental trends. For example, the replacement of thatch by corrugated iron roofs, while offering advantages of increased weather-proofing, may increase exposure to extreme heat, particularly when the eaves are closed, thus reducing ventilation. In a study in The Gambia, metal-roofed houses with closed eaves and poorly fitting doors had a higher density of malaria-transmitting mosquitoes and were hotter than those with thatched roofs. The higher temperature increases sweating and probably CO_2 levels from increased breathing rates, both of which attract mosquitoes. Simple design changes in a metal-roofed house, such as screening the doors and adding a screened window directly under the roof, were shown to prevent mosquito ingress and to make

Table 9.2. *Housing features, sustainability, and health in low-income settings. Chapter 8 discusses reductions in short-lived climate pollutants from clean household fuels and lighting and Chapter 5 discusses adaptation and resilience to environmental change.*

Housing features	Potential health outcomes affected	Mechanism of effects	Potential trade-offs and constraints
Lack of screened housing, ceilings and open eaves	Malaria and other VBDs; fly-borne diseases e.g. trachoma; diarrhoeal diseases	Prevention of house entry by insect vectors	Reduced ventilation, increased thermal stress and household air pollution
Lack of efficient, low-emission cookstoves or clean fuels (e.g. LPG/biogas) or reliable electricity	ARI in children, COPD, IHD, burns/scalds, etc.	Exposure to products of incomplete combustion (particulates, CO, PAHs, etc.); accidents with fire	Barriers include costs and intermittent supply
Lack of safe and clean (electric) lighting	Burns and household air pollution from kerosene and other lamps	Indirect effects through inability to study, lack of physical security	Lack of reliable electricity, initial costs of solar lamps
Lack of ventilation, inappropriate albedo (e.g. white paint may increase reflection of sunlight and reduce heating)	Heat-related mortality and morbidity (converse in high-altitude/latitude sites)	Thermal stress/cold exposure	See above
Fragile or inappropriate structure for location	Risk from extreme weather. Also, injuries, sexual violence, mental ill-health, VBDs	Robbery, physical attack, susceptibility to landslide, flood, storm, etc. Elevated housing may also protect against some VBDs	Cost may be prohibitive, ventilation may be reduced by smaller and more secure windows
Lack of clean water supply, washing facilities, toilet	Diarrhoeal disease, trachoma, intestinal parasites, ARIs. Improving WaSH provision can reduce undernutrition due to diarrhoea and intestinal parasites	Ingestion of pathogens; poor hygiene	Poorly maintained latrines or inadequate drainage may provide opportunities for mosquito and fly breeding

ARI: acute respiratory infections; CO: carbon monoxide; COPD: chronic obstructive pulmonary disease; IHD: ischaemic heart disease; LPG: liquefied petroleum gas; PAHs: polycyclic aromatic hydrocarbons; VBD: vector-borne diseases; WaSH: water, sanitation, and hygiene

Source: Haines A, Bruce N, Cairncross S, et al. Promoting health and advancing development through improved housing in low-income settings. *Journal of Urban Health.* 2013;90:810–31.

the house as cool as a thatched-roofed house at night by encouraging airflow through the house, although it remained hotter during the day (90).

A study in Tanzania showed that building houses with low-cost materials based on designs from Southeast Asia could greatly reduce the numbers of *Anopheles gambiae* mosquitoes, which transmit malaria, and reduce indoor temperatures by over 2 °C compared with modified and traditional Sub-Saharan African houses. Houses were built with walls made of lightweight permeable materials (bamboo, shade net, or timber) with bedrooms elevated from the ground and with screened windows (91). Two-storey houses seemed to provide the greatest reduction in mosquito numbers. These results are promising but based on relatively small numbers of comparisons so there is a need for large-scale experiments.

These examples from The Gambia and Tanzania illustrate the value of careful design – in these cases balancing the needs for ventilation, vector control, and temperature control. Several other principles for sustainable, low-income housing can be identified. First, use materials that are durable and to the extent possible sourced locally and sustainably. Second, use materials that are acceptable to the poor and do not contribute to stereotyping of poor communities. In many cases these will be similar to materials used in middle class communities. Third, involve communities as far as possible in the design of houses tailored to their needs and resources. Fourth, pre-fabrication and standardization can keep costs low, and speed up construction, particularly with mass production using industrialized building systems (92). Care must be taken however to avoid monotonous developments with little character, little community input, and inadequate infrastructure and green space.

In high-income settings the challenges include the imperative to reduce energy requirements for heating and cooling, to minimize the build-up of harmful pollutants, and to address profound inequities in the quality of housing. Since the rate of new house construction is relatively low in many countries, the retrofitting of older houses to reduce energy demands and achieve greater comfort is central to achieving both health and environmental goals.

In the UK for example, housing is responsible for about 20% of CO_2 emissions, about half of these due to space heating (93). There is also a high burden of cold weather mortality attributed to poor housing, together with fuel poverty (94). In order to meet ambitious national GHG reduction targets there will need to be a national programme of retrofitting housing with insulation and other improvements to increase energy efficiency. The retrofits would need to comprise: double or triple-glazed windows; cavity wall, solid wall, and loft insulation; new condensing gas boilers or increasingly heat pumps or electrical central heating. These upgrades should be combined with effective draught-proofing of leaky dwellings, together with repair of non-operational extractor fans and installation of window trickle fans. These interventions improve health through reduced household air pollution and improved temperature control, provided they include improved ventilation (95). Inadequate ventilation can permit the build-up of harmful pollutants, such as second-hand tobacco smoke, PM from cooking, formaldehyde and volatile organic compounds from furnishings and other sources, and radon emitted from rocks and soils in some regions (96). Ventilation systems with heat recovery can reduce heat loss and therefore energy requirements while maintaining a comfortable household temperature. Unfortunately, some green building codes, which mandate energy efficiency, do not reflect the exposures to potential indoor pollutants and need revision to harmonize health and environmental standards.

If multiple interventions are combined to retrofit existing housing stock in the UK, including increased air tightness, improved insulation, increased ventilation (with particulate filters for the houses in the top 20% of air tightness) and fuel switching to electricity, then GHG emissions could be reduced by 0.6 megatonnes of CO_2 per million population (about 36% reduction in baseline space-heating emissions from housing). Additional reductions in GHG emissions could be achieved by decarbonizing the electricity supply. Together, these interventions could avert about 5400 premature deaths annually, mainly from improved ventilation control with air filtering and fuel switching (97). Cold-related deaths, which are particularly high in the UK, could also be reduced, together with a reduction in fuel poverty because of lowered energy bills. Sadly though, to date the British government has failed to fund housing refurbishment adequately, with the consequence that vast amounts of energy are wasted and GHG emissions reductions have stalled at a time when deep cuts are sorely needed (93). At the same time, the opportunity to ensure that new housing is built to the highest possible standard of energy efficiency is also partly untapped; although standards have improved they could be considerably higher with modest additional cost (estimated at £4800 to provide an air source heat pump and ultra-high levels of fabric efficiency in a newly constructed home (93)). The costs of retrofitting inadequately built new housing with such measures are much greater – about £26,000 on average. This is important because the UK plans to build about 1.5 million new homes by 2022. **Figure 9.5** summarizes the measures needed for existing and new homes in the UK context; many of the points are widely applicable to housing in high/middle-income settings.

Gas boilers are powered by natural gas, a fossil fuel. They are therefore sources of both CO_2 and NO_2 emissions and need to be replaced over time, e.g. by heat pumps powered by zero-carbon renewable electricity. Biomethane offers an alternative to natural gas in some locations where biogas is readily available, but it still produces NO_2 and other combustion products. The Committee on Climate Change suggests that in the UK no new houses should be connected to the gas grid by 2025.

Net-zero buildings are becoming increasingly feasible. They generate the energy they use, typically with solar panels. They use natural lighting, passive solar gain, maximum insulation, natural ventilation, and window louvres and overhangs to optimize summer shade and winter sunlight, and are designed to encourage walking between floors rather than the use of elevators. Buildings that generate more energy than they use, and return that energy to the grid, can offset some of the embodied emissions in the building materials (98).

Multiple stakeholders are involved in the housing sector, and the building sector more broadly, with different interests and motivations. Participatory techniques based on system dynamics modelling can help elucidate complex relationships, build shared understanding, and guide decisions (99, 100). In these exercises, causal loop diagrams (CLDs) may be used to depict key processes, including cycles of cause and effect. There are two kinds of feedback loop: reinforcing loops (R) that reinforce patterns of system behaviour; and balancing loops (B) that can dampen trends. Specially designed software can be used to develop an initial set of CLDs which can be further refined through stakeholder workshops.

What does a low-carbon, sustainable home look like?

Current technology, and measures aimed at preparing for the impacts of climate change, can help new and existing homes to become low-carbon and ultra-efficient as well as adapted to flooding, heat and water scarcity.

Existing homes

Improving existing homes can help existing house-holders meet the challenges of climate change

1. **Insulation**
   in lofts and walls (cavity and solid)

2. **Double or triple glazing with shading**
   (e.g. tinted window film, blinds, curtains and trees outside)

3. **Low-carbon heating**
   with heat pumps or connections to district heat networks

4. **Draught proofing**
   of floors, windows and doors

5. **Highly energy efficient appliances**
   (e.g. A++ and A+++ rating)

6. **Highly water efficient devices**
   with low-flow showers and taps, insulated tanks and hot water theromstats

7. **Green space (e.g. gardens and trees)**
   to help reduce the risk and impacts of flooding and overheating

8. **Flood resilience and resistence**
   with removable air brick covers, relocated appliances (e.g installing washing machines upstairs), treated wooden floors

New build homes

New build homes can and should meet even more ambitious standards in some areas

High levels of airtightness A

More fresh air B
with mechanical ventilation and heat recovery, and passive cooling measures such as openable windows

Triple glazed windows and extenal shading C
especially on south and west faces

Low-carbon heating D
and no new homes on the gas grid by 2025 at the latest

Water management and cooling E
more ambitious water efficiency standards, green roofs and reflective walls

Flood resilience and resistance F
e.g. raised electricals, concrete floors and greening your garden

Construction site and planning G
timber frames, sustainable transport options (such as cycling)

Figure 9.5. Features of a low-carbon home.

Source: Committee on Climate Change. *UK Housing: Fit for the Future?* London: Committee on Climate Change; 2019. www.theccc.org.uk/publication/uk-housing-fit-for-the-future/.

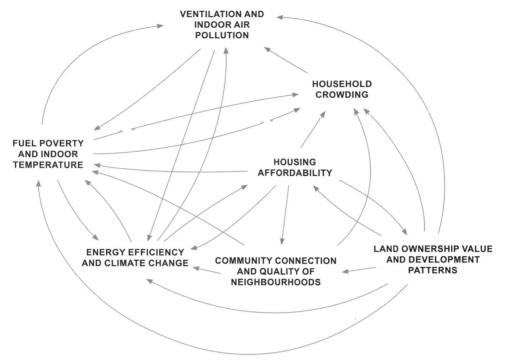

Figure 9.6. Seven themes used to develop causal loop diagrams to support decision making on housing, energy, well-being, and health.
Source: Macmillan A, Davies M, Shrubsole C, et al. Integrated decision-making about housing, energy and wellbeing: a qualitative system dynamics model. *Environmental Health*. 2016;15(1):S37.

In addition, interviews can be used to make explicit people's understanding of complex linkages, such as between housing, energy use, well-being, and health, a process called cognitive mapping. This can facilitate decision making in the face of trade-offs and conflicts between different policies. In one such effort, a CLD oriented around seven themes depicted the links between housing, well-being, and energy use (99) (**Figure 9.6**). Approaches such as these help to integrate diverse priorities and perspectives.

Tackling Consumption-Based Emissions

Although a growing number of cities aim to tackle GHG emissions directly within their territorial boundaries, few have plans to tackle consumption-based emissions, which as discussed above are major contributors to climate change. The C40 group of 94 major cities aims to address consumption-based emissions from six sectors (food, buildings and infrastructure, private transport, aviation, clothing and textiles, and electronics and household appliances) in order to contribute to achieving the ambitious 1.5 °C target. Currently per capita annual GHG emissions of member cities range from 1.2 to 39.7 tCO$_2$eq (101). Of the total urban emissions arising from goods and services consumed in C40 cities (comprising public transport, utilities, food, stationary fuel/energy, personal transport, and purchased goods and services), 85% is imported from elsewhere. Consumption emissions are still rising in C40 cities and, if unchecked, are projected to nearly double by 2050. However, by 2030 urban policies to influence consumption could substantially reduce GHG emissions. For example, in the category of clothing and textiles, reducing the amount of new clothing bought and reducing supply chain waste could cut emissions by 39%, and in the category of food, dietary changes and waste reduction could cut emissions by 36%. Many of these policies would involve the application of circular economy principles (see Chapter 8). Sectoral policies in the transport, housing, and food sectors (Chapter 10) have major potential for GHG reductions and health co-benefits.

Water and Sanitation

Water is essential to a city's function, and water constraints pose growing threats to urban development. An urban water system consists of the natural and built parts of a city's water cycle, including sources (rainfall and groundwater, and surface water), natural freshwater ecosystems, drinking water and wastewater plants, distribution and storage facilities, and sewer networks. All these elements are intrinsically linked by the movement of water, and the fluxes of energy and materials, and each interacts with the others (102). However, traditional approaches to water services have been siloed into four categories: water resources (quantity), drinking water (quality), wastewater, and stormwater. Integrated solutions increasingly aim to deliver an ample supply of high-quality water, with recovery of water, energy, waste products and nutrients, and with resilience to perturbations such as drought and extreme rainfall (103).

Improved drinking water sources should, but do not always, provide safe drinking water. They include piped household water (the norm in wealthy cities), public standpipes, boreholes, protected dug wells, protected springs, and rainwater collection. Although the provision of safe water in cities has increased and overall 96% of urban populations have access to improved water supplies (104), many informal settlements are supplied with water that has high levels of microbial contamination. Even in wealthy countries, legacy infrastructure, deferred maintenance, and poor management can result in chemical contamination of drinking water, as famously occurred in Flint, Michigan, USA, beginning in 2014 (105), and water supplies can be routinely contaminated by disinfection by-products, pharmaceuticals and personal care products (106), and other contaminants.

There are major inefficiencies in current urban water resource management. For example, phosphorus is removed from wastewater to prevent eutrophication of water bodies, using expensive technologies, and disposed of as a waste product (107). However, phosphorus – dubbed 'life's bottleneck' – is a valuable resource, essential in phosphate-based fertilizers, and the global supply is finite (108). This waste stream could therefore be a valuable source of fertilizer (109, 110), exemplifying a circular economy approach. Another inefficiency is the entry of emerging biological and chemical pollutants into water supplies – metals, pharmaceuticals, persistent organics, microplastics, and more. These pose ill-defined but concerning threats to the health of humans and of natural systems requiring costly treatment facilities to reduce the risks. Preventing contamination in the first place would be far less costly and more effective.

Urban water systems also have substantial energy costs and environmental impacts. Moving and purifying water are energy intensive; in the USA, for example, about 35% of municipal energy use is devoted to water and wastewater facilities and electricity use accounts for 80% of the budget for drinking water treatment and provision (103). As cities expand, and as they outgrow or deplete proximate sources, water must often be transported for longer distances with consequent increases in energy demand. Many urban water systems are designed for high water flows necessary for fire fighting, but water does not need to be potable for use in extinguishing fires. Thus dual systems, which separate drinking water from non-potable water using reclaimed wastewater, are a more efficient alternative. Such systems could be used for domestic supply since many household applications, such as flushing toilets or watering gardens, do not require potable water.

Increasingly impervious surfaces in cities cause higher stormwater run-offs which may be amplified by climate change. This can result in flooding and discharge of untreated stormwater into surface water bodies which can increase contamination and reduced water quality. And excessive groundwater withdrawals can lead to land subsidence, a problem confronting cities from Mexico City to Jakarta.

Systems approaches are essential to understanding and managing urban water systems (111). These integrated solutions include analytical frameworks such as integrated water resource management (112) and the ecosystem services framework (102, 113). While these approaches differ in emphasis, both explicitly address complexity, and attempt to identify trade-offs and synergies operating within complex systems (114). Observers have noted that implementing the findings of such analyses can be difficult (114, 115) – more a reflection of the messy complexity of urban political and social life than an inherent flaw of complex analytical approaches.

Many solutions to urban water challenges are technical. The concept of a smart water grid, analagous to the smart electric grid, relies on information and communications technology (ICT) to monitor supply and demand, manage flow, and optimize energy use, water supply and quality (116). A different approach involves decentralized water systems, which have increasingly been proposed as sustainable, reliable alternatives to costly and energy-consuming centralized systems (117, 118). These systems include on-site water collection and storage, and point-of-use treatment of collected rainwater and/or of wastewater. They avoid the energy use required for water transport in large systems. Using smart monitoring and control approaches, such systems could be networked to provide added resilience and flexibility in the face of changing demands. Household water treatment and safe storage can reduce diarrhoeal disease incidence by about 60% in those whose water supply has microbial contamination, but many technologies have limited effectiveness. A recent WHO evaluation showed that 23 of 30 products met basic WHO performance criteria but only five provided comprehensive protection (including bacteria, protozoa, and viruses), including three of four solar disinfection products (119).

Another household-level intervention that could improve efficiency is waste collection. Human waste contains potentially useful resources that could be re-used; urine contributes about 80% nitrogen, 50% phosphorus, and 90% potassium in domestic wastewater (120). Toilets that separate urine and faeces can simplify recovery of nutrients from waste. The Bill and Melinda Gates Foundation (BMGF) has made considerable investments in developing 'toilets of the future' which combine affordability with efficient recovery of nutrients and minimize the risks of diarrhoeal disease transmission. An evaluation of different toilet technologies using forward osmosis funded by the BMGF has shown that recovering nutrients only from urine had the greatest environmental and economic benefit (121). 'Resource-oriented' toilets, which enable the recovery of nutrients, are generally costlier to construct but have lower operational costs compared with conventional toilets. The forward osmosis process requires considerable energy but this could be provided by photovoltaics or from biogas, and the production of fertilizer from human waste offsets the environmental and economic costs.

Nature-based solutions offer many human, environmental, and economic benefits (122). For example, greywater (domestic non-sewage water) and stormwater can be treated with low energy and capital expenditure by wetlands which can also act as carbon sinks. Their ability to act as carbon sinks is due not just to photosynthesis within the wetlands but also to their capacity to remove carbon from greywater and stormwater. Wetlands can also provide impressive capacity to provide denitrification of run-off water, support biodiversity conservation, and reduce drought risks and urban heat island effects. Nature trails around wetlands can provide the benefits of green space (discussed below). However, care must be taken to ensure that wetlands do not provide breeding sites for disease-transmitting mosquitoes in locations where these can thrive.

In China the concept of sponge cities aims to increase the capacity to absorb surface water discharges by 20%, to retain 70% of urban stormwater and to re-use 80% by 2030 in 30 cities (123). The measures include protection and enhancement of wetlands and other bodies of water in cities, together with permeable pavements, green roofs, and green spaces. Planners of urban roadways are required to maintain green space between roads, cycle ways, and pavements to absorb excess water (see also Chapter 5).

Nature-based approaches to urban water are especially relevant given increasing pressure on, and degradation of, the watersheds that supply cities. Globally, population density in urban source watersheds increased by a factor of 5.4 from a median of 22 to 124 people/km^2 between 1900 and 2005, with the fastest growth occurring in the late twentieth century (124). About one-third of the world's largest cities depend on protected forests for good-quality drinking water (125). A study of 309 large cities linking watershed degradation from human activities to water quality showed that between 1900 and 2005, average pollutant yield of urban source watersheds increased by 40% for sediment, 47% for phosphorus, and 119% for nitrogen, because of watershed degradation. High nitrogen and phosphorus levels lead to growth of algae which may be toxic. Water treatment costs increased in about 30% of cities as a result of watershed degradation and the increases were substantial, averaging over 50% for operation and maintenance and over 40% for replacement of treatment plants (124). Policies to protect watersheds are increasingly recognized to yield a range of benefits: conserving water, reducing microbial contamination and sedimentation, and saving resources.

Strategies to protect watersheds and restore forest cover are being employed in many locations. These include payments to communities living on or owning land on watersheds to protect their integrity. Payment for ecosystem services (PES) initiatives have had mixed results depending on the financial value of the compensation, targeting of payments, and types of trees used in reforestation programmes (some types of trees have high water requirements). Whilst payments may be beneficial for land cover and biodiversity, their impact on poverty and livelihoods is variable and the additional effects of payments over and above other strategies is often unclear (126). In some cases they increased income inequities because the poorest communities and individuals possess little or no land and are therefore ineligible for payments. Some schemes have shown 'spillover' effects because damaging activities are displaced outside the area covered. Local participation in the design of PES initiatives and respect for local cultural values help to promote success and preliminary evidence suggests that PES initiatives financed by those who use the services they provide, rather than central government, may enhance the chances of success (127). Like many such interventions PES cannot be considered a 'silver bullet', but with appropriate design and implementation can contribute to the achievement of environmental and socioeconomic objectives.

Much of the literature on urban water centres on technologies, from smart monitors to nature-based solutions. However, any such intervention needs to be accepted and used by the communities whose water it provides. Research points again and again to an 'implementation gap'; what is known to work is not put in place, due to a range of social and institutional barriers (128). Successful urban water management therefore requires effective governance, management, and community participation, using processes that foster social capital, inter-sectoral professional development, and inter-organizational coordination (129).

Green Space

Contact with nature – or the absence of such contact – is an important feature of city life. Nature contact may come through large, iconic urban parks such as London's Hampstead Heath or New York's Central Park. It may come from smaller neighbourhood-scale parks, from linear parks that also feature waterways and/or pathways for cycling and walking, or from street trees. It may even come from vegetated vacant lots, or from such small-scale amenities as flowerboxes and planters.

Nature contact is measured in various ways. A common metric of nature contact is the normalized difference vegetation index (NDVI), which assesses the density of photosynthetically active biomass based on satellite imagery (130). This is commonly combined with geospatial data to assess the amount of green space within a given distance of where people live. Land use data – say, municipal maps indicating the presence of forest, grassland, buildings, and so on – are also used as an exposure metric. Some studies use subjective reports of nature contact as the exposure metric, and increasingly researchers are using street-view data from sources such as Google to quantify the presence of vegetation from a perspective that urban dwellers would see (131).

Urban nature offers a range of health benefits. Some of these are direct, while others function through a variety of indirect pathways. The literature on these benefits has expanded greatly in recent years and has been extensively reviewed (131–136). While most research has been carried out in wealthy nations, there is an increasing body of evidence from LMICs as well (137).

The direct benefits of nature contact extend to both mental health and physical health. Some of the strongest evidence pertains to mental health and cognitive function (138–141). Nature contact has been associated with reduced risk of anxiety and depression across the lifespan. It reduces the risk of attention-deficit hyperactivity disorder and of psychiatric disorders in children, improves attention and academic performance in children, and slows loss of cognitive function in older adults (142). Nature contact may also predict well-being more broadly. A systematic review of 50 studies assessing the links between different aspects of green space and well-being made the important distinction between happiness and life satisfaction (hedonic well-being) and fulfilment, functioning, and purpose in life (eudaimonic well-being) (143). The review found that the amount of nearby green space was associated with life satisfaction (hedonic well-being), but not with eudaimonic well-being. There was insufficient evidence of which types of green space were particularly beneficial and on accessibility, number of visits, and viewing patterns to guide policymakers seeking to optimize the benefits. The mechanisms by which green space results in improved mental health may vary according to the specific condition. For example, it may enhance psychological restoration and counteract the effects of urban noise or crowding.

There is also considerable evidence linking green space with physical health. A systematic review of residential green space and mortality identified 12 eligible studies conducted in North America, Europe, and Oceania. Although the studies varied greatly in design, study population, green space assessment, and the extent to which they adjusted for potential confounding, the majority showed a small but significant reduced risk of cardiovascular disease (CVD) mortality in areas with higher residential greenness (144). There is also some evidence that green space exposure reduces the risk of diabetes (145) and hyperlipidaemia (146). These effects are not large, and need to be carefully distinguished from the effects of reductions in other urban exposures such as to noise and air pollution (147, 148). But because cardiovascular diseases are so prevalent, even relatively small risk reductions could translate to large public health benefits over time. This was vividly illustrated by a study in Toronto of the association between street tree density and cardiometabolic conditions (hypertension, high blood glucose, overweight, high cholesterol, myocardial infarction, heart disease, stroke, and diabetes), controlling for socioeconomic and demographic factors (149). Each additional 11 trees on a city block, on average, was associated with a reduction in cardiometabolic conditions equivalent to that of a US$20,000 higher annual income, moving to a neighbourhood with US$20,000 higher median income, or being 1.4 years younger.

Other reported health benefits of nature contact include improved birth outcomes (150), greater physical activity (151), and longer life expectancy (152). Green space is also associated with improved social capital (153, 154) and with reduced crime and aggressive behaviour (155–157).

To optimize the health benefits of urban nature, parks and green space should be as accessible as possible – distributed throughout the city, especially in deprived neighbourhoods where access may be historically limited, not separated from where people live by major thoroughfares, and with porous borders between neighbourhoods and natural areas. Programming such as gardening classes and organized sports activities helps increase use, as do good maintenance, lighting, restroom facilities, and safety. Community engagement in planning and caring for parks is a key strategy.

Urban nature contact does not invariably improve health. For example, while some studies show an association between tree canopy and reduced asthma and allergic symptoms, others show the reverse effect (158, 159). Part of the explanation may lie with the allergenicity of different tree species, with seasonal effects, and/or with other confounders such as social class, air pollution, and home environments. For example, birch pollen is a common cause of allergic symptoms and birch trees are used for ornamental planting in some cities (160). In some circumstances trees may block airflow that would otherwise disperse pollutants, worsening local air quality (161). Trees also emit biogenic volatile organic compounds (VOCs) that act as ozone precursors, with the levels varying by species (162). For example, a study in Denver showed that planting a million trees that emit low levels of biogenic VOCs compared with, for example, a million English oak trees that are high emitters, would be equivalent to preventing emissions from up to 490,000 cars (163). Trees require water, and the demand for water must be balanced against the benefits of trees, especially in arid regions. Careful consideration of urban vegetation strategies is essential in order to fully realize ecosystem services (164, 165), and to realize the potential for urban nature to advance the SDGs (166). Trees may be also be perceived as a nuisance in low-income neighbourhoods, in part because they are poorly maintained and sometimes vandalized, and in part because they may be seen as providing concealment for criminals (165). Importantly, the addition of trees or other natural amenities to low-income neighbourhoods may push up property values and displace the people who live there, a process known as green gentrification (167, 168). This can threaten the health of people who are displaced from their neighbourhoods (169). To optimize health benefits, then, natural elements in cities need to be carefully planned, considering factors from species selection to unintended social consequences.

Urban nature may also offer indirect health benefits – by cooling cities (see also Chapter 5), by reducing air pollution, and by controlling stormwater flow (165). Trees and other vegetation cool cities through evapotranspiration and by providing shade – an effect that can operate at the small scale of individual blocks, as well as across large sections of cities (170). A study using data from over 300 cities in 22 countries showed higher heat-related mortality associated with increases in population density, $PM_{2.5}$, GDP, and Gini index (a measure of income inequality), whereas higher levels of green space were associated with a decreased effect of heat (171). In Greater London lower daily average temperatures were recorded in areas with more tree and vegetation cover, and heat-related deaths were lower in these areas after adjusting for a range of confounding factors (172).

Buildings that are shaded, in turn, require less energy for air conditioning during hot weather, reducing overall energy demand in some regions (173, 174). Trees and other vegetation may reduce both particulate and gaseous air pollution as their leaves capture and remove pollutants from air (158). While published quantitative estimates vary widely, this effect is generally small, on the order of a single-digit percentage reduction in pollutant levels (175). The effectiveness of trees in cleaning the air depends on the species, the morphology of streets (and therefore of air movement), and other factors.

Strategies for optimizing health benefits of urban green space, identified through a WHO review, include the provision of multiple green spaces rather than a few large green areas within a city, provision of green space that can be used for a range of purposes targeted to different age and ethnic groups, assuring community engagement to promote appropriate design and equitable use of green space, and finally the avoidance of allergenic species (136).

Conclusions

Urban growth will continue over coming decades, especially in rapidly urbanizing Africa and South Asia. There is considerable opportunity to learn and apply the lessons from past experience. The provision of basic services in informal settlements is a priority – including safe water, sanitation, clean energy (increasingly from renewable sources), waste collection, and recycling. In high-income settings the challenges are to increase resilience to environmental change at much lower environmental footprints than exist today through zero-carbon energy, electrification of the transport system with enhanced opportunities for active travel and universal access to affordable public transport, implementation of circular economy principles and more energy-efficient buildings. The latter will require large-scale retrofitting programmes with improved ventilation and insulation of buildings, together with passive cooling where feasible. The provision of green space and tree planting can deliver substantial benefits for health and the environment but careful design and selection are needed to minimize unintended adverse consequences. Nature-based solutions such as wetlands, coastal mangroves, and watershed protection can also enhance the ability of cities to withstand extreme events such as floods and droughts. Growing urban networks provide an opportunity for shared learning to scale up potential solutions rapidly using cross-sectoral strategies integrating environmental sustainability with health protection and promotion.

References

1. UN DESA. World Urbanization Prospects: The 2018 Revision. New York: United Nations Department of Economic and Social Affairs, Population Division; 2018. Contract No. ST/ESA/SER.A/420. Available from https://population.un.org/wup.
2. Satterthwaite D. Cities' contribution to global warming: notes on the allocation of greenhouse gas emissions. *Environment and Urbanization*. 2008;20(2):539–49. doi: 10.1177/0956247808096127.

3. Moran D, Kanemoto K, Jiborn M, et al. Carbon footprints of 13 000 cities. *Environmental Research Letters*. 2018;13(6):064041. doi: 10.1088/1748-9326/aac72a.

4. Dodman D. Blaming cities for climate change? An analysis of urban greenhouse gas emissions inventories. *Environment and Urbanization*. 2009;21(1):185–201. doi: 10.1177/0956247809103016.

5. Muñiz I, Dominguez A. The impact of urban form and spatial structure on per capita carbon footprint in U.S. larger metropolitan areas. *Sustainability*. 2020;12(1):389. www.mdpi.com/2071-1050/12/1/389.

6. Milner J, Davies M, Wilkinson P. Urban energy, carbon management (low carbon cities) and co-benefits for human health. *Current Opinion in Environmental Sustainability*. 2012;4(4):398–404. http://dx.doi.org/10.1016/j.cosust.2012.09.011.

7. UN DESA. The Sustainable Development Goals Report 2018. New York: United Nations Department of Economic and Social Affairs; 2018. Available from www.un.org/development/desa/publications/the-sustainable-development-goals-report-2018.html.

8. United Nations Human Settlements Programme. *The Challenge of Slums: Global Report on Human Settlements 2003*. London: Routledge; 2003.

9. Milner J, Harpham C, Taylor J, et al. The challenge of urban heat exposure under climate change: an analysis of cities in the Sustainable Healthy Urban Environments (SHUE) database. *Climate*. 2017;5(4):93. www.mdpi.com/2225-1154/5/4/93.

10. Pal JS, Eltahir EAB. Future temperature in southwest Asia projected to exceed a threshold for human adaptability. *Nature Climate Change*. 2015;6:197–200. doi: 10.1038/nclimate2833.

11. Akbari H, Cartalis C, Kolokotsa D, et al. Local climate change and urban heat island mitigation techniques – the state of the art. *Journal of Civil Engineering and Management*. 2016;22(1):1–16. doi: 10.3846/13923730.2015.1111934.

12. Bhamare DK, Rathod MK, Banerjee J. Passive cooling techniques for building and their applicability in different climatic zones – the state of art. *Energy and Buildings*. 2019;198:467–90. https://doi.org/10.1016/j.enbuild.2019.06.023.

13. Pichler P-P, Zwickel T, Chavez A, et al. Reducing urban greenhouse gas footprints. *Scientific Reports*. 2017;7(1):14659. doi: 10.1038/s41598-017-15303-x.

14. Wackernagel M, Kitzes J, Moran D, Goldfinger S, Thomas M. The Ecological Footprint of cities and regions: comparing resource availability with resource demand. *Environment and Urbanization*. 2006;18(1):103–12. doi: 10.1177/0956247806063978.

15. Minx J, Baiocchi G, Wiedmann T, et al. Carbon footprints of cities and other human settlements in the UK. *Environmental Research Letters*. 2013;8(3):035039. http://stacks.iop.org/1748-9326/8/i=3/a=035039.

16. Global Covenant of Mayors for Climate & Energy. 2020. Available from www.globalcovenantofmayors.org/.

17. Shank M, Partin J, Plastrik P, Cleveland J. Game Changers: Bold Actions by Cities to Accelerate Progress Toward Carbon Neutrality. Carbon Neutral Cities Alliance; 2018. Available from https://carbonneutralcities.org/initiatives/game-changers/.

18. Kona A, Melica G, Koffi Lefeivre B, et al. Covenant of Mayors: Greenhouse Gas Emissions Achievements and Projections. European Union; 2016. Available from https://ec.europa.eu/jrc/en/publication/eur-scientific-and-technical-research-reports/covenant-mayors-greenhouse-gas-emissions-achievements-and-projections.

19. Frumkin H. Urban sprawl and public health. *Public Health Reports*. 2002;117(3):201–17.

20. Spence JC, Kim Y-B, Lamboglia CG, et al. Potential impact of autonomous vehicles on movement behavior: a scoping review. *American Journal of Preventive Medicine*. 2020;58(6):e191–9. doi: 10.1016/j.amepre.2020.01.010.

21. Rojas-Rueda D. New transport technologies and health. In Nieuwenhuijsen MJ, Khreis H, editors. *Advances in Transportation and Health*: Elsevier; 2020. pp. 225–37.

22. Ramaswami A, Russell AG, Culligan PJ, Sharma KR, Kumar E. Meta-principles for developing smart, sustainable, and healthy cities. *Science*. 2016;352(6288):940–3. doi: 10.1126/science.aaf7160.

23. Shiva Nagendra SM, Reddy YP, Narayana MV, Khadirnaikar S, Rani P. Mobile monitoring of air pollution using low cost sensors to visualize spatio-temporal variation of pollutants at urban hotspots. *Sustainable Cities and Society*. 2019;44:520–35. https://doi.org/10.1016/j.scs.2018.10.006.

24. Rossi B. Is It Time to Nationalise Data? Information Age. 6 June 2017. Available from www.information-age.com/time-nationalise-data-123466627/.

25. Lorinc J. Promise and Peril in the Smart City: Local Government in the Age of Digital Urbanism. University of Toronto Institute on Municipal Finance & Governance; 2019.

26. Gupta A, Panagiotopoulos P, Bowen F. An orchestration approach to smart city data ecosystems. *Technological Forecasting and Social Change*. 2020;153:119929. https://doi.org/10.1016/j.techfore.2020.119929.

27. Taylor J, Haines A, Milner J, Davies M, Wilkinson P. A comparative analysis of global datasets and initiatives for urban health and sustainability. *Sustainability*. 2018;10(10). doi: 10.3390/su10103636.

28. Carbon Disclosure Project. Over 100 global cities get majority of electricity from renewables; 2018. Available from www.cdp.net/en/articles/cities/over-100-global-cities-get-majority-of-electricity-from-renewables.

29. Falk J, Gaffney O, Bhowmik AK, et al. Exponential Roadmap 1.5. Stockholm: Future Earth; 2019. Available from https://exponentialroadmap.org.

30. Shoup D. *The High Cost of Free Parking (Updated Edition)*. Chicago: Taylor & Francis; 2011.

31. Bates J, Leibling D. Spaced Out: Perspectives on Parking Policy. London: RAC Foundation; 2012. Available from www.racfoundation.org/research/mobility/spaced-out-perspectives-on-parking.

32. Fulton L, Mason J, Meroux D. Three Revolutions in Urban Transportation. Davis, CA: UC Davis Institute of Transportation Studies and Institute for Transportation & Development Policy; 2017. Available from https://steps.ucdavis.edu/wp-content/uploads/2017/05/STEPS_ITDP-3R-Report-5-10-2017-2.pdf.

33. OECD. Shared Mobility: Innovation for Liveable Cities. OECD International Transport Forum; 2016. Available from www.itf-oecd.org/shared-mobility-innovation-liveable-cities.

34. Gardner C. We are the 25%: looking at street area percentages and surface parking; 2011. Available from https://oldurbanist.blogspot.com/2011/12/we-are-25-looking-at-street-area.html.

35. Gössling S. Why cities need to take road space from cars – and how this could be done. *Journal of Urban Design*. 2020;25(4):443–8. doi: 10.1080/13574809.2020.1727318.

36. Chester M, Fraser A, Matute J, Flower C, Pendyala R. Parking infrastructure: a constraint on or opportunity for urban redevelopment? A study of Los Angeles County parking supply and growth. *Journal of the American Planning Association*. 2015;81(4):268–86. doi: 10.1080/01944363.2015.1092879.

37. Karagulian F, Belis CA, Dora CFC, et al. Contributions to cities' ambient particulate matter (PM): a systematic review of local source contributions at global level. *Atmospheric Environment*. 2015;120:475–83. http://dx.doi.org/10.1016/j.atmosenv .2015.08.087.

38. Titos G, Lyamani H, Drinovec L, et al. Evaluation of the impact of transportation changes on air quality. *Atmospheric Environment*. 2015;114:19–31. https://doi.org/10 .1016/j.atmosenv.2015.05.027.

39. Peel JL, Klein M, Flanders WD, Mulholland JA, Tolbert PE. Impact of improved air quality during the 1996 Summer Olympic Games in Atlanta on multiple cardiovascular and respiratory outcomes. *Research Reports (Health Effects Institute)*. 2010(148):3–23; discussion 25–33.

40. Hawken P, editor. *Drawdown: The Most Comprehensive Plan Ever Proposed to Reverse Global Warming*. London and New York: Penguin; 2017.

41. Guthold R, Stevens GA, Riley LM, Bull FC. Worldwide trends in insufficient physical activity from 2001 to 2016: a pooled analysis of 358 population-based surveys with 1.9 million participants. *The Lancet Global Health*. 2018;6(10): e1077–86. doi: 10.1016/S2214-109X(18)30357-7.

42. Lee IM, Shiroma EJ, Lobelo F, et al. Effect of physical inactivity on major non-communicable diseases worldwide: an analysis of burden of disease and life expectancy. *The Lancet*. 2012;380(9838):219–29. doi: 10.1016/S0140-6736(12)61031-9.

43. Ekelund U, Tarp J, Steene-Johannessen J, et al. Dose-response associations between accelerometry measured physical activity and sedentary time and all cause mortality: systematic review and harmonised meta-analysis. *BMJ*. 2019;366:l4570. doi: 10.1136/bmj.l4570.

44. Raza W, Forsberg B, Johansson C, Sommar JN. Air pollution as a risk factor in health impact assessments of a travel mode shift towards cycling. *Global Health Action*. 2018;11(1):1429081. doi: 10.1080/16549716.2018.1429081.

45. Doorley R, Pakrashi V, Ghosh B. Quantifying the health impacts of active travel: assessment of methodologies. *Transport Reviews*. 2015;35(5):559–82. doi: 10.1080/ 01441647.2015.1037378.

46. Maizlish N, Woodcock J, Co S, et al. Health cobenefits and transportation-related reductions in greenhouse gas emissions in the San Francisco Bay area. *American Journal of Public Health*. 2013;103(4):703–9. doi: 10.2105/ajph.2012.300939.

47. Rojas-Rueda D, de Nazelle A, Andersen ZJ, et al. Health impacts of active transportation in Europe. *PLoS One*. 2016;11(3):e0149990. doi: 10.1371/journal.pone.0149990.

48. Tainio M, de Nazelle AJ, Gotschi T, et al. Can air pollution negate the health benefits of cycling and walking? *Preventive Medicine*. 2016;87:233–6. doi: 10.1016/j. ypmed.2016.02.002.

49. Sa TH, Tainio M, Goodman A, et al. Health impact modelling of different travel patterns on physical activity, air pollution and road injuries for Sao Paulo, Brazil. *Environment International*. 2017;108:22–31. doi: 10.1016/j.envint.2017.07.009.

50. Chang KM, Hess JJ, Balbus JM, et al. Ancillary health effects of climate mitigation scenarios as drivers of policy uptake: a review of air quality, transportation and diet co-benefits modeling studies. *Environmental Research Letters*. 2017;12(11):113001. http://stacks.iop.org/1748-9326/12/i=11/a=113001.

51. Giallouros G, Kouis P, Papatheodorou S, Woodcock J, Tainio M. The long-term mortality impact of restricting cycling and walking during high air pollution days on all-cause mortality: health impact assessment study. *Environment International*. 2020. https://doi.org/10.17863/CAM.51003.

52. Patterson R, Webb E, Millett C, Laverty AA. Physical activity accrued as part of public transport use in England. *Journal of Public Health*. 2018;41(2):222–30. doi: 10.1093/pubmed/fdy099.
53. Hirsch JA, DeVries DN, Brauer M, Frank LD, Winters M. Impact of new rapid transit on physical activity: a meta-analysis. *Preventive Medicine Reports*. 2018;10:184–90. doi: 10.1016/j.pmedr.2018.03.008.
54. Rissel C, Curac N, Greenaway M, Bauman A. Physical activity associated with public transport use – a review and modelling of potential benefits. *International Journal of Environmental Research and Public Health*. 2012;9(7):2454–78. doi: 10.3390/ijerph9072454.
55. Miller HJ, Tribby CP, Brown BB, et al. Public transit generates new physical activity: evidence from individual GPS and accelerometer data before and after light rail construction in a neighborhood of Salt Lake City, Utah, USA. *Health & Place*. 2015;36:8–17. doi: 10.1016/j.healthplace.2015.08.005.
56. Recio A, Linares C, Banegas JR, Diaz J. Road traffic noise effects on cardiovascular, respiratory, and metabolic health: an integrative model of biological mechanisms. *Environmental Research*. 2016;146:359–70. doi: 10.1016/j.envres.2015.12.036.
57. Dzhambov MA, Lercher P. Road traffic noise exposure and depression/anxiety: an updated systematic review and meta-analysis. *International Journal of Environmental Research and Public Health*. 2019;16(21). doi: 10.3390/ijerph16214134.
58. Casey JA, Morello-Frosch R, Mennitt DJ, et al. Race/ethnicity, socioeconomic status, residential segregation, and spatial variation in noise exposure in the contiguous United States. *Environmental Health Perspectives*. 2017;125(7):077017. doi: 10.1289/ehp898.
59. Kenworthy JR, Laube FB. Patterns of automobile dependence in cities: an international overview of key physical and economic dimensions with some implications for urban policy. *Transportation Research Part A: Policy and Practice*. 1999;33:691–723.
60. Newman P. Density, the sustainability multiplier: some myths and truths with application to Perth, Australia. *Sustainability*. 2014;6(9):6467–87. www.mdpi.com/2071-1050/6/9/6467.
61. Newman P, Kenworthy JR. *Cities and Automobile Dependence: A Sourcebook*. Aldershot, UK and Brookfield, VT: Gower Technical; 1989.
62. Ewing R, Hamidi S, Tian G, et al. Testing Newman and Kenworthy's theory of density and automobile dependence. *Journal of Planning Education and Research*. 2018;38(2):167–82. doi: 10.1177/0739456X16688767.
63. Ewing R, Cervero R. Travel and the built environment. *Journal of the American Planning Association*. 2010;76(3):265–94. doi: 10.1080/01944361003766766.
64. Creatore MI, Glazier RH, Moineddin R, et al. Association of neighborhood walkability with change in overweight, obesity, and diabetes. *JAMA*. 2016;315(20): 2211–20. doi: 10.1001/jama.2016.5898.
65. Project Drawdown. The Drawdown Review 2020: Climate Solutions for a New Decade. San Francisco: Project Drawdown; 2020. Available from www.drawdown.org/drawdown-framework/drawdown-review-2020.
66. Johnson M, Rose G. Extending life on the bike: electric bike use by older Australians. *Journal of Transport & Health*. 2015;2(2):276–83. https://doi.org/10.1016/j.jth.2015.03.001.
67. Jones T, Harms L, Heinen E. Motives, perceptions and experiences of electric bicycle owners and implications for health, wellbeing and mobility. *Journal of Transport Geography*. 2016;53:41–9. https://doi.org/10.1016/j.jtrangeo.2016.04.006.

68. Requia WJ, Mohamed M, Higgins CD, Arain A, Ferguson M. How clean are electric vehicles? Evidence-based review of the effects of electric mobility on air pollutants, greenhouse gas emissions and human health. *Atmospheric Environment*. 2018;185: 64–77. doi: 10.1016/j.atmosenv.2018.04.040.

69. Amato F, Cassee FR, Denier van der Gon HAC, et al. Urban air quality: the challenge of traffic non-exhaust emissions. *Journal of Hazardous Materials*. 2014;275:31–6. doi: 10.1016/j.jhazmat.2014.04.053.

70. Holmatov B, Hoekstra AY. The environmental footprint of transport by car using renewable energy. *Earth's Future*. 2020;8(2):e2019EF001428. doi: 10.1029/ 2019EF001428.

71. Erhardt GD, Roy S, Cooper D, et al. Do transportation network companies decrease or increase congestion? *Science Advances*. 2019;5(5):eaau2670. doi: 10.1126/sciadv. aau2670.

72. Graehler MJ, Mucci RA, Erhardt GD, editors. Understanding the Recent Transit Ridership Decline in Major US Cities: Service Cuts or Emerging Modes? Transportation Research Board 98th Annual Meeting, Washington, DC; 2019.

73. Clewlow RR, Mishra GS. Disruptive Transportation: The Adoption, Utilization, and Impacts of Ride-Hailing in the United States. Davis, CA: UC Davis Institute of Transportation Studies; 2017.

74. Duarte F, Ratti C. The impact of autonomous vehicles on cities: a review. *Journal of Urban Technology*. 2018;25(4):3–18. doi: 10.1080/10630732.2018.1493883.

75. Larson W, Zhao W. Self-driving cars and the city: effects on sprawl, energy consumption, and housing affordability. *Regional Science and Urban Economics*. 2020;81:103484. https://doi.org/10.1016/j.regsciurbeco.2019.103484.

76. Harris JE. The Subways Seeded the Massive Coronavirus Epidemic in New York City. National Bureau of Economic Research Working Paper No. 27021; 2020. Available from http://web.mit.edu/jeffrey/harris/HarrisJE_WP2_COVID19_NYC_24-Apr-2020.pdf.

77. McLaren J. Racial Disparity in COVID-19 Deaths: Seeking Economic Roots with Census Data. National Bureau of Economic Research; Working Paper No. 27407; 2020. Available from www.nber.org/papers/w27407.

78. Sadik-Khan J, Solomonow S. Fear of Public Transit Got Ahead of the Evidence. The Atlantic. 14 June 2020. Available from www.theatlantic.com/ideas/archive/2020/06/ fear-transit-bad-cities/612979/.

79. Chinn D, Lotz C, Speksnijder L, Stern S. Restoring Public Transit Amid COVID-19: What European Cities Can Learn from One Another. McKinsey; 2020 [updated 5 June 2020]. Available from www.mckinsey.com/industries/travel-logistics-and-transport-infrastructure/our-insights/restoring-public-transit-amid-covid-19-what-european-cities-can-learn-from-one-another.

80. Muller J. Cities are Retooling Public Transit to Lure Riders Back. Axios. 5 June 2020. Available from www.axios.com/public-transit-coronavirus-0767f3fb-22ce-4102-bf30-f7907f5ac857.html.

81. Nurse A, Dunning R. Is COVID-19 a turning point for active travel in cities? *Cities & Health*. 2020:1–3. doi: 10.1080/23748834.2020.1788769.

82. McIlwain JK, Hammerschmidt S, Simpson M. Housing in America: Integrating Housing, Health, and Resilience in a Changing Environment. Washington, DC: Urban Land Institute; 2014.

83. Satterthwaite D, Archer D, Colenbrander S, et al. Building resilience to climate change in informal settlements. *One Earth*. 2020;2(2):143–56. doi: 10.1016/j. oneear.2020.02.002.

84. Makantasi A-M, Mavrogianni A. Adaptation of London's social housing to climate change through retrofit: a holistic evaluation approach. *Advances in Building Energy Research*. 2016;10(1):99–124. doi: 10.1080/17512549.2015.1040071.

85. WHO. WHO Housing and Health Guidelines. Geneva: World Health Organization; 2018. Available from www.who.int/sustainable-development/publications/housing-health-guidelines/en/.

86. Ezeh A, Oyebode O, Satterthwaite D, et al. The history, geography, and sociology of slums and the health problems of people who live in slums. *The Lancet*. 2017;389(10068):547–58. doi: 10.1016/S0140-6736(16)31650-6.

87. Tusting LS, Bisanzio D, Alabaster G, et al. Mapping changes in housing in Sub-Saharan Africa from 2000 to 2015. *Nature*. 2019;568:391–4. doi: 10.1038/s41586-019-1050-5.

88. Jones A, Grigsby-Toussaint DS. Housing stability and the residential context of the COVID-19 pandemic. *Cities & Health*. 2020:1–3. doi: 10.1080/23748834.2020.1785164.

89. Haines A, Bruce N, Cairncross S, et al. Promoting health and advancing development through improved housing in low-income settings. *Journal of Urban Health*. 2013;90(5):810–31. doi: 10.1007/s11524-012-9773-8.

90. Jatta E, Jawara M, Bradley J, et al. How house design affects malaria mosquito density, temperature, and relative humidity: an experimental study in rural Gambia. *The Lancet Planetary Health*. 2018;2(11):e498–508. doi: 10.1016/S2542-5196(18)30234-1.

91. von Seidlein L, Ikonomidis K, Mshamu S, et al. Affordable house designs to improve health in rural Africa: a field study from northeastern Tanzania. *The Lancet Planetary Health*. 2017;1(5):e188–99. https://doi.org/10.1016/S2542-5196(17)30078-5.

92. Husain SF, Shariq M, editors. Low cost modular housing – a review. International Conference on Advances in Construction Materials and Structures (ACMS-2018), Roorkee, Uttarakhand, India; 2018.

93. Committee on Climate Change. UK Housing: Fit for the Future? London: Committee on Climate Change; 2019. Available from www.theccc.org.uk/publication/uk-housing-fit-for-the-future/.

94. Marmot Review Team. The Health Impacts of Cold Homes and Fuel Poverty. London: Friends of the Earth; 2011. Available from www.instituteofhealthequity.org/projects/the-health-impacts-of-cold-homes-and-fuel-poverty.

95. Hamilton I, Milner J, Chalabi Z, et al. Health effects of home energy efficiency interventions in England: a modelling study. *BMJ Open*. 2015;5(4):e007298. doi: 10.1136/bmjopen-2014-007298.

96. Institute of Medicine Committee on the Effect of Climate Change on Indoor Air Quality and Public Health. *Climate Change, the Indoor Environment, and Health*. Washington, DC: The National Academies Press; 2011.

97. Wilkinson P, Smith KR, Davies M, et al. Public health benefits of strategies to reduce greenhouse-gas emissions: household energy. *The Lancet*. 2009;374(9705):1917–29. doi: 10.1016/s0140-6736(09)61713-x.

98. Kneifel J, O'Rear E, Webb D, O'Fallon C. An exploration of the relationship between improvements in energy efficiency and life-cycle energy and carbon emissions using the BIRDS low-energy residential database. *Energy and Buildings*. 2018;160:19–33. doi: 10.1016/j.enbuild.2017.11.030.

99. Macmillan A, Davies M, Shrubsole C, et al. Integrated decision-making about housing, energy and wellbeing: a qualitative system dynamics model. *Environmental Health*. 2016;15(1):S37. doi: 10.1186/s12940-016-0098-z.

100. Eker S, Zimmermann N, Carnohan S, Davies M. Participatory system dynamics modelling for housing, energy and wellbeing interactions. *Building Research & Information*. 2018;46(7):738–54. doi: 10.1080/09613218.2017.1362919.

101. C40 Cities. The Future of Urban Consumption in a 1.5 °C World. London: C40 Cities, Arup and the University of Leeds; 2019. Available from www.c40.org/press_releases/new-research-shows-how-urban-consumption-drives-global-emissions.

102. Garcia X, Barceló D, Comas J, et al. Placing ecosystem services at the heart of urban water systems management. *Science of The Total Environment*. 2016;563–4:1078–85. doi: 10.1016/j.scitotenv.2016.05.010.

103. Ma X, Xue X, González-Mejía A, Garland J, Cashdollar J. Sustainable water systems for the city of tomorrow – a conceptual framework. *Sustainability*. 2015;7(9): 12071–105. doi: 10.3390/su70912071.

104. UNICEF and WHO. 25 Years: Progress on Sanitation and Drinking Water. 2015 Update and MDG Assessment. Geneva: World Health Organization; 2015.

105. Hanna-Attisha M, LaChance J, Sadler RC, Champney Schnepp A. Elevated blood lead levels in children associated with the Flint drinking water crisis: a spatial analysis of risk and public health response. *American Journal of Public Health*. 2015;106(2):283–90. doi: 10.2105/AJPH.2015.303003.

106. Wilkinson J, Hooda PS, Barker J, Barton S, Swinden J. Occurrence, fate and transformation of emerging contaminants in water: an overarching review of the field. *Environmental Pollution*. 2017;231(Pt 1):954–70. doi: 10.1016/j.envpol.2017.08.032.

107. Bunce JT, Ndam E, Ofiteru ID, Moore A, Graham DW. A review of phosphorus removal technologies and their applicability to small-scale domestic wastewater treatment systems. *Frontiers in Environmental Science*. 2018;6(8). doi: 10.3389/fenvs.2018.00008.

108. Cordell D, White S. Life's bottleneck: sustaining the world's phosphorus for a food secure future. *Annual Review of Environment and Resources*. 2014;39(1):161–88. doi: 10.1146/annurev-environ-010213-113300.

109. Theregowda RB, González-Mejía AM, Ma X, Garland J. Nutrient recovery from municipal wastewater for sustainable food production systems: an alternative to traditional fertilizers. *Environmental Engineering Science*. 2019;36(7):833–42. doi: 10.1089/ees.2019.0053.

110. Grant SB, Saphores JD, Feldman DL, et al. Taking the 'waste' out of 'wastewater' for human water security and ecosystem sustainability. *Science*. 2012;337(6095): 681–6. doi: 10.1126/science.1216852.

111. Xue X, Schoen ME, Ma X, et al. Critical insights for a sustainability framework to address integrated community water services: technical metrics and approaches. *Water Research*. 2015;77:155–69. doi: https://doi.org/10.1016/j.watres.2015.03.017.

112. Rahaman MM, Varis O. Integrated water resources management: evolution, prospects and future challenges. *Sustainability: Science, Practice and Policy*. 2005;1(1): 15–21. doi: 10.1080/15487733.2005.11907961.

113. Kandulu JM, MacDonald DH, Dandy G, Marchi A. Ecosystem service impacts of urban water supply and demand management. *Water Resources Management*. 2017;31(15):4785–99. doi: 10.1007/s11269-017-1778-3.

114. Cook BR, Spray CJ. Ecosystem services and integrated water resource management: different paths to the same end? *Journal of Environmental Management*. 2012;109:93–100. https://doi.org/10.1016/j.jenvman.2012.05.016.

115. Giordano M, Shah T. From IWRM back to integrated water resources management. *International Journal of Water Resources Development*. 2014;30(3):364–76. doi: 10.1080/07900627.2013.851521.

116. Lee SW, Sarp S, Jeon DJ, Kim JH. Smart water grid: the future water management platform. *Desalination and Water Treatment*. 2015;55(2):339–46. doi: 10.1080/19443994.2014.917887.

117. Piratla KR, Goverdhanam S. Decentralized water systems for sustainable and reliable supply. *Procedia Engineering*. 2015;118:720–6. https://doi.org/10.1016/j.proeng.2015.08.506.

118. Arora M, Malano H, Davidson B, Nelson R, George B. Interactions between centralized and decentralized water systems in urban context: a review. *WIREs Water*. 2015;2(6):623–34. doi: 10.1002/wat2.1099.

119. WHO. Results of Round II of the WHO International Scheme to Evaluate Household Water Treatment Technologies. Geneva: World Health Organization; 2019. Available from www.who.int/water_sanitation_health/publications/results-round-2-scheme-to-evaluate-houshold-water-treatment-tech/en/.

120. Larsen TA, Peters I, Alder A, et al. Re-engineering the toilet for sustainable wastewater management. *Environmental Science & Technology*. 2001;35(9):192A–7A. doi: 10.1021/es012328d.

121. Shi Y, Zhou L, Xu Y, Zhou H, Shi L. Life cycle cost and environmental assessment for resource-oriented toilet systems. *Journal of Cleaner Production*. 2018;196:1188–97. https://doi.org/10.1016/j.jclepro.2018.06.129.

122. Oral HV, Carvalho P, Gajewska M, et al. A review of nature-based solutions for urban water management in European circular cities: a critical assessment based on case studies and literature. *Blue-Green Systems*. 2020;2(1):112–36. doi: 10.2166/bgs.2020.932.

123. Chan FKS, Griffiths JA, Higgitt D, et al. 'Sponge City' in China – a breakthrough of planning and flood risk management in the urban context. *Land Use Policy*. 2018;76:772–8. https://doi.org/10.1016/j.landusepol.2018.03.005.

124. McDonald RI, Weber KF, Padowski J, Boucher T, Shemie D. Estimating watershed degradation over the last century and its impact on water-treatment costs for the world's large cities. *Proceedings of the National Academy of Sciences*. 2016;113(32):9117–22. doi: 10.1073/pnas.1605354113.

125. Secretariat of the Convention on Biological Diversity. Protected Areas in Today's World: Their Values and Benefits for the Welfare of the Planet. Montreal: UN Environment Programme and Convention on Biological Diversity; 2008. Available from www.cbd.int/doc/publications/cbd-ts-36-en.pdf.

126. Calvet-Mir L, Corbera E, Martin A, Fisher J, Gross-Camp N. Payments for ecosystem services in the tropics: a closer look at effectiveness and equity. *Current Opinion in Environmental Sustainability*. 2015;14:150–62. https://doi.org/10.1016/j.cosust.2015.06.001.

127. Börner J, Baylis K, Corbera E, et al. The effectiveness of payments for environmental services. *World Development*. 2017;96:359–74. https://doi.org/10.1016/j.worlddev.2017.03.020.

128. Brown RR, Farrelly MA. Delivering sustainable urban water management: a review of the hurdles we face. *Water Science and Technology*. 2009;59(5):839–46. doi: 10.2166/wst.2009.028.

129. van de Meene SJ, Brown RR, Farrelly MA. Towards understanding governance for sustainable urban water management. *Global Environmental Change*. 2011;21(3):1117–27. https://doi.org/10.1016/j.gloenvcha.2011.04.003.

130. Gascon M, Cirach M, Martínez D, et al. Normalized difference vegetation index (NDVI) as a marker of surrounding greenness in epidemiological studies: the case of

Barcelona city. *Urban Forestry & Urban Greening*. 2016;19:88–94. http://dx.doi .org/10.1016/j.ufug.2016.07.001.

131. Frumkin H, Bratman GN, Breslow SJ, et al. Nature contact and human health: a research agenda. *Environmental Health Perspectives*. 2017. doi: 10.1289/EHP1663.

132. Kabisch N, van den Bosch M, Lafortezza R. The health benefits of nature-based solutions to urbanization challenges for children and the elderly – a systematic review. *Environmental Research*. 2017;159:362–73. doi: 10.1016/j.envres.2017.08.004.

133. Kondo MC, Fluehr JM, McKeon T, Dranas CC. Urban green space and its impact on human health. *International Journal of Environmental Research and Public Health*. 2018;15(3):445. doi: 10.3390/ijerph15030445.

134. Fong KC, Hart JE, James P. A review of epidemiologic studies on greenness and health: updated literature through 2017. *Current Environmental Health Reports*. 2018;5:77–87. doi: 10.1007/s40572-018-0179-y.

135. Lai H, Flies EJ, Weinstein P, Woodward A. The impact of green space and biodiversity on health. *Frontiers in Ecology and the Environment*. 2019;17(7):383–90. doi: 10.1002/fee.2077.

136. WHO. Urban Green Space Interventions and Health: A Review of Impacts and Effectiveness. Copenhagen: World Health Organization Regional Office for Europe; 2017. Available from www.euro.who.int/__data/assets/pdf_file/0010/ 337690/FULL-REPORT-for-LLP.pdf.

137. Shuvo FK, Feng X, Akaraci S, Astell-Burt T. Urban green space and health in low and middle-income countries: a critical review. *Urban Forestry & Urban Greening*. 2020;52:126662. https://doi.org/10.1016/j.ufug.2020.126662.

138. Collins RM, Spake R, Brown KA, et al. A systematic map of research exploring the effect of greenspace on mental health. *Landscape and Urban Planning*. 2020;201:103823. https://doi.org/10.1016/j.landurbplan.2020.103823.

139. McCormick R. Does access to green space impact the mental well-being of children: a systematic review. *Journal of Pediatric Nursing*. 2017;37:3–7. doi: 10.1016/j. pedn.2017.08.027.

140. Norwood MF, Lakhani A, Fullagar S, et al. A narrative and systematic review of the behavioural, cognitive and emotional effects of passive nature exposure on young people: evidence for prescribing change. *Landscape and Urban Planning*. 2019;189:71–9. https://doi.org/10.1016/j.landurbplan.2019.04.007.

141. Tillmann S, Tobin D, Avison W, Gilliland J. Mental health benefits of interactions with nature in children and teenagers: a systematic review. *Journal of Epidemiology and Community Health*. 2018;72(10):958–66. doi: 10.1136/jech-2018-210436.

142. de Keijzer C, Gascon M, Nieuwenhuijsen MJ, Dadvand P. Long-term green space exposure and cognition across the life course: a systematic review. *Current Environmental Health Reports*. 2016;3(4):468–77. doi: 10.1007/s40572-016-0116-x.

143. Houlden V, Weich S, Porto de Albuquerque J, Jarvis S, Rees K. The relationship between greenspace and the mental wellbeing of adults: a systematic review. *PLoS One*. 2018;13(9):e0203000. doi: 10.1371/journal.pone.0203000.

144. Gascon M, Triguero-Mas M, Martínez D, et al. Residential green spaces and mortality: a systematic review. *Environment International*. 2016;86:60–7. http://dx.doi.org/ 10.1016/j.envint.2015.10.013.

145. den Braver NR, Lakerveld J, Rutters F, et al. Built environmental characteristics and diabetes: a systematic review and meta-analysis. *BMC Medicine*. 2018;16(1):12. doi: 10.1186/s12916-017-0997-z.

146. Kim H-J, Min J-Y, Kim H-J, Min K-B. Parks and green areas are associated with decreased risk for hyperlipidemia. *International Journal of Environmental Research and Public Health.* 2016;13(12):1205. Available from www.mdpi.com/1660-4601/13/12/1205.

147. Yitshak-Sade M, James P, Kloog I, et al. Neighborhood greenness attenuates the adverse effect of PM(2.5) on cardiovascular mortality in neighborhoods of lower socioeconomic status. *International Journal of Environmental Research and Public Health.* 2019;16(5). doi: 10.3390/ijerph16050814.

148. Klompmaker JO, Janssen NAH, Bloemsma LD, et al. Associations of combined exposures to surrounding green, air pollution, and road traffic noise with cardiometabolic diseases. *Environmental Health Perspectives.* 2019;127(8):87003. doi: 10.1289/ehp3857.

149. Kardan O, Gozdyra P, Misic B, et al. Neighborhood greenspace and health in a large urban center. *Scientific Reports.* 2015;5:11610. doi: 10.1038/srep11610.

150. Akaraci S, Feng X, Suesse T, Jalaludin B, Astell-Burt T. A systematic review and meta-analysis of associations between green and blue spaces and birth outcomes. *International Journal of Environmental Research and Public Health.* 2020;17(8). doi: 10.3390/ijerph17082949.

151. O'Donoghue G, Perchoux C, Mensah K, et al. A systematic review of correlates of sedentary behaviour in adults aged 18–65 years: a socio-ecological approach. *BMC Public Health.* 2016;16:163. doi: 10.1186/s12889-016-2841-3.

152. Rojas-Rueda D, Nieuwenhuijsen MJ, Gascon M, Perez-Leon D, Mudu P. Green spaces and mortality: a systematic review and meta-analysis of cohort studies. *The Lancet Planetary Health.* 2019;3(11):e469–77. doi: 10.1016/S2542-5196(19)30215-3.

153. Putra I, Astell-Burt T, Cliff DP, et al. The relationship between green space and prosocial behaviour among children and adolescents: a systematic review. *Frontiers in Psychology.* 2020;11:859. doi: 10.3389/fpsyg.2020.00859.

154. Goldy SP, Piff PK. Toward a social ecology of prosociality: why, when, and where nature enhances social connection. *Current Opinion in Psychology.* 2020;32:27–31. doi: 10.1016/j.copsyc.2019.06.016.

155. Shepley M, Sachs N, Sadatsafavi H, Fournier C, Peditto K. The impact of green space on violent crime in urban environments: an evidence synthesis. *International Journal of Environmental Research and Public Health.* 2019;16(24):5119. Available from www.mdpi.com/1660-4601/16/24/5119.

156. Bogar S, Beyer KM. Green space, violence, and crime. *Trauma, Violence, & Abuse.* 2015;17(2):160–71. doi: 10.1177/1524838015576412.

157. Burley BA. Green infrastructure and violence: do new street trees mitigate violent crime? *Health & Place.* 2018;54:43–9. https://doi.org/10.1016/j.healthplace.2018.08.015.

158. Eisenman TS, Churkina G, Jariwala SP, et al. Urban trees, air quality, and asthma: an interdisciplinary review. *Landscape and Urban Planning.* 2019;187:47–59. https://doi.org/10.1016/j.landurbplan.2019.02.010.

159. Lambert KA, Bowatte G, Tham R, et al. Residential greenness and allergic respiratory diseases in children and adolescents – a systematic review and meta-analysis. *Environmental Research.* 2017;159:212–21. doi: 10.1016/j.envres.2017.08.002.

160. Grundström M, Dahl Å, Ou T, Chen D, Pleijel H. The relationship between birch pollen, air pollution and weather types and their effect on antihistamine purchase in two Swedish cities. *Aerobiologia.* 2017;33(4):457–71. doi: 10.1007/s10453-017-9478-2.

161. Vos PE, Maiheu B, Vankerkom J, Janssen S. Improving local air quality in cities: to tree or not to tree? *Environmental Pollution.* 2013;183:113–22. doi: 10.1016/j.envpol.2012.10.021.

162. Grote R, Samson R, Alonso R, et al. Functional traits of urban trees: air pollution mitigation potential. *Frontiers in Ecology and the Environment*. 2016;14(10):543–50. doi: 10.1002/fee.1426.

163. Curtis AJ, Helmig D, Baroch C, Daly R, Davis S. Biogenic volatile organic compound emissions from nine tree species used in an urban tree-planting program. *Atmospheric Environment*. 2014;95:634–43. https://doi.org/10.1016/j.atmosenv.2014.06.035.

164. Bodnaruk EW, Kroll CN, Yang Y, et al. Where to plant urban trees? A spatially explicit methodology to explore ecosystem service tradeoffs. *Landscape and Urban Planning*. 2017;157:457–67. http://dx.doi.org/10.1016/j.landurbplan.2016.08.016.

165. Salmond JA, Tadaki M, Vardoulakis S, et al. Health and climate related ecosystem services provided by street trees in the urban environment. *Environmental Health*. 2016;15(1):95–111. doi: 10.1186/s12940-016-0103-6.

166. Maes MJA, Jones KE, Toledano MB, Milligan B. Mapping synergies and trade-offs between urban ecosystems and the Sustainable Development Goals. *Environmental Science & Policy*. 2019;93:181–8. https://doi.org/10.1016/j.envsci.2018.12.010.

167. Anguelovski I, Connolly JJT, Pearsall H, et al. Opinion: Why green 'climate gentrification' threatens poor and vulnerable populations. *Proceedings of the National Academy of Sciences*. 2019;116(52):26139–43. doi: 10.1073/pnas.1920490117.

168. Cole HVS, Triguero-Mas M, Connolly JJT, Anguelovski I. Determining the health benefits of green space: does gentrification matter? *Health & Place*. 2019;57:1–11. https://doi.org/10.1016/j.healthplace.2019.02.001.

169. Schnake-Mahl AS, Jahn JL, Subramanian SV, Waters MC, Arcaya M. Gentrification, neighborhood change, and population health: a systematic review. *Journal of Urban Health*. 2020;97(1):1–25. doi: 10.1007/s11524-019-00400-1.

170. Rahman MA, Stratopoulos LMF, Moser-Reischl A, et al. 2020. Traits of trees for cooling urban heat islands: a meta-analysis. *Building and Environment*. 170:106606. doi: https://doi.org/10.1016/j.buildenv.2019.106606.

171. Sera F, Armstrong B, Tobias A, et al. How urban characteristics affect vulnerability to heat and cold: a multi-country analysis. *International Journal of Epidemiology*. 2019;48(4):1101–12. doi: 10.1093/ije/dyz008.

172. Murage P, Kovats S, Sarran C, et al. What individual and neighbourhood-level factors increase the risk of heat-related mortality? A case-crossover study of over 185,000 deaths in London using high-resolution climate datasets. *Environment International*. 2020;134:105292. doi: 10.1016/j.envint.2019.105292.

173. Hwang WH, Wiseman PE, Thomas VA. Enhancing the energy conservation benefits of shade trees in dense residential developments using an alternative tree placement strategy. *Landscape and Urban Planning*. 2017;158:62–74. http://dx.doi.org/10.1016/j.landurbplan.2016.09.022.

174. Wang Z-H, Zhao X, Yang J, Song J. Cooling and energy saving potentials of shade trees and urban lawns in a desert city. *Applied Energy*. 2016;161:437–44. https://doi.org/10.1016/j.apenergy.2015.10.047.

175. Nowak DJ, Hirabayashi S, Bodine A, Greenfield E. Tree and forest effects on air quality and human health in the United States. *Environmental Pollution*. 2014;193:119–29. http://dx.doi.org/10.1016/j.envpol.2014.05.028.

10

Food Systems and Land Use

Providing equitable access to nutritious and affordable food for a growing world population in the face of multiple environmental changes is one of the greatest challenges facing humanity. Despite steep increases in agricultural yields beginning in the mid-twentieth century, the global food system is failing to provide nutritious food for much of the world's population. The food system also imposes a heavy burden on Earth systems; it is a major driver of land use change, biodiversity loss, freshwater depletion, air and water pollution and climate change, as outlined in Chapters 1–3. Current food systems are also grossly inefficient, with, for example, overuse of nitrogen and phosphorus in some regions and underuse in others, together with high levels of food loss and waste, such that about 30% of food produced is never eaten (1). This chapter focuses on potential strategies to improve nutrition and health while reducing the environmental footprint of food systems, with the aim of staying within planetary boundaries.

The State of the World's Food Systems

Food systems include 'all the elements (e.g. environment, people, inputs, processes, infrastructures, and institutions) and activities that relate to the production, processing, distribution, preparation and consumption of food' (2). Food systems have evolved since the dawn of civilization to meet the changing needs and demands of humanity, shaped by local, and increasingly global, environmental challenges and influenced by powerful interests, particularly commercial companies. History reminds us that no civilization has sustained itself over long periods without a reliable supply of food through agriculture and trade. Furthermore, the current affluence of Western societies is rooted in millennia of agrarian change. Food surpluses were instrumental in the rise of cities and provided essential income for national treasuries. More recently, the impressive economic success of many Asian countries, such as the Philippines, Indonesia, and India, was driven in part by marked increases in agricultural yields (3).

The Green Revolution of the 1960s increased yields of major staple crops through a combination of breeding of improved varieties, management practices, hybrid crop development, increased application of fertilizers, and enhanced investment in research. However, the Green Revolution yielded numerous unintended consequences: increasing dependence on costly pesticides and seeds controlled by multinational companies, impoverishment of small farmers, loss of biodiversity associated with monocropping, and excessive freshwater use (4). Nutrient-rich fruits and vegetables have often been overlooked in favour of high-calorie grains providing energy but limited nutrient quality and little diversity.

Recent large-scale trends in the global food system include a loss of diversity, a shift towards animal-source and processed foods, increased globalization, a growing role for large multinational companies, contributions to global environmental changes, and vulnerability to these changes. Seventy-five per cent of the world's food is produced from only twelve plant and five animal species. Just three plants – rice, maize, and wheat – contribute nearly 60% of calories and proteins. For comparison, estimates of the number of edible plant species range from 12,000 to 75,000 (5).

With increasing prosperity has come a shift towards Western diets. These feature high levels of processed foods, sugar, fats, and salt, and relatively lower levels of whole grains, fruits, vegetables, and nuts. A study of 171 countries showed that South Korea, China, and Taiwan experienced the largest changes in food supply over the past five decades. In these countries the supply of animal-source foods, sugar, vegetables, seafood, and oil crops all increased. In many Western countries, by contrast, over the same period the supply of animal-source foods and sugar declined, albeit from a high baseline. Meanwhile, the food supply in Sub-Saharan Africa countries showed little change (6). In many countries unprocessed fresh foods are being replaced by highly processed, ready-made meals such as pizzas and soups, increasingly supplied by large food manufacturers.

The world's food system is highly globalized. Countries in Latin America, East Africa, and South Asia are generally net food exporters, while most countries in the rest of Asia and Africa are net food importers (7). The food industry is increasingly concentrated, with the largest companies growing most rapidly. Fifty manufacturers account for 50% of global food sales (8). The import and export of agricultural commodities are dominated by four companies: Archer Daniels Midland (ADM), Bunge, Cargill, and the Louis Dreyfus Company. Together they are known simply as 'ABCD'. The first three are USA-based and Louis Dreyfus is headquartered in Amsterdam. Cargill, the largest company, holds one-quarter of the global trade in palm oil and is the only one directly involved in meat production and marketing.

On a global scale though, small farms are still a major source of food. An analysis of the global production levels of 41 crops, 7 livestock species, and 14 aquaculture and fish products showed that 51–77% of nearly all commodities and nutrients assessed were produced by small- and medium-sized farms of less than 50 hectares. The nutrients comprised vitamin A, vitamin B_{12}, folate, iron, zinc, calcium, calories, and protein (9). There are major regional differences: large farms dominate production in the Americas, Australia, and New Zealand, while in Sub-Saharan Africa, Southeast Asia, South Asia, and China, farms smaller than 20 hectares produce more than 75% of most food commodities. Evidence from China suggests that small farms are less efficient than larger farms in terms of fertilizer and pesticide use, and in terms of labour and economic productivity (10).

In settings with access to global and national markets, farms are getting larger and are increasingly dominated by agribusiness, but in Asia and Africa farms have become smaller as a result of population growth. In Asia increasing urban employment opportunities are reversing that trend but in Africa that process has not yet resulted in sufficient urban migration to offset the declines in farm size (11). Women make up about 40% of the agricultural workforce worldwide but there is large variation between continents, ranging from about 20% in the Americas and about 35% in South Asia to almost 50% in East and Southeast Asia and 50% in Sub-Saharan Africa (12).

Despite the trend of declining agrobiodiversity, there is still an important role for wild and diverse foods. As discussed in Chapter 3, about 1 billion people use wild foods in their diets, often to provide essential nutrients. These can be harvested sustainably as indigenous communities around the world have long demonstrated, but overexploitation can contribute to biodiversity loss. A review showed that agricultural and forager communities in 22 countries of Asia and Africa (36 studies) use on average 90–100 species per location. In countries such as India, Ethiopia, and Kenya, 300–800 species may be used (13). Indigenous communities in both industrialized and developing countries use about on average 120 species per community. But the distinction between wild and cultivated foods may be artificial (13). Many wild foods are actively managed, suggesting that in some contexts agricultural products and wild foods overlap: hunter-gatherers and foragers farm and manage their environments, and cultivators may use many wild plants and animals.

Food supply and nutritional quality are also affected by the dramatic environmental changes discussed in Chapters 1 and 2. For example, climate change is projected to cause declines in yields of staple crops, particularly in tropical and sub-tropical regions; loss of pollinators could contribute to reductions in nutrition, including vitamin A intake; and fishery declines also place many low-income populations at increased risk of undernutrition. Multiple environmental changes also reduce the yields of fruit, vegetables, nuts, and seeds (14, 15) which have important nutritional benefits, including reducing the risks of non-communicable diseases. The nutritional value of many of these plants is reduced, for example, by carbon dioxide fertilization (16, 17).

Over coming decades, global population growth, together with increasing prosperity leading to changing food preferences, will necessitate large increases in food availability to meet the increasing demand. One analysis suggested that under a 'business as usual' scenario, a 56% gap would exist between crop calories produced in 2010 and those needed in 2050 (18). The corresponding gap between cropland in 2010 and cropland needed in 2050 was 593 million-hectares – an area nearly twice the size of India – and the agricultural GHG emissions in 2050 would be 11 gigatonnes higher than needed to meet a global 2 °C target. Fundamental transformation of the food system will be necessary to meet these challenges while improving health, but effective strategies need to be tailored to the diverse contexts of food production and consumption. A combination of approaches will be needed – including halting agricultural expansion, closing 'yield gaps' on underperforming lands, increasing cropping efficiency, shifting diets, and reducing food loss and waste (19).

Food Systems and Nutrition

Malnutrition in all its forms refers to 'an abnormal physiological condition caused by inadequate, unbalanced, or excessive consumption of macronutrients or micronutrients' (20). It includes undernutrition (wasting, stunting, underweight); caloric excess leading to overweight, obesity, and associated non-communicable diseases; and inadequate vitamin or mineral intake (micronutrient deficiencies). These affect billions of people worldwide.

Undernutrition remains a widespread problem. After decades of decline, the numbers of hungry people plateaued in around 2015 and began to rise, reaching over 820 million globally by 2018 (21). Over 150 million children are stunted and about 2 billion are deficient in micronutrients, particularly iron deficiency, causing anaemia, together with zinc, vitamin A, and iodine deficiency. At the same time, the prevalence of obesity and overweight has increased in many countries over recent decades, so that by 2016 nearly 2 billion adults worldwide (39% of adult men and women) were overweight (BMI $\geq$ 25 kg/m^2) and, of these, more than half a billion were obese (BMI $\geq$ 30 kg/m^2) (22). This latter trend is related to lack of access to nutritious food, and the increasing availability of ultra-processed energy-dense foods, as well as declines in physical activity.

Cereal grains are the largest source of energy in most diets. Consumption of high levels of whole grains has however been associated with reduced coronary disease and diabetes risk and lower total mortality. Refining grains results in loss of micronutrients and fibre, with adverse consequences for health.

Adequate consumption of vegetables, fruit, nuts, and seeds improves health by reducing the burden of non-communicable diseases, including coronary heart disease and stroke, and reducing overall risks of premature death (23). In the case of fruit and non-starchy vegetables, consumption of five portions daily (80 g per portion) confers major benefits, with additional portions probably conferring modest additional health benefits.

In contrast, based on evidence largely from high-income countries, consumption of red and processed meat is associated with higher risks of premature mortality from heart disease, diabetes, and from all causes considered together (23, 24). The extent of this risk remains a subject of debate (25–27). However, several conclusions seem warranted. First, the health risks associated with meat consumption are lower than those associated with inadequate consumption of vegetables and fruit. Second, processed meats confer higher risk than unprocessed meat. Third, red meat confers higher risk than fish and poultry. Fourth, plant-based proteins are substantially healthier than animal-source proteins (28). Finally, meat consumption has a markedly higher environmental footprint, including GHG emissions, land use, and water use, than do plant-based diets.

Figure 10.1 summarizes the health impacts, in terms of both deaths and disability-adjusted life-years (DALYs), attributable to various dietary risk factors.

A *Lancet* Commission has described undernutrition, climate change, and obesity/overweight as a syndemic (29). The term was originally defined as two or more diseases that co-occur in time and place; interact with each other at biological, psychological, or societal levels; and share common underlying societal drivers (30). The *Lancet* Commission extended the concept to encompass global trends in health and the environment. The

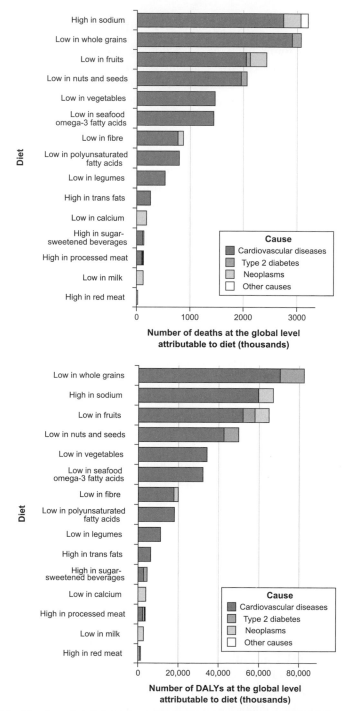

Figure 10.1. Number of global deaths and DALYs attributable to individual dietary risks in 2017. DALY: disability-adjusted life-year.

Source: Afshin A, Sur PJ, Fay KA, et al. Health effects of dietary risks in 195 countries, 1990–2017: a systematic analysis for the Global Burden of Disease Study 2017. *The Lancet*. 2019;393 (10184):1958–72. doi: 10.1016/S0140-6736(19)30041-8.

drivers of the syndemic arise from the food system but also include urban planning and transport systems. This conceptual approach is closely aligned with Planetary Health. For example, climate and other environmental changes reduce crop yield and nutritional quality of food, undermining food security and increasing risks of undernutrition (see Chapter 2). Ultra-processed foods, whose consumption contributes to obesity, are responsible for substantial greenhouse gas emissions according to some analyses. Studies from Australia estimate that consumption of 'discretionary foods' (those that are unnecessary for nutrition, such as ultra-processed snacks) accounts for about 30% of the total food-related GHG emissions, as well as about 35% of water use, 39% of energy use, 33% of carbon dioxide equivalent emissions, and 35% of land use (31, 32). Replacing consumption of these high-impact, low-nutritional-value foods with alternatives such as nuts, seeds, fruit, and whole grains offers an opportunity to improve health and achieve environmental benefits.

The additional food demands of obese and overweight people and increased transport-related GHG emissions related to overweight make a modest additional contribution to societal GHG emissions. The Commission is careful not to amplify stigmatization by attributing climate change to obesity. It is clear that there are many complex inter-relationships between the food system, socioeconomic realities, and environmental change. One aspect not fully explored by the Commission is the strong social class gradient of obesity, with a higher prevalence amongst the least educated and most disadvantaged (33) – those with the fewest opportunities to consume healthy food and to be physically active in congenial surroundings. As we have seen, it is generally the wealthiest in societies and amongst nations who have the highest per capita greenhouse gas emissions.

Environmental Impacts of Food Systems

Food production has far-reaching impacts on many planetary systems, as discussed in Chapters 2 and 3. The impacts include driving climate change, (over)exploiting freshwater, converting land, depleting soil, altering nitrogen and phosphorus cycling, contributing to atmospheric aerosol formation, and driving biodiversity loss. Through these effects, food systems propel humanity towards many of the planetary boundaries described in Chapter 1 (34).

Accurate reckoning of food system impacts on the environment requires comprehensive assessment of its many components, from production through distribution, packaging, and consumption, to waste. Life cycle analysis (LCA) is an internationally recognized approach to estimating inputs, outputs, and environmental impacts. LCA comprises a four-phase process including goal/scope definition; inventory analysis; impact assessment; and interpretation (35).

Using an LCA approach that includes pre- and post-production activities, the global food system accounts for about 21–37% of total net anthropogenic GHG emissions (36, 37). Numerous pathways operate across the life cycle, from pre-production (e.g. fertilizer manufacturing) to production (e.g. nitrous oxide release from tilled soils, methane production by livestock) and post-production (e.g. refrigeration, manure management, and packaging).

Figure 10.2 shows the life cycle GHG emissions associated with different food groups based on data from the EU. An important conclusion is that emissions from land use and from on-farm sources are much larger than those from transport, retail, and packaging. In most cases it is the choice of food items rather than whether they are locally produced that is the key determinant of GHG emissions.

A study using data from 28 high-income and 9 middle-income countries, representing over 60% of the global population, showed that GHG emissions of average national diets increase with income. Per capita diet-related GHG emissions in high-income nations are more than double those in lower-middle-income countries, with upper-middle-income nations having intermediate per capita emissions (38). Animal products (meat, fish, and dairy) account for 22%, 65%, and 70% of emissions in the diets of lower-middle-, upper-middle-, and high-income nations, respectively. There is wide diversity of GHG emissions between countries in the same income group. For example, Australia and Brazil have emissions about three-fold larger than the average of their respective income groups, driven by high levels of red meat consumption, including a preponderance of grass-fed beef. Norway has about 40% higher dietary GHG emissions than average because of its high fish consumption and associated energy requirements for its fishing fleet.

Beef has the highest GHG emission profile of major foods, as shown in Figure 10.2. The emissions occur along the production life cycle. Major sources include the deforestation that creates cropland to produce cattle feed, and the methane that cattle (and other ruminants such as goats and sheep) emit. Can beef be produced sustainably? An interesting debate exists on this point (see also **Box 10.3**). One view holds that carefully managed grazing, with livestock moved regularly from one field to another to permit grass recovery and prevent soil loss, can result in robust grasslands with net carbon sequestration (39). However, a meta-analysis of production systems found that grass-fed beef needs more land than grain-fed beef, and is responsible for similar GHG emissions (40). Another meta-analysis found that grassland grazing could either increase or decrease soil carbon sequestration, depending on contextual factors such as soil type, grass type, and rainfall (41). And a modelling study found that, in the USA, meeting current beef demand with grass-finished cattle would require 30% more cattle and a several-fold increase in grass production (implying increased demands on land), and would result in increased GHG emissions (42). Even if grass-fed beef could achieve net-zero or even negative emissions in some settings, it is clear that limits on grassland availability would impose stringent limits on production. In general, plant-based foods have the lowest environmental impacts per unit of food across all environmental indicators and nutritional units, with intermediate impacts for eggs, dairy, pork, poultry, non-trawling fisheries, and non-recirculating aquaculture. Meat has the highest impacts; strikingly, ruminant meat has impacts as much as 100 times those of plant-based food (40). This suggests that heavy beef consumption, as currently practised in such countries as the USA, Argentina, Australia, and Brazil, is unsustainable on a global basis, and supports the need for a shift from meat- to plant-based diets.

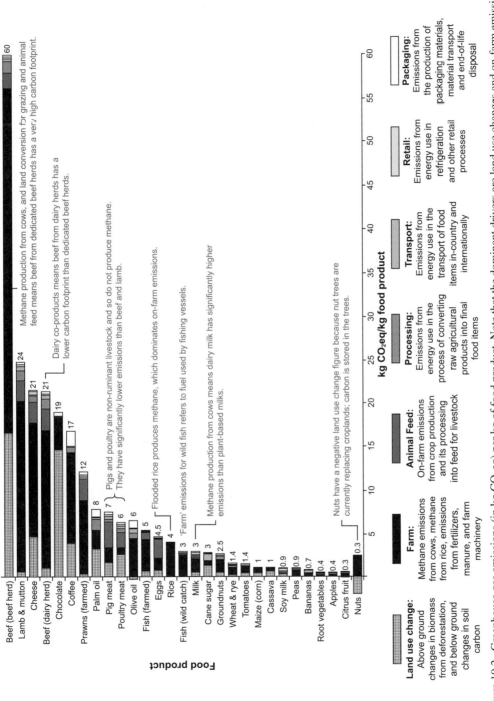

Figure 10.2. Greenhouse gas emissions (in kg CO_2eq) per kg of food product. Note that the dominant drivers are land use changes and on-farm emissions (especially methane emissions in meat production). GHG emissions from most plant-based products are 10–50 times lower than for most animal-based products, so the food type is far more important in determining emissions than such factors as travel distance and packaging.

Source: Data from Poore J, Nemecek T. Reducing food's environmental impacts through producers and consumers. *Science*. 2018;360(6392):987–92. doi:10.1126/science. aaq0216; image adapted from https://ourworldindata.org/food-choice-vs-eating-local).

Food production entails major changes in land use – an impact closely linked with GHG emissions. Land is both a source and a sink of GHG emissions. Recent decades have featured a net negative balance (that is, more CO_2 emission than sequestration) due largely to deforestation. Agriculture occupies about 40% of global land and contributes substantially to tropical deforestation, thus destroying biodiversity and reducing the uptake of carbon dioxide by carbon sinks (23). Globally, total pasture area has declined at a rate of 8.2 million hectares per year, totalling 140 million hectares over the past 17 years, with little conversion of land to other agricultural uses (21). The reasons include inadequate investment, low economic returns, competition for water and human resources, and soil degradation. Pressure on land has had particular implications for indigenous people, who hold rights over about a quarter of Earth's land surface (**Box 10.1**).

In addition to changes in land use, agriculture can threaten the very soil on which it depends. While some soil erosion occurs naturally, improper farming practices can greatly accelerate erosion and soil loss. In some locations this is occurring at rates up to 100 times greater than soil formation, and an estimated half of the planet's topsoil has been lost in the past 150 years (46). Loss of soil compromises farm production, clogs downstream waterways with sediment, pollutes waterways, and compromises dams (47). In addition to soil loss, soil quality is compromised by overfertilization (which acidifies soil) and by overuse of pesticides. Scientists are only beginning to understand the complexities of the soil microbiome and its role in the biological function of soil (48). Finally, and of critical importance in the context of climate change, healthy soil not only provides food, it is also a significant carbon sink (49, 50) (although the global magnitude of potential carbon sequestration in soil is debated (51)). This ecosystem service is undermined by soil degradation.

Agriculture is a major user of freshwater, including both 'blue water' (ground- and surface water) and 'green water' (rainfall), accounting for an estimated 70% of global freshwater use. Expanding irrigation is depleting aquifers in many locations, from northwest India (52) to the central USA (53), threatening the sustainability of agriculture on a large scale. Different crops, and the animals that are fed these crops, vary greatly in their water requirements, so different food items have widely varying water footprints, as shown in **Figure 10.3**. There is considerable variation depending on where crops are grown. In India, for example, rice-based diets have higher associated GHG emissions and green-water footprints, but wheat-based diets have higher blue-water footprints (54). Overall a dietary pattern characterized by high meat and rice consumption has the highest environmental impact.

Agriculture is also a major source of PM air pollution. This problem is well recognized in many parts of Asia, where seasonal crop burning, especially of rice stubble, drives extremes in pollution levels (55, 56). But agriculture in high-income countries also contributes to PM pollution (57). For example, in Germany agriculture is responsible for about 45% of $PM_{2.5}$-related excess mortality. Ammonia (NH_3) plays a central role in the formation of secondary particles and is the predominant air pollutant from fertilizer use and animal husbandry in the EU. On a global scale about one-fifth of PM-related deaths could be prevented by eliminating agricultural emissions (58).

These many environmental impacts of the food system could increase by 50–90% between 2010 and 2050, driven by population growth and increased demand for animal

---

Box 10.1. **Territories of Life: Indigenous People, Land Rights, and Health**

Throughout human history communities have shaped the diversity of their local natural surroundings. In many cases indigenous communities have been custodians of landscapes, protecting them from degradation and overexploitation in order to safeguard their integrity for current and future inhabitants. The central role of indigenous people in the conservation of nature has become increasingly recognized, particularly over the last 20 years, although some Western conservation organizations may still see their role as protecting biodiversity from people, including those with traditional and customary rights to live in biodiverse areas. Indigenous peoples manage or have tenure rights over about 25% of the Earth's land surface, intersecting 40% of the terrestrial protected areas and ecologically intact landscapes. Many of these territories are under increasing threat because of illegal encroachment including logging, mining, and agriculture. Such threats can undermine community cohesion, health, and well-being; when violence accompanies land expropriation, indigenous leaders can pay with their lives (43).

Community-conserved areas (CCAs) have been called 'territories of life' in recognition of how they host and protect rich biodiversity These areas are characterized by (a) a deep connection to place rooted in history, cultural identity, spirituality, and/or resource dependence; (b) custodial decisions about the land made and implemented through a governance institution; and (c) contributing through this governance both to stewardship of nature and to community well-being. Conservation is rarely an explicit objective for the concerned community but usually results from efforts to protect spiritually or culturally significant locations such as sacred forests, or to provide livelihoods. Indeed, the rate of degradation of nature is slower in territories governed by indigenous people than in other comparable areas. However even in these territories the situation is deteriorating with increasing pressure from extractive industries, agriculture, and other activities (44).

Many, but not all, CCAs are protected areas and are listed in the World Database of Protected Areas (www.protectedplanet.net/). This resource is used to monitor progress towards the Aichi Biodiversity targets and other global goals, as well as supporting accountability of local and national decision makers. The land area covered by these 'territories of life' is estimated to equal or exceed that covered by government protected areas, which underlines their vital role in protecting nature, in addition to protecting cultural diversity and sustainable livelihoods. There is no single optimum governance system for all CCAs, but a set of principles has been articulated that can help to determine whether a governance system can support the achievement of key objectives such as conservation, livelihood protection, equity, and cultural preservation (45). A self-strengthening process that deepens the communities' knowledge of their own history and customs, improves governance systems, links with other such communities nationally and internationally, and increases resilience to external threats, can help enhance ecological integrity and support community well-being and livelihoods. Respect for indigenous knowledge and belief systems with support for their rights to govern territories of life is a key step in safeguarding Planetary Health.

---

products that accompanies economic growth, unless there are effective mitigation measures and new technologies (59). Currently the global agricultural system produces insufficient fruits, vegetables, and protein to meet the nutritional needs of today's population, but delivers excess grains, fats, and sugars. According to one study, correcting this imbalance (using the Harvard Healthy Eating Plate as a benchmark) would reduce the amount of

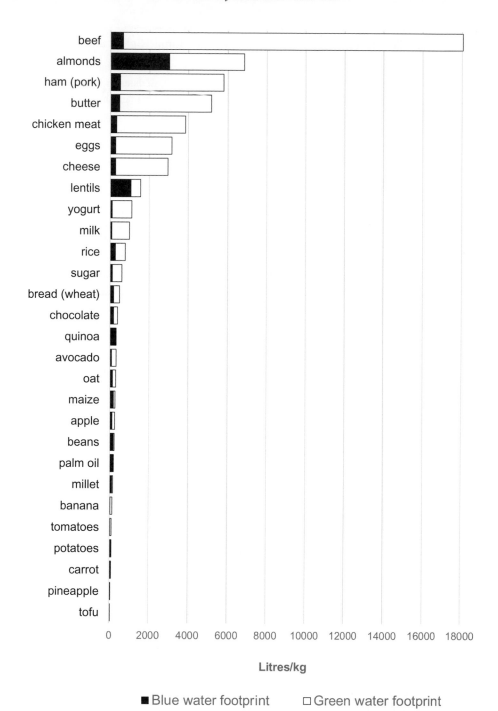

Figure 10.3. Global average water footprints (litres/kg) of various food items.

Source: Data from Mekonnen MM, Hoekstra AY. The green, blue and grey water footprint of crops and derived crop products. *Hydrology and Earth System Sciences*. 2011;15:1577–600. https://doi.org/10 .5194/hess-15-1577-2011; and Mekonnen MM, Hoekstra AY. A global assessment of the water footprint of farm animal products. *Ecosystems*. 2012;15(3):401–15. doi: 10.1007/s10021-011-9517-8. Figure prepared by Francesca Harris.

arable land used to produce grain, fat and oil, and sugar, but increase the arable land needed to produce livestock, fruits, and vegetables, for a net reduction of 51 million hectares globally (60). However, if pastureland used to produce livestock is also considered, then the total land used for agriculture would increase by 407 million hectares, and GHG emissions would increase. The study suggested that, to provide a nutritionally balanced diet for a growing global population, conserve land, and reduce GHG emissions, a transition to diets higher in plant-based protein is needed, with overall increases in consumption and production of fruits and vegetables.

Waste and Inefficiency

Waste and inefficiency bedevil the global food system. Food loss and food waste are distinguishable concepts. Both refer to diminished quality or quantity of food along the supply chain; food loss occurs at the level of producers (at harvest, slaughter, or catch) and suppliers, while food waste occurs at the level of retailers and consumers (1). About one-third of global food production, amounting to about 1.3 billion tonnes (1.6 billion tonnes if inedible waste is included) annually is lost or wasted. Almost 1.4 billion hectares (14 million km²) of agricultural land and 250 km^3 of ground- and surface water are used to grow food that is produced but not eaten. Country-level food waste increases with development and such trends need to be reversed. In low-income countries food loss tends to dominate and, as countries become more developed, the wastage moves further up the supply chain. This loss of food contributes to hunger and food insecurity, and also has adverse environmental impacts. Food loss and waste are responsible for about 3.3 GtCO$_2$eq of GHG emissions (without considering land use change), ranking as the third largest GHG emitter worldwide after the USA and China (61, 62).

Another form of inefficiency relates to the production and use of certain kinds of food. About 36% of the calories from the world's crops are fed to animals and, because of inefficiencies in conversion of feed crops into animal products, only about 12% ultimately contribute to human intake. In the USA about two-thirds of the total calorie production is used for animal feed and about one-third is delivered direct to the food system, while only about 27% of plant protein produced is available for consumption as plant or animal protein. In the case of Brazil about 50% of the total calorie production is delivered to the food system, thus croplands could feed 10.6 people per hectare but only feed 5.2 people per ha. The need for dietary diversity and micronutrient sufficiency should also be taken into account and the estimate should be considered illustrative. Of the total plant protein produced only about 50% is delivered to the food system as plant or animal protein. Additionally, the proportion of crops that could be eaten by humans but were used instead for biofuels increased from 1 to 4% between 2000 and 2010 (63). If crops were grown for direct human consumption globally an additional 4 billion people could be fed – more than sufficient to cope with projected population growth. While this may be unachievable in practice, and environmental changes pose additional challenges, these statistics indicate the transformative potential of re-orienting agricultural policy to focus on providing healthy equitable choices, including increased consumption of plant-based foods.

Potential Solutions

Developing and implementing policies, technologies, and interventions to improve access to healthy and more sustainable dietary choices in the face of population growth, changing demands, and pervasive environmental changes is imperative. Options for reducing the environmental impacts of the food system include dietary changes towards healthier, more plant-based diets, improvements in agricultural technologies and practices, and reductions in food loss and waste. To drive these, both policy changes and behaviour changes are needed. No single measure will suffice to achieve this objective; a combination of approaches will be needed (59).

The relative contributions of different options to reduce the environmental impacts of the global food system sufficiently to stay within the planetary boundaries related to food production are shown in **Figure 10.4**. This summarizes the reduction in environmental impacts from measures to address food loss and waste, changes in practices and technology, and dietary change for medium ambition and high ambition scenarios. This analysis

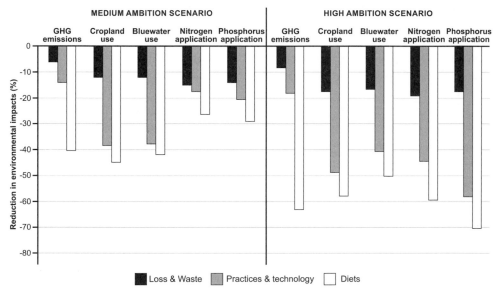

Figure 10.4. Options for keeping the food system within environmental limits, through reducing food loss and waste, technological change, and altered diets. The left-hand panel represents a 'medium ambition' scenario, and the right-hand panel represents a 'high ambition' scenario.

Source: Springmann M, Clark M, Mason-D'Croz D, et al. Options for keeping the food system within environmental limits. *Nature*. 2018;562:519–25. doi: 10.1038/s41586-018-0594-0.

assessed the impacts of moving towards global dietary guidelines for the consumption of red meat, sugar, fruits, vegetables, and energy intake and a more ambitious scenario of more plant-based (flexitarian) diets. There are uncertainties (not shown on the figure) reflecting the differences between development pathways that are more optimistic (higher income and lower population growth) and more pessimistic (lower income and higher population growth).

It is conventional to divide potential solutions into supply and demand-side interventions; there needs to be alignment between the food produced and the food consumed. In many cases both must be implemented in an integrated way to accelerate food systems transformation. For example, producing healthy, low environmental impact food will need to be accompanied by increasing consumer demand, particularly from those who can benefit most, often the disadvantaged. Increasing equity of access to healthy sustainable food choices is a prerequisite to achieving optimal outcomes for Planetary Health. The current incentives that drive the production of food are not consistent with either population health or environmental sustainability and the metrics used to assess agricultural productivity fail to reflect the paramount importance of nourishing the world's population within planetary boundaries.

The following sections address the three major categories of solutions: dietary changes, changes in agricultural technologies and practices, and reductions in food loss and waste. In each case, we focus on solutions that simultaneously deliver benefits for human health and for the planet.

Sustainable Healthy Diets

Individual dietary choices have major implications for the environmental footprint of the food system. Diets that aim to lower environmental impacts are sometimes referred to as sustainable diets, defined by the UN Food and Agriculture Organization as those 'which are healthy, have a low environmental impact, are affordable, and culturally acceptable' (64). Here we address the characteristics of sustainable, healthy diets, as well as how to promote the adoption of such diets.

An important contextual issue for healthy, sustainable diets is the imperative to increase the income of the global poor. Rising prosperity brings with it a demand for more animal-source protein, which poses challenges to sustainability. An example from India illustrates the scale of the challenge. If the entire Indian population adopted the diet of the wealthiest 25%, then GHG emissions, green and blue water use, and land use would increase by 19–36% (65). In contrast, if India's whole population adopted sustainable, healthy diets, there would be minimal increases in environmental impacts.

A systematic review of different types of sustainable diets showed variable effects on GHG emissions, land use, and water use. Most of the studies were undertaken in high-income countries. Reductions of over 70% in GHG emissions and land use, and more than 50% in water use, could in theory be achieved by shifting typical Western diets to more environmentally sustainable dietary patterns, but there was considerable variation between studies. Medians of these impacts across all studies suggest possible reductions of 20–30%.

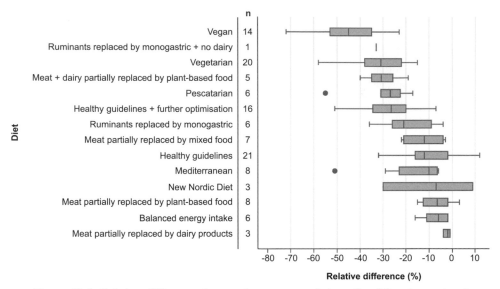

Figure 10.5. Relative differences in greenhouse gas emissions (kg CO_2eq/person/year) between current average diets and alternative dietary patterns. The vertical lines within the shaded areas represent the medians, dots are outliers, and n denotes the number of studies.
Source: Aleksandrowicz L, Green R, Joy EJM, Smith P, Haines A. The impacts of dietary change on greenhouse gas emissions, land use, water use, and health: a systematic review. *PLoS One.* 2016;11(11): e0165797. https://doi.org/10.1371/journal.pone.0165797.

The largest reductions in GHG emissions (**Figure 10.5**) and land use were from vegan diets; other dietary changes had more modest effects (66). A few (~5%) effects were not in the expected direction – when meat was replaced by dairy products, and when the fruits and vegetables consumed had a large environmental footprint – underlining the importance of local context. Pescatarian diets may be advantageous nutritionally but are infeasible in many locations because of the decline in fish catches and the unsustainability of some aquaculture practices. There is a dearth of studies addressing the effects of different diets on other key environmental endpoints, such as soil carbon, which makes comprehensive assessments difficult (67).

In general, healthy diets and sustainable diets are aligned. An analysis of large prospective cohort studies in the UK on adherence to the Eatwell Guide (EWG), a set of dietary guidelines, supports the proposition that healthy dietary choices can also have lower environmental footprints. Participants with 'intermediate-to-high adherence' to EWG recommendations had a reduced risk of death relative to those with 'very low' adherence, after adjusting for potential confounding factors. Adherence to the recommendation on fruit and vegetable consumption was associated with the largest reduction in total mortality risk, whereas adherence to the recommendations on red and processed meat consumption was associated with the largest reduction in GHG emissions and blue water footprint (-1.48 kg CO_2eq/day and -22.5 L/day, respectively) (68).

Dietary changes can significantly reduce agricultural water use. A modelling study optimized typical dietary patterns in an Indian population sample to meet projected decreases in the availability of water per person for irrigation due to population growth up to 2050. The optimized diet contained lower amounts of wheat, dairy, and poultry, and increased amounts of legumes, relative to baseline. It reduced the dietary blue water footprint (WF) by up to 30%, and yielded moderate reductions in mortality from reduced NCD risks (69). Cereals with lower irrigation dependency than wheat include maize, sorghum, and millet (70).

A systematic review of studies of dietary water footprints showed that, on average, European (170 estimates) and Oceanian (18 estimates) dietary patterns have the highest green WFs (median per capita: 2999 L/day and 2924 L/day, respectively), whereas Asian dietary patterns (98 estimates) have the highest blue WFs (median per capita: 382 L/day). Foods of animal origin are major contributors to the green WFs of diets, whereas cereals, fruits, nuts, and oils are major contributors to the blue WFs of diets (71).

So dietary patterns can have a substantial impact on environmental sustainability. Can these insights be combined with nutritional considerations to point the way to a food system that is both sustainable and healthy?

The EAT–Lancet Commission

One of the most extensive assessments to date of how to feed the global population nutritiously while remaining within planetary boundaries was the EAT–*Lancet* Commission (**Box 10.2**) (23). It framed its analysis around the need to address the Paris Climate Agreement and the Sustainable Development Goals, while meeting the needs of a growing population for a healthy, nutritious diet by mid-century. The Commission

---

Box 10.2. Characteristics of Healthy, Low Environmental Impact Diets

- Diversity – a wide variety of foods eaten.
- Balance achieved between energy intake and energy needs.
- Based around: minimally processed tubers and whole grains; legumes; fruits and vegetables – particularly those that are field grown, 'robust' (less prone to spoilage) and less requiring of rapid and more energy-intensive transport modes. Meat, if eaten, in moderate quantities – and all animal parts consumed.
- Dairy products or alternatives (e.g. fortified milk substitutes and other foods rich in calcium and micronutrients) eaten in moderation.
- Unsalted seeds and nuts.
- Small quantities of fish and aquatic products sourced from certified fisheries.
- Very limited consumption of foods high in fat, sugar or salt and low in micronutrients, e.g. crisps, confectionery, sugary drinks.
- Oils and fats with a beneficial Omega 3:6 ratio such as rapeseed and olive oil.
- Tap water in preference to other beverages – particularly soft drinks.

Source: Adapted from Garnett T. *Changing What We Eat: A Call for Research and Action on Widespread Adoption of Sustainable Healthy Eating.* Oxford, UK: Food Climate Research Network; 2014.

concluded that it is possible to provide healthy sustainable diets for about 10 billion people within planetary boundaries.

The EAT–*Lancet* Commission analysis shows that transformation to healthy diets by 2050 will require substantial dietary shifts, including a more than 50% reduction in global consumption of red meat and sugar, and a more than 100% increase in consumption of nuts, fruits, vegetables, and legumes. However, the magnitude and type of changes needed vary greatly by region and do not apply to children aged less than 2 years (who consume a small proportion of total food intake). After extensive review of the evidence linking the consumption of different food groups to health outcomes and environmental impacts, the EAT–*Lancet* Commission proposed a global average 'Planetary Health diet' (**Figure 10.6**) which would benefit both health and the environment. These broad recommendations would of course need to be adapted to local circumstances and diverse cultures and were not intended to be prescriptive. The optimal energy intake for healthy body weight will depend on individual characteristics such as height and level of physical activity.

Following such a diet on a global scale was estimated to reduce premature deaths by about 11 million annually by 2050 compared with current diets, amounting to 19–24% of adult deaths. While the emphasis is on dietary changes, achieving these ambitious goals will require major transformation of the food system – not only dietary changes, but also reduced food loss and waste and substantial improvements in technology, with a particular emphasis on improved access to technologies by low-income farmers. The food system must shift from maximizing the quantity of food to production of healthy food more equitably distributed, with increased diversity rather than reliance on a small range of food crops with little genetic diversity.

Upon its release in 2019, the Commission report triggered a number of significant but sometimes misplaced criticisms. Critics charged that the report was a 'one size fits all' prescription that failed to consider cultural appropriateness in different settings, such as the importance of livestock to many poor communities, including pastoralists. Critics also cited problems with affordability, and argued that it failed to address the nutritional needs of vulnerable groups, including children. In fact, the report explicitly addressed many of these issues. Affordability did emerge as a serious concern. One analysis (72) estimated the median daily cost of the EAT–*Lancet* diet to be US\$2.84 (in 2011 dollars), of which the largest share was the cost of fruits and vegetables (31%), with substantial contributions from legumes and nuts (19%), meat, eggs, and fish (15%), and dairy products (13%). This cost exceeded the daily household income of about 1.6 billion people. The analysis was inevitably simplified and does not take into account urban/rural differences in costs, differences in nutritional requirements, or the availability of produce from gardens and smallholdings. Nevertheless, it is a powerful reminder of the need to consider affordability when making dietary recommendations.

A counter-campaign to the findings of the EAT–*Lancet* report erupted at the time of its release, with substantial support from the meat industry. This counter-campaign disseminated caricatures and negative perceptions of the report, centred on advocacy of meat consumption (using the #yes2meat Twitter hashtag). Over the weeks following the launch of the report, the number of negative tweets exceeded those mentioning the report itself,

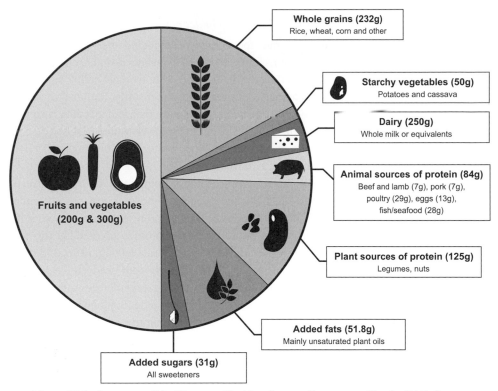

Figure 10.6. Summary of the Planetary Health reference diet proposed by the EAT–*Lancet* Commission. Daily intakes are shown in grams; the sizes of each 'slice' reflect their volumes (for fruits and vegetables) or their caloric content (for all other food types). The daily intakes shown in each food group are global averages; local and regional variation is expected, and the original report provides ranges for each food type. Processing of foods including partial hydrogenation of oils, refining of grains, and addition of salt and preservatives can substantially affect health. Unsaturated oils are 20% each of olive, soybean, rapeseed, sunflower, and peanut oil. The quantities of wheat, rice, dry beans, and lentils refer to dry, raw product. The mix and amount of grains can vary to maintain isocaloric intake. Beef and lamb are exchangeable with pork and vice versa. Chicken and other poultry are exchangeable with eggs, fish, or plant protein sources. Dairy intake is equivalent to half a pint of milk. Legumes, peanuts, tree nuts, seeds, and soy are interchangeable.

Source: Adapted from Willett W, Rockström J, Loken B, et al. Food in the Anthropocene: the EAT–*Lancet* Commission on healthy diets from sustainable food systems. *The Lancet*. 2019;393 (10170):447–92. doi: 10.1016/S0140-6736(18)31788-4. Graphic by Emanuel Santos.

and commentators who were initially ambivalent about the Commission findings appeared to become more sceptical over time, influenced by the countermovement (73). This is but one example of how vested interests can stir dissent when their interests are threatened. We discuss disinformation – a significant impediment to Planetary Health – in Chapter 12.

While more work is needed to tailor broad recommendations to the needs of specific groups, the EAT–*Lancet* Commission report represents a significant contribution to the debate about how to feed a growing world population within planetary boundaries.

Food-Based Dietary Guidelines

International and national bodies generate food-based dietary guidelines (FBDGs). The WHO has issued dietary recommendations, directed to policymakers and the general public (74). Many governments have also produced FBDGs adapted to national eating habits and cultural preferences (75). These guidelines usually advocate more fruit, nut, and vegetable consumption and fewer animal products, refined carbohydrates, fats, and oils than currently consumed in each nation's average diets. Widespread compliance with FBDGs would in general yield environmental benefits as well as health benefits, including reduced GHG emissions, reduced land use, and reduced eutrophication (38, 76).

National dietary guidelines have at least four notable limitations. First, because of uncertainties in nutritional science, the evidence to support unequivocal recommendations is sometimes unavailable (77–79). Second, FBDGs rarely consider the environmental benefits associated with nutritious diets, let alone optimize for both health and sustainability. In one study of the 83 countries identified as having official dietary guidelines, only four (Germany, Brazil, Sweden, and Qatar) addressed environmental sustainability in their main messaging (although several additional countries – the Nordic countries, the Netherlands, France, Estonia, and the UK – included sustainability considerations in quasi-official guidance (75)). Adherence to the majority of FBDGs would be incompatible with the Paris Climate Agreement and other environmental targets (76). Compared with WHO dietary recommendations, adherence to the EAT–*Lancet* recommendations would result in 34% greater reductions in premature mortality, more than three-fold reductions in greenhouse gas emissions, and other environmental benefits. Third, evidence for the impact of FBDGs is equivocal; it is not clear that they drive dietary choices (80). Fourth, the development of FBDGs can be subverted by powerful interests. In the USA, the Department of Agriculture undertakes a five-yearly process of reviewing dietary guidelines overseen by an independent Dietary Guidelines Advisory Committee (DGAC). In 2015 the DGAC published its advisory report which made the case for including sustainability considerations in the guidelines. Industry lobbyists, largely representing the meat industry, successfully argued that this was beyond the scope of the guidelines but, significantly, did not contest the science behind the recommendations (81).

Despite these limitations, FBDGs provide frameworks within which private sector firms produce and market food, and within which civil society works to influence dietary patterns. Incorporating sustainability into FBDGs, and advancing the implementation of FBDGs on both supply and demand sides, could potentially contribute to advancing both health and environmental goals.

Promoting Behaviour Change Towards Healthy, Sustainable Diets

If FBDGs are not well established as a method to alter people's eating patterns, are there other more effective approaches? This is a well-explored area with regard to improving health, less so with regard to environmental sustainability.

Individual behaviour change techniques (BCTs), as used in clinical settings, appeal to people's conscious awareness of their food choices, and aim to support their making healthy choices. An umbrella review of such approaches in the context of non-communicable diseases (NCDs) found some evidence supporting a cluster of 'self-regulatory' intervention techniques: identifying barriers to change, problem solving, goal setting, self-monitoring, and relapse prevention training (82). A systematic review of dietary BCTs among young people showed a number of effective interventions, particularly directed at fruit and vegetable consumption. The most effective BCTs were habit formation (for example prompting participants to fill half of their plate with cooked vegetables or salad each night at dinner) (100%), salience of consequences (83%), and adding objects to the environment (70%). These trials were largely in high-income countries and their applicability to other settings is unknown (83).

Population approaches aim to influence dietary choices on a much broader scale than clinical approaches, with an emphasis on environmental and social interventions. Moreover, they have the potential to be effective, as they rely on 'automatic processes' as opposed to conscious, deliberative choice. These approaches may also be more cost-effective. This approach accords with psychological and neuroscientific evidence that much human behaviour is driven not by conscious deliberation, but is unconsciously cued by environmental stimuli (84).

Prices provide a powerful signal. If healthy, sustainable food is less costly than the alternative, it is more likely to be preferred, purchased, and consumed. Many techniques are used, including taxes, discounts, and subsidies at the point of production and/or sale. A systematic review of price signals as a dietary strategy found that much of the evidence comes from modelling studies rather than empirical observations and from the USA, and uses highly variable methods, but did offer support for the role of price signals (85). Another systematic review and meta-analysis found that in pooled analyses a 10% decrease in price (i.e. subsidy) increased consumption of healthful foods (including fruits and vegetables) by 12% (95% CI = 10±15%) whereas a 10% increase in price (i.e. tax) decreased consumption of unhealthful foods by 6% (95% CI = 4±8%) (86). Increases in food prices have a larger effect in low-income countries than high-income nations, and within countries a greater effect on consumption by the poor than by the rich, so care must be taken not to further disadvantage the poor (87).

Choice architecture is another set of approaches to motivating healthy, sustainable dietary choices. Coined by behavioural economists Richard Thaler and Cass Sunstein (88, 89), it refers to efforts to nudge people towards certain decisions – in this case healthy and sustainable food choices – without changing their values, using psychological principles. Pricing may be considered a form of choice architecture, but other examples include displaying preferred foods at prominent positions on restaurant menus and food store shelves, manipulating plate, cutlery, and/or portion sizes, labelling foods (see Chapter 12), meat-free Mondays, and signage and other programming in workplaces. The evidence is mixed on many architecture options (90–92) – no surprise in view of the complexity of human decision making, the wide range of behaviours, interventions, outcomes, and settings studied, but some reports, such as on nudges to reduce meat consumption, are very promising (93). For example, observational and experimental studies in the University of Cambridge college cafeterias showed that when the proportion of vegetarian meals offered was doubled, vegetarian sales increased by between 41% and 79% (94).

Marketing strategies can also increase demand for plant-based foods. Key points include avoiding terms such as meat-free, vegetarian or vegan, and emphasizing instead flavour or origin (e.g. 'Cumberland-spiced'), or using more appealing words for plant-based options (e.g. 'field-grown' or 'garden') which significantly increase sales of the target vegetarian dishes (95).

Four broad principles can be applied to dietary behaviour change efforts (96):

1. Minimize disruption, for example producing affordable, recognizable, and tasty plant-based alternatives;
2. Sell a compelling benefit, such as improved taste, health benefits, or reduced cost;
3. Maximize awareness, which can be informed by traditional food marketing strategies such as putting plant-based foods at the top of a menu;
4. Help shift norms so that plant-based foods become the default rather than seen as a fringe behaviour.

Improved Agricultural Technologies and Practices

Meeting global needs for healthy, sustainable food also requires a range of agricultural technologies and practices. Some of these are land use practices, with the aims of protecting biodiversity, reducing GHG emissions, and promoting carbon sequestration, and shifting agricultural land from tobacco (a health-destroying crop) to nutritious food crops. Water management is equally important. The use of dangerous pesticides needs to be reduced. Other needed practices fall under the rubric of sustainable intensification, and include agroforestry and optimizing fertilizer use. Still other practices relate to the types of food that are produced. Options range from rediscovering traditional and indigenous types of food to devising innovative new foods. Finally, better indicators of progress, and better policies, are needed to drive technological change and improved practices.

Land Area Required to Protect Biodiversity

There are major trade-offs between the exploitation of land to feed the world population and to support other types of economic activity on the one hand, and the imperative to rapidly reduce greenhouse gas emissions and biodiversity loss on the other. The tensions between biodiversity conservation and food production triggered debate between 'land sparing' and 'land sharing' approaches. Land sparing aims to protect land for conservation purposes whereas land sharing aims to ensure that farming supports diverse wildlife. There are numerous points of disagreement between commentators, including the relative importance of food production compared with broader approaches to address food security, such as poverty reduction (97). Those who value biodiversity have different perspectives than those who put more weight on economic growth. Differing spatial and temporal scales of analysis further complicate the picture. Moving beyond polarized views requires looking afresh at how to reconcile multiple objectives, including health, equity, and livelihoods, within planetary boundaries, including in the context of land scarcity. In addressing these complex interconnected challenges we can learn from the critical analysis of diverse perspectives.

Several research groups have proposed maximum levels of land use for crops and livestock and minimum levels of forest cover for specific biomes compatible with preserving the integrity of the biosphere. Estimates suggest that 12.6–15.2% of global ice-free land is the maximum area of cropland compatible with biosphere integrity. However, a recent analysis using 90% of naturally occurring local species abundance (number of individuals) and 80% of naturally occurring local species richness (number of species) as the acceptable levels of biodiversity suggests a lower maximum of 4.6–11.2% of the global ice-free land that can be allocated to cropland (and 7.9–15.7% to pasture) (98). The 90% abundance level was set as a precautionary measure and may be lower in some cases, but even so the findings illustrate the constraints on land use to feed human populations. Assumptions about future dietary patterns significantly affect the results. If current dietary patterns are assumed, the lower end of the range for projected cropland would apply because of the need for more pasture, but if there is a move to lower consumption of animal products then more land is available to grow crops and the upper end of the range is applicable.

While there may be resistance to acquiring abandoned agricultural land, if for example it has cultural and livelihood importance to specific communities, restoring uncontested abandoned and degraded agricultural land could be a major opportunity for conservation without much political opposition. Land with low purchase and transaction costs is particularly attractive for conservation purposes (99). Assessing the potential social and health benefits of such land restoration for biodiversity conservation will be an important task for research and action.

Land Management Options for GHG Removal

Rising emissions, and rising atmospheric levels of CO_2 and methane, confirm the need to sequester substantial amounts of carbon. The two most important GHG removal (GGR) technologies are bioenergy with carbon capture and storage (BECCS) and afforestation (planting a new forest) or reforestation (replanting a former forest). Each of these has major land use implications, and requires substantial trade-offs (101). Other approaches to GGR include wetland restoration and soil carbon sequestration (SCS) (102).

Growing and burning biofuels represents a seemingly straightforward intersection between land use and GGR. Growing biofuel plants would, in theory, capture the CO_2 that is later released with combustion, achieving no net release. However, the theory rarely translates into practice. Consider the case of wood pellets from forest biomass, which are misleadingly classified as carbon-neutral renewable energy sources. Currently about 10 million tonnes of wood pellets are traded, largely between the USA and Europe, but also increasingly to Asia, to be substituted for coal in electricity generation. The net effect of such policies on CO_2 levels and thus on the global climate depends on how

---

Box 10.3. **Is Grass-Fed, Free-Range Beef a Solution?**

The disadvantages of factory farming of beef have been well documented. They include air and water pollution, risk of antibiotic resistance and infectious disease propagation, and animal cruelty. These problems, together with the large methane emissions of beef, have led many to advocate reduced beef and dairy consumption.

A counterpoint argues that beef can be produced sustainably through grazing. According to this view, some land is unsuitable for any other kind of agriculture, grazing by ruminants is a long-standing 'natural' process (e.g. the huge buffalo herds on the American plains before European settlement), and when well managed, livestock grazing can enhance carbon sequestration in grazed land soil through manure deposition.

Resolving this issue is a matter of numbers. While ruminant milk and meat currently contribute 13 g protein/person/day, amounting to about half of the world's terrestrial animal protein supply (27 g protein/person/day), grazing systems account for less than 5% of overall terrestrial animal protein supply globally, about 1 g/person/day (100). There may be some potential for increase in grazing lands, but it is limited and could not provide a substantial proportion of global protein needs. Additional deforestation to create more grazing land would accelerate biodiversity loss and climate change. Moreover, the potential benefits of grazing in terms of carbon sequestration decline over time and are context-specific, depending for example on livestock density, local plant species, the presence of wild herbivores, and soil type. Poor grazing management can result in large soil carbon losses.

Overall, even with optimistic assumptions about soil carbon sequestration, grazing could only make a tiny contribution to climate change mitigation. For some communities, grazing of cattle and other ruminants provides an important source of essential nutrients and livelihoods. In these cases, good grazing management, including avoiding overstocking, can improve soil carbon, particularly where soils are depleted, and conserve biodiversity.

---

quickly the carbon released by burning the biomass is absorbed by the trees that are grown to replace those used for fuel. The carbon from burning enters the atmosphere in minutes but the rate of uptake depends on the growth rate of the tree, which in turn depends on the species and the growing conditions. Only leftovers from traditional processes such as timber, board and paper processing, or rapidly decaying wood from trees affected by diseases or fire damage, have payback periods in the order of years. If trees are harvested specifically for processing stem wood into wood pellets there may be a net increase in CO_2 levels for decades or centuries, depending on the scenario (103). Some studies conclude that wood pellet emissions could be double those of coal, and even when 65% of the wood is from residues and 35% is from harvesting additional trees, the CO_2 emissions could exceed those from coal burning. Burning wood is also an important contributor to PM air pollution. Under current regulations these crucial issues are ignored and wood pellets receive large subsidies as 'carbon-neutral' sources of energy. These sums can be considerable – in the UK for example one power station (Drax) received £789 million in 2018.

The amount of electricity that can be produced from a given area of land using photovoltaics is at least 50–100 times greater than that from biomass burning, which also produces harmful air pollution. The biomass industry should publish full life cycle assessment of CO_2 and air pollution emissions, and regulations should ensure the use of genuine agricultural or forestry residues where payback times could be compatible with the Paris Agreement. The most useful contribution of forests to carbon management is likely to be through the provision of timber and other forest products that can lock away carbon for long periods, replacing carbon-intensive materials in the construction industry, rather than by burning healthy trees for fuel.

A more complex approach is BECCS, which generates bioenergy and incorporates carbon capture at the point of combustion. Some IPCC scenarios foresee massive deployment of BECCS, with annual removal rates as high as 15 $GtCO_2eq$/year by 2100, and providing up to 50% of primary energy requirements (compared with ~10% today). The GHG emissions reductions from BECCS technology could in theory reduce global atmospheric carbon concentrations close to pre-industrial levels by the end of the present century (104). However, BECCS at this scale would require massive conversion of land from pastureland, grassland, and other crops to growing biofuels. Indeed, scenarios that contemplate a major role for BECCS in meeting climate targets require heroic assumptions about the scale of biofuel cropland expansion (105). Critics have noted that this process would not only put pressure on agricultural land needed for food, but would likely transgress planetary boundaries for land use. It would also require vast amounts of water and fertilizer, with implications for water, nitrogen, and phosphorus planetary boundaries (106).

Alternatives include the use of microalgae for livestock feed, which could in theory free up to 2 billion hectares of land currently used for pasture and feed crops (107). By one estimate, forest plantations established on this land could potentially meet 50% of global primary energy demand.

Serious questions about the technical feasibility and effectiveness of BECCS remain. Land-intensive bioenergy can lead to substantial carbon emissions from land use change and from the production, harvesting, and transport of the biofuels (108). Even the reduced carbon emissions claimed by existing BECCS projects have been overstated, underlining the need for rigorous, consistent indicators (109). And with the cost of solar and wind energy declining rapidly, the economics of BECCS as an energy source are increasingly unfavourable (110). While the future of BECCS is uncertain, land use conversion at the scale contemplated by some scenarios is unlikely.

Forests are highly effective carbon sinks. According to one estimate, the world's forests, which cover 30% of global land surface, now sequester 8.3 Gt CO_2/year. To meet the 2 °C climate target, global forest cover would need to double to 60% to sequester an additional 7.8 Gt CO_2/year. This would require a substantial decrease in red meat consumption, to enable grazing land and cropland used for livestock feed to be forested (111). While afforestation and reforestation at that scale are unlikely, they can make an important contribution to CO_2 removal. Forests yield other ecosystem services as well, including water and soil conservation, water filtration, and health. Intact forests are associated with reduced incidence of diarrhoeal disease in their watersheds (112) and protected forest areas have reduced incidence of malaria, acute respiratory infection, and diarrhoeal disease (113). Forests exemplify the links among land use, food, energy, health, and climate change.

Wetlands currently cover less than 9% of the global landscape and about half have already been lost (102). They have the potential to deliver 2.7 $GtCO_2$eq additional climate change mitigation annually and to reduce ocean acidification, because of their ability to take up large amounts of CO_2 from the atmosphere. The CO_2 absorbed more than compensates for the increased methane emissions that can arise when wetlands are restored (102). Wetlands can also filter pollutants and sediments, thus providing clean water for human populations, as well as protecting important ecosystems such as coastal coral reefs. Flood defences and support for livelihoods through, for example, shrimp and fish farming are further examples of potential direct and indirect health benefits from protecting and restoring wetlands. Trade-offs, such as reduced biodiversity from excessive nutrient loading resulting from overuse of fertilizers that cause dominance of aggressive plant species, need to be anticipated and addressed.

A final aspect of land management for GGR relates to soil. Globally, soil is a massive carbon repository. Recent research has clarified the extent to which land use changes have undermined this function, reducing the stock of soil carbon (114) and the potential to sequester more carbon with appropriate management (115). Strategies include (a) optimizing the primary productivity of crops; (b) reducing soil disturbance and managing soil physical properties through techniques such as no-till agriculture; (c) minimizing deliberate carbon removal and erosion; (d) addition of carbon from external sources such as manure and biochar; and (e) additional carbon inputs within the cropping system such as agroforestry and cover cropping (116). Many of these practices form part of regenerative agriculture; they are feasible and affordable, even for small-scale farmers, and they deliver improved crop yields. However, for soil carbon sequestration to make a significant contribution to balancing the Earth's carbon budget, it must be implemented on a vast scale, overcoming substantial social, economic, and technical barriers (51). For example, tenant farmers may have little incentive to invest the effort and money required. Moreover, on a heating planet, soil stores CO_2 less effectively (117). Other concerns relate to the need for extensive inputs of nitrogen and phosphorus, which cause their own environmental problems (see Chapter 3) including emissions of nitrous oxide, a powerful GHG (118). Increasing soil carbon is good for farmers but forests and wetlands are likely to play more substantial roles in carbon sequestration.

Assessment of the health effects of GGR strategies through pathways such as food and nutrition, freshwater quantity and quality, air pollution, poverty and livelihoods, and infectious disease transmission has been relatively neglected and will be a key topic for more research. Nevertheless, it is clear that they will have important health effects, both positive and negative, and by taking these into account in design and implementation the balance of benefits and harms could be substantially positive.

Reclaiming Land from Tobacco Farming

Another land-based strategy for advancing Planetary Health is to end land uses that undermine health. There is no better example than tobacco farming, which accounts for about 1% of global arable land. Tobacco farming has increased in a number of countries such as Malawi, Tanzania, and China where it has doubled since the 1960s. The tobacco industry has encouraged increases in land under tobacco cultivation in a number of LMICs. Although this brings short-term economic benefits for some farmers, it causes a range of long-term disbenefits from adverse effects on health, the environment, and in some cases on local food production (119). Tobacco farming is often dependent on higher use of pesticides, fertilizers, and growth regulators than other crops and can deplete soil nutrients by taking up more phosphorus, nitrogen, and potassium than other crops. Once soils are depleted, tobacco farming may move to more fertile areas, thus undermining the prospects for future food production. In 66 tobacco-growing LICs, tobacco farming was estimated to account for nearly 5% of deforestation between 1990 and 1995 and output has increased since then (120). Tobacco farming requires substantial amounts of wood for a variety of purposes, including curing and for barn construction, amounting to about 20 stacked cubic metres per tonne of tobacco. The total post-consumption solid waste from the 6 trillion cigarettes smoked annually was estimated to be at least 2 million tonnes – greater than the annual plastic waste from mineral water bottles (121). Trillions of non-biodegradable cigarette butts containing cellulose acetate are discarded annually and are the most frequent item of litter on beaches and water edges worldwide. Action to curb tobacco smoking can therefore have substantial environmental as well as health benefits.

Water Management

Freshwater use forms one of the planetary boundaries, driven largely by the irrigation requirements of food production. Local and regional availability varies greatly. Global redistribution of crops from regions of low to high freshwater availability would in theory feed an additional 825 million people while reducing overall water consumption by over 10%, but would be difficult to achieve (122). It assumes a transition from smallholder farming to large-scale commercial farms with widespread negative impacts on rural livelihoods. Redistribution within national borders is more feasible. In China, for example, the breadbasket has shifted towards water-scarce northern regions. Irrigation would need to be reduced by 18% in these water scarcity 'hotspots' to achieve sustainable use of freshwater, with a potential 21% loss of grain production. However, this loss can be reduced to about 8% by improving yields in water-rich regions (123).

In parts of Africa where malaria transmission is unstable and where people have little or no immunity to malaria parasites, such as the African highlands and desert fringes, irrigation can increase malaria incidence by providing habitats for the breeding of vector species. However the risks can be addressed by better management practices (124).

Sustainable Intensification

Sustainable intensification refers to efforts to increase agricultural production, yields, and/or income per unit of land area through greater investment of financial or human resources and more efficient use of inputs such as fertilizer or pesticides (125). It can take many forms depending on climate and land, socioeconomic circumstances, individual choice, and market forces. A comprehensive assessment of approaches is beyond the scope of this book but we discuss some of the options in subsequent sections, with a particular emphasis on the implications for nutrition and health. Sustainable intensification has synergies with 'Climate-Smart' agriculture, which aims to enhance adaptation and resilience to climate change by supporting diversified farming systems, local adaptation planning, responsive governance systems, and enhancing leadership skills (126). Sustainable intensification has sometimes been equated with large-scale industrial agriculture but the principles can also be applied in regions such as Sub-Saharan Africa (SSA), where yields are low and 75% of the total land has degraded or highly degraded soil (127).

Sustainable intensification comprises ecological, socioeconomic, and genetic intensification (127). Examples of ecological intensification are integrated pest management (IPM), which depends on replacing pesticides by natural enemies, and no-till conservation farming to encourage the accumulation of organic matter in the soil. Interspersing herbaceous crops with perennial trees or shrubs is a form of intercropping used in agroforestry. With deeper roots, trees can often access water and nutrients that are inaccessible to the crop. Socioeconomic intensification includes improved access to markets, value chains, affordable finance, and information about weather conditions and innovations in agriculture. Genetic intensification of desirable traits can be achieved by conventional breeding which in recent years has been augmented by biotechnologies – cell and tissue culture, marker-assisted selection, and genetic engineering (see below for discussion).

A relatively neglected topic in sustainable intensification is how to sustain the health of subsistence farmers in the face of climate change. Major challenges loom; for example, by later this century extreme heat could prevent a billion people from working outdoors in the hottest months, even in the shade (128). This will make physical labour more difficult and necessitate increasing mechanization as well as changes in working practices to reduce heat stress exposure and ensure adequate hydration.

According to a meta-analysis of data from 38,000 farms worldwide producing 40 types of agricultural products, farms producing similar products can differ by as much as 50-fold in environmental impacts (GHG emissions, land use, terrestrial acidification, eutrophication, and scarcity-weighted freshwater withdrawals) (129). While some of the differences may be due to geographical factors and resource availability, this finding suggests the potential for transfer of knowledge between low and high environmental impact farms to support the rapid scale-up of more sustainable farming practices. Emissions from deforestation and cultivated organic soils drive on average 42% of the variance in each product's agricultural GHG emissions, indicating the importance of reversing deforestation and the exploitation of peatlands. Diversified cropping and improvements in degraded pasture deliver significant reductions in both land use and GHG emissions. Methane emissions from flooded rice paddies can be reduced by shallower and shorter flooding.

Communicating the benefits of sustainable farm practices to processors, distributors, retailers, and ultimately consumers, by reference to specified environmental standards, could motivate rapid uptake of more sustainable products. However, these standards can be circumvented if producers can market their products to less discerning purchasers, as in the case of 'sustainable palm oil'. Not only has the value of this designation been challenged (130) but there is virtually no demand for this option in importing countries such as China, Indonesia, and India. Products with multiple supply outlets and invisible use in processed foods may be particularly challenging to regulate (131). Here the power of large retailers could in theory achieve change at scale.

Agroforestry

Agroforestry – the integrated management of trees and shrubs with crops and livestock – is increasingly advanced as a solution to interconnected challenges of food insecurity, climate change, biodiversity loss, declining freshwater availability, and air pollution from land-scape fires. For example, 71% of African countries have committed to using agroforestry for climate change adaptation and/or mitigation in their Nationally Determined Contributions (NDCs) submitted to the United Nations Framework Convention on Climate Change. More than half of Sub-Saharan African countries have committed to ecological restoration of a total of 1,130,000 km^2 of land, in many cases relying on tree planting (132). Agroforestry features less, but still significantly, in the NDCs from the Americas (34%), Asia (21%), and Oceania (7%). A systematic review of 207 case studies in SSA showed that agroforestry increased yields of crops or livestock products in 68% of studies (133). There are many pathways by which such approaches could benefit human health if they are well designed and implemented, although systematic evaluation of agroforestry for health and wider social outcomes has not yet been undertaken. The potential benefits could include reductions in poverty, water and air pollution, infectious disease transmission, and heat stress together with improvements in nutrition, NCD risks, and mental health (134).

Some trees are able to fix nitrogen and obviate the need for fertilizers. For example *Faidherbia albida* is a nitrogen-fixing acacia, indigenous to Africa, which sheds its leaves during the monsoon season when maize is sown, making it especially suited to intercropping with maize. This approach, which has been widely used in regreening of the Sahel in Niger as well as in Zambia and Malawi, encourages carbon sequestration, weed suppression, and drought tolerance (135).

Optimizing Fertilizer Use

In Sub-Saharan Africa the low level of fertilizer use is leading to nutrient mining, whereby soils are losing nutrients – at a level exceeding 30 kg/hectare/year in 80% of SSA countries. The technique of microdosing has been developed to minimize the application of and excessive reliance on inorganic fertilizer. It both improves efficiency of nutrient use and protects against drought (127).

Ammonia (NH_3) from animal husbandry and fertilizer use plays a key role in the formation of airborne particulate matter. A range of approaches can be used to reduce NH_3 emissions, including improved manure storage and processing, more appropriate dosage and timing of urea fertilizer application, and replacement with other fertilizers that release less NH_3. A 50% reduction in agricultural emissions in the EU would reduce

---

Box 10.4. **Does Organic Food Advance Planetary Health?**

Organic agriculture is promoted for its environmental and health benefits, including those from avoiding use of synthetic pesticides. A number of studies have assessed the benefits and trade-offs from organic agriculture. A meta-analysis that assessed the environmental impacts of organic agriculture found that although it uses less energy, GHG emissions are similar to those of conventional agriculture and both land requirements and eutrophication are greater for organic agriculture (40). Another study modelling the implications of transition to 100% organic agriculture in England and Wales found that, although domestic GHG emissions are reduced with organic farming, when increased overseas land use to compensate for shortfalls in domestic supply are included, net emissions are greater than in conventional agriculture (136). Enhanced soil carbon sequestration could offset only a small part of the higher overseas emissions. There may be other benefits not captured by this study, including lack of pesticide residues in food and the environment as well as greater biodiversity, but organic agriculture, at least using current practices in industrialized countries, is not by itself an effective method for reducing the environmental footprint of food systems.

   The putative health benefits of organic foods have also been questioned. Observational studies are difficult to interpret because people who consume predominantly organic food tend to have healthier lifestyles in general than those who do not. Two systematic reviews showed no clear health benefits or differences in nutritional quality, but many of the studies reviewed were relatively small and extended over short periods of follow up (137, 138). A third systematic review also found no clear nutritional benefits from consuming organic food, but there was evidence of reduced exposure to pesticide residues and antibiotic-resistant bacteria (139, 140). Studies measuring the urinary concentration of dialkylphosphate (DAP), a metabolite of organophosphorus pesticides (OPs), showed that these are lower in people who regularly consume organic foods (140). However, there is no gold standard for measuring exposure because there are large differences in toxicity between OPs, and children may be particularly sensitive to low-level exposures. Overall then the picture is nuanced with potential benefits such as reduced pesticide exposure, but there are trade-offs because of lower yields and therefore larger overall land use requirements.

$PM_{2.5}$-related deaths by 18% (141) and also reduce eutrophication of water. These measures could generate net annual economic benefits of US\$89–177 billion in the EU, as measured using the Valuation of a Statistical Life approach, for 50% and 75% emission reductions, respectively. Implementation of such measures is more cost-effective on larger than smaller farms and the benefits increase disproportionately once 50% reductions are exceeded. Decrease in consumption of animal products could contribute to reduced NH_3 emissions including through reduced manure production.

   Phosphorus is a finite resource, with known deposits in only a few countries and depletion a real threat (142, 143). At the same time, excess application of phosphorus in many places is wasteful and contributes to eutrophication of waterways. One proposed solution is phosphorus recycling, relying primarily on manure (144, 145). Happily, some of the places most appropriate for phosphorus recycling, because of a plentiful local supply of manure relative to need, are in India, China, Southeast Asia, and parts of Africa, where fertilizer has had to be imported. However, phosphorus recycling has a substantial carbon footprint (146), so careful management of the process, and weighing of competing priorities, is essential.

Methane Management

Methane is a GHG, a precursor of tropospheric ozone which is hazardous to health (see Chapter 3) and is also toxic to crops, particularly in humid rain-fed and irrigated areas of major wheat-producing countries, where it reduces yields by about 10% (147). Measures to reduce methane emissions, e.g. by alternating wetting and drying in rice paddies, will therefore also yield air pollution benefits for health (148) and improve crop yields.

Safer Use of Pesticides

Pesticides are widely used in agriculture to protect crops against pests, diseases, and weeds (see also Chapter 3). They may also be used for other applications such as in public green spaces and in public health protection programmes, for example to kill mosquito vectors that carry diseases. Exposure can be through contact with skin, by inhalation, and by ingestion. Occupational exposures have been a particular cause for concern but human exposure is ubiquitous through residues found in food and drinks. In the UK for example, testing of over 3000 samples of 38 types of food (including 376 pesticides in fruit and vegetables and 100 in animals) showed that 47% contained a residue and 3% of the samples contained levels above the limit set by law (149).

Several reviews have described the effects of pesticides on a range of health outcomes (150), particularly based on studies of occupational exposures of farmers and others with relatively prolonged, high-dose exposure. Principal toxic effects include endocrine-disruptor activity (151), neurotoxicity (152), and increased risks of various cancers including bladder cancer, hormone-related cancers such as breast cancer, and haematopoietic (blood) cancers and non-Hodgkin lymphoma (153). Pesticides also have a range of adverse impacts on ecosystems. For example, the use of neonicotinoids has led to concerns about effects on bee populations as well as on human health, including suggestive but not conclusive adverse effects on developmental and neurological outcomes in exposed populations (154).

There are inevitable scientific uncertainties but there is a strong case for major reductions in pesticide use and tighter regulation of pesticides given the probable pervasive effects on health and the environment.

Genetic Strategies to Increase Crop Yield

In the 1980s scientific advances resulted in the first introduction of bioengineered genes into plant genomes. Genetic modification has yielded plants that resist diseases, that have enhanced nutritional value, and that are adapted to changing climate conditions such as heat and drought. Innovations that may benefit farmers and consumers include apples that resist browning, drought-tolerant rice, and bananas and rice fortified with pro-vitamin A. Derivative benefits have flowed as well: resistance to herbicides and to insects enables decreased tillage and insecticide use, respectively. However, these techniques can lead to increasing selection for resistance in weeds and insect pests. Loss of genetic diversity has been an unpredicted consequence of developing high-yield crops with advanced cultivation practices; this in turn lowers resilience to climate extremes and other environmental stressors. Acceptance has been variable, and in the EU the cultivation of genetically modified (GM) crops is restricted. Africa has failed to benefit as much as Asia from these technological advances, with food insecurity still rife. This will necessitate building capacity to regulate, evaluate, and implement beneficial GM crops. There is a need for objective assessment not only of the health implications of GM crops but also the benefits and risks more generally (155).

GM crops may offer benefits in some circumstances but not in others, for example if they compel farmers to become dependent on costly seeds and pesticides. An assessment of the controversy around Golden Rice, promoted as a solution to vitamin A deficiency, illustrates the issues. Evaluative studies by natural scientists, many commercially funded, focused on biophysical plant–consumer interactions, and were overwhelmingly positive. In contrast, social science research addressed a wider range of sustainability perspectives including participation, equity, ethics, and biodiversity, and was more likely to emphasize the downsides of Golden Rice (156). Assessment should encompass the natural sciences, public health sciences, and the social sciences, as well as the perspectives of affected farmers and communities, providing a holistic overview of the range of impacts consistent with a Planetary Health perspective. The benefits and risks should be assessed, considering the likely harm of making no intervention to address the problem that the new product aims to address (157). Reviews have not identified any clear adverse effects on human health for common GM crops such as soybeans, rice, corn/maize, and wheat, but there is a dearth of evidence about long-term effects including on tests of mutagenicity, teratogenicity, and carcinogenicity (158).

The distinction between GM and non-GM crops may become increasingly blurred. For example, with the advent of genome editing with the CRISPR–Cas system it will become impossible to tell whether a crop variety arose from conventional breeding and/or use of new methods. It will therefore become more difficult to maintain a blanket ban on GM and other emerging technologies.

In coming decades advances in agricultural technology are needed to address multiple challenges simultaneously by providing higher nutrient quality, increased dietary diversity, and greater resilience to environmental change, while reducing the environmental footprint of the food system (GHG emissions, freshwater, nitrogen, phosphorus, land use, etc.) (159). This may involve a range of approaches, including but not limited to GM crops, such as gene editing and selective breeding. It will be essential to breed crops with a diversity of resistance genes and/or to cultivate a wider diversity of varieties to minimize the risks that pathogens become resistant and to enhance resilience to abiotic stressors such as climate extremes. Newly engineered traits that confer resistance to disease-causing organisms can be transferred to elite, high-performing crops. Genes that result in the development of pattern recognition to different disease-causing organisms can be transferred from one species to another, unrelated one. The use of wild relatives of domesticated crops to recover disease-resistant genes, which may have been lost during domestication or have evolved independently in different plant lineages, opens new possibilities for enhancing the resilience of major food crops to emerging diseases and environmental change. The stacking of multiple disease-resistant genes with different abilities to recognize disease-causing agents can help to engender resistance to a range of pathogens. Caution is also needed because some genetically engineered crops have demonstrated unanticipated vulnerability to disease. This emphasizes the need for ongoing evaluation of impacts and avoiding uncritical acceptance of new technologies.

A range of genes protect against climate-related factors such as flooding. Even crops such as rice, which are relatively resistant to immersion, can suffer yield losses as a result of prolonged flooding. Approaches to increased flood tolerance include manipulation of traits that promote root tolerance to waterlogging, and genes that modify the tendency of pods of oil crops to shatter prematurely when exposed to heavy rainfall. The flood-resistant SUB1 gene, when transferred into popular rice varieties such as high yielding Swarna rice using a technique known as 'marker-assisted backcrossing', allows them to survive up to two weeks of flooding (160, 161). The modified rice retains the characteristics that make it popular with farmers in India and Bangladesh, including low fertilizer requirements.

Manipulation of transporter genes can reduce the accumulation of sodium in the leaves of crops exposed to salination as a result of irrigation or seawater ingress into coastal freshwater. Similar approaches can also be used to reduce uptake of toxic elements such as aluminium (maize) or boron (wheat) (159). The yields of four major crops – maize, wheat, soy, and rice – are reduced by increasing temperatures (160). Sensitivity to heat is increased during reproduction leading to reduced male fertility and seed quality and protective responses may reduce yields. Between 100 and 400 molecules of water are lost for each atom of carbon that is fixed by photosynthesis. This means that there is a trade-off between carbohydrate production and water use. Carbon dioxide fertilization can lead to reduced size of stomatal pores through which water is lost, thus increasing water use efficiency. However, many plants have weak stomatal responses and increasing understanding of how genetic diversity of stomatal control and number could be harnessed to optimize the balance between carbon dioxide uptake and water use.

Additional challenges that could be addressed through genetic and other technologies include overuse of fertilizers, which suppress beneficial natural associations between micro-organisms and root systems. Many plant species associate with arbuscular mycorrhizal fungi that greatly expand the nutrient uptake by the plant root system by actively mining immobilized phosphates from the soil. These natural associations could be capitalized on with improved knowledge to increase yield responses to moderate phosphate use.

There are several obstacles to implementation of new knowledge for the benefit of low-income populations. Restrictive patent legislation that gives control to large corporations intent on profit maximization rather than public benefit is a significant challenge. So is public resistance in some countries. The topic is contentious, with entrenched views from both opponents and proponents (155). GM and other technologies could increase the dependence of poor farmers on expensive seeds and pesticides but could also increase yields, thus offsetting these costs. In some cases the assessment depends on the perspective taken and on the interpretation of incomplete evidence. The empirical evidence of benefits of GM agriculture to the poor is still fairly limited. A study of insect-resistant GM cotton adoption in rural India showed, for example, that early adopters benefitted from increased income, increased calorie consumption, and improved dietary quality, with an overall reduction in food insecurity of 15–20% (163). Improving access to effective technological advances by disadvantaged populations is imperative to reduce inequities in health and wealth. It is clear that GM and other technologies need to be assessed on a case by case basis when considering their implications for equitable development. Evaluations should encompass a full range of potential benefits and harms. When developed and implemented appropriately they can yield major benefits through increased resistance to insect pests, nutritional fortification, tolerance to high temperatures and to salinity (164).

Wild Foods and Indigenous Crop Varieties

Wild foods provide essential micronutrients (e.g. iron) (165) and are an enduring source of dietary diversity for many hundreds of millions, when used sustainably. Consumption of wild foods not only perpetuates cultural traditions, it can be highly adaptive to disruptions in other food supplies. For example, overfishing by the fleets of EU nations off the West African coast reduced fish availability for local fishers, and led to increased hunting of bushmeat (166). Conservation of biodiversity, for example through forest protection, can therefore address nutritional challenges, by conserving local species that provide dietary diversity and a range of essential nutrients, particularly for indigenous communities.

Currently over 20% of the population in South Africa have inadequate access to food and this proportion remains stubbornly high. Stunting affects over 25% of children but at the same time nearly 70% of women are overweight or obese. Current agricultural policies encourage the cultivation of energy-dense, nutrient-poor cash crops which contribute to the current burden of nutrition-related ill-health. Several underutilized traditional crops including cowpeas, amaranth, pumpkin leaves, and Bambara groundnut are nutrient-rich and adaptable to marginal conditions, thus providing potentially sustainable and accessible food sources for low-income populations (167). However, traditional and indigenous crops are currently excluded from policy priorities and, as a consequence, subsistence farmers, many of them women, are disadvantaged, while deprived populations, who could particularly benefit from greater dietary diversity, are unable to do so.

South Africa provides an example of how the dominant agro-industrial food system co-exists with an alternative informal system, including subsistence farmers who lack access to modern technologies, markets, finance, and fertilizers, and as a result are vulnerable to climate shocks. These subsistence farmers are often custodians of indigenous crop varieties that have been overlooked in large-scale systems, sometimes because their cultivation and use require traditional knowledge (167, 168). Ensuring support for subsistence farmers and integrating underused traditional and indigenous crops into national food systems can increase resilience and enhance dietary diversity. Because they can survive in harsh conditions and require less disturbance of landscapes for their cultivation, these crops can also advance environmental goals – reducing freshwater use, eutrophication, GHG emissions, and biodiversity loss. However, it is unclear to what extent these indigenous crops can be scaled up in markets that require competing with mass advertising of ultra-processed foods. Policy options to increase cultivation include campaigns to counter stigma due to perceptions that they are foods for the poor or elderly, marketing to middle- and high-income consumers on health grounds (as has been done with quinoa), and training of subsistence farmers to increase yields.

Promoting Healthy, Sustainable Crops

The production of fruit, vegetables, legumes, nuts, and seeds is threatened by climate change and freshwater declines (see also Chapter 2) (14, 15). Thus, increasing the availability and affordability of fruits, vegetables, nuts, and seeds, particularly for low-income populations, is a key strategy for safeguarding health in the Anthropocene. But within these broad categories there is considerable variability in both environmental impacts and climate sensitivity. For example, although global nut production has increased by about one-quarter since 2007–8 (169), some nuts, such as cashews, have low yields and others, such as almonds, are water, pesticide, and fertilizer intensive. About 74% of irrigated nuts are produced under blue water stress. To meet the recommendations of the EAT–*Lancet* Commission, nut varieties with low water requirements should be prioritized (170). More generally, policies should selectively prioritize the cultivation of crops that address both health and environmental requirements.

Novel Foods

A growing array of innovations is being developed to respond to the challenges of improving and sustaining nutrition at lower levels of environmental impact than current food systems.

Insect-based food is one such innovation. The ability of some insects to convert food waste and agricultural by-products into food together with their high nutrient density suggests that they may have potential to contribute to food security and nutrition. About 2 billion people live in countries where insects are eaten by at least some of the population, and worldwide over 2000 insect species are consumed by humans (171). Currently about 90% are harvested from the wild but overharvesting can undermine the integrity of local ecosystems. A recent review suggested how production of 'mini-livestock' through insect agriculture could help address nutritional and environmental challenges posed by current food systems (172). First, many species are rich in nutrients including protein, micronutrients, and polyunsaturated fatty acids, and their nutrient content can be influenced by their feedstock towards healthier composition. Insects may be a source of dietary fibre in the form of chitin from their exoskeletons, which may have beneficial effects on microbial flora (173). Second, insects can be grown with relatively low capital and technology requirements, with the potential to use food waste as feedstock. They can yield innovative products such as protein powder and fertilizer, as well as providing animal feed. Third, insects also have a lower environmental footprint than many conventional foods. Wide acceptance of insect-based foods must overcome the 'yuck' factor in some cultures (174).

Duckweed, an edible aquatic plant, is increasingly recognized as a source of essential nutrients with a low environmental impact (175) and seems relatively acceptable to consumers (176). Mankai (a newly developed high-protein strain of *Wolffia globosa*, a type of duckweed) provides all nine essential amino acids, dietary fibre, polyphenols, iron, zinc, and vitamin B12, thus offering a potential high-quality replacement for animal protein (177).

Microbial protein production offers particularly exciting prospects. Recent work has shown that protein-rich biomass can be produced by direct air capture of carbon dioxide with the help of bacteria that can oxidize hydrogen, powered by renewable energy in a closed system (178).

Manufactured meat alternatives are another food innovation. This group of products includes cultured meat (produced by culturing animal muscle cells in appropriate media), meat substitutes made from plant-based sources such as beets, tofu, and tempeh, and products made from micro-organisms (such as Quorn, a commercial product made from the soil mould *Fusarium venenatum*). Food with the taste and mouth feel of meat can help ease the transition to plant-based diets for dedicated carnivores. However, the benefits of alternative meats, for both health and the environment, need to be documented. A study compared the land requirements and GHG emissions for conventional animal products, insects, plant-based meat alternatives, and cultured meat (179). It showed that cultured meat did not offer significant land use benefits over poultry and eggs and had higher energy requirements. Meat alternatives and insects have the highest land use benefits although not much higher than eggs and poultry meat. Comparison of two leading alternative meat products (Beyond Meat and Impossible Burgers) with lean ground beef has revealed that the alternatives are similar to the beef in calories, saturated fat, and protein content, much lower in cholesterol, and about five-fold higher in sodium – all told, not a compelling health advantage (180). Future formulations should aim to improve the health profile, but in any event the emphasis should not be on consumption of ultra-processed food.

A study compared the environmental impacts and nutritional profiles of nine 'future foods' with conventional plant (PSF) and animal (ASF) source foods (181). The future foods comprised both terrestrial foods (cultured meat, mycoprotein (*Fusarium venenatum*), black soldier fly larvae (*Hermetia illucens*), housefly larvae (*Musca domestica*), and mealworm larvae (*Tenebrio molitor*)) and aquatic foods (chlorella (*Chlorella vulgaris*), spirulina (*Arthrospira platensis*), sugar kelp (*Saccharina latissima*), and mussels (*Mytilus* spp.)). All the future foods studied, except sugar kelp, had a comparable or higher dry-matter protein content than PSF and ASF, and provide essential amino acids. Although there was considerable diversity in their nutritional profile they were a better replacement for ASF than PSF. A judicious mix of future foods as part of a diverse diet could contribute to meeting micro- and macronutrient requirements. For example adequate calcium can be provided by sugar kelp or black soldier fly larvae, iron by chlorella and spirulina (in very high amounts suggesting that recommended intakes should be limited), and zinc by sugar kelp, all insect species, and mussels. More research is needed on bioavailability of nutrients and digestibility but current indications suggest this is unlikely to be a barrier to use. Unlike GHG emissions from ASF and fertilizer use, GHG emissions associated with producing future foods come largely from fossil fuel energy sources and will decline when renewable energy sources are used. Future foods also require less land than ASF although this is dependent on the feedstock used.

Novel foods can play significant roles in improving nutrition with reduced environmental impacts but new technologies and food sources are not yet ready for scaling up to address the needs of 10 billion people and face barriers of acceptability and integration with existing dietary choices.

Aquatic Food Products

Captured and farmed aquatic food products provide only 1.2% of total calorie supply and 6.7% of total protein supply but are nevertheless vitally important sources of protein and micronutrients to many low-income populations. About 3 billion people source 20% of their animal protein intake from fish. Fish provide 60% of dietary protein in some West African countries (e.g. The Gambia, Ghana, Sierra Leone) and over 50% in some small island states and Asian countries such as Bangladesh, Indonesia, and Sri Lanka (182). About 10% of the global population could face micronutrient and fatty acid deficiencies over coming decades as a result of fishery decline, especially in equatorial locations (183). Consumption is profoundly inequitable, such that high-income countries consume over 70% of total available seafood exports.

Wild capture fisheries have declined in recent years because of overfishing. Aquaculture has been responsible for most of the increase in fish production since the 1990s, with an annual average growth rate of over 8%, and is projected to increase substantially in the future to address growing fish demand and falling wild fish supply (182). However, aquaculture is still responsible for only a small proportion of global calorie and protein supply. Over 90% of aquaculture is in Asia; in China, farmed fish provide the second highest supply of farmed meat after pork.

Fish convert feed into body mass much more efficiently than do land animals. Fish also have much lower nitrogen and phosphorus requirements than beef and pork, although slightly higher than poultry. They are also responsible for lower GHG emissions than land animals. However, aquaculture is overly dependent on unsustainable feed sources such as krill; there is currently much research on plant-based feedstocks. Carnivorous fish such as salmon can consume up to five times more fish as feed than they produce (184), and cannot therefore make a major sustainable contribution to food security. The use of fishmeal 'to feed the fish' has declined from an average of 23% (26 million tonnes/year) in the 1990s to 10% in 2012 (16 million tonnes/year) as a result of increased use of fishmeal replacers, including plant proteins, waste products from fish and terrestrial animals, and the use of improved breeds with better feed conversion. Those farmed aquatic species that are low on the aquatic food chain (including freshwater species such as tilapia, carp, and omnivorous catfish species) are not dependent on using fishmeal and fish oil (185). Shellfish such as mussels and oysters are not feed-dependent but are threatened by ocean acidification, eutrophication, and harmful algal blooms in some locations.

Recommendations to increase the sustainable consumption of fish include the removal of harmful subsidies encouraging overfishing, policies to support small-scale fisheries and aquaculture, use of improved feedstocks that do not directly compete with human foods, improvement of key genetic traits contributing to nutrition, and using agroecological models of production in aquaculture (182). Self-sufficiency of aquatic food supply needs emphasis over imports of fish and other aquatic products by rich nations. Aquaculture has the potential to increase resilience of food supplies to environmental change in the face of increasing food demands but the achievement of this potential depends crucially on the implementation of appropriate regulation and governance systems (186) (see Chapter 12).

Reducing Food Loss and Waste

Given the magnitude of food loss and food waste, described above, addressing these challenges has the potential to reduce environmental impacts of food systems, improve food security, and conserve resources. Relatively more staple crops are wasted than animal products, so the benefits of reduced waste and loss are greater for land use, blue water use, nitrogen, and phosphorus than for GHG emissions (59).

Policies to address food loss and waste vary according to the causes (23, 61). In low-income countries enhanced storage and transport infrastructure together with 'cold chains' (refrigeration) and improved processing (drying and packaging) facilities will contribute to reduced losses. Education and training of subsistence and other low-income farmers in good animal husbandry and improved storage techniques can play a significant role in reducing food loss. In high-income countries consumers are responsible for a high proportion of waste, and potential strategies to reduce waste from this source include better marketing of visually imperfect but edible produce, legislation to encourage supermarkets to provide food reaching its 'use by' date to food banks or other outlets, and reduced portion size. Food waste can also be used to feed insects for high-value protein for animal feed. SDG 12.3 aims to halve per capita global food waste at the retail and consumer levels by 2030 and reduce food losses along production and supply chains, including post-harvest losses. Two new indices have been created to monitor progress. The Food Waste Index will measure tonnes of wasted food per capita, from processing to consumption. The Food Loss Index has already been created by FAO, to assess food loss along supply activities including production, handling and storage, and processing.

Redefining Metrics of Food and Agricultural Progress

In Chapter 6 we discussed the need for improved metrics of economic progress – shifting from GDP to metrics that account for human well-being, equity, and quality of life. The same is true for the global food system. To track progress towards a food system that supplies ample nutritious food to all, while remaining within planetary boundaries, we need redefined metrics of agricultural productivity. Conventional metrics such as tonnes of product/hectare need to give way to metrics such as people nourished/hectare (63). Summary metrics could include environmental endpoints, such as GHG emissions, freshwater use, land use, and nitrogen and phosphorus for a given quantity of food, as defined in terms of energy, protein, or micronutrients. Still other metrics could address the diversity of diets and the proportion of a population that is well nourished. Research in LMICs has consistently shown that dietary diversity scores are associated with adequate nutrient intake and nutritional status among women and young children (187). Such measures of environmental impact, dietary quality, and food-related health outcomes could complement standard measures of agricultural productivity.

Taxes and Subsidies: Getting the Prices Right and Advancing Equity

As we have seen for energy policies there are large costs arising from food systems that are not reflected in prices. A recent estimate suggests the hidden costs of the world food system are about US$12 trillion annually – compared to a market value of about US$10 trillion. The costs of the health effects make up over 50% (about US$6.6 trillion) and environmental costs account for about 25% (US$3.1 trillion) (188).

Greenhouse gas taxes on food, could, if well designed, have the potential to improve health and reduce food system GHG emissions. But badly designed policies could disadvantage the poor and increase ill-health. According to a modelling study, a tax rate of about US$50/tonne CO_2eq would raise the price of beef by about US$2.80/kg and of lamb by about US$1.30/kg but would have little effect on the prices of vegetables, fruits, legumes, grains, and roots (189). To avoid negative impacts on food security, such a policy might apply no taxes to healthy food groups, selectively compensate for income losses associated with tax-related price increases, and invest a portion of tax revenues in health promotion. Implemented in each world region, a tax structure with these features could avert approximately 500,000 premature deaths annually, and reduce GHG emissions by nearly 1 $GtCO_2$eq.

Increasing the price of high sugar snacks by 20% was estimated to reduce prevalence of obesity in the UK by 2.7% after one year, particularly in low-income households classified as obese, and was twice as effective as a similar price increase on sugar sweetened drinks (190). However, the overall effect on GHG emissions from the diet is unclear because sugar has low emissions and there is a danger that such taxes could displace demand towards higher GHG emission foods. Combining GHG-related taxes on food with taxes on sugar-containing snacks and beverages offers a way forward that could optimize both health and GHG benefits.

Subsidy policy can also play a role, as discussed in depth in Chapter 12. Government support for agriculture and fisheries currently totals over US$700 billion a year worldwide but only about 15% is targeted at public goods. Redirecting these subsidies towards paying farmers and fishers to produce the healthy food, reduce GHG emissions, and protect nature would have major benefits (188).

Future projections of hunger and food security under climate change show how different pathways of socioeconomic development can dramatically influence the prospects for addressing undernutrition and food insecurity. For example, shared socioeconomic pathways (SSPs) (described in Chapter 2) that project low population growth and more equitable food distribution, wealth, and education result in the virtual elimination of hunger by the end of the century. In contrast, SSPs that project high population growth, inequitable economic growth, and economic stagnation in low-income countries result in increased hunger (191). Notably, two of the relatively 'optimistic' SSPs (1 and 5) also project higher areas of primary forest cover than current levels, with benefits for biodiversity and carbon sequestration. Clearly, economic and social policy must complement changes in what we eat, how we produce food, and how we waste food, if we are to feed the world's population and remain within planetary boundaries.

Box 10.5. **Food Sovereignty**

Food sovereignty is a paradigm that has emerged from struggles over many years of social movements, often linked to indigenous groups working in mutual solidarity to define principles that aim to underpin progress towards a future without hunger. At the 1996 World Food Summit, *La Via Campesina*, an organization representing about 200 million small farmers in 73 countries, presented seven principles that underpin food sovereignty: (a) access to safe, nutritious food as a basic human right; (b) agrarian reform that confers land ownership and control, together with access to credit and technologies to local farmers, and returns traditional lands to indigenous communities; (c) conservation of biodiversity and protection of the health of soils; (d) food trade reform focusing on the achievement of national self-sufficiency and prioritization of domestic food production with support of smallholder farmers, together with food pricing that reflects the full economic costs of food production; (e) better regulation of transnational food trade; (f) social peace that aims to reduce violence towards and discrimination against poor and indigenous communities; and (g) democratic control and participation, particularly by rural women, supported by access to accurate information to ensure good governance and accountability.

These principles build on the experience of small farmers in often difficult circumstances, whose rights are challenged by powerful interests and whose scarce technical and financial resources make it difficult to compete with agro-industrial corporations. A coalition of NGOs, including Via Campesina, has been instrumental in the persuading the UN Human Rights Council to begin negotiations on a new international human rights instrument that if adopted will recognize the rights of rural working populations to land, seeds, biodiversity protection, livelihoods, and food sovereignty (192).

The global food trade and large-scale production offer efficiency and keep food prices low but do so at substantial costs to health and the environment. Learning from indigenous knowledge, including of local food systems, can offer the prospects of mutual benefit, but the deterioration of nature, even in many indigenous territories, suggests that experience of the past may increasingly prove an inadequate guide to the future.

Conclusion: A Menu of Potential Solutions

Achieving the goals of healthy, sustainable food systems and land use patterns necessitates addressing multiple objectives simultaneously, including environmental, socioeconomic, and health outcomes. However, much of the research evidence examines subsets of these outcomes in isolation. It is only through integrating knowledge from different disciplinary and geographical perspectives that we can advance our understanding of the pros and cons of different potential 'solutions'.

The pursuit of greater equity in access to healthy food in the face of environmental change must be a cornerstone of food system policies for the Anthropocene. An example is the paramount importance of making vegetables, legumes, fruit, nuts, and seeds affordable to those on low incomes.

The increasing concentration of power and market dominance in a few companies means that any policies to improve sustainability will have to address their role. Multiple strategies are needed, including regulation, taxation, land policy, and consumer influence, to shift them to a more sustainable, healthy, and equitable trajectory.

Small farms including subsistence farmers still provide a major proportion of the world's food yet they face major challenges. They tend to be less productive than larger farms; they will face increasing heat and chaotic weather; and plot sizes are declining in regions such as SSA as populations increase. The practices of sustainable intensification and improved access to market, climate, and agronomic information can help to enhance their performance. While there is much to learn from indigenous agricultural and ecological knowledge, the rapidity of environmental change in the Anthropocene implies that knowledge founded on centuries of experience will need to be supplemented by access to tools that enable rapid adaptation to evolving environmental threats. Although innovation based on science and technology may sometimes be perceived as incompatible with traditional wisdom, there are opportunities to learn from this diverse range of approaches about what works and in which context, particularly to improve nutrition and livelihoods of the most vulnerable populations on the frontline of environmental change.

The inefficiency of current food systems cannot be sustained. Reducing food loss and waste will require interventions targeted along the supply chain from farms, through processing and transport to households. Inefficient use of water and fertilizer can be tackled through combinations of regulation, technology, pricing, and other strategies.

Dietary change is inevitably difficult and controversial. People are attached to the foods they grew up eating, and that their cultures value. People value personal choice. But much can be done to make transparent the environmental impacts of food systems and to increase consumer demand for healthy, low environmental impact products. Appropriate tax and subsidy regimes can help to achieve universal access to healthy and affordable dietary choices.

Novel foods are an area of rapidly evolving knowledge driven by the imperative of producing healthy food choices with minimal environmental impact. Which of these will provide enduring solutions is still unclear but many of them show real promise.

Food systems are currently dysfunctional, contributing to malnutrition in its different forms, to unsustainable land exploitation and conversion, and to the dramatic environmental threats confronting humanity. Re-orienting them to address the challenges of Planetary Health offers the potential to address many of the health and sustainability challenges of the Anthropocene.

References

1. FAO. The State of Food and Agriculture. Moving Forward on Food Loss and Waste Reduction. Rome: FAO; 2019.
2. UN. Updated Comprehensive Framework for Action. UN High Level Task Force on the Global Food Security Crisis; 2010.
3. Frankema E. Africa and the green revolution: a global historical perspective. *NJAS – Wageningen Journal of Life Sciences*. 2014;70:17–24.
4. Pingali PL. Green revolution: impacts, limits, and the path ahead. *Proceedings of the National Academy of Sciences*. 2012;109(31):12302–8. doi: 10.1073/pnas. 0912953109.
5. Bioversity International. *Mainstreaming Agrobiodiversity in Sustainable Food Systems: Scientific Foundations for an Agrobiodiversity Index*. Rome: Bioversity International; 2017.
6. Bentham J, Singh GM, Danaei G, et al. Multidimensional characterization of global food supply from 1961 to 2013. *Nature Food*. 2020;1(1):70–5.
7. UNCTAD. Key Statistics and Trends in International Trade 2018; 2019.
8. Alliot C, Bartz D, Becheva S, et al. *Agrifood Atlas*. 2017. Available from www.boell .de/en/agrifood-atlas.
9. Herrero M, Thornton PK, Power B, et al. Farming and the geography of nutrient production for human use: a transdisciplinary analysis. *The Lancet Planet Health*. 2017;1(1):e33–42. doi: 10.1016/S2542-5196(17)30007-4.
10. Ren C, Liu S, van Grinsven H, et al. The impact of farm size on agricultural sustainability. *Journal of Cleaner Production*. 2019;220:357–67.
11. Masters WA, Djurfeldt AA, De Haan C, et al. Urbanization and farm size in Asia and Africa: implications for food security and agricultural research. *Global Food Security*. 2013;2(3):156–65. doi: 10.1016/j.gfs.2013.07.002.
12. Doss C. and the SOFA Team. The Role of Women in Agriculture. Rome: FAO; 2010.
13. Bharucha Z, Pretty J. The roles and values of wild foods in agricultural systems. *Philosophical Transactions of the Royal Society B: Biological Sciences*. 2010; 365(1554):2913–26. doi: 10.1098/rstb.2010.0123.
14. Alae-Carew C, Nicoleau S, Bird FA, et al. The impact of environmental changes on the yield and nutritional quality of fruits, nuts and seeds: a systematic review. *Environmental Research Letters*. 2020;15(2):023002. doi: 10.1088/1748-9326/ab5cc0.
15. Scheelbeek PFD, Bird FA, Tuomisto HL, et al. Effect of environmental changes on vegetable and legume yields and nutritional quality. *Proceedings of the National Academy of Sciences*. 2018;115(26):6804–9. doi: 10.1073/pnas.1800442115.
16. Zhu C, Kobayashi K, Loladze I, et al. Carbon dioxide (CO_2) levels this century will alter the protein, micronutrients, and vitamin content of rice grains with potential health consequences for the poorest rice-dependent countries. *Science Advances*. 2018;4(5):1–9. doi: 10.1126/sciadv.aaq1012.
17. Beach RH, Sulser TB, Crimmins A, et al. Combining the effects of increased atmospheric carbon dioxide on protein, iron, and zinc availability and projected climate change on global diets: a modelling study. *The Lancet Planet Health*. 2019;3(7):e307–17.
18. Searchinger T, Waite R, Hanson C, et al. *World Resources Report: Creating a Sustainable Food Future*. World Resources Institute, EcoAgiculture Partners; 2018.

19. Foley JA, Ramankutty N, Brauman KA, et al. Solutions for a cultivated planet. *Nature*. 2011;478(7369):337–42. doi: 10.1038/nature10452.
20. FAO, IFAD, WFP. *The State of Food Insecurity in the World. Meeting the 2015 International Hunger Targets: Taking Stock of Uneven Progress*. Rome: FAO; 2015.
21. FAO, IFAD, UNICEF, WFP, WHO. *The State of Food Security and Nutrition in the World: Transforming Food Systems for Affordable Healthy Diets. The State of the World*. Rome: FAO; 2020.
22. World Health Organization. *Overweight and Obesity*. Global Health Observatory (GHO) data.
23. Willett W, Rockström J, Loken B, et al. Food in the Anthropocene: the EAT–*Lancet* Commission on healthy diets from sustainable food systems. *The Lancet*. 2019; 6736:3–49. doi: 10.1016/S0140-6736(18)31788-4.
24. Zhong VW, Van Horn L, Greenland P, et al. Associations of processed meat, unprocessed red meat, poultry, or fish intake with incident cardiovascular disease and all-cause mortality. *JAMA Internal Medicine*. 2020;180(4):503–12.
25. Johnston BC, Zeraatkar D, Han MA, et al. Unprocessed red meat and processed meat consumption: dietary guideline recommendations from the nutritional recommendations (NUTRIRECS) consortium. *Annals of Internal Medicine*. 2019;171(10): 756–64. doi: 10.7326/m19-1621.
26. Zeraatkar D, Johnston BC, Bartoszko J, et al. Effect of lower versus higher red meat intake on cardiometabolic and cancer outcomes. *Annals of Internal Medicine*. 2019;171(10):721–31. doi: 10.7326/m19-0622.
27. Zeraatkar D, Han MA, Guyatt GH, et al. Red and processed meat consumption and risk for all-cause mortality and cardiometabolic outcomes. *Annals of Internal Medicine*. 2019;171(10):703–10. doi: 10.7326/m19-0655.
28. Naghshi S, Sadeghi O, Willett WC, Esmaillzadeh A. Dietary intake of total, animal, and plant proteins and risk of all cause, cardiovascular, and cancer mortality: systematic review and dose-response meta-analysis of prospective cohort studies. *BMJ*. 2020;370:m2412. doi: 10.1136/bmj.m2412.
29. Swinburn BA, Kraak VI, Allender S, et al. The global syndemic of obesity, undernutrition, and climate change: the *Lancet* Commission report. *The Lancet*. 2019; 393(10173):791–846. doi: 10.1016/S0140-6736(18)32822-8.
30. Singer M, Clair S. Syndemics and public health: reconceptualizing disease in biosocial context. *Medical Anthropology Quarterly*. 2003;17(4):423–41.
31. Hendrie GA, Baird D, Ridoutt B, Hadjikakou M, Noakes M. Overconsumption of energy and excessive discretionary food intake inflates dietary greenhouse gas emissions in Australia. *Nutrients*. 2016;8(11):690. doi: 10.3390/nu8110690.
32. Hadjikakou M. Trimming the excess: environmental impacts of discretionary food consumption in Australia. *Ecological Economics*. 2017;131:119–28.
33. Loring B, Robertson A. Obesity and Inequities: Guidance for Addressing Inequities in Overweight and Obesity. World Health Organization Europe; 2014 pp.1–6.
34. Steffen W, Richardson K, Rockström J, et al. Planetary boundaries: guiding human development on a changing planet. *Science*. 2015;347(6223):1259855.
35. Cucurachi S, Scherer L, Guinée J, Tukker A. Life cycle assessment of food systems. *One Earth*. 2019;1(3):292–7. doi: 10.1016/j.oneear.2019.10.014.
36. Shukla PR, Skea J, Calvo Buendia E, et al. *Climate Change and Land: An IPCC Special Report on Climate Change, Desertification, Land Degradation, Sustainable Land Management, Food Security, and Greenhouse Gas Fluxes in Terrestrial Ecosystems. Summary For Policymakers*; 2019. Available from www .ipcc.ch/srccl/.

37. Vermeulen S, Park T, Khoury CK, et al. Changing Diets and Transforming Food Systems. Working Paper No. 282. Wageningen, the Netherlands: CGIAR Research Program on Climate Change, Agriculture and Food Security (CCAFS); 2019.

38. Behrens P, Kiefte-de Jong JC, Bosker T, et al. Evaluating the environmental impacts of dietary recommendations. *Proceedings of the National Academy of Sciences.* 2017;114(51):201711889. doi: 10.1073/pnas.1711889114.

39. Teague WR, Apfelbaum S, Lal R, et al. The role of ruminants in reducing agriculture's carbon footprint in North America. *Journal of Soil and Water Conservation.* 2016;71(2):156–64. doi: 10.2489/jswc.71.2.156.

40. Clark M, Tilman D. Comparative analysis of environmental impacts of agricultural production systems, agricultural input efficiency, and food choice. *Environmental Research Letters.* 2017;12(6):064016. doi: 10.1088/1748-9326/aa6cd5.

41. Mcsherry ME, Ritchie ME. Effects of grazing on grassland soil carbon: a global review. *Global Change Biology.* 2013;19(5):1347–57. doi: 10.1111/gcb.12144.

42. Hayek MN, Garrett RD. Nationwide shift to grass-fed beef requires larger cattle population. *Environmental Research Letters.* 2018;13(8):84005.

43. Scheidel A, Del Bene D, Liu J, et al. Environmental conflicts and defenders: a global overview. *Global Environmental Change.* 2020;63:102104.

44. Díaz S, Settele J, Brondízio ES, et al., editors. *Summary for Policymakers of the Global Assessment Report on Biodiversity and Ecosystem Services of the Intergovernmental Science-Policy Platform on Biodiversity and Ecosystem Services.* IPBES Secretariat; 2019.

45. Borrini-Feyerabend G, Dudley N, Jaeger T, et al. Governance of Protected Areas: From Understanding to Action. Vol. 20, Best Practice Protected Area Guideline Series No. 20. IUCN; 2013.

46. Shukla PR, Skea J, Calvo Buendia E, et al. IPCC 2019: Summary for Policymakers. In *Climate Change and Land: An IPCC Special Report on Climate Change, Desertification, Land Degradation, Sustainable Land Management, Food Security, and Greenhouse Gas Fluxes in Terrestrial Ecosystems.* IPCC; 2020. pp. 1–36. Available from www.ipcc.ch/srccl/.

47. Montgomery DR. Soil erosion and agricultural sustainability. *Proceedings of the National Academy of Sciences.* 2007;104(33):13268–72. doi: 10.1073/pnas.0611508104.

48. Flandroy L, Poutahidis T, Berg G, et al. The impact of human activities and lifestyles on the interlinked microbiota and health of humans and of ecosystems. *Science of The Total Environment.* 2018;627:1018–38.

49. Kritee K, Nair D, Zavala-araiza D, et al. Changing diets and transforming food systems. *Land Use Policy.* 2019;9(Working Paper no. 282):1–6.

50. Lal R. Soil carbon sequestration to mitigate climate change. *Geoderma.* 2004; 123(1–2):1–22. doi: 10.1126/science.1097396.

51. Amundson R, Biardeau L. Opinion: Soil carbon sequestration is an elusive climate mitigation tool. *Proceedings of the National Academy of Sciences.* 2018;115(46): 11652–6. doi: 10.1073/pnas.1815901115.

52. Harris FB, Green R, Joy E, Haines A, Dangour A. The water use of diets in India. *Annals of Global Health.* 2017;83(1):89–90.

53. Steward DR, Bruss PJ, Yang X, et al. Tapping unsustainable groundwater stores for agricultural production in the High Plains Aquifer of Kansas, projections to 2110. *Proceedings of the National Academy of Sciences.* 2013;110(37):E3477–86.

54. Green RF, Joy EJM, Harris F, et al. Greenhouse gas emissions and water footprints of typical dietary patterns in India. *Science of The Total Environment.* 2018;643: 1411–18. doi: 10.1016/j.scitotenv.2018.06.258.

55. Gadde B, Bonnet S, Menke C, Garivait S. Air pollutant emissions from rice straw open field burning in India, Thailand and the Philippines. *Environmental Pollution.* 2009;157(5):1554–8. doi: 10.1016/j.envpol.2009.01.004.

56. Wang Z, Zhao J, Xu J, et al. Influence of straw burning on urban air pollutant concentrations in northeast China. *International Journal of Environmental Research and Public Health.* 2019;16(8):1379. doi: 10.3390/ijerph16081379.

57. Bauer SE, Tsigaridis K, Miller R. Significant atmospheric aerosol pollution caused by world food cultivation. *Geophysical Research Letters.* 2016;43(10):5394–400.

58. Lelieveld J, Evans JS, Fnais M, Giannadaki D, Pozzer A. The contribution of outdoor air pollution sources to premature mortality on a global scale. *Nature.* 2015;525(7569):367–71. doi: 10.1038/nature15371.

59. Springmann M, Clark M, Mason-D'Croz D, et al. Options for keeping the food system within environmental limits. *Nature.* 2018;562(7728):519–25.

60. Krishna Bahadur K, Dias GM, Veeramani A, et al. When too much isn't enough: does current food production meet global nutritional needs? *PLoS One.* 2018;13(10):e0205683.

61. FAO. *Food Wastage Footprint. Impacts on Natural Resources.* Rome: FAO; 2013.

62. Ishangulyyev R, Kim S, Lee SH. Understanding food loss and waste: why are we losing and wasting food? *Foods.* 2019;8(8). doi: 10.3390/foods8080297.

63. Cassidy ES, West PC, Gerber JS, Foley JA. Redefining agricultural yields: from tonnes to people nourished per hectare. *Environmental Research Letters.* 2013;8(3).

64. Burlingame B, Dernini S, eds. *Sustainable Diets and Biodiversity: Directions and Solutions for Policy, Research and Action. Proceedings of the International Scientific Symposium, Biodiversity and Sustainable Diets United against Hunger*, 3–5 November, 2020. Rome: Food and Agriculture Organization of the United Nations and Bioversity International, 2010. Available from www.bioversityinternational.org/e-library/publications/detail/sustainable-diets-and-biodiversity/.

65. Aleksandrowicz L, Green R, Joy EJM, et al. Environmental impacts of dietary shifts in India: a modelling study using nationally-representative data. *Environment International.* 2019;126:207–15. doi: 10.1016/j.envint.2019.02.004.

66. Aleksandrowicz L, Green R, Joy EJM, Smith P, Haines A. The impacts of dietary change on greenhouse gas emissions, land use, water use, and health: a systematic review. *PLoS One.* 2016;11(11):e0165797. doi: 10.1371/journal.pone.0165797.

67. Ridoutt BG, Hendrie GA, Noakes M. Dietary strategies to reduce environmental impact: a critical review of the evidence base. *Advances in Nutrition.* 2017;8(6):933–46.

68. Scheelbeek P, Green R, Papier K, et al. Health impacts and environmental footprints of diets that meet the Eatwell Guide recommendations: analyses of multiple UK studies. *BMJ Open.* 2020;10:37554. doi: 10.1136/bmjopen-2020-037554.

69. Milner J, Joy EJM, Green R, et al. Projected health effects of realistic dietary changes to address freshwater constraints in India: a modelling study. *The Lancet Planetary Health.* 2017;1(1):e26–32. doi: 10.1016/S2542-5196(17)30001-3.

70. Kayatz B, Harris F, Hillier J, et al. 'More crop per drop': exploring India's cereal water use since 2005. *Science of The Total Environment.* 2019;673:207–17.

71. Harris F, Moss C, Joy EJM, et al. The water footprint of diets: a global systematic review and meta-analysis. *Advances in Nutrition.* 2020;11(2):375–86.

72. Hirvonen K, Bai Y, Headey D, Masters WA. Affordability of the EAT–*Lancet* reference diet: a global analysis. *The Lancet Global Health.* 2020;8(1):e59–66.

73. Garcia D, Galaz V, Daume S. EATLancet vs yes2meat: the digital backlash to the planetary health diet. *The Lancet.* 2019;394(10215):2153–4.

74. World Health Organization. A Healthy Diet Sustainably Produced. 2018. Available from www.who.int/publications/i/item/WHO-NMH-NHD-18.12

75. Gonzalez Fischer C, Garnett T. *Plates, Pyramids and Planets. Developments in National Healthy and Sustainable Dietary Guidelines: A State of Play Assessment.* Rome: FAO; 2016.

76. Springmann M, Spajic L, Clark MA, et al. The healthiness and sustainability of national and global food based dietary guidelines: modelling study. *BMJ.* 2020;370: 2322. doi: 10.1136/bmj.m2322.

77. Blake P, Durão S, Naude CE, Bero L. An analysis of methods used to synthesize evidence and grade recommendations in food-based dietary guidelines. *Nutrition Reviews.* 2018;76(4):1–11. doi: 10.1093/nutrit/nux074.

78. Hedin B, Katzeff C, Eriksson E, Pargman D. A systematic review of digital behaviour change interventions for more sustainable food consumption. *Sustain.* 2019;11(9).

79. Katz DL, Meller S. Can we say what diet is best for health? *Annual Review of Public Health.* 2014;35:83–103. doi: 10.1146/annurev-publhealth-032013-182351.

80. Webb D, Byrd-Bredbenner C. Overcoming consumer inertia to dietary guidance. *Advances in Nutrition.* 2015;6(4):391–6. doi: 10.3945/an.115.008441.

81. Merrigan K, Griffin T. Building a case, over time, for adding sustainability to nutritional guidelines. The Conversation. 13 October 2015. Available from https://theconversation.com/building-a-case-over-time-for-adding-sustainability-to-nutritional-guidelines-48556.

82. Browne S, Minozzi S, Bellisario C, Sweeney MR, Susta D. Effectiveness of interventions aimed at improving dietary behaviours among people at higher risk of or with chronic non-communicable diseases: an overview of systematic reviews. *European Journal of Clinical Nutrition.* 2019;73:9–23. doi: 10.1038/s41430-018-0327-3.

83. Ashton LM, Sharkey T, Whatnall MC, et al. Effectiveness of interventions and behaviour change techniques for improving dietary intake in young adults: a systematic review and meta-analysis of RCTs. *Nutrients.* 2019;11(4):825.

84. Marteau TM, Hollands GJ, Fletcher PC. Changing human behavior to prevent disease: the importance of targeting automatic processes. *Science.* 2012;337(6101): 1492–5. doi: 10.1126/science.1226918.

85. Shemilt I, Hollands GJ, Marteau TM, et al. Economic instruments for population diet and physical activity behaviour change: a systematic scoping review. *PLoS One.* 2013;8(9):e75070. doi: 10.1371/journal.pone.0075070.

86. Afshin A, Peñalvo JL, Gobbo L, et al. The prospective impact of food pricing on improving dietary consumption: a systematic review and meta-analysis. *PLoS One.* 2017;12(3). doi: 10.1371/journal.pone.0172277.

87. Green R, Cornelsen L, Dangour AD, et al. The effect of rising food prices on food consumption: systematic review with meta-regression. *BMJ.* 2013;347(7915):1–9.

88. Thaler R, Sunstein CR. *Nudge: Improving Decisions about Health, Wealth, and Happiness.* New York: Penguin Books; 2009.

89. Thaler R. *Misbehaving: The Making of Behavioral Economics.* New York: W.W. Norton; 2015.

90. Skov LR, Lourenço S, Hansen GL, Mikkelsen BE, Schofield C. Choice architecture as a means to change eating behaviour in self-service settings: a systematic review. *Obesity Reviews.* 2013;14(3):187–96. doi: 10.1111/j.1467-789X.2012.01054.x.

91. Lehner M, Mont O, Heiskanen E. Nudging – a promising tool for sustainable consumption behaviour? *Journal of Cleaner Production.* 2016;134:166–77.

92. Marteau TM, Ogilvie D, Roland M, Suhrcke M, Kelly MP. Judging nudging: can nudging improve population health? *BMJ.* 2011;342(7791):263–5.

93. Rust NA, Ridding L, Ward C, et al. How to transition to reduced-meat diets that benefit people and the planet. *Science of The Total Environment.* 2020;718:137208.

94. Garnett EE, Balmford A, Sandbrook C, Pilling MA, Marteau TM. Impact of increasing vegetarian availability on meal selection and sales in cafeterias. *Proceedings of the National Academy of Sciences*. 2019;116(42):20923–9. doi: 10.1073/pnas.1907207116.

95. Bacon L, Wise J, Attwood S, Vennard D. The language of sustainable diets: a field study exploring the impact of renaming vegetarian dishes on U.K. café menus. World Resources Institute; 2018.

96. Ranganathan J, Vennard D, Waite R, et al. Shifting diets for a sustainable food future. Installment 11 of Creating a Sustainable Food Future. World Resources Institute; 2016.

97. Fischer J, Abson DJ, Butsic V, et al. Land sparing versus land sharing: moving forward. *Conservation Letters*. 2014;7(3):149–57. doi: 10.1111/conl.12084.

98. Usubiaga-Liaño A, Mace GM, Ekins P. Limits to agricultural land for retaining acceptable levels of local biodiversity. *Nature Sustainability*. 2019;2(6):491–8.

99. Xie Z, Game ET, Hobbs RJ, et al. Conservation opportunities on uncontested lands. *Nature Sustainability*. 2019;3:9–15.

100. Garnett T, Godde C, Muller A, et al. Grazed and confused? Ruminating on cattle, grazing systems, methane, nitrous oxide, the soil carbon sequestration question – and what it all means for greenhouse gas emissions. Food Climate Research Network, University of Oxford; 2017.

101. Roe S, Streck C, Obersteiner M, et al. Contribution of the land sector to a 1.5 °C world. *Nature Climate Change*. 2019;9(11):817–28. doi: 10.1038/s41558-019-0591-9.

102. Smith P, Adams J, Beerling DJ, et al. Land-management options for greenhouse gas removal and their impacts on ecosystem services and the Sustainable Development Goals. *Annual Review of Environment and Resources*. 2019;7:40.

103. Norton M, Baldi A, Buda V, et al. Serious mismatches continue between science and policy in forest bioenergy. *GCB Bioenergy*. 2019;11(11):1256–63.

104. Walsh BJ, Rydzak F, Palazzo A, et al. New feed sources key to ambitious climate targets. *Carbon Balance Management*. 2015;10(1).

105. Turner PA, Field CB, Lobell DB, Sanchez DL, Mach KJ. Unprecedented rates of land-use transformation in modelled climate change mitigation pathways. *Nature Sustainability*. 2018;1(5):240–5. doi: 10.1038/s41893-018-0063-7.

106. Fuhrman J, McJeon H, Patel P, et al. Food–energy–water implications of negative emissions technologies in a +1.5 °C future. *Nature Climate Change*. 2020;10:920–7.

107. Madeira MS, Cardoso C, Lopes PA, et al. Microalgae as feed ingredients for livestock production and meat quality: a review. *Livestock Science*. 2017;205:111–21.

108. Fajardy M, Koberle A, Mac Dowell N, Fantuzzi A. BECCS Deployment: A Reality Check. Grantham Institute: Briefing Paper 28; 2019.

109. West TAP, Börner J, Sills EO, Kontoleon A. Overstated carbon emission reductions from voluntary REDD+ projects in the Brazilian Amazon. *Proceedings of the National Academy of Sciences*. 2020;117(39):202004334.

110. Fridahl M, Lehtveer M. Bioenergy with carbon capture and storage (BECCS): global potential, investment preferences, and deployment barriers. *Energy Research and Social Science*. 2018;42:155–65. doi: 10.1016/j.erss.2018.03.019.

111. Mader S. Plant trees for the planet: the potential of forests for climate change mitigation and the major drivers of national forest area. *Mitigation and Adaptation Strategies for Global Change*. 2019;25(4):519–36. doi: 10.1007/s11027-019-09875-4.

112. Herrera D, Ellis A, Fisher B, et al. Upstream watershed condition predicts rural children's health across 35 developing countries. *Nature Communications*. 2017;8(1):811.

113. Bauch SC, Birkenbach AM, Pattanayak SK, Sills EO. Public health impacts of ecosystem change in the Brazilian Amazon. *Proceedings of the National Academy of Sciences*. 2015;112(24):7414–19. doi: 10.1073/pnas.1406495111.

114. Deng L, Zhu GY, Tang ZS, Shangguan ZP. Global patterns of the effects of land-use changes on soil carbon stocks. *Global Ecology and Conservation*. 2016;5:127–38.

115. Lal R. Digging deeper: a holistic perspective of factors affecting soil organic carbon sequestration in agroecosystems. *Global Change Biology*. 2018;24(8):3285–301.

116. Sykes AJ, Macleod M, Eory V, et al. Characterising the biophysical, economic and social impacts of soil carbon sequestration as a greenhouse gas removal technology. *Global Change Biology*. 2020;26(3):1085–108. doi: 10.1111/gcb.14844.

117. Follett RF, Stewart CE, Pruessner EG, Kimble JM. Effects of climate change on soil carbon and nitrogen storage in the US Great Plains. *Journal of Soil and Water Conservation*. 2012;67(5):331–42. doi: 10.2489/jswc.67.5.331.

118. Aneja VP, Schlesinger WH, Li Q, Nahas A, Battye WH. Characterization of atmospheric nitrous oxide emissions from global agricultural soils. *SN Applied Sciences*. 2019;1(12):1–11. doi: 10.1007/s42452-019-1688-5.

119. Lecours N, Almeida GEG, Abdallah JM, Novotny TE. Environmental health impacts of tobacco farming: a review of the literature. *Tobacco Control*. 2012;21(2):191–6.

120. Geist HJ. Global assessment of deforestation related to tobacco farming. *Tobacco Control*. 1999;8:18–28.

121. Novotny TE, Bialous SA, Burt L, et al. The environmental and health impacts of tobacco agriculture, cigarette manufacture and consumption. *Bulletin of the World Health Organization*. 2015;93(12):877–80. doi: 10.2471/blt.15.152744.

122. Davis KF, Rulli MC, Seveso A, D'Odorico P. Increased food production and reduced water use through optimized crop distribution. *Nature Geoscience*. 2017;10(12):919–24.

123. Huang J, Ridoutt BG, Sun Z, et al. Balancing food production within the planetary water boundary. *Journal of Cleaner Production*. 2020;253:119900.

124. Ijumba JN, Lindsay SW. Impact of irrigation on malaria in Africa: paddies paradox. *Medical and Veterinary Entomology*. 2001;15(1):1–11.

125. Godfray HCJ, Garnett T. Food security and sustainable intensification. *Philosophical Transactions of the Royal Society B: Biological Sciences*. 2014;369(1639): 20120273. doi: 10.1098/rstb.2012.0273.

126. Campbell BM, Thornton P, Zougmoré R, van Asten P, Lipper L. Sustainable intensification: what is its role in climate smart agriculture? *Current Opinion in Environmental Sustainability*. 2014;8:39–43. doi: 10.1016/j.cosust.2014.07.002.

127. Montpelier Panel. Sustainable Intensification: A New Paradigm for African Agriculture. Ag4Impact; 2013.

128. Andrews O, Le Quéré C, Kjellstrom T, Lemke B, Haines A. Implications for workability and survivability in populations exposed to extreme heat under climate change: a modelling study. *The Lancet Planetary Health*. 2018;2(12):e540–7.

129. Poore J, Nemecek T. Reducing food's environmental impacts through producers and consumers. *Science*. 2018;360(6392):987–92. doi: 10.1126/science.aaq0216.

130. Cazzolla Gatti R, Velichevskaya A. Certified 'sustainable' palm oil took the place of endangered Bornean and Sumatran large mammals habitat and tropical forests in the last 30 years. *Science of The Total Environment*. 2020;742:140712.

131. Waldman KB, Kerr JM. Limitations of certification and supply chain standards for environmental protection in commodity crop production. *Annual Review of Resource Economics*. 2014;6(1):429–49.

132. Rosenstock TS, Wilkes A, Jallo C, et al. Making trees count: measurement and reporting of agroforestry in UNFCCC national communications of non-Annex I countries. *Agriculture, Ecosystems & Environment*. 2019;284:106569.

133. Kuyah S, Öborn I, Jonsson M, et al. Trees in agricultural landscapes enhance provision of ecosystem services in Sub-Saharan Africa. *International Journal of Biodiversity Science*. 2016;12(4):255–73.

134. Rosenstock TS, Dawson IK, Aynekulu E, et al. A planetary health perspective on agroforestry in Sub-Saharan Africa. *One Earth*. 2019;1(3):330–44.

135. Garrity DP, Akinnifesi FK, Ajayi OC, et al. Evergreen agriculture: a robust approach to sustainable food security in Africa. *Food Security*. 2010;2(3):197–214.

136. Smith LG, Kirk GJD, Jones PJ, Williams AG. The greenhouse gas impacts of converting food production in England and Wales to organic methods. *Nature Communications*. 2019;10(1):1–10. doi: 10.1038/s41467-019-12622-7.

137. Dangour AD, Dodhia SK, Hayter A, et al. Nutritional quality of organic foods: a systematic review, *American Journal of Clinical Nutrition*. 2009;90(3):680–5.

138. Dangour AD, Lock K, Hayter A, et al. Nutrition-related health effects of organic foods: a systematic review. *American Journal of Clinical Nutrition*. 2010;92(1):203–10.

139. Smith-Spangler C, Brandeau ML, Hunter GE, et al. Are organic foods safer or healthier than conventional alternatives? A systematic review. *Annals of Internal Medicine*. 2012;157:348–66. doi: 10.7326/0003-4819-157-5-201209040-00007.

140. Curl CL, Beresford SAA, Fenske RA, et al. Estimating pesticide exposure from dietary intake and organic food choices: the multi-ethnic study of atherosclerosis (MESA). *Environmental Health Perspectives*. 2014;123(5):475–83.

141. Giannadaki D, Giannakis E, Pozzer A, Lelieveld J. Estimating health and economic benefits of reductions in air pollution from agriculture. *Science of The Total Environment*. 2018;622–3:1304–16. doi: 10.1016/j.scitotenv.2017.12.064.

142. Scholz RW, Ulrich AE, Eilittä M, Roy A. Sustainable use of phosphorus: a finite resource. *Science of The Total Environment*. 2013;461–2:799–803.

143. Cordell D, Neset TSS. Phosphorus vulnerability: a qualitative framework for assessing the vulnerability of national and regional food systems to the multi-dimensional stressors of phosphorus scarcity. *Global Environmental Change*. 2014;24(1):108–22.

144. Schneider KD, Thiessen Martens JR, Zvomuya F, et al. Options for improved phosphorus cycling and use in agriculture at the field and regional scales. *Journal of Environmental Quality*. 2019;48(5):1247–64. doi: 10.2134/jeq2019.02.0070.

145. Powers SM, Chowdhury RB, MacDonald GK, et al. Global opportunities to increase agricultural independence through phosphorus recycling. *Earth's Future*. 2019;7(4):370–83. doi: 10.1029/2018EF001097.

146. Golroudbary SR, El Wali M, Kraslawski A. Environmental sustainability of phosphorus recycling from wastewater, manure and solid wastes. *Science of The Total Environment*. 2019;672:515–24. doi: 10.1016/j.scitotenv.2019.03.439.

147. Mills G, Sharps K, Simpson D, et al. Ozone pollution will compromise efforts to increase global wheat production. *Global Change Biology*. 2018;24(8):3560–74.

148. Shindell D, Kuylenstierna JCI, Vignati E, et al. Simultaneously mitigating near-term climate change and improving human health and food security. *Science*. 2012;335(6065):183–9. doi: 10.1126/science.1210026.

149. Health and Safety Executive. The Expert Committee on Pesticide Residues in Food (PRiF) Annual Report 2017; 2017.

150. Nicolopoulou-Stamati P, Maipas S, Kotampasi C, Stamatis P, Hens L. Chemical pesticides and human health: the urgent need for a new concept in agriculture. *Frontiers in Public Health*. 2016;4:148. doi: 10.3389/fpubh.2016.00148.

151. Bergman Å, Heindel JJ, Jobling S, Kidd KA, Zoeller TR. The State of Science of Endocrine Disrupting Chemicals 2012. Summary for Decision-Makers. Inter-organization programme for the Sound Management of Chemicals. UNEP, WHO; 2012.

152. Richardson JR, Fitsanakis V, Westerink RHS, Kanthasamy AG. Neurotoxicity of pesticides. *Acta Neuropathologica*. 2019;138:343–62. doi: 10.1007/s00401-019-02033-9.

153. Kim K-H, Kabir E, Jahan SA. Exposure to pesticides and the associated human health effects. *Science of The Total Environment*. 2017;575:525–35.

154. Cimino AM, Boyles AL, Thayer KA, Perry MJ. Effects of neonicotinoid pesticide exposure on human health: a systematic review. *Environmental Health Perspectives.* 2017;125(2):155–62. doi: 10.1289/ehp515.

155. Whitty CJM, Jones M, Tollervey A, Wheeler T. Biotechnology: Africa and Asia need a rational debate on GM crops. *Nature.* 2013;497:31–3. doi: 10.1038/497031a.

156. Kettenburg AJ, Hanspach DJ, Abson J, Fischer J. From disagreements to dialogue: unpacking the Golden Rice debate. *Sustainable Science.* 2018;13:1469–82.

157. Hunter J, Duff G. GM crops – lessons from medicine. *Science.* 2016;353(6305):1187.

158. Domingo JL. Safety assessment of GM plants: an updated review of the scientific literature. *Food and Chemical Toxicology.* 2016;95:12–18.

159. Bailey-Serres J, Parker JE, Ainsworth EA, Oldroyd GED, Schroeder JI. Genetic strategies for improving crop yields. *Nature.* 2019;575:109–18.

160. Septiningsih EM, Pamplona AM, Sanchez DL, et al. Development of submergence-tolerant rice cultivars: the Sub1 locus and beyond. *Annals of Botany.* 2009;103(2): 151–60.

161. Sarkar RK, Bhattacharjee B. Rice genotypes with SUB1 QTL differ in submergence tolerance, elongation ability during submergence and re-generation growth at re-emergence. *Rice.* 2011;5(1).

162. Zhao C, Liu B, Piao S, et al. Temperature increase reduces global yields of major crops in four independent estimates. *Proceedings of the National Academy of Sciences.* 2017;114(35):9326–31. doi: 10.1073/pnas.1701762114.

163. Qaim M, Kouser S. Genetically modified crops and food security. *PLoS One.* 2013;8(6):e64879. doi: 10.1371/journal.pone.0064879.

164. Godfray HCJ, Beddington JR, Crute IR, et al. Food security: the challenge of feeding 9 billion people. *Science.* 2010;327(5967):812–18.

165. Golden CD, Fernald LCH, Brashares JS, Rasolofoniaina BJR, Kremen C. Benefits of wildlife consumption to child nutrition in a biodiversity hotspot. *Proceedings of the National Academy of Sciences.* 2011;108(49):19653–6.

166. Brashares JS, Arcese P, Sam MK, et al. Bushmeat hunting, wildlife declines, and fish supply in West Africa. *Science.* 2004;306(5699):1180–3.

167. Mabhaudhi T, Chibarabada TP, Chimonyo VGP, et al. Mainstreaming underutilized indigenous and traditional crops into food systems: a South African perspective. *Sustainability.* 2018;11(1). doi: 10.3390/su11010172.

168. Whitmee S, Haines A, Beyrer C, et al. Safeguarding human health in the Anthropocene Epoch: report of The Rockefeller Foundation–*Lancet* Commission on Planetary Health. *The Lancet.* 2015;386:1973–2028. doi: 10.1016/S0140-6736(15)60901-1.

169. International Nut & Dried Fruit Council. Nuts & Dried Fruits Statistical Yearbook 2017/2018; 2018.

170. Vanham D, Mekonnen MM, Hoekstra AY. Treenuts and groundnuts in the EAT-Lancet reference diet: concerns regarding sustainable water use. *Global Food Security.* 2020;24:100357. doi: 10.1016/j.gfs.2020.100357.

171. Van Huis A. Edible insects are the future? In Conference on 'The Future of Animal Products in the Human Diet: Health and Environmental Concerns'. *Proceedings of the Nutrition Society.* 2016;75(3):294–305.

172. Stull V, Patz J. Research and policy priorities for edible insects. *Sustainability Science.* 2019;15:633–45. doi: 10.1007/s11625-019-00709-5.

173. Stull VJ, Finer E, Bergmans RS, et al. Impact of edible cricket consumption on gut microbiota in healthy adults, a double-blind, randomized crossover trial. *Science Reports.* 2018;8(1):1–13. doi: 10.1038/s41598-018-29032-2.

174. Tuorila H, Hartmann C. Consumer responses to novel and unfamiliar foods. *Current Opinion in Food Science.* 2020;33:1–8. doi: 10.1016/j.cofs.2019.09.004.

175. Appenroth KJ, Sowjanya Sree K, Bog M, et al. Nutritional value of the duckweed species of the Genus *Wolffia* (Lemnaceae) as human food. *Frontiers in Chemistry*. 2018;6.

176. de Beukelaar MFA, Zeinstra GG, Mes JJ, Fischer ARH. Duckweed as human food. The influence of meal context and information on duckweed acceptability of Dutch consumers. *Food Quality and Preference*. 2019;71:76–86.

177. Kaplan A, Zelicha H, Tsaban G, et al. Protein bioavailability of *Wolffia globosa* duckweed, a novel aquatic plant – a randomized controlled trial. *Clinical Nutrition*. 2019;38(6):2576–82.

178. Sillman J, Nygren L, Kahiluoto H, et al. Bacterial protein for food and feed generated via renewable energy and direct air capture of CO_2: can it reduce land and water use? *Global Food Security*. 2019;22:25–32. doi: 10.1016/j.gfs.2019.09.007.

179. Alexander P, Brown C, Arneth A, et al. Could consumption of insects, cultured meat or imitation meat reduce global agricultural land use? *Global Food Security*. 2017;15:22–32. doi: 10.1016/j.gfs.2017.04.001.

180. Gelsomin E. Impossible and beyond: how healthy are these meatless burgers? Harvard Health blog. 2019. Available from www.health.harvard.edu/blog/impossible-and-beyond-how-healthy-are-these-meatless-burgers-2019081517448.

181. Parodi A, Leip A, De Boer IJM, et al. The potential of future foods for sustainable and healthy diets. *Nature Sustainability*. 2018;1(12):782–9.

182. High Level Panel of Experts on World Food Security. *Sustainable Fisheries and Aquaculture for Food Security and Nutrition*. Rome: FAO; 2014 (June):1–119.

183. Golden C. Fall in fish catch threatens human health. *Nature*. 2016;534:317–20.

184. Naylor RL, Hardy RW, Bureau DP, et al. Feeding aquaculture in an era of finite resources. *Proceedings of the National Academy of Sciences*. 2009;106(36): 15103–10. doi: 10.1073/pnas.0905235106.

185. Tacon AGJ, Metian M. Feed matters: satisfying the feed demand of aquaculture. *Reviews in Fisheries Science and Aquaculture*. 2015;23(1):1–10.

186. Troell M, Naylor RL, Metian M, et al. Does aquaculture add resilience to the global food system? *Proceedings of the National Academy of Sciences*. 2014;111(37): 13257–63. doi: 10.1073/pnas.1404067111.

187. Caswell BL, Talegawkar SA, Siamusantu W, West KP, Palmer AC. A 10-food group dietary diversity score outperforms a 7-food group score in characterizing seasonal variability and micronutrient adequacy in rural Zambian children. *Journal of Nutrition*. 2018;148(1):131–9.

188. Food and Land Use Coalition. Growing Better: Ten Critical Transitions to Transform Food and Land Use. The Global Consultation Report of the Food and Land Use Coalition; 2019.

189. Springmann M, Mason-D'Croz D, Robinson S, et al. Mitigation potential and global health impacts from emissions pricing of food commodities. *Nature Climate Change*. 2017;7(1):69–74. doi: 10.1038/nclimate3155.

190. Scheelbeek PFD, Cornelsen L, Marteau TM, Jebb SA, Smith RD. Potential impact on prevalence of obesity in the UK of a 20% price increase in high sugar snacks: modelling study. *BMJ*. 2019;366. doi: 10.1136/bmj.l4786.

191. Hasegawa T, Fujimori S, Takahashi K, Masui T. Scenarios for the risk of hunger in the twenty-first century using Shared Socioeconomic Pathways. *Environmental Research Letters*. 2015;10(1):14010. doi: 10.1088/1748-9326/10/1/014010.

192. European Coordination Via Campesina. Food Sovereignty Now! A Guide to Food Sovereignty; 2018.

11

The Role of Health Professionals in Fostering Planetary Health

Environmental change will pose numerous challenges to health systems, as described in earlier chapters. They will need to become more resilient to shocks, including extreme events, and be able to detect and respond to changing patterns of disease (see Chapter 5). This chapter describes four major ways in which health professionals can catalyse rapid decarbonization of the economy and support moves to live within planetary boundaries while protecting health: reducing the burden of preventable ill-health; reducing the environmental impact of health care; contributing to slowing population growth; and providing broader societal leadership.

Reducing the Burden of Preventable Ill-Health

The clearest way to reduce the need for health care, and therefore to avoid the environmental impact associated with delivering that care, is to keep people well in the first place. Prevention of disease and suffering begins with addressing root causes – poverty and social inequities together with the environmental drivers of ill-health (1). Many other preventive strategies occur upstream from the health sector; examples include reducing air pollution (Chapter 8), overhauling urban transportation systems (Chapter 9), and reforming agricultural practices and food policy (Chapter 10). Other preventive strategies are pursued within health care settings, as described below (2). While there are compelling ethical and financial reasons for preventing illness, such efforts also help reduce the environmental footprint of health care.

Primary Health Care, Prevention, and the Environment

Achieving universal health coverage (UHC) (SDG 3) is fundamental to Planetary Health, and primary health care is the essential platform on which UHC depends. The original conception of primary health care (PHC), as articulated in the landmark Alma Ata declaration of 1978, encompassed addressing economic and social development to attain good health for all, reducing health inequities within and between countries, supporting international collaboration, as well as providing access to essential health care. PHC was envisioned as integrating promotive, preventive, curative, and rehabilitative services.

The declaration also addressed the imperative of technology transfer, economic justice, and the regulation of health-harming activities by transnational corporations. At a minimum PHC encompassed 'education concerning prevailing health problems and the methods of preventing and controlling them; promotion of food supply and proper nutrition; an adequate supply of safe water and basic sanitation; maternal and child health care, including family planning; immunization against the major infectious diseases; prevention and control of locally endemic diseases; appropriate treatment of common diseases and injuries, and provision of essential drugs' (3). Progress was hampered by the political divisions of the Cold War era and from the outset there was contestation between those promoting the comprehensive approach outlined at Alma Ata and a selective primary care approach based on a limited number of priority interventions (4).

In the latest iteration of PHC in 2018, when the 40th anniversary of PHC was celebrated at Nur-Sultan, Kazakhstan (then known as Astana), much of the compelling language of the original Alma Ata declaration and its inspiring vision of PHC was lost. The focus was on health care delivery, reflecting priorities in many middle- and high-income countries (5).

Nevertheless, strengthening PHC will be essential to addressing the challenges of the Anthropocene, not least because it provides care close to where people live and aims to reduce the need for specialist care with its attendant high costs and environmental impacts. A well-functioning primary health care system provides the first point of contact with the health care system in most settings. It integrates care of patients suffering from a range of conditions with preventive activities such as immunization, screening, and health promotion. Depending on the level of resources, primary health care (often shortened to 'primary care' in high- and middle-income settings to reflect the focus on clinical care) may be delivered by varying combinations of community health workers, nurses, and family doctors, often working in teams.

There are many potential roles of primary care professionals in addressing the challenges of climate change and Planetary Health (6). One is the promotion of 'low carbon' and healthy behaviours such as healthy diets, active travel, and the uptake of clean energy. Evidence for the impact of behaviour change intervention in primary care is equivocal; effectiveness is better documented in patients with pre-existing disease such as hypertension and diabetes than in the general population (7, 8). However, there is some evidence that if health care providers themselves model healthy behaviours, their counselling is more effective (9, 10); this could suggest a rationale for physicians modelling sustainable and healthy practices in their own lives. Digital behaviour-change strategies may have a role in promoting healthy, sustainable choices, but a recent review of their use in promoting more sustainable food consumption shows that many studies are poorly designed (11). Interventions will need to be complemented by a supportive policy environment that incentivizes the desired behaviour changes, together with empowerment of people to have greater control over making healthy decisions (12). Many of these roles are expressed in a proposed Planetary Health pledge for health professionals, published in 2020 (**Box 11.1**).

Box 11.1. **A Planetary Health Pledge for Health Professionals in the Anthropocene**

- I solemnly pledge to dedicate my life to the service of humanity, and to the protection of natural systems on which human health depends.
- The health of people, their communities, and the planet will be my first consideration and I will maintain the utmost respect for human life, as well as reverence for the diversity of life on Earth.
- I will practise my profession with conscience and dignity and in accordance with good practice, taking into account Planetary Health values and principles.
- To do no harm, I will respect the autonomy and dignity of all persons in adopting an approach to maintaining and creating health which focuses on prevention of harm to people and planet.
- I will respect and honour the trust that is placed in me and leverage this trust to promote knowledge, values, and behaviours that support the health of humans and the planet.
- I will actively strive to understand the impact that direct, unconscious, and structural bias may have on my patients, communities, and the planet, and for cultural self-awareness in my duty to serve.
- I will advocate for equity and justice by actively addressing environmental, social, and structural determinants of health while protecting the natural systems that underpin a viable planet for future generations.
- I will acknowledge and respect diverse sources of knowledge and knowing regarding individual, community, and Planetary Health such as from Indigenous traditional knowledge systems while challenging attempts at spreading disinformation that can undermine Planetary Health.
- I will share and expand my knowledge for the benefit of society and the planet; I will also actively promote transdisciplinary, inclusive action to achieve individual, community, and Planetary Health.
- I will attend to my own health, well-being, and abilities in order to provide care and serve the community to the highest standards.
- I will strive to be a role model for my patients and society by embodying Planetary Health principles in my own life, acknowledging that this requires maintaining the vitality of our common home.
- I will not use my knowledge to violate human rights and civil liberties, even under threat; recognizing that the human right to health necessitates maintaining Planetary Health.
- I make these promises solemnly, freely, and upon my honour. By taking this pledge, I am committing to a vision of personal, community, and Planetary Health that will enable the diversity of life on our planet to thrive now and in the future.

Source: Wabnitz K-J, Gabrysch S, Guinto R, et al. A pledge for planetary health to unite health professionals in the Anthropocene. *The Lancet*. 2020. doi: 10.1016/S0140-6736(20)32039-0.

An effective health care system, grounded in primary care, can also help reduce the use of ineffective treatments – another key strategy that advances both health and environmental performance. This means identifying and eliminating medicines and surgical procedures that do no good, using the methods of clinical epidemiology (13). Such treatments are far too often incentivized by the profit motive, which rewards treatment rather than health maintenance (albeit far more in some countries than in others) (14–16). The high carbon emissions associated with some pharmaceuticals and surgical procedures (see below) indicate the potential for environmental and health benefits from reducing unnecessary interventions. Inpatient end-of-life care is resource- and energy-intensive; palliative care provides better symptom control, coordination of care and improved communication between professionals and the patient and family, at lower cost (17, 18). As with primary prevention, there are compelling humanitarian and economic reasons to provide palliative care, and the environmental benefits provide additional rationale.

The Environmental Footprint of Health Care

Health care is a large economic sector, accounting for about US\$7.2 trillion in annual spending, or about 10% of global GDP, and it has a large environmental footprint. A recent report from the NGO Health Care Without Harm (HCWH) and the engineering and design firm Arup assessed the GHG emissions from health care based on data from 43 countries up to 2014, the most recent year for which data were available (19). The study used multi-regional input–output (MRIO) analysis, which tracks flows of goods and services from different sectors of the economy into the health sector, monetizes these flows, links monetary accounts to GHG emissions in each sector, and allocates 'embedded' carbon emissions to the health sector (20). It found that the health sector accounts for about 4.4% of global net emissions, or 2 GtCO$_2$eq – equivalent to the emissions of about 514 coal-powered plants. If the global health sector were a country it would rank fifth in GHG emissions after China, the USA, India, and Russia. The US health care sector is the highest GHG emitter in both per capita and absolute terms, and the top three emitters, the USA, China, and the EU countries, contribute 56% of estimated global health care emissions.

A more recent analysis considered not only GHG emissions but also other environmental impacts of health care, using MRIO analysis and data for 189 countries (21). It found that health care accounted for 4.4% of global GHG emissions, consistent with the HCWH/Arup report. It also found that health care accounted for 2.8% of global PM emissions, 3.4% of NO$_x$ emissions, and 3.6% of SO$_2$ emissions, and found an increasing trend for all emissions between 2000 and 2015. It also calculated three other footprints: malaria risk, the release of reactive nitrogen into waterways, and the use of scarce water, and found that health care accounted for 0.7%, 1.8%, and 1.5%, respectively.

The impact of such emissions on health can be calculated. A life cycle analysis of Canadian health care, for example, showed that the health system is responsible for annual emissions of about 33 million tonnes of CO$_2$eq, 4.6% of the national total, as well as over 200,000 tonnes of other pollutants. An estimated 23,000 disability-adjusted life-years (DALYs) (uncertainty range 4500–610,000 DALYs) are lost annually from direct exposures to hazardous pollutants and from environmental changes caused by this pollution (22). This illustrates the paradox of adverse impacts on health and the environment from the delivery of health care.

The sources of GHG emissions in health care are diverse, including energy consumption, transport, and product manufacture, use, and disposal. The HCWH/Arup report assessed GHG emissions from all three 'scopes' of the Greenhouse Gas Protocol (23):

- Scope 1: Direct emissions from sources owned or controlled by a health care facility, such as emissions from combustion in hospital boilers and furnaces; health system vehicles; and emissions from chemical sterilizing equipment.
- Scope 2: Indirect emissions from the generation of electricity purchased and consumed by the health care facility.
- Scope 3: All other indirect emissions associated with the health care facility's activities but that occur outside the facility's direct control, such as emissions in the production or transport of medications, supplies, equipment, and food.

Strikingly, Scope 1 emissions account for just 17% of health care GHG emissions, and Scope 2 for 12%; Scope 3 emissions make up 71% of the total. Of the total emissions, about 40% could be attributed to electricity and gas for energy (in both Scope 2 and Scope 3), and 13% are due to operational emissions from health care facilities (Scope 1) (**Figure 11.1**). Other significant sources of emissions include agriculture (9%, e.g. from catering at health facilities, growing cotton for surgical gowns, etc.), pharmaceuticals and chemicals (not including the energy used to produce them, 5%), transport (7%), and waste treatment (3%).

A study of 36 OECD countries, representing 54% of global population and 78% of GDP, used somewhat different methods than the HCWH study, but it also yielded an estimate of about 5% of national GHG emissions from the health care sector. It found the key determinants of emissions to be the carbon intensity of the national energy system, the energy intensity of the national economy, and the level of health care expenditure, which together explain half of the variance in per capita health sector carbon footprints (24). These findings, together with the importance of Scope 3 emissions in the overall health sector carbon footprint, emphasize that significant emissions reductions will require not only changes in health care delivery, but also system-wide transformation, including in the energy, manufacturing, and transportation sectors.

That said, research has highlighted specific practices in health care facilities that account for substantial emissions and that are amenable to change. Examples include:

- *Anaesthetic gases*: Halogenated inhalation anaesthetics such as isoflurane, desflurane, and sevoflurane are potent GHGs (25, 26). These gases are released into the atmosphere during manufacturing and use (patients exhale them virtually unchanged during and after anaesthesia), and their atmospheric concentrations, as measured in diverse settings globally, have increased significantly. Nitrous oxide, another commonly used anaesthetic, is a less potent GHG than the halogenated compounds, but it has an atmospheric lifetime of over a century, amplifying its impact (27). A limited estimate based on UNFCCC data from 31 countries suggests that anaesthetics account for 0.6% of health care's global climate footprint, or nearly 2.5 Mt CO_2eq (19).
- *Metered-dose inhalers* (MDIs) for asthma use hydrofluorocarbons as propellants; these are 1480–2900 times more powerful than CO_2 in their ability to trap heat in the

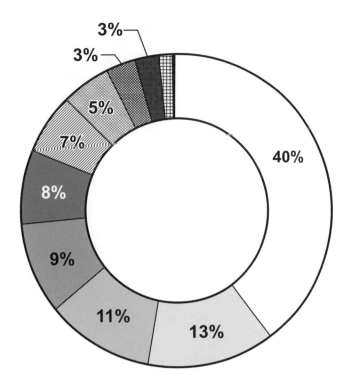

Figure 11.1. Global health care emissions by production sector.

Source: Karliner J, Slotterback S, Boyd R, Ashby B, Steele K. *Health Care's Carbon Footprint: How the Health Sector Contributes to the Global Climate Crisis and Opportunities for Action*. Health Care Without Harm and Arup; 2019.

atmosphere (28). They currently account for about 0.3% of health care's global climate footprint (19).

- *Energy use*: Health care facilities are energy intensive. For example, the energy intensity of US hospitals ranges from 640.7 kWh/m^2 in the hottest climates to 781.1 kWh/m^2 in the coldest climates, with an average of 738.5 kWh/m^2 – approximately 2.6 times higher than other commercial buildings (29). Moreover, the sites of highest energy use within hospitals can be identified. Operating theatres are 3–6 times more energy intensive than the hospital as a whole, largely due to heating, air conditioning, and ventilation requirements. A study of three operating theatres in referral hospitals in Canada, the USA, and the UK showed that the average GHG emissions per case ranged from 146 kg CO_2eq to 232 kg CO_2eq (30). Anaesthetic gases accounted for a substantial part of the emissions, but energy use accounted for half again as much.

- *Travel*: Health- and care-related travel accounts for around 5% of all road travel in England each year (31), corresponding to about 6.25 MtCO_2eq in 2017 (32). Some travel demand varies with specific therapeutic choices (33), while other travel demand depends on the need for face-to-face patient encounters – a need that has been called into question in some circumstances, as highlighted by the COVID-19 pandemic.

Reducing the Environmental Impacts of Health Care

Substantial progress is possible in reducing GHG emissions from health systems, even as the volume of health care delivered grows (34). Some of this progress comes from changes in the way clinical care is delivered, some comes from changes in the electricity used by the facilities (Scope 2 emissions), and some comes from upstream and downstream (Scope 3) interventions.

Potential solutions to each of the example problems highlighted above include:

- Anaesthetists can choose less polluting gases (e.g. substituting sevoflurane for desflurane, which yields a 15-fold reduction in global warming potential); implement waste gas scavenging technology; use low-flow anaesthesia; use regional and intravenous anaesthesia where possible; and use less nitrous oxide (35, 36).

- Metered dose inhalers can often be replaced by dry powder inhalers that do not require the use of hydrofluorocarbons as propellants (37). In 2017, 70% of all inhalers sold in England were MDIs, whereas in Sweden only 13% were MDIs and the remainder were powder inhalers. Applying the Swedish distribution of inhaler use to England would result in a reduction of 550 kt CO_2eq annually (28).

- Energy use, and dependence on the energy grid, can be reduced at the level of individual facilities. For example, the University Hospital of the North Midlands, UK, was able to attract £336,000 of community investment to place 1000 rooftop solar (photovoltaic) panels on hospital buildings at no cost to the National Health Service (NHS). This reduced the hospital's CO_2 emissions by about 2300 tonnes annually, and yielded financial returns for the investors from feed-in tariffs, which pay for electricity entering the grid. Additional income of £300,000 was directed to reducing local fuel poverty – hence the programme name, 'Saving Lives with Solar' – nicely demonstrating both health and socioeconomic co-benefits of well-designed initiatives (38).

- With regard to health care-related travel, studies have increasingly demonstrated the potential of telehealth to reduce the environmental impact of travel while delivering high-quality care (and saving costs and time). Evidence is available from general practice settings in the UK (39), from rehabilitation services in Sweden (40), from large academic medical centres in California (41) and Toronto (42), and from specialty services such as dermatology in the Catalan region of Spain (43) and head and neck cancer care in the Shetland Islands (44). Over a 5-year period the University of California, Davis provided more than 13,000 outpatient consultations in about 30 different medical specialties. This was calculated to have saved 4.7 million miles of travel and an average of 6 hours of patient travel time per visit, with fuel savings corresponding to 1700 tonnes of reduced carbon emissions (41). Not surprisingly, the emissions reductions are greatest where travel distances are longest and mass transit options least available. More research is needed to define which conditions and patients are suitable for telehealth approaches and which require face-to-face consultations. In addition to travel reduction, solutions are emerging to reduce many other operational aspects of clinical care, from materials selection to waste management (45).

There is also opportunity for reducing Scope 2 emissions by improving the sourcing and use of energy. As the GHG intensity of national grids declines, so do health sector Scope 2 emissions. The NHS in England reduced GHG emissions by an estimated 18.5% between 2007 and 2017, reflecting in part the decarbonization of the electrical grid in the UK occurring over that period. This was due to growth in wind generation, and increased transport and energy efficiency (31). In a study of the health care systems in OECD countries, China, and India, between 2005 and 2014, the carbon footprint fell in 14 of the countries studied, mostly due to decreasing emissions intensities in those countries' energy sectors (24). In the UK, energy measures alone resulted in savings to the health system of £1.85 billion over an 8-year period from 2007. These savings were retained in the local health economy to be spent on health care delivery (46).

Overall, however, progress in reducing health sector GHG emissions has lagged behind what is needed to contribute to reaching global targets. A majority of emissions reside in the supply chain, requiring widespread innovation. In addition, care pathways and health technologies need to be designed with both health and sustainability in mind. Health technology assessment programmes that provide evidence to guide the use of appropriate technologies in health care should include assessments of environmental impact as well as cost-effectiveness (47).

Health research also offers opportunities for reducing GHG emissions. The Sustainable Trials Study group estimated that a large, 5-year randomized clinical trial of the effects of corticosteroids following head injury (CRASH 1) was responsible for about 630 tonnes of CO_2 emissions. The coordinating centre was responsible for about 40% of emissions, with distribution of medications and documents accounting for 28% and travel for 23%. In a subsequent trial (CRASH 2) the investigators were able to reduce emissions by more efficient recruitment (48, 49). Similar efforts in laboratory research are also timely (50). Another area of opportunity is medical conferences and meetings, which in the aggregate have large travel footprints. They can be held virtually with considerable avoided emissions even while increasing attendance (51, 52).

The health sector, as noted above, has substantial environmental impacts other than GHG emissions. One is water use. Studies have assessed strategies to conserve water in hospitals, for example by recycling reject water that remains after kidney dialysis (which consumes 120–800 litres of water per treatment session) for grey use such as watering lawns or flushing toilets. Recycling such water is less costly and probably less energy intensive than reverse osmosis of seawater in arid regions (53). The use of spring-loaded foot-controlled taps can greatly reduce water use for surgical scrubbing (54). Knee-operated taps can also reduce water use compared with elbow-operated taps. Changing all surgical sinks in the UK from elbow- to knee-activated taps would save about 3,000,000 kWh in energy for heating water annually (55).

Addressing Population Growth

The global population has expanded dramatically since the turn of the twentieth century – a cardinal feature of the Great Acceleration. By 2050 it is expected to be roughly ten times larger than it was for most of the nineteenth century – 10 billion as opposed to 1 billion (56). The role of population in Planetary Health is explored in **Box 11.2**.

The landmark 1994 International Conference on Population and Development reaffirmed a rights-based approach to population: 'All couples and individuals have the basic right to decide freely and responsibly the number and spacing of their children and to have information, education and means to do so' (66). The inclusion of 'responsibly' has potential resonances in the Anthropocene that were probably largely unforeseen by those drafting the statement, but could be interpreted to refer to the responsibility not to allow today's population to undermine the potential for future generations to flourish.

In this context, a principal role for the health sector is in the provision of reproductive health services. Access to affordable reproductive health, including family planning, is facilitated by provision of universal health coverage, including of primary care services. The coverage of family planning can be defined as the ratio of users of family planning to the total demand, including the unmet need. But desired family size is influenced by social norms such that living in countries with high total fertility may raise desired family size and vice versa (61). Thus current estimates of 'unmet need' that do not take into account social norms in high fertility populations may underestimate the potential requirements if norms change rapidly. However, there is a substantial unmet need for family planning even without taking into account the influence of such social norms, because of lack of universal access to modern contraception.

Globally, the proportion of women of reproductive age whose need for modern methods of family planning is met has continued to increase slowly, from 74% in 2000 to 76% in 2019. Although adolescent births have declined globally, they have remained persistently

Box 11.2. **Population and Planetary Health**

Population growth implies growth in resource and energy use. Historical studies show that CO_2 emissions from energy use respond almost proportionately to population growth (57). Other demographic changes, such as ageing and urbanization, also have significant effects, but they are less than proportional. Future projections, based on the relationship of population growth and energy use, suggest that by the end of the century the UN low population projection (5.5 billion) would yield emissions 40% lower than the UN medium population growth projection (9.1 billion) (57). These projections are of course dependent on assumptions about economic growth, dependence on fossil fuels, consumption patterns, and technology changes.

As discussed in Chapters 6 and 8, consumption patterns are critical in determining environmental impacts. Per capita GHG emissions are over 50 times higher in the USA, Canada, and Australia than in Burundi. Low-income, high-fertility countries of the Global South have contributed little to global carbon emissions to date, but emissions are increasing with economic development. Population growth in a wealthy country such as the USA, although contributing little overall to world population, will contribute substantially to GHG emissions because of the high environmental footprint of each individual. Therefore, reducing consumption, and reducing the carbon intensity of economic activity, are essential to reducing GHG emissions.

But slowing, and subsequently halting population growth is also essential. This is not just true for GHG emissions; population pressure drives many other planetary changes, such as deforestation (58), freshwater depletion, and biodiversity loss (since population growth rates are especially high in biodiversity hotspots) (59).

The carrying capacity of the Earth has been the subject of much speculation. A review of past estimates of the population that could be supported by the Earth showed an extraordinarily wide range of estimates, which varied by two orders of magnitude (60). This reflected a wide range of assumptions about consumption patterns, environmental constraints, technology innovation, and other factors. Clearly there are major uncertainties but a recent notional estimate based on the ecological footprint (Chapter 1), assuming a per capita income of US$20,000 (the income above which there is no demonstrable increase in happiness), suggests a global carrying capacity of about 1.8 billion people, similar to that of a century ago (61). Whilst the number is open to discussion, there is no doubt that a substantially smaller world population than today's would ease both adaptation to and mitigation of planetary changes.

Some observers have noted that population is conspicuous by its absence in much of the climate change discourse (62, 63). This may be due to misperceptions that population growth is no longer a problem, that population policies are ineffective, that population has little impact on climate, that population policies are coercive, and/or that population policy is too controversial to succeed (62). It may be that those who work on climate change are simply too segregated by disciplinary and organizational boundaries from those who work on population (56). And it may be that polarization impedes constructive debate – with some, usually in the Global North, strongly emphasizing the role of population (64) (especially in poor countries), and others, especially in the Global South, pointing to the role of overconsumption in wealthy countries, the historic north–south imbalance of emissions, and the legacy of colonialism, perceiving hypocrisy in calls by the wealthy for slowed population growth, and asserting the right to procreate (65).

high in Sub-Saharan Africa, at 101 births per 1000 adolescent women (67). Providing universal access to modern contraception yields major near-term health benefits. For example, according to estimates by the Guttmacher Institute (68), such access would cut the annual number of maternal deaths from unplanned pregnancies from 96,000 to 8000. Such access would also approximately halve the number of neonatal deaths and stillbirths globally (69), improve childhood nutritional status, and reduce stunting (70).

Of the 1.6 billion women and girls aged 15–49 living in Africa, Asia, Latin America and the Caribbean, and Oceania (excluding Australia, Japan, and New Zealand), about half (885 million) want to avoid or delay a pregnancy. About three-quarters of these (671 million) are using modern contraceptives, with the needs of the remainder (214 million) unmet (68). The definition of modern contraception includes hormonal methods, IUDs, surgical sterilization, male and female condoms, diaphragms, and some other methods (68). Women with an unmet need for modern contraception account for 84% of all unintended pregnancies in these regions. The prevalence of unmet need is highest in Africa (21%) but absolute numbers are highest in Asia (70 million). Current use of modern contraceptives prevents about 308 million unintended pregnancies annually.

The costs of providing universal access to modern contraception are modest. A recent estimate suggests that current direct and indirect annual costs of providing access to modern contraception for 671 million women in LMICs total US$6.3 billion (68). Scaling up coverage to meet the remaining unmet needs would cost about US$12.1 billion. Because it is far cheaper to prevent an unplanned pregnancy than to deliver maternity care, there would be average net savings of US$2.20 for every additional US$1 invested above current levels.

At least three caveats are necessary with regard to reproductive health services. First, they should comprise more than the distribution of contraceptives; they should be accompanied by the empowerment of women. Second, they should be accompanied by investments in education, to build human capital. Third, reducing the population growth rate will not yield an immediate reduction in population size.

The empowerment of women is associated with more rapid economic and social development, with better community health, and with greater adoption of family planning (71–73).

Building human capital through education, especially the education of girls, is critical. This priority arises in the context of debate about the 'demographic dividend' – the economic growth that may occur in low-income countries after family planning is implemented, fertility falls, there are fewer young mouths to feed, and the working-age population 'bulge' is large – when, in other words, the 'dependency ratio' is low. This dividend is not, however, inevitable; it is more likely to emerge if concurrent policies advance health care, education, and equitable access to opportunity (74). Recent analyses suggest that human capital, particularly through better education, is a more important predictor of economic dividends than population age structure (75, 76). Integrated policies should therefore address both reproductive health rights and improving human capital.

Some of the benefits of reducing population growth will accrue over time. Even if it were possible to transition very rapidly to a one child policy, the world population in 2100 would still be similar to today's. An analysis by the Institute for Health Metrics and

Evaluation modelled the 'completed cohort fertility at age 50 years', or CCF50, through 2100, as a function of educational attainment and contraceptive met need. It projected a peak in the global population in 2064 at 9.73 billion (95% uncertainty interval: 8.84–10.9) with a decline to 8.79 billion (6.83–11.8) in 2100. Many countries were forecasted to experience substantial population declines by the end of the century – China by 48%, and Japan, Thailand, Spain, and 20 other countries by more than 50%, relative to the 2017 population (77). The recession following the COVID-19 pandemic is likely to reduce birth rates further, a typical effect of economic downturns (78). But even a hypothetical catastrophic event in mid-century leading to the mass deaths of 2 billion people could still leave the world with a population of 8.5 billion by the end of the century (79).

While increased ageing combined with population decline in some countries may pose challenges of caring for the elderly, overall the benefits to the environment, and thus to long-term sustainability are likely to be substantial. For health professionals, then, the task is to advocate for reproductive health services as part of health services more generally, and to deliver these services equitably and effectively, in the context of advancing women's empowerment, education, and social equity.

Health Professional Leadership in the Anthropocene

The need to promote Planetary Health, reproductive health, women's empowerment, education, and social equity introduces a fourth role for health professionals in the Anthropocene: advocacy. Throughout this book we have emphasized that action is needed in many sectors other than health – in energy, agriculture, transportation, urban planning, and more – as well as in the health sector. Health professionals are well positioned to frame and deliver key messages, and to serve as leading voices for healthful, sustainable policies in their communities and nations.

Human thriving in the Anthropocene will require policy changes and behavioural changes, as explored throughout this book. Policymakers and publics need to embrace these changes; otherwise they will not occur at the necessary scale and pace to reduce the projected risks to humanity. This embrace depends on 'clear and consistent messages, repeated often, from a variety of trusted sources' – fundamental strategies of social marketing, as honed in campaigns to reduce smoking, encourage condom use, and promote other health behaviours (80). Health professionals are highly trusted voices in many cultural settings (**Figure 11.2**) – a salient point at a time when public trust in many institutions is declining, and 'fake news' spreads rapidly and widely on social media (81). Health professionals are well positioned to inform members of the public, and policymakers, about the consequences of unsustainable practices and the benefits to health and well-being of careful stewardship of resources.

There is a substantial history of health professional advocacy at the global scale. During the Cold War, health professionals from the USA, the USSR, and many other countries convened around opposition to nuclear weapons, played a major role in moving their governments towards arms control treaties, and were awarded the 1985 Nobel Peace

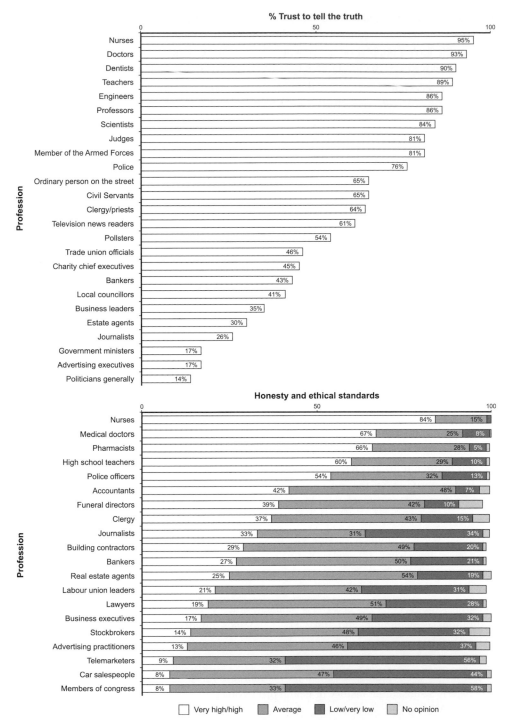

Figure 11.2. Trust in various professions in the UK (upper panel) and USA (lower panel). Sources: Ipsos MORI Veracity Index, 2019. www.ipsos.com/ipsos-mori/en-uk/trust-politicians-falls-sending-them-spiralling-back-bottom-ipsos-mori-veracity-index and Gallup Poll, 2018. https://news .gallup.com/poll/245597/nurses-again-outpace-professions-honesty-ethics.aspx.

Prize – an apt model for addressing health threats in the Anthropocene (82–84). There have been many calls for physicians and other health professionals to speak forthrightly on the climate crisis and other planetary changes (85–88). Many medical and nursing organizations have taken strong positions, especially regarding climate change, and many health professionals have emerged as leading voices. One potentially high-impact example is the declaration by the World Organization of Family Doctors (WONCA), representing 500,000 members, of the unique role they can play in advancing Planetary Health and the Sustainable Development Goals (89).

Whether in their capacity as private citizens, in clinical settings, as community leaders, as elected officials, and/or as professional leaders, health professionals can help advance progress towards the solutions described in this book.

Conclusion: Health and Population in the Anthropocene

For many low- and middle-income countries, achieving universal health coverage (SDG 3) is clearly of paramount importance, particularly in the face of growing environmental threats to health. Universal health care must be based on effective primary health care to ensure equity, efficiency, and cost-effectiveness. Current patterns of health care in high-income countries seem increasingly unsustainable, financially and environmentally. Health care costs continue to escalate in many countries and yield diminishing returns. While the quest for biomedical advances will continue and no doubt bring some genuine progress, sustaining health in the Anthropocene will necessitate health systems that are not only universal and resilient to environmental change, but also that have a far lower environmental footprint than today's. Reducing the environmental footprint of health care will require working with the private sector, such as pharmaceutical and medical equipment industries, to incentivize more efficient use of resources and zero-carbon supply chains. Providing universal access to modern family planning and supporting reproductive health rights can help address population growth and improve neonatal and maternal health outcomes in the near term. Health professionals can potentially reach the whole population of the planet and could have a transformative role in communicating how individuals, families, and communities can improve their own health and the state of natural systems on which our health depends.

References

1. Marmot M. *The Health Gap: The Challenge of an Unequal World*. New York and London: Bloomsbury Press; 2015.
2. Fries JF, Koop CE, Beadle CE, et al. Reducing health care costs by reducing the need and demand for medical services. *New England Journal of Medicine*. 1993;329(5): 321–5. doi: 10.1056/nejm199307293290506.
3. WHO. Declaration of Alma-Ata International Conference on Primary Health Care, Alma-Ata, USSR, 6–12 September 1978.
4. Walsh JA, Warren KS. Selective primary health care: an interim strategy for disease control in developing countries. *New England Journal of Medicine*. 1979;301(18): 967–74. doi: 10.1056/nejm197911013011804.

5. Sanders D, Nandi S, Labonté R, Vance C, Van Damme W. From primary health care to universal health coverage – one step forward and two steps back. *The Lancet*. 2019;394(10199):619–21. doi: 10.1016/s0140-6736(19)31831-8.

6. Xie E, de Barros EF, Abelsohn A, Stein AT, Haines A. Challenges and opportunities in planetary health for primary care providers. *The Lancet Planetary Health*. 2018;2(5): e185–7. doi: 10.1016/s2542-5196(18)30055-x.

7. Ebrahim S, Taylor F, Ward K, et al. Multiple risk factor interventions for primary prevention of coronary heart disease. *Cochrane Database of Systematic Reviews*. 2011(1). doi: 10.1002/14651858.CD001561.pub3.

8. Lobelo F, Young DR, Sallis R, et al. Routine assessment and promotion of physical activity in healthcare settings: a scientific statement from the American Heart Association. *Circulation*. 2018;137(18):e495–522. doi:10.1161/CIR.0000000000000559.

9. Lobelo F, de Quevedo IG. The evidence in support of physicians and health care providers as physical activity role models. *American Journal of Lifestyle Medicine*. 2016;10(1):36–52. doi: 10.1177/1559827613520120.

10. Frank E, Dresner Y, Shani M, Vinker S. The association between physicians' and patients' preventive health practices. *Canadian Medical Association Journal/Journal de l'Association medicale canadienne*. 2013;185(8):649–53. doi: 10.1503/cmaj.121028.

11. Hedin B, Katzeff C, Eriksson E, Pargman D. A systematic review of digital behaviour change interventions for more sustainable food consumption. *Sustainability*. 2019;11(9):2638. Available from www.mdpi.com/2071-1050/11/9/2638.

12. Laverack G. The challenge of behaviour change and health promotion. *Challenges*. 2017;8(2):25. Available from www.mdpi.com/2078-1547/8/2/25.

13. Sackett DL, Haynes RB, Tugwell P. *Clinical Epidemiology: A Basic Science for Clinical Medicine*, 2nd ed. Boston: Little, Brown; 1991.

14. Kassirer JP. *On the Take: How Medicine's Complicity with Big Business Can Endanger Your Health*. Oxford and New York: Oxford University Press; 2005.

15. Doran T, Maurer KA, Ryan AM. Impact of provider incentives on quality and value of health care. *Annual Review of Public Health*. 2017;38(1):449–65. doi: 10.1146/annurev-publhealth-032315-021457.

16. Gray BH. *For-Profit Enterprise in Health Care*. Washington, DC: Institute of Medicine Committee on Implications of For-Profit Enterprise in Health Care; 1986.

17. Marik PE. The cost of inappropriate care at the end of life: implications for an aging population. *American Journal of Hospice and Palliative Medicine*. 2015;32(7):703–8. doi: 10.1177/1049909114537399.

18. Smith S, Brick A, O'Hara S, Normand C. Evidence on the cost and cost-effectiveness of palliative care: a literature review. *Palliative Medicine*. 2014;28(2):130–50. doi: 10.1177/0269216313493466.

19. Karliner J, Slotterback S, Boyd R, Ashby B, Steele K. Health Care's Carbon Footprint: How the Health Sector Contributes to the Global Climate Crisis and Opportunities for Action. Health Care Without Harm and Arup; 2019. Available from www.arup.com/perspectives/publications/research/section/healthcares-climate-footprint.

20. Wiedmann T. A review of recent multi-region input–output models used for consumption-based emission and resource accounting. *Ecological Economics*. 2009;69(2):211–22. https://doi.org/10.1016/j.ecolecon.2009.08.026.

21. Lenzen M, Malik A, Li M, et al. The environmental footprint of health care: a global assessment. *The Lancet Planetary Health*. 2020;4(7):e271–9. doi: 10.1016/S2542-5196(20)30121-2.

22. Eckelman MJ, Sherman JD, MacNeill AJ. Life cycle environmental emissions and health damages from the Canadian healthcare system: an economic-environmental-

epidemiological analysis. *PLoS Medicine*. 2018;15(7):e1002623. doi: 10.1371/journal. pmed.1002623.

23. Ranganathan J, Corbier L, Bhatia P, et al. *The Greenhouse Gas Protocol: A Corporate Accounting and Reporting Standard (Revised Edition)*. Geneva and Washington, DC: World Business Council on Sustainable Development (WBCSD) and World Resources Institute (WRI); 2004. Available from https://ghgprotocol.org/corporate-standard.

24. Pichler P-P, Jaccard IS, Weisz U, Weisz H. International comparison of health care carbon footprints. *Environmental Research Letters*. 2019;14(6):064004. doi: 10.1088/ 1748-9326/ab19e1.

25. Vollmer MK, Rhee TS, Rigby M, et al. Modern inhalation anesthetics: potent greenhouse gases in the global atmosphere. *Geophysical Research Letters*. 2015;42(5): 1606–11. doi: 10.1002/2014gl062785.

26. Ryan SM, Nielsen CJ. Global warming potential of inhaled anesthetics: application to clinical use. *Anesthesia and Analgesia*. 2010;111(1):92–8. doi: 10.1213/ ANE.0b013e3181e058d7.

27. Parker NW, Behringer EC. Nitrous oxide: a global toxicological effect to consider. *Anesthesiology*. 2009;110(5):1195; author reply 6. doi: 10.1097/ ALN.0b013e31819faca9.

28. Janson C, Henderson R, Löfdahl M, et al. Carbon footprint impact of the choice of inhalers for asthma and COPD. *Thorax*. 2020;75(1):82–4. doi: 10.1136/thoraxjnl-2019-213744.

29. Bawaneh K, Nezami FG, Rasheduzzaman M, Deken B. Energy consumption analysis and characterization of healthcare facilities in the United States. *Energies*. 2019;12(19). https://doi.org/10.3390/en12193775.

30. MacNeill AJ, Lillywhite R, Brown CJ. The impact of surgery on global climate: a carbon footprinting study of operating theatres in three health systems. *The Lancet Planetary Health*. 2017;1(9):e381–8. doi: 10.1016/S2542-5196(17)30162-6.

31. NHS. Reducing the Use of Natural Resources in Health and Social Care. Cambridge, UK: National Health Service Sustainable Development Unit; 2018. Available from www.sduhealth.org.uk/policy-strategy/reporting/natural-resource-footprint-2018.aspx.

32. BEIS. 2018 UK Greenhouse Gas Emissions, Provisional Figures. London: UK Department for Business, Energy & Industrial Strategy; 2019.

33. Zander A, Niggebrugge A, Pencheon D, Lyratzopoulos G. Changes in travel-related carbon emissions associated with modernization of services for patients with acute myocardial infarction: a case study. *Journal of Public Health*. 2011;33(2):272–9. doi: 10.1093/pubmed/fdq048.

34. Salas RN, Maibach E, Pencheon D, Watts N, Frumkin H. A pathway to net zero emissions for healthcare. *BMJ*. 2020;371:m3785. doi: 10.1136/bmj.m3785.

35. Alexander R, Poznikoff A, Malherbe S. Greenhouse gases: the choice of volatile anesthetic does matter. *Canadian Journal of Anesthesia/Journal canadien d'anesthésie*. 2018;65(2):221–2. doi: 10.1007/s12630-017-1006-x.

36. Charlesworth M, Swinton F. Anaesthetic gases, climate change, and sustainable practice. *The Lancet Planetary Health*. 2017;1(6):e216–17. doi: 10.1016/s2542-5196(17)30040-2.

37. Wilkinson AJK, Braggins R, Steinbach I, Smith J. Costs of switching to low global warming potential inhalers. An economic and carbon footprint analysis of NHS prescription data in England. *BMJ Open*. 2019;9(10):e028763. doi: 10.1136/bmjopen-2018-028763.

38. Sustainability West Midlands. 'Saving Lives with Solar' Community Energy Scheme 2017. Available from www.sustainabilitywestmidlands.org.uk/resources/saving-lives-with-solar-community-energy-scheme/.

39. Wootton R, Tait A, Croft A. Environmental aspects of health care in the Grampian NHS region and the place of telehealth. *Journal of Telemedicine and Telecare.* 2010;16(4):215–20. doi: 10.1258/jtt.2010.004015.

40. Holmner A, Ebi KL, Lazuardi L, Nilsson M. Carbon footprint of telemedicine solutions – unexplored opportunity for reducing carbon emissions in the health sector. *PLoS One.* 2014;9(9):e105040. doi: 10.1371/journal.pone.0105040.

41. Yellowlees PM, Chorba K, Burke Parish M, Wynn-Jones H, Nafiz N. Telemedicine can make healthcare greener. *Telemedicine and e-Health.* 2010;16(2):229–32. doi: 10.1089/tmj.2009.0105.

42. Masino C, Rubinstein E, Lem L, Purdy B, Rossos PG. The impact of telemedicine on greenhouse gas emissions at an academic health science center in Canada. *Telemedicine and e-Health.* 2010;16(9):973–6. doi: 10.1089/tmj.2010.0057.

43. Vidal-Alaball J, Franch-Parella J, Lopez Seguí F, et al. Impact of a telemedicine program on the reduction in the emission of atmospheric pollutants and journeys by road. *International Journal of Environmental Research and Public Health.* 2019;16(22):4366. Available from www.mdpi.com/1660-4601/16/22/4366.

44. Dorrian C, Ferguson J, Ah-See K, et al. Head and neck cancer assessment by flexible endoscopy and telemedicine. *Journal of Telemedicine and Telecare.* 2009;15(3): 118–21. doi: 10.1258/jtt.2009.003004.

45. WHO. Towards Environmentally Sustainable Health Systems in Europe: A Review of the Evidence. Copenhagen: WHO Regional Office for Europe; 2016. Available from www.euro.who.int/en/health-topics/environment-and-health/Climate-change/publications/2016/towards-environmentally-sustainable-health-systems-in-europe.-a-review-of-the-evidence-2016.

46. NHS. Securing Healthy Returns: Realising the Financial Value of Sustainable Development. Cambridge, UK: National Health Service Sustainable Development Unit; 2016.

47. Marsh K, Ganz ML, Hsu J, Strandberg-Larsen M, Gonzalez RP, Lund N. Expanding health technology assessments to include effects on the environment. *Value in Health.* 2016;19(2):249–54. doi: 10.1016/j.jval.2015.11.008.

48. Sustainable Trials Study G. Towards sustainable clinical trials. *BMJ.* 2007;334(7595): 671–3. doi: 10.1136/bmj.39140.623137.BE.

49. Subaiya S, Hogg E, Roberts I. Reducing the environmental impact of trials: a comparison of the carbon footprint of the CRASH-1 and CRASH-2 clinical trials. *Trials.* 2011;12:31. doi: 10.1186/1745-6215-12-31.

50. Zak JD, Wallace J, Murthy VN. How neuroscience labs can limit their environmental impact. *Nature Reviews Neuroscience.* 2020;21(7):347–8. doi: 10.1038/s41583-020-0311-5.

51. Zotova O, Pétrin-Desrosiers C, Gopfert A, Van Hove M. Carbon-neutral medical conferences should be the norm. *The Lancet Planetary Health.* 2020;4(2):e48–50. doi: 10.1016/S2542-5196(20)30003-6.

52. Coroama VC, Hilty LM, Birtel M. Effects of Internet-based multiple-site conferences on greenhouse gas emissions. *Telematics and Informatics.* 2012;29(4):362–74. https://doi.org/10.1016/j.tele.2011.11.006.

53. Tarrass F, Benjelloun M, Benjelloun O. Recycling wastewater after hemodialysis: an environmental analysis for alternative water sources in arid regions. *American Journal of Kidney Diseases.* 2008;52(1):154–8. doi: 10.1053/j.ajkd.2008.03.022.

54. Jones EE. Water use in the surgical scrub: surgeons can reduce their environmental footprint. *ANZ Journal of Surgery.* 2009;79(5):319–20. doi: 10.1111/j.1445-2197.2009.04881.x.

55. Somner JE, Stone N, Koukkoulli A, et al. Surgical scrubbing: can we clean up our carbon footprints by washing our hands? *Journal of Hospital Infection.* 2008;70(3):212–15. doi: 10.1016/j.jhin.2008.06.004.

56. Stephenson J, Crane SF, Levy C, Maslin M. Population, development, and climate change: links and effects on human health. *The Lancet.* 2013;382(9905):1665–73. doi: 10.1016/s0140-6736(13)61460-9.

57. O'Neill BC, Liddle B, Jiang L, et al. Demographic change and carbon dioxide emissions. *The Lancet.* 2012;380(9837):157–64. http://dx.doi.org/10.1016/S0140-6736(12)60958-1.

58. Bologna M, Aquino G. Deforestation and world population sustainability: a quantitative analysis. *Scientific Reports.* 2020;10(1):7631. doi: 10.1038/s41598-020-63657-6.

59. Cincotta RP, Wisnewski J, Engelman R. Human population in the biodiversity hotspots. *Nature.* 2000;404(6781):990–2. doi: 10.1038/35010105.

60. Cohen JE. *How Many People Can the Earth Support?* New York: W.W. Norton; 1995.

61. Dasgupta A, Dasgupta P. Population overshoot. In Bykvist K, Campbell T, editors. *Oxford Handbook of Population Ethics.* Oxford: Oxford University Press; forthcoming.

62. Bongaarts J, O'Neill BC. Global warming policy: is population left out in the cold? *Science.* 2018;361(6403):650–2. doi: 10.1126/science.aat8680.

63. O'Sullivan JN. The social and environmental influences of population growth rate and demographic pressure deserve greater attention in ecological economics. *Ecological Economics.* 2020;172:106648. https://doi.org/10.1016/j.ecolecon.2020.106648.

64. Murtaugh PA, Schlax MG. Reproduction and the carbon legacies of individuals. *Global Environmental Change.* 2009;19(1):14–20. https://doi.org/10.1016/j.gloenvcha.2008.10.007.

65. Ramesh R. Leave population out of climate talks, Indian minister says. *Guardian.* 28 August 2009.

66. UNFPA. Programme of Action. Adopted at the International Conference on Population and Development, Cairo, 5–13 September 1994. New York: UNFPA; 1994.

67. UN Secretary-General. Progress towards the Sustainable Development Goals. Report of the Secretary-General. New York: United Nations Economic and Social Council; 2019. Contract No. E/2019/68. Available from https://unstats.un.org/sdgs.

68. Guttmacher Institute. Adding It Up: Investing in Contraception and Maternal and Newborn Health, 2017. New York: Guttmacher Institute; 2017. Available from www.guttmacher.org/fact-sheet/adding-it-up-contraception-mnh-2017.

69. Bhutta ZA, Das JK, Bahl R, et al. Can available interventions end preventable deaths in mothers, newborn babies, and stillbirths, and at what cost? *The Lancet.* 2014;384(9940):347–70. doi: 10.1016/S0140-6736(14)60792-3.

70. de Onis M, Dewey KG, Borghi E, et al. The World Health Organization's global target for reducing childhood stunting by 2025: rationale and proposed actions. *Maternal and Child Nutrition.* 2013;9(Suppl. 2):6–26. doi: 10.1111/mcn.12075.

71. Pratley P. Associations between quantitative measures of women's empowerment and access to care and health status for mothers and their children: a systematic review of evidence from the developing world. *Social Science & Medicine.* 2016;169:119–31. doi: 10.1016/j.socscimed.2016.08.001.

72. Varkey P, Kureshi S, Lesnick T. Empowerment of women and its association with the health of the community. *Journal of Women's Health.* 2010;19(1):71–6. doi: 10.1089/jwh.2009.1444.

73. Afshar H, editor. *Women and Empowerment: Illustrations from the Third World.* New York: St. Martin's Press; 1998.

74. National Research Council. *Population Growth and Economic Development: Policy Questions*. Washington, DC: The National Academies Press; 1986.

75. Lutz W, Crespo Cuaresma J, Kebede E, et al. Education rather than age structure brings demographic dividend. *Proceedings of the National Academy of Sciences*. 2019;116(26):12798–803. doi: 10.1073/pnas.1820362116.

76. Rentería E, Souto G, Mejía-Guevara I, Patxot C. The effect of education on the demographic dividend. *Population and Development Review*. 2016;42(4):651–71. doi: 10.1111/padr.12017.

77. Vollset SE, Goren E, Yuan C-W, et al. Fertility, mortality, migration, and population scenarios for 195 countries and territories from 2017 to 2100: a forecasting analysis for the Global Burden of Disease Study. *The Lancet*. 2020;396(10258):1285–306. doi: 10.1016/S0140-6736(20)30677-2.

78. Sobotka T, Skirbekk V, Philipov D. Economic recession and fertility in the developed world. *Population and Development Review*. 2011;37(2):267–306. doi: 10.1111/j.1728-4457.2011.00411.x.

79. Bradshaw CJA, Brook BW. Human population reduction is not a quick fix for environmental problems. *Proceedings of the National Academy of Sciences*. 2014;111(46):16610–15. doi: 10.1073/pnas.1410465111.

80. Maibach E. Increasing public awareness and facilitating behavior change: two guiding heuristics. In Hannah L, Lovejoy T, editors. *Climate Change and Biodiversity*, 2nd ed. New Haven, CT: Yale University Press; 2019. pp. 336–46.

81. Vosoughi S, Roy D, Aral S. The spread of true and false news online. *Science*. 2018;359(6380):1146–51. doi: 10.1126/science.aap9559.

82. Nusbaumer MR, DiIorio JA. The medicalization of nuclear disarmament claims. *Peace & Change*. 1985;11(1):63–73. doi: 10.1111/j.1468-0130.1985.tb00073.x.

83. Williamson J. Nuclear war, climate change, and medical activism. *The Lancet Planetary Health*. 2020;4(6):e221–2. doi: 10.1016/S2542-5196(20)30127-3.

84. Haines A, Hartog M. Doctors and the test ban: 25 years on. *BMJ*. 1988;297(6645):408–11. doi: 10.1136/bmj.297.6645.408.

85. McCally M, Cassel CK. Medical responsibility and global environmental change. *Annals of Internal Medicine*. 1990;113(6):467–73.

86. Schwartz BS, Parker C, Glass TA, Hu H. Global environmental change: what can health care providers and the environmental health community do about it now? *Environmental Health Perspectives*. 2006;114(12):1807–12. doi: 10.1289/ehp.9313.

87. Jameton A. The importance of physician climate advocacy in the face of political denial. *AMA Journal of Ethics*. 2017;19(12):1222–37. doi: 10.1001/journalofethics.2017.19.12.sect1-1712.

88. Parker CL. Slowing global warming: benefits for patients and the planet. *American Family Physician*. 2011;84(3):271–8. Available from www.aafp.org/afp/2011/0801/p271.pdf.

89. WONCA. WONCA Statement on Planetary Health and Sustainable Development Goals. World Organization of National Colleges, Academies and Academic Associations of General Practitioners/Family Physicians; 2017. Available from www.globalfamilydoctor.com/News/PlanetaryHealthandSustainableDevelopmentGoals.aspx.

12

Sustaining Planetary Health in the Anthropocene

The Anthropocene Epoch confronts humanity with unprecedented challenges. Meeting these challenges demands fundamentally different modes of thought, institutions, technologies, policies, values, and governance systems than those that propelled the Great Acceleration. Humanity is at a crossroads. With environmental stressors intensifying, and with the world population growing, we need integrated solutions across sectors to address today's challenges and to reduce future risks to a minimum. The scale of change required is dramatic.

Chapter 6 introduced three categories of Planetary Health challenges: conceptual and empathy failures (imagination challenges), knowledge failures (research and information challenges), and implementation failures (governance challenges) (1). In this chapter, building on that framework, we explore some of the transformations needed in the Anthropocene, not only to safeguard health, but to assure, so far as possible, that all people have the opportunity to thrive.

We ground this discussion in a human rights perspective. The notion that health is a human right dates from the founding of the World Health Organization in 1948, and is enshrined in The International Covenant on Economic, Social and Cultural Rights:

Health is a fundamental human right indispensable for the exercise of other human rights. Every human being is entitled to the enjoyment of the highest attainable standard of health conducive to living a life in dignity. (2)

This does not necessarily connote a right to *be healthy* since some ailments are unavoidable, and since individuals should have agency, including the freedom to control their own bodies (including making unhealthy choices). But it does connote a right to access 'to a system of health protection which provides equality of opportunity for people to enjoy the highest attainable level of health'. Importantly, the right to health extends beyond health care; the UN General Assembly committee that drafted Article 12 stated that health is

... an inclusive right extending not only to timely and appropriate health care but also to the underlying determinants of health, such as access to safe and potable water and adequate sanitation, an adequate supply of safe food, nutrition and housing, healthy occupational and environmental conditions, and access to health-related education and information, including on sexual and reproductive health. (2)

Rights cannot exist without responsibilities. The COVID-19 pandemic has served as a reminder that collective responsibility – to wear masks, to maintain social distance from others, to wash hands – is necessary for societal function, and to achieve shared goals. In the broader context of Planetary Health, individuals are responsible for making healthy, sustainable choices – such as in what they eat, how they travel, and how they consume – to the extent they are free to make these choices. Institutions, such as private firms and governments, are also responsible for implementing policies and practices needed to achieve Planetary Health.

We also ground this discussion in the ethical mandate to achieve equity and justice. There are profound inequities between nations, as well as within nations, that in many cases severely constrain the achievement of health through the life course. Furthermore, in general, wealthy populations profit disproportionately from exploitation of the global commons (air, water, land, and ecosystems) and the poor confront disproportionate risks. Accordingly, the wealthy bear a special responsibility and must shoulder much of the burden of change.

Relatedly, gender equity is key to achieving equitable, just solutions. The Women and Gender Constituency (WGC), one of the nine stakeholder groups at the UNFCCC, offers a set of criteria to assess to what extent climate change projects are gender-just and equitable (3). Examples include that the projects should:

1. Provide equal access to benefits for women, men and youth;
2. Aim to alleviate and/or not add additional burden to women's workload;
3. Promote women's democratic rights and participation; and
4. Produce results that can be shared, spread and scaled up.

These criteria should be widely used to assess and shape policies and investments for environmental sustainability.

The need for profound change is another theme that runs through this chapter. All is not well with the world. While many indicators of human health and well-being have improved in recent decades, these improvements have relied heavily on unsustainable practices that cannot continue. Poverty and suffering are far too common. Even in wealthy countries, strains are showing – extreme concentrations of wealth among a small elite while the prospects of the majority stagnate, illustrated by plateauing (and, in the face of COVID-19, declining) healthy life expectancies as seen in the USA and UK (4). Loneliness, suicide, and overuse of prescription and illegal drugs and of alcohol are commonplace. Populist movements are on the rise in many countries. Countries with substantial human, technical, and financial resources such as the USA, the UK, and Brazil have failed spectacularly to keep death rates low when confronted with the COVID-19 crisis (5). Increasing automation and the advent of artificial intelligence and machine learning threaten the viability of many jobs, across the spectrum of skill levels.

These realities call for major course corrections. Increasing the resilience of societies to environmental threats is necessary but not sufficient. History suggests that even resilient societies have collapsed, unable to change at the required scale and pace (6). Geographer and archaeologist Karl Butzer wrote that what is needed is transformability: '. . . after overcoming initial, ideological dissonance, people can indeed come together to support change'. Reframing progress as the achievement of equitable health and well-being within finite environmental limits could become a politically attractive priority, if it can be communicated in a compelling manner.

Overcoming Conceptual and Empathy Failures

We need new ways of thinking (albeit sometimes drawn from long-standing and even ancient wisdom) – conceptual frameworks that situate humans in the context of thriving ecosystems, that define well-being based on authentic human needs, that extend over the time span of generations, that privilege fairness, solidarity, and empathy over conflict and competition. Economic activity must be re-envisioned as a means of fulfilling human needs while remaining within planetary boundaries, rather than simply as a set of efficient markets. The magnitude of change needed is indicated by such terms as 'Great Transition' (7) and 'Great Turning' (8).

Here we explore several examples of such conceptual frameworks, beginning with the Earth Charter (2000), the Sustainable Development Goals (2015), and the Canmore Declaration (2018), proceeding to statements of Planetary Health ethics, and turning next to economic frameworks aligned with Planetary Health. Finally, we discuss consumerism – the relationship with material goods, especially in wealthy countries, that has helped drive excessive demands on the planet, without commensurate improvements in health, well-being, or happiness. This discussion is not exhaustive; the examples we present are intended to illustrate promising efforts to meet imagination challenges and their limitations.

The Earth Charter

The Earth Charter originated with the 1987 World Commission on Environment and Development (the Brundtland Commission) (9), which called for a 'new charter' based on 'new norms' for sustainable development. Following the Earth Summit in Rio de Janeiro five years later, the summit's Secretary General, Canadian businessman and diplomat Maurice Strong, joined former Soviet president Mikhail Gorbachev to develop an Earth Charter as a civil society initiative. A lengthy drafting process included extensive consultation, and the Earth Charter was finally released at UNESCO headquarters in Paris in 2000 (earthcharter .org). The Charter begins with the statement 'We stand at a critical moment in Earth's history, a time when humanity must choose its future' – a statement that is even more true today. Its 16 principles (**Box 12.1**) provide an ethically grounded vision of sustainable development. Strengths include its recognition of the interconnections among environmental protection, human rights, equitable human development, and peace. It provides a suitable framework for safeguarding health in the Anthropocene, including specific references to universal access to health care, reproductive health rights, poverty reduction, and the use of the precautionary approach in addressing environmental challenges. The Earth Charter has been cited as a statement of principles and used in teaching, but it has not been translated into binding action – a goal that has been attempted, thus far without success, by a UN General Assembly effort beginning in 2017 to create a Global Pact for the Environment.

The Sustainable Development Goals

Another major statement of principles relevant to Planetary Health is the Sustainable Development Goals, adopted in 2015 by the UN member states. The SDGs are discussed

---

Box 12.1. **Earth Charter Key Principles**

I. Respect and Care for the Community of Life

 1. Respect Earth and life in all its diversity.
 2. Care for the community of life with understanding, compassion, and love.
 3. Build democratic societies that are just, participatory, sustainable, and peaceful.
 4. Secure Earth's bounty and beauty for present and future generations.

II. Ecological Integrity

 5. Protect and restore the integrity of Earth's ecological systems, with special concern for biological diversity and the natural processes that sustain life.
 6. Prevent harm as the best method of environmental protection and, when knowledge is limited, apply a precautionary approach.
 7. Adopt patterns of production, consumption, and reproduction that safeguard Earth's regenerative capacities, human rights, and community well-being.
 8. Advance the study of ecological sustainability and promote the open exchange and wide application of the knowledge acquired.

III. Social and Economic Justice

 9. Eradicate poverty as an ethical, social, and environmental imperative.
 10. Ensure that economic activities and institutions at all levels promote human development in an equitable and sustainable manner.
 11. Affirm gender equality and equity as prerequisites to sustainable development and ensure universal access to education, health care, and economic opportunity.
 12. Uphold the right of all, without discrimination, to a natural and social environment supportive of human dignity, bodily health, and spiritual well-being, with special attention to the rights of indigenous peoples and minorities.

IV. Democracy, Nonviolence, and Peace

 13. Strengthen democratic institutions at all levels, and provide transparency and accountability in governance, inclusive participation in decision making, and access to justice.
 14. Integrate into formal education and life-long learning the knowledge, values, and skills needed for a sustainable way of life.
 15. Treat all living beings with respect and consideration.
 16. Promote a culture of tolerance, nonviolence, and peace.

Source: Earth Charter International. Paris: Earth Charter Commission; 2000. https://earthcharter.org/.

---

in detail in Chapter 7. Notably, health both appears as its own SDG (SDG 3), focusing on universal health coverage, and is embedded in many other SDGs (Table 7.3).

The COVID-19 pandemic poses grave challenges to achieving the SDGs. In early 2021 at the time of writing, as the pandemic's global death toll surpassed 2.5 million, the world confronted a severe economic recession, widespread protests against structural inequality and racism, geopolitical tensions, and continued climate-related disasters. Commentators pointed out that COVID-19 had left two enabling conditions for achieving the SDGs – economic expansion and globalization – in 'shreds' (10). The UN's annual SDG report for

2020, usually an upbeat document, had a sombre tone; it noted that even before the pandemic, progress had been uneven and the world had not been on track to meet the SDGs by 2030 – and that the pandemic had curtailed progress even further, especially in the world's poorest places (11). There were warnings about the impacts on health and sustainability (12). Some called for narrowing the ambitions of the SDGs (10), while others called for redoubling efforts to achieve the SDGs (11, 13). Time will tell how the ambitions of the SDGs, and the intellectual scaffolding that supports them, weather the pandemic storm. In some respects the SDGs exemplify a systems approach that integrates planetary processes with human thriving, but despite their ambitious scope, they do not explicitly allow observers to assess whether a country is achieving progress within planetary boundaries. The successor of the SDGs will therefore need substantial rethinking as the target date of 2030 approaches.

The Canmore Principles

The Canmore Declaration (**Box 12.2**) emerged from a 2018 meeting in Canmore, Alberta, Canada, of inVIVO Planetary Health, an academic network (14). The full statement of these principles contains much to support; strengths include the aspirational quality and the emphasis on deep changes in contemporary ways of thinking. These principles integrate core ideas of human and ecosystem health with a grounding in values. They recognize the need for personal commitment to change and for social justice – robust responses to imagination challenges. However, the principles do not explicitly recognize the need to live within planetary limits, and they provide little concrete guidance on how to effect change – on how to adjudicate difficult trade-offs, on what governance arrangements are necessary, on how to drive attitude and behaviour change at scale. Operationalizing these principles will require a broad portfolio of actions, as discussed below.

Box 12.2. **Canmore Planetary Health Principles, from the inVIVO Network. The full statement of principles includes explanatory text for each principle.**

1. Sustainable vitality for all systems
2. Values and purpose
3. Integration and unity
4. Narrative health (Deploying traditional knowledge and science in solutions-focused narratives)
5. Planetary consciousness
6. Nature relatedness
7. Biopsychosocial interdependence
8. Advocacy
9. Countering elitism, social dominance and marginalization
10. Personal commitment to shaping new normative behaviours

Source: Prescott LS, Logan CA, Albrecht G, et al. The Canmore Declaration: Statement of Principles for Planetary Health. *Challenges*. 2018;9(2). doi:10.3390/challe9020031.

Planetary Health Ethics

Planetary Health implies the need for an ethical framework – one that broadens traditional biomedical ethics and environmental ethics to address ethical obligations to local and global communities, to those who are underserved, to future generations, and to the natural world. One formulation, called 'biome ethics', argues that ethics in the Anthropocene calls for 'focusing on the interconnectedness of all life forms and creating sensibilities of care, compassion, and shared suffering' (15). Others have called for an ethical critique of exploitative economic models, leading to a shift 'away from neoliberal and anthropocentric belief systems towards a more ecologically aware perspective on life' (16).

Foundations for such a Planetary Health ethics can be found in the extensive literature on climate ethics (17–20). Another foundation can be found in traditional indigenous belief systems, which in many cases embody ethical principles applicable to the Anthropocene (21–23). Planetary Health ethics has begun to emerge as a distinct focus, with a set of principles (24). A challenge of developing Planetary Health ethics is reconciling universal principles with heterogeneous cultural traditions. Planet-wide ethical principles cannot be imposed by small unrepresentative groups but must be shaped by ongoing debates involving those who are most affected as well as those whose consumption patterns have driven many of the changes we are witnessing. Such discussions and debates should be cultivated using inclusive approaches to ensure gender balance and representation of those whose voices are often unheard, including indigenous people, the disadvantaged, those in the Global South, and youth, who will have to live with a situation they have not created.

Conceptual and Empathy Failures in Economics: Beyond Neoliberalism

Imagination challenges call not only for conceptual frameworks that accommodate socio-environmental complexity, and for ethical approaches fit for the Anthropocene; they also require economic thinking that envisions commerce in the context of both human needs and planetary boundaries.

Much contemporary economic thinking dates from the aftermath of World War II. In 1947 Frederick Hayek, the renowned economist, invited 39 eminent economists, historians, and philosophers to a meeting in the Hotel du Parc, in the picturesque Swiss village of Mont Pèlerin, with panoramic views of Lake Geneva. Following the defeat of fascism, there were real concerns about the potential for communism to undermine the ascendancy of liberal thought. Participants were motivated by the imperative of confronting and defeating this growing threat. The meeting led to the formation of the Mont Pèlerin Society (MPS), which remains in existence today (25). Its Statement of Aims began with a clarion call of alarm:

The central values of civilization are in danger. Over large stretches of the Earth's surface the essential conditions of human dignity and freedom have already disappeared. In others they are under constant menace ... The position of the individual and the voluntary group are progressively undermined by extensions of arbitrary power. Even that most precious possession of Western Man, freedom of thought and expression, is threatened ... (26)

The statement went on to identify declining faith in private property and the competitive market as a particular concern. The MPS has had enduring impact; among its members have been eight economics Nobel Prize winners, whose work helped establish neoliberal ideas as economic orthodoxy (27). In the era of Margaret Thatcher and Ronald Reagan in the 1980s, neoliberal thinking edged out the Keynesian economics that had prevailed in the decades after World War II, and came to dominate much of modern economic policy.

Neoliberal economics places great faith in free markets. Proponents support privatization, deregulation, limited government spending, and free trade. It prizes individual liberty but not collective responsibility. It requires unending (and unsustainable) growth without consideration of ecological limits. It has little to say about inequities. Complete efficiency – the lodestar of free market theory (even if practice often fails to conform to theory) – reduces institutions' ability to adapt. Institutions that optimize efficiency in a particular context often fail when circumstances change. This has been illustrated by the failure of countries to stockpile sufficient protective equipment and other supplies to deal with the COVID-19 pandemic. These omissions have helped drive our current crisis.

However, there are lessons to be learned from neoliberalism's rapid rise from the periphery of economic thinking to political orthodoxy. It identified a clear threat (communism and the undermining of private property), it offered a clear solution (unregulated markets), it appealed to high values (such as individual liberty), it offered (and marketed) a vision of a good life (centred on acquisition of consumer goods), it cultivated powerful financial and political support (business interests and their lobbyists), and it identified a clear metric of success (GDP growth).

Those who aspire to transformative solutions will do well to learn from, if not necessarily emulate, some of the effective strategies used to promote neoliberalism. This alternative vision must be couched in Planetary Health terms, focusing on human health and well-being within planetary boundaries. Economics should aim to meet the needs of all members of the human family, to increase the resilience of societies (including of health systems) to environmental change, to reduce humanity's environmental footprint rapidly, and simultaneously to move towards much greater equity of consumption. Clearly government policies can be important levers for change, but excessive government power is also risky. Pluralistic systems with strong civil society voices will be essential to preserve freedom of thought and enhance sustainability. The critique of neoliberal economics should not preclude the use of market-based mechanisms to effect change which will necessitate constructive engagement of the private sector in driving transformation, incentivized by reforms in taxes, subsidies, and regulation.

The COVID-19 pandemic has posed a profound challenge to a canon of neoliberal thinking: opposition to state intervention to prop up failing businesses. A recent quote from conservative legal scholar Richard Epstein of the Hoover Institution is telling:

In ordinary times, the classical liberal approach favours strong property rights and limited government. But it is less widely known that this same theory, like virtually every other general political approach, advocates strong government controls in any emergency situation that poses an immediate peril to life and health. As a classical liberal, I often begin my analysis with the Roman maxim '*Salus populi suprema lex esto*' – the health (or well-being) of the population is the highest good. That principle is widely understood to invoke a second maxim, which is that ordinary property rights are suspended in times of necessity, but only so long as the necessity lasts. (28)

In August 2020, as the pandemic entered its second six months, the conservative US media company Fox News fielded a poll, whose results were also telling. Respondents were asked to choose one of two messages they would like to send to the Federal government: 'lend me a hand' or 'leave me alone'. A 57% majority opted for 'lend me a hand' – an increase from 34% 18 months earlier. Over the same period the proportion opting for 'leave me alone' dropped from 55% to 36% (29). Just as, according to the old aphorism, there are no atheists in the trenches, perhaps there are no pure neoliberal economists in pandemics. Fears of the 'nanny state' may give way to openness to collective action, through well-functioning and open systems of government, in the face of adversity. The left–right polarization that characterizes much of contemporary political discourse is proving unproductive. New approaches are needed to break the ideological logjam, including the constraints of left–right polarization, focusing on how best to accelerate progress towards a more equitable society within planetary boundaries.

The Anthropocene Epoch, with its attendant risks to life and health, provides ample motivation for new economic approaches. In recent years, triggered by both the economic and environmental crises of the Anthropocene, many observers have questioned the viability of capitalism as we know it. Some have argued that capitalism should be managed and reformed, but not eliminated (30, 31); others have pointed to intrinsic contradictions and shortcomings, and have argued for a profound overhaul of economic arrangements (32, 33). A full consideration of these issues is beyond the scope of this book (and of the authors' expertise), but clearly, any economic system that does not put the equitable achievement of human health and well-being, together with environmental sustainability at its core is doomed to fail.

Natural Capital as an Economic Fundamental

One economic concept that helps address imagination challenges is natural capital. Different stocks of capital are fundamental to any economic thinking. Various authors have proposed various typologies of capital, ranging from three to as many as twelve (34, 35). Financial capital consists of money in various forms, the dominant means of economic exchange. Manufactured, or produced, capital includes roads, buildings, machines, equipment, and other physical infrastructure. Human capital encompasses knowledge, education, skills, health, and aptitude. Social capital consists of networks of relationships among people, characterized by trust and reciprocity, that enable social functioning. (Others have written of spiritual, cultural, experiential, and symbolic capitals.) Here we focus on a relative newcomer to the discourse on capitals: natural capital.

As discussed in Chapter 6, natural capital refers to 'the living and non-living components of ecosystems – other than people and what they manufacture – that contribute to the generation of goods and services of value for people' (36). Nature has real value, and provides services to people that can be incorporated into economic thinking, but it is often excluded from economic calculations because its exploitation is a free good (37). This omission leads to externalities and other fundamental errors in valuation (38). The concept of natural capital, and the related concept of ecosystem services, were highlighted by the Millennium Ecosystem Assessment in 2005 (39). A subset of natural capital is sometimes called critical capital – natural assets such as freshwater that are essential for human survival and irreplaceable. Economic thinking in the Anthropocene must properly value the assets nature provides – and that are in many cases being depleted – as a foundation for careful stewardship of these resources, and for health and well-being (40).

Inclusive wealth is a measure of the aggregate value of all capital assets, including natural capital (41). Because it is comprehensive, and reflects the stock of assets that will be available to future generations, it helps assess whether society is on a sustainable development trajectory. The Inclusive Wealth Index (IWI) was launched in 2012 by the UN University and the UN Environment Programme, with the intention that it would replace the GDP and the Human Development Index (42).

A criticism of the IWI is that a nation's IWI can be maintained or grow even if its natural capital and/or critical capital is being depleted. Suppose that a country cuts down US$100 billion worth of forest and invests US$100 billion in infrastructure. This type of policymaking does not lead to strong sustainability because natural capital is being depleted. However, the country's IWI would remain constant. If substitution across the three forms of capital is allowed this leads to what is called weak sustainability. In fact, this is the current global pattern, a reflection of prevailing economic priorities that value produced capital above natural capital (**Figure 12.1**). The 2018 Inclusive Wealth Report (42) shows that 123 of 140 countries for which data were available experienced decreasing natural capital between 1990 and 2014, while wealth increased. This analysis included non-renewable resources as a positive natural capital asset, rather than a negative one, although if the social costs of carbon emissions such as those from air pollution are considered, they would be considered liabilities. Ten countries experienced growth in natural capital and in GDP. These are mainly former Soviet states that have improved their forest management since Soviet times and in which population has declined. Although the inclusive wealth approach is an improvement on GDP alone it does not fully reflect the challenge of growing human capital, while maintaining, and where necessary restoring, natural capital.

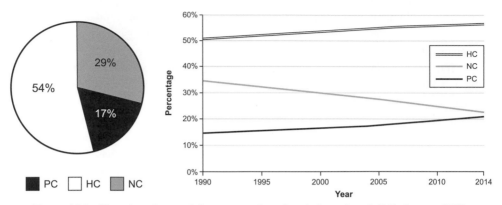

Figure 12.1. Changing shares of three categories of capital, produced (PC), human (HC), and natural (NC) – from 1990 to 2014. The percentages in the pie chart refer to averages of the three types of capital over the time period.

Source: Adapted from UNEP. *Inclusive Wealth Report*. London: UN Environment Programme; 2018. www.unenvironment.org/resources/report/inclusive-wealth-report-2018.

Doughnut Economics

An influential proposal for ensuring that the two interlinked objectives of environmental and social sustainability are addressed in a coherent and integrated way is the concept of the doughnut economy proposed by Kate Raworth (43) (**Figure 12.2**, redrawn as a lifebuoy instead of a doughnut because doughnuts – with between 200 and 400 empty calories each – are hardly symbols of health!). In this conceptualization, the planetary boundaries represent a ceiling that humanity should not transgress, and the essentials for satisfying and healthy lives – human needs – represent a floor, below which humanity should not descend. The space for achieving equitable human development exists between these two sets of limits. The twelve dimensions of the social foundation Raworth proposes are derived from internationally agreed minimum social standards, as embodied in the Sustainable Development Goals. This is a powerful formulation, showing the combined necessity of planetary and human priorities.

Like all simplifying diagrams, this one has limitations. The social foundations in the figure comprise a heterogeneous collection of domains; some are endpoints of development (such as human health and improved education), some reflect the basic necessities for development (such as access to clean water), some are enablers of progress (such as networks), and others reflect the effectiveness of governance systems (such as political voice). Nevertheless, it is a useful compass to orient our thinking about how to address the interlinked challenges of the Anthropocene.

Some ecological economists have taken strong positions against growth (44, 45) and others have advocated 'degrowth' – not only the abolition of economic growth as a social objective, but also a planned and equitable economic contraction in wealthy nations, eventually reaching a steady state that operates within planetary boundaries (46, 47). It is clear that if growth is to continue, it must be a very different growth than what has prevailed until now (48, 49) – with radically different technologies, with absolute

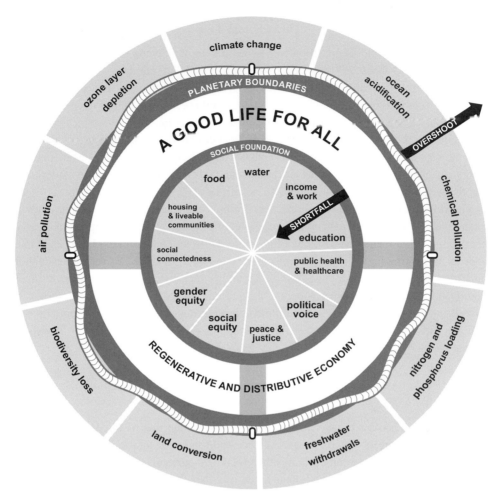

Figure 12.2. The Lifebuoy Economy, redrawn from Kate Raworth's image of the doughnut economy (www.kateraworth.com/doughnut/) (because doughnuts are used for gorging and contribute to obesity, while lifebuoys are used for rescue and contribute to survival!). Graphic by Emanuel Santos.

decoupling of growth from the use of energy and resources (which may not be plausible (50)) and with different metrics of growth commonly utilized (as discussed in Chapter 6). While Raworth is agnostic about growth, she describes a range of potential approaches to creating a steady state regenerative economy. Examples include replacing profit-based shareholder dividends with payment of a share of income of an enterprise to investors, allowing investors to benefit from a business that is maintaining a steady state. Another approach is demurrage, a small fee incurred for holding money, analogous to a negative interest rate. This could increase the incentive for regenerative investments such as reforestation schemes.

Thriving with Less

About 10% of the world's population was living in extreme poverty (less than US$1.90 per day) in 2015, and about 23% was living on less than US$3.20 per day – still far too many, but a major improvement over the preceding decades, albeit one that may be reversed by the economic recession triggered by COVID-19. For these people, an improved standard of living, with increased use of energy and resources, is essential.

But among the world's wealthy, the opposite pattern prevails: compulsive consumerism, with per capita ecological footprints substantially larger than the planet can sustain (51, 52). A key imagination challenge of the Anthropocene is this: Can we reverse this pattern, and can the wealthy learn to live – *contentedly* – with less?

This transformation requires a new social ethic, one that replaces consumption with other markers of success (53, 54). Because consumption is often a matter of identity signalling, and of social competition, a transition to different attitudes and behaviours must address identity formation (55–57). A positive vision of a good life, one that depends more on strong networks of family and friends, economic security, freedom and justice, meaningful work, and health, must supplant the current model – relentlessly promoted by advertising and media imagery – of fulfilment through materialism (58). This cultural shift will depend on a range of factors, from modelling by respected thought leaders to championing by religious authorities.

Overcoming Knowledge Failures

Throughout this book, we have flagged gaps in knowledge that, if filled, would provide better understanding of the challenges of the Anthropocene, and enable more effective solutions. Currently environmental health research in general, and particularly research directed at global environmental challenges and health, are deprived of funds. The Wellcome Trust, the major research funder in this domain, spent about £70 million over 5 years beginning in 2015 on its Our Planet Our Health programme (59). In contrast the global pharmaceutical industry invested over US$1.36 trillion in R&D in the decade from 2007 to 2016, and an annual investment of US$181 billion is projected by 2022 (60). An analysis published in 2016 showed that the ten largest public and philanthropic funding organizations together funded US$37.1 billion of research annually (with the US National Institutes of Health, the largest single funder, spending US$26.1 billion), about 40% of all public and philanthropic health research globally (61). It is likely that less than 0.1% of the annual total of public and philanthropic health R&D is spent on research directly relevant to the challenges of the Anthropocene Epoch.

The Rockefeller Foundation–*Lancet* Commission on Planetary Health identified five priority domains for research and provides a more detailed discussion (1):

1. The mechanisms through which environmental change affects human health.
2. Strategies to reduce environmental damage and harmful emissions including assessment of co-benefits/harms.

3. Strategies and technologies to promote resilience and support adaptation to environmental change.
4. Improved indicators of human welfare and the integrity of underpinning natural systems.
5. Translational research and implementation science to address the on-the-ground realities of what is feasible and relevant in the settings facing the greatest threats to Planetary Health.

Building on these categories, we describe several directions for Planetary Health research, followed by general comments about how to advance this research. A detailed research agenda is beyond the scope of this chapter.

Research on the mechanisms through which environmental change affects human health addresses the questions, 'How is the planet changing, and how does this affect people's health and well-being?' The first of these questions is a matter for earth and atmospheric scientists, ecologists, marine biologists, and others who characterize environmental processes. Examples of such work include quantifying trends in glacial mass (62), insect abundance (63), and fish populations (64). Changing patterns of human activity such as population ageing and urbanization, also part of how the planet is changing, are the domains of demographers, sociologists, and other social scientists. The second part of the question – how these changes affect people – is a matter for epidemiologists, other biomedical scientists, and social scientists. Examples of such work include explicating the associations between land use change and infectious diseases (65), Amazon fires and respiratory health (66), and changing plant chemistry and nutritional status (67). Transdisciplinary collaborations between these two groups of scientists can advance knowledge about the attribution of health impacts to anthropogenic environmental change.

The next sets of questions correspond to prevention and treatment, respectively. Strategies to reduce environmental damage and harmful emissions are strategies for primary prevention. A wide range of scientists develop and test these strategies. Engineers are refining renewable energy systems that will displace fossil fuels; agronomists are testing farming practices that conserve soil, reduce the need for inputs, and yield plenty of nutritious food; and urban planners and architects are designing cities that facilitate low-carbon lifestyles. For each such intervention, it is necessary to document both the reduced environmental impacts, and the impacts on people. Happily, effective environmental strategies are generally also effective health strategies – yielding health co-benefits through such pathways as cleaner air and more physical activity – but this is not invariable, and must be demonstrated empirically. Evidence of improved health can be a powerful driver of progressive environmental policies (68).

Strategies and technologies to promote resilience and support adaptation to environmental change correspond to the public health concepts of disaster preparedness and secondary prevention. Evidence of what approaches are effective, safe, and feasible is essential to developing protective strategies. Does an urban heat wave preparedness plan save lives, and if so, which elements of the plan are most efficacious and cost-effective (69)? Is an innovative crop resistant to heat and drought, does it provide an acceptable yield of nutritious food, and do consumers enjoy eating it (70)? Do pre-existing community bonds protect against the mental health impacts of disasters, and what are the best ways to build such bonds (71)?

Underlying all these research areas is the need for improved indicators of human welfare and the integrity of underpinning natural systems. In Chapter 6 we discussed the importance of replacing the GDP with more suitable indicators of human well-being and societal progress (72), and earlier in this chapter we discussed the Inclusive Wealth Index. It is equally important to develop robust indicators of biodiversity, of soil health, of freshwater availability, and other parameters of Earth systems and the ecosystem services they provide. Research is needed to develop and validate indicators, and to generate the data that populate the indicators. A comprehensive 'Planetary Health watch' would aggregate health and environmental data from multiple sources including 'top down' from satellites and 'bottom up' from monitoring sites in diverse locations, align the data in temporal and spatial scales, and make it readily available to researchers and decision makers (73).

Across all these areas, research must do more than document associations and validate solutions. Because of the implementation challenges discussed earlier – the 'know-do gap' – it is essential to understand the barriers to effective change, and how to overcome them. These are matters for behavioural and political scientists, economists, and allied researchers. What fiscal policies are most effective in shifting manufacturing practices towards the circular economy? What are the determinants of choosing low-carbon life-styles, and are people satisfied when they do so (74)? What is the role of disinformation in slowing progress, and how can it be countered? (See **Box 12.3**.) Once solutions are identified, how best can they be scaled up in different social and cultural contexts? Translational research and implementation science are needed to address the on-the-ground realities of what is feasible and relevant in advancing Planetary Health.

Across the breadth of Planetary Health, numerous research agendas have been proposed. Examples range from agroecology (88) to biodiversity (89), from migration (90) to nature contact (91), from transportation (92) to urban health (93, 94). While some directly address questions of human health and well-being, others are more limited to earth and environmental sciences. It is important for such agendas to be future oriented, as many challenges of the Anthropocene are dynamic; solutions centred, as Planetary Health sciences need to help solve urgent problems; co-produced with input from scientists, decision makers, and affected communities; and transdisciplinary, as no consequential Planetary Health problem can be solved within the confines of a conventional discipline.

Finally, it is essential that research funders – in government, foundations and charities, and the private sector – embrace the breadth and complexity of Planetary Health, and make grants accordingly. Far too many funders operate in silos, supporting environmental *or* health *or* social science and economic research, but not research that spans these domains.

Overcoming Implementation Failures

The final set of challenges for the Anthropocene grows out of implementation failures. Too often we know what to do, but don't do it. In the original Planetary Health Commission report (1) these were identified as governance challenges. Improved governance is certainly a major part of the solution, but other domains are relevant as well. Here we begin by briefly discussing the governance principles required, then specific policies, subsequently moving

---

Box 12.3. **Disinformation and How to Counter It**

Recent decades have seen intensive efforts by vested interests to obfuscate science and spread disinformation, with the goal of blocking efforts to protect public health and the environment. These efforts have now been well documented (75–77). *Misinformation* may be defined as misleading information irrespective of intent to deceive, while *disinformation* is misleading information deliberately created and spread with intent to deceive (78). Such practices are long-standing, but reached unprecedented levels in the hands of the tobacco industry beginning in the 1970s. Similar efforts – often carried out by the same public relations firms, think tanks, and hired scientists – have been deployed by the asbestos industry, the chemical industry (in defence of pesticides, lead, and plastics, to name a few), the fossil fuel industry (in an effort to undermine climate science and slow decarbonization of the energy sector), and other industries. Much of the disinformation focuses on climate change, but other bodies of evidence, such as on biodiversity (79) and sustainable diets (80), are also targeted. Within private firms and governments, scientists are commonly blocked from publicly communicating their data and conclusions (81). Key strategies include creating confusion about established science, exaggerating the level of scientific uncertainty, purveying conspiracy theories, inflating cost estimates of regulations, and deceptively downplaying both the risks of various hazards and the benefits of controlling the hazards. In an age of social media, such disinformation can rapidly go viral (78, 82) – a broader challenge that has bedevilled public health efforts to promote vaccinations and to combat HIV, COVID-19, and other diseases (83). Disinformation on social media does damage well beyond health and the environment, such as vilifying members of minority groups and undermining fair elections. Disinformation is used as a tool to sow discord within societies, manipulate democratic processes, widen existing fault lines, and weaken trust in science, institutions, and conventional media.

Several strategies to counter disinformation have been proposed (83–86). Circulating pre-emptive statements may be effective because initial information about a topic tends to be more influential in shaping beliefs than subsequent messages. Conveying the strength of scientific consensus on an issue may also be effective. Some evidence suggests that high-profile rebuttals may attract additional attention to disinformation and be counterproductive. However, other evidence suggests that correcting disinformation may be more effective than forewarnings; this is especially true if rebuttals come from a credible source and appeal to coherence rather than simply correct facts. Attempts to persuade the most committed deniers and conspiracy theorists are likely a waste of time. Instead, people who are less entrenched in their beliefs can be encouraged to use 'slow thinking', engaging their rational faculties rather than knee-jerk responses (87).

---

on to technology solutions, and ending with consideration of large-scale, integrated implementation strategies.

Managing the Commons

In traditional economic thinking, rational self-interest is the universal driver of economic choices, and the market is the ideal site of exchange of private resources. This unidimensional view of human motivation led to a classic analysis, ecologist Garret Hardin's 1968 'Tragedy of the Commons' (95). Faced with a shared resource such as a pasture or a fishery, Hardin wrote, each individual would be motivated to exploit the resource for maximum personal gain – an approach that would rapidly deplete the shared resource, subverting the common good. External authorities were needed to manage common resources, relegating individuals to the roles of worker, consumer, and voter.

Eleanor Ostrom, the Nobel Prize winning economist, challenged Hardin's analysis. She argued that it failed to recognize 'the wide diversity of institutional arrangements that humans craft to govern, provide, and manage public goods and common-pool resources', and she provided extensive empirical evidence of such arrangements. She argued for the importance of self-governing management of shared resources, beyond that exercised by the state and the market (96).

Ostrom observed that effective rules are often adaptive or negotiable, allowing them to evolve with changing circumstance. Instead of universal rules, she proposed a series of design principles that inform the probability of long-term collective management success **(Figure 12.3)**. Although polycentric management of shared resources through multiple institutions without a clear hierarchy may appear chaotic, it can achieve efficiencies of scale whilst avoiding the bungling of large-scale centralized systems.

Inevitably there have been criticisms of Ostrom's concepts, particularly her suggestion that the state has little part to play at the local level, and some have warned that her proposals could lead to privatization of common resources (97). Nevertheless, she highlighted a key insight: that people can be motivated by more than immediate personal gain, and that collective stewardship of shared resources is possible. This insight is highly salient in the Anthropocene, a time that demands such stewardship. In particular, it is relevant to the role of NGOs, local civil society groups, and indigenous communities as they contest the expropriation of common resources by powerful interests governed by market forces.

An important implication of Ostrom's work is that public policy should support the development of institutions that nurture positive behaviours including adaptability, trustworthiness, cooperation, innovativeness. This is in contrast to many current policies that seek to nudge self-interested individuals to achieve better outcomes as often defined by central authorities and sometimes shaped by vested interests (98). Nudges may have a role in changing behaviour but they are tactical rather than strategic and cannot by themselves re-orient society on a sustainable path.

Current economic policies and practices rarely steward the commons equitably and sustainably. A wide range of economic policies has been proposed to address these challenges in the Anthropocene. Here we consider several: ending harmful subsidies, pricing carbon and other approaches to internalizing the risks of fossil fuels, universal basic income, fighting corruption, and greater transparency, focusing on addressing implementation challenges. We then turn briefly to regulation and litigation, two other tactics for advancing Planetary Health.

Ending Harmful Subsidies

The pervasive nature and harmful impact of perverse subsidies were introduced in Chapter 8. These persist as a result of varying combinations of political inertia, influential

 Strong group identity and understanding of purpose

Fair distribution of costs and benefits

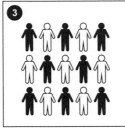 Fair and inclusive decision making

Monitoring agreed upon behaviours

 Graduated sanctions for misbehaviours

Fast and fair conflict resolution

 Authority to self-govern

Appropriate relations with other groups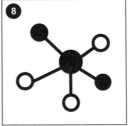

Figure 12.3. Eight principles for managing a commons.
Source: Content from Ostrom E. *Governing the Commons: The Evolution of Institutions for Collective Action.* Cambridge, UK: Cambridge University Press; 1990. Graphic by Emanuel Santos.

vested interests, corruption, lax regulation, and lack of transparency (99). A number of steps can be taken to shift money out of fossil fuels **(Figure 12.4)**.

There are numerous success stories that illustrate how harmful subsidies can be phased out. Redirection of fossil fuel subsidies to support social programmes, including universal health coverage, may be particularly relevant in countries where high fossil fuel subsidies and poor health care coverage co-exist. In Indonesia, for example, fossil fuel subsidies accounted for almost 20% of the government budget between 2009 and

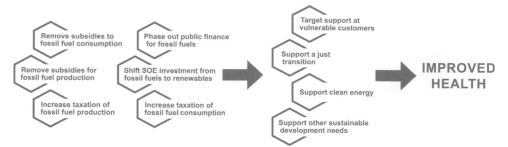

Figure 12.4. Approaches to shifting money out of fossil fuels. SOE refers to state-owned enterprises.
Source: Adapted from Gerasimchuk I, Whitley S, Beaton C, et al. Stories from G20 Countries: Shifting public money out of fossil fuels. Winnipeg: IISD; 2018. www.iisd.org/publications/stories-g20-coun tries-shifting-public-money-out-fossil-fuels.

2013. President Joko Widodo reformed petrol and diesel subsidies in 2015, saving US$15.6 billion, and redirected the funds to social programmes, including free health-care entitlement for low-income households not previously covered by health insurance schemes. In Iran fossil fuel subsidies were cut by nearly 20% in 2010, and about 50% of the funds were redirected to social assistance programmes for the poor, including universal health coverage (100). This helped to gain public support, an important element of successful subsidy and tax reform. A vital aspect of subsidy withdrawal should be to provide support to communities that depend on fossil fuels and are facing impoverishment as a result of the transition to a zero-carbon economy. An example of such support is the Just Transition Fund, established in Europe to blunt the impact of fossil fuel phaseout in EU member countries.

Some cautionary tales show how persistent subsidies can be. For example, in Canada successes have been partial and progress hard fought. That nation possesses the world's second largest reserves of non-conventional oil and has a long history of generous subsidies. Canada reformed seven tax exemptions for oil, gas, and coal exploration during the decade of the 2010s, saving about US$260 million annually, but continues to provide substantial support. It provided US$1.7 billion in April 2020 to fund the cleanup of orphan and abandoned oil and gas wells to provide jobs and prevent bankruptcies in fossil fuel companies reeling from oil price cuts and a dramatic fall in demand triggered by the COVID pandemic. In an attempt to soften criticism and fulfil its climate obligations it also set up a US$750 million Emission Reduction Fund that will provide employment through measures to reduce methane emissions (101). This shows how progress is often non-linear, with advances offset by continuing subsidies driven by political expediency and the desire to protect employment, even in polluting and unproductive industries.

A study of subsidy reform in Iran, India, and Nigeria concluded that four measures were needed to ensure public support for subsidy withdrawal:

1. developing a comprehensive reform policy and public engagement plan, including clear communication to the public before price increases;
2. phasing in price adjustments over a period of time to reduce the impacts;

3. alleviating financial impacts on low- to middle-income households, e.g. by providing a targeted compensatory cash transfer; and

4. capitalizing on favourable global macroeconomic conditions, for example when oil prices have fallen and the economy is buoyant (102).

Failure to heed these lessons can result in social disruption, as seen by the rise of the *gilets jaunes* in France and the violence in Iran when subsidies were suddenly withdrawn during a period of economic hardship.

Carbon Pricing

Carbon pricing has long been advocated by economists as an efficient way of decarbonizing the economy but has only made limited progress. There is a range of approaches to the pricing of carbon including emissions trading systems, carbon taxes, results-based climate finance, carbon offsets, and internal carbon prices set by organizations. As outlined in Chapter 8, only about 21% of carbon emissions worldwide are covered by carbon pricing mechanisms and less than 5% of those are adequately priced to reach the temperature goals of the Paris Agreement (103). To guide the implementation of carbon pricing, the World Bank has proposed FASTER principles – **F**airness, **A**lignment with other policies and objectives, **S**tability and predictability, **T**ransparency, **E**fficiency and cost effectiveness, **R**eliability and environmental integrity (104). These principles acknowledge the potential health co-benefits of such policies but suggest that these are best achieved by local policies because of different levels of air pollution and other contextual factors.

Few countries have set a carbon tax at a level sufficient to decrease emissions rapidly enough to achieve the Paris climate goals. The global average carbon price (explicit and implicit) needed to achieve the long-term climate stabilization target is between US$40/$tCO_2$ and US$80/$tCO_2$ by 2020 and between US$80 and US$120 in 2030, according to a range of studies (105). Much higher prices would be needed to achieve the 1.5 °C target together with other complementary policies to accelerate decarbonization, such as ambitious energy efficiency standards, investment in research and development, and subsidies for renewables. Sweden has a carbon price of US$127/tonne but it only covers about 40% of national emissions. Conversely, California and Quebec cover about 85% of their emissions but at a low carbon price of US$16/tonne.

Emissions trading systems (ETSs) allow flexible pricing but, by putting a cap on emissions and issuing limited numbers of permits, aim to provide certainty about the environmental impact. Their intention can be undermined by issuing too many permits and thus driving down the costs, as a result of lobbying by industries dependent on fossil fuels. The EU ETS for example, which covers about 45% of GHG emissions of member states, had to be reformed to improve its effectiveness, including by reducing the cap on the total permitted emissions by 2.2% annually (106). The sectors at highest risk of relocating their production outside the EU receive full free allocation of permits, which is a disincentive to decarbonization, but this will be gradually phased out from 2026. At the time of writing the EU carbon price has fallen as a result of decreased fossil fuel demand from the effects of the COVID-19 pandemic on the economy.

Although ETSs and carbon taxes are often thought to be equivalent, a carbon tax may be simpler, faster to implement, and more transparent in practice. Taxes have the added advantage that they also send a stronger public message about the social cost of pollution. Additionally, they can raise substantial revenues for the public sector, which can be used to reinforce the positive aspects of the tax or counteract its potential negative effects, for example on low-income groups (107). Although the prospect of new taxes is rarely attractive to policymakers, emphasizing the health and wider co-benefits that will result from decreased air pollution and the potential to redistribute the income could make them much more acceptable. As is the case for subsidy removal (see above) revenue from carbon taxes can provide financial support for universal health coverage, affordable energy-efficient housing, or healthy food choices.

Although carbon pricing has attracted most attention, similar principles could be used for other health and environmentally damaging activities such as pollution, biodiversity loss, and the underlying driving forces causing them. In general, shifting taxes from labour to resource use would encourage a transition from making more consumer goods to supporting a circular economy.

However, while putting a price on carbon (or other environmental harms) is theoretically attractive, taxes are unpopular with the public and with policymakers. While it is possible to insulate poor households from economic pain, fear of such impacts is easily exploited by opponents. To be effective, prices need to rise rapidly – which may increase political resistance. So, while carbon pricing is likely to remain an important arrow in the quiver, the needed transition will likely rely on a range of other strategies.

Other Approaches to Internalizing the Risk of Fossil Fuels

Carbon pricing is a mechanism for internalizing the identifiable costs of fossil fuel use, such as health and environmental damage. However, there is an additional set of risks and liabilities – future damages. The fossil fuel industry confronts enormous future costs, such as cleaning up improperly managed wastes and spills. Moreover, as the industry sunsets, there are financial liabilities associated with the lost value of fossil fuel reserves, coal-fired power plants, and similar 'stranded assets' (108–111). In addition, litigation may impose costs on industries that contributed to climate change or other environmental damages. The risk was exemplified by the bankruptcy filing of California's largest utility in 2019, based on liabilities from massive wildfires, in what the *Wall Street Journal* characterized as 'the first climate-change bankruptcy' (112). Historically, the associated costs have often been borne by governments – that is, by taxpayers. Shielding the public from such costs, and internalizing them to the responsible industries, must be policy priorities.

Governments have a range of policy options: bonding, insurance, letters of credit, corporate guarantees, and surcharge-based trust funds. Such policies require firms to hedge against future obligations and risks, with cash or enforceable assurances. Any such policy, to function effectively, requires standardized ways of assessing and quantifying risk, and transparent disclosure of the findings (113). Reliable information is critical as many firms conceal or minimize the climate risks they face in communicating with investors, regulators, and the public (114), and as markets have tended to undervalue catastrophic risk (115). In 2015, the G20's Financial Stability Board created a Task Force on Climate-related Financial Disclosures (TCFD), in an effort to achieve more complete and consistent disclosure of climate-related risks. The TCFD issued recommendations for voluntary reporting in 2017 (116); the impact of these recommendations has yet to be fully evaluated but if they were made obligatory they could trigger a rapid exit from fossil fuel investments.

Universal Basic Income

A far-reaching proposal, which can be traced to such varied thinkers as Sir Thomas More, Thomas Paine, John Stuart Mill, and Bertrand Russell, and revived in recent years, is Universal Basic Income (UBI) – the unconditional payment, to all members of a society, of enough funds to meet their basic needs. The philosophical argument for UBI is based on its capacity to increase human freedom (117). Practical arguments for UBI cite its capacity to increase gender equality (118), to reduce poverty and inequality (119), and to buffer people from the employment impacts of robotization, automation, and artificial intelligence (120). Another argument in favour of UBI is highly relevant to Planetary Health: rapid decarbonization will be disruptive for many workers whose livelihoods are linked to the fossil fuel economy, and UBI could allay fears and ease the transition. Arguments against UBI centre on its cost, on the morality of cash transfers that are unrelated to need, and on the possibilities that UBI would reduce the incentive to work and/or that people would spend their income in 'unworthy' ways (121–123).

Experiments have tested various forms of UBI in both wealthy and low-income settings, generally on a localized scale. Examples include a nationwide sample of 2000 beneficiaries in Finland (124), the town of Dauphin, Manitoba, Canada (125, 126), and the US state of Alaska (through its Permanent Fund, ironically derived from petroleum revenues) (127), as well as the village of Otjivero, Namibia (128–130), 63 villages in Kenya (131), and 8 villages in the Indian state of Madhya Pradesh (132–134). Evaluations of these initiatives have generally shown little if any reduction in motivation to work, little or no increase (and often a decrease) in spending on 'temptation goods' such as tobacco and alcohol (135), and a range of social and health benefits, sometimes substantial. For example, a study of Namibia's BIG (Basic Income Grant) pilot revealed reductions in poverty, child malnutrition, crime, and women's economic dependency on men, and improvements in school attendance, economic activity, and health (129). However, few UBI programmes have been sustained or evaluated over a long term. Interestingly, the economic catastrophe triggered by the COVID pandemic has moved many governments to provide UBI-like cash payments to their citizens. Spain accelerated implementation of a planned UBI initiative in response to the pandemic (136), in what is the world's largest such effort to date. If UBI becomes more widely accepted and implemented, it could well ease the disruptive economic transitions the Anthropocene will require, but must be accompanied by rigorous evaluation to assess whether the intended benefits are achieved.

Fighting Corruption and Tax Evasion

Governments across the world are subject to improper influence by private interests that work to block progress towards decarbonization and sustainable practices, and lobby for support for harmful industries. One example is that of Kelcey Warren, CEO of Energy Transfer Partners, the firm responsible for building the Dakota Access pipeline. Warren donated US$100,000 to Donald Trump's Presidential campaign and US$250,000 for his Presidential inauguration. Trump rapidly approved the pipeline once in office and Warren's net worth surged from US$1.7 billion to US$4.5 billion (137). More generally, the role of 'dark money' from industries ranging from tobacco to fossil fuels in subverting science, distorting public discourse, purchasing politicians, and blocking regulation, has been well documented (75, 76, 138–141).

In the Brazilian Amazon, deforestation is driven by criminal gangs that have been dubbed 'rainforest mafias' (142). These operate with impunity, often compromising the public officials responsible for enforcing laws. During the decade of the 2010s, more than 300 people were killed in that country in the context of conflicts over the use of land and resources in the Amazon – many of them by people involved in illegal logging (142). This tragic pattern is not limited to Brazil; environmental activists, especially indigenous people, are murdered in high numbers throughout Latin America and globally, especially in countries in which the rule of law has been subverted (143–145). From decisions on individual projects, to the passage and enforcement of laws, and even to acts of political violence, corruption bedevils attempts to rein in industries whose activities threaten the planet.

Another form of corruption that threatens sustainability and equity – and therefore health – is tax evasion. Oxfam estimated in 2013 that wealthy individuals had stashed an estimated US$18.5 trillion in tax havens worldwide, representing US$156 billion in lost tax revenue annually (146). The Tax Justice Network, using different methods, estimated hidden assets (including non-financial assets such as gold) to amount to as much as US$46 trillion as of 2015 (147). Forfeited taxes from corporate tax avoidance are even higher, roughly US$500 billion annually (148). When adjusted for the size of national economies, the greatest losses from corporate tax avoidance occur in LMICs in Sub-Saharan Africa, South Asia, and Latin America and the Caribbean; these losses exceed the amount such countries receive in development assistance. Recouping unpaid taxes could make a major contribution to funding a just transition to a low-carbon economy and end extreme poverty (149).

A full discussion of pathways to overcoming corruption is beyond the scope of this book. It involves strengthening the rule of law, often driven by civil society – both local groups and international NGOs such as Transparency International and Global Witness, with support from governments and foundations committed to more open, effective, and honest government (150, 151). Good government is essential to earning and maintaining public trust, to fairly adjudicating the competing claims arising during the needed transitions of the Anthropocene, to controlling the excesses and abuses of markets, and to assuring that all elements of society have fair access to voice and power.

Transparency

Transparency is essential not only in rooting out corruption; but also in tracking progress towards environmental and health goals. Reporting is a standard feature of public health systems and environmental protocols globally, and is useful in tracking progress, in holding countries and businesses to account, and in motivating improvements in performance. With regard to carbon emissions, for example, nations report their annual emissions under the terms of the UN Framework Convention on Climate Change. Civil society plays an equally important role. For example, the NGO CDP Worldwide (formerly the Carbon Disclosure Project) supports the disclosure by private firms of carbon emissions data, as well as data on water and forest impacts, to enable investors to take this information into account.

Another form of transparency is product labelling. In many countries, food is routinely labelled with nutritional information, allergy information, and with the presence or absence of genetically modified organisms. There is a Globally Harmonized System of Classification and Labelling of Chemicals (152), and in some countries, additional requirements provide for chemical hazard labels for both workers and members of the public. One recent proposal called for point-of-sale health warnings on fossil fuels (153). There are voluntary labelling systems that document labour practices, fair trade provisions, and environmental aspects of products (the latter called 'eco-labels').

Extending these practices, some have called for integrated product labelling that combines nutritional (in the case of food), environmental, and social reporting (154, 155). While such transparency aligns well with Planetary Health goals, the track record of labelling is mixed. Food labelling does influence food *producers*, leading to healthier food formulations, but systematic reviews show equivocal impacts on *consumer* behaviour (156–158). Multidimensional product labels – what might be called Planetary Health labels – would confront the challenges of conveying complex information in understandable form, and of establishing consistent labelling standards across a wide range of products and geographies. Moreover, as retail purchasing increasingly shifts to the Internet, customer interaction with labels will evolve, and new studies to document efficacy will be needed. Finally, unfortunately, there is evidence of cheating and gaming some of the certification schemes that underlie current labelling schemes (159); provisions to prevent such abuses are needed. Planetary Health labelling is a promising tool, but needs careful design underpinned by realistic expectations and evidence of impact (160).

Regulation

Direct regulation is increasingly used to effect changes needed to protect health and the environment, and to achieve the transition to sustainability. Examples range from emissions standards for motor vehicles and factories, to restrictions on cutting forests or on resource extraction. In the USA, more than half the state governments have promulgated Renewable Portfolio Standards, which require utilities to obtain a specified percentage of the electricity they sell from renewable sources. (As of 2020, 14 states had requirements of 50% or greater.) While regulation is politically controversial, and powerful vested interests can effectively block them, they represent a direct and targeted way of shifting practices.

Litigation

A growing trend in recent years has been legal challenges by private citizens and NGOs against governments or private firms, to compel action to protect public health and the environment (161, 162). While some lawsuits claim non-enforcement of environmental laws, others utilize innovative legal approaches. In many cases these lawsuits reflect frustration with failures of governance and policymaking. Climate change is the most common focus; other litigation goals include protecting forests, rivers, or wildlife, and stopping the emission of harmful pollutants. This litigation began as a trickle in the 1980s and accelerated rapidly after 2005; by 2020, 1587 climate lawsuits had been brought worldwide – 1213 in the USA and 374 in 36 other countries and eight regional or international jurisdictions (162). Of these cases, 37 were brought in the Global South – 16 in Asia, 7 in Africa and 14 in Latin America. Examples are shown in **Box 12.4** (163, 164).

These lawsuits often employ novel legal theories, or apply established legal principles in novel ways. These theories include public trust doctrine (the principle that the state must hold and manage certain natural resources for public benefit), human rights law (167), and even legal entitlements for nature (168).

Technologies for a Healthy Anthropocene

Innovative technologies will be essential and many of these have been discussed in preceding chapters. Examples in the energy sector include replacing fossil fuels with renewable sources of power (especially wind and solar), achieving large-scale battery storage for times when solar and wind fall short, and modernizing electrical grids. In the transport sector, electric or hydrogen-powered motor vehicles, efficient, affordable mass transit, and providing opportunities for active travel, including e-bikes, are needed. In the building sector, buildings need to be highly energy-efficient, fully electrified (so that oil, gas, coal, and biomass are no longer used for heating and cooking), and made of sustainably sourced materials. In food and agriculture, innovative technologies range from remote sensing and geospatial data to guide precision application of inputs, to biotechnologies that yield resilient, nutritious foods.

A cross-cutting and vitally important category of technology innovation is information and communications technology (ICT). Universal access to digital technologies and broadband, and implementation of strategies for digital inclusion, accompanied by protection of privacy and digital identity, can reduce inequities and deliver multiple benefits in sectors such as health and education. Some principles of digital transformation for sustainable development are summarized in **Table 12.1**.

The rapid penetration of mobile phones represents a communications revolution of unprecedented scale and speed. As of late 2020 there were 5.16 billion unique mobile

Box 12.4. **Examples of Climate Litigation in Recent Years**

- In 2015, the Dutch environmental group Urgenda and 900 Dutch citizens sued the Dutch government to require it to do more to meet that nation's fair contribution towards the goal of keeping global temperature increases within two degrees Celsius of pre-industrial levels. The court found for the plaintiffs, citing Article 21 of the Dutch Constitution ('It shall be the concern of the authorities to keep the country habitable and to protect and improve the environment'); EU emissions reduction targets; the European Convention on Human Rights; the 'no harm' principle of international law; the doctrine of hazardous negligence; the principle of fairness; the precautionary principle; and the UNFCCC sustainability principle, among others. The decision was appealed through successive levels of Dutch courts, and finally upheld by the Supreme Court in 2019, with an order that the government cut the nation's GHG emissions by 25% from 1990 levels by the end of 2020. This was the first time a nation was required by its courts to act against climate change for reasons other than statutory mandates (165).
- In 2015, a group of 21 youth plaintiffs, represented by the NGO Our Children's Trust, sued the US government and several government officials for infringing on their constitutional and public trust rights to life and liberty. Although the initial outcome was positive for the plaintiffs, with the judge holding that 'a climate system capable of sustaining human life' was a fundamental right, in 2020 a higher court dismissed the case, on the basis that it was beyond the court's power to create and oversee a comprehensive remedy for the plaintiff's complaint (166).
- In 2019, the NGO Friends of the Irish Environment sued the Irish government, arguing that the government's approval of a National Mitigation Plan did not comply with Ireland's Climate Action and Low Carbon Development Act and violated rights protected by the European Convention on Human Rights.
- In 2020, a group of indigenous tribes in the US states of Alaska and Louisiana submitted a complaint to the United Nations alleging the US government had violated their rights to life, health, housing, water, sanitation, a healthy environment and food, resulting in forced displacement, by failing to address climate change.
- In 2020, two Houses of the Wet'suwet'en tribe of British Columbia filed a suit in the Canadian Federal Court alleging that the government's approach to climate change had violated their constitutional and human rights, causing significant warming effects on their territories and negative health impacts.
- In 2020, the French NGOs Notre Affaire à Tous, Sherpa, Zea, and Les EcoMaires joined with more than a dozen French local governments in filing a complaint in the Nanterre District Court against oil major Total. The complaint claimed that Total had violated the 2017 French Law of Vigilance. That law requires companies to formulate 'plans of vigilance' that identify and seek to mitigate risks to human rights, fundamental freedoms, the environment, and public health that could result from their operations. The plaintiffs sought a court order forcing Total to issue a new vigilance plan analysing the risks related to global warming beyond 1.5 °C and Total's contributions to those risks, and aligning its activities with warming of no more than 1.5 °C.

At the time of this writing, the last four cases remain in litigation and have not been resolved.

Table 12.1. *Principles for digital transformation*

Principles	Examples
Enabling digital infrastructure	• Universal access to high-quality, low-cost mobile broadband
Online services	• Online governance to support public services and participation • Online finance and payments to facilitate trade and business services • Regulatory security for online identity and privacy • Online national systems (or 'platforms') for health care and education
Digital systems to increase efficiency of resource use	• Smart grids and Internet of Things for sustainable cities
Instruments for a sustainable digital revolution	• Income redistribution to address income inequalities arising from digital scale-up • Tax and regulatory systems to avoid monopolization of Internet services • Democratic oversight of cutting-edge technologies (biotech, nanotech, artificial intelligence, big data, autonomous systems) • Universal access to high-quality, low-cost mobile broadband education to avoid new digital divides and to develop capacities for sustainable digitalization • Alignment of the emerging digital technologies and infrastructures with human norms and the paradigm of sustainable development

Source: TWI2050. *The World in 2050: Transformations to Achieve the Sustainable Development Goals.* Laxenburg, Austria: International Institute for Applied Systems Analysis (IIASA); 2018. www .twi2050.org.

phone users in the global population of 7.77 billion, a penetration of 66%. There were 4.57 billion Internet users, and 3.81 billion social media users (169). This access has enabled millions of low-income households to benefit from services such as banking and vital information such as weather forecasts for farmers and personal health information. One example of a digital success story is Kenya, which boasts a higher percentage of the population with access to the Internet than in other African countries, and the highest use of mobile phones for financial and other transactions (170). This is a result of sustained policies to build skills, capacity and infrastructure, foster employment, and encourage digital entrepreneurship.

Smartphones offer environmental advantages. They have low energy requirements – between 2.2 watts of power in standby and about 5 watts in use – with the potential to replace a range of other devices that require up to a hundred times more power to deliver similar services (171). A smartphone contains about 25 times less embodied energy than the devices it can replace, with a proportional reduction in GHG emissions.

Mobile phones, together with low-cost sensors, drone technology, and geo-positioning systems (GPS), offer new prospects for monitoring processes relevant to Planetary Health.

One example is the use of GPS to develop fine-scale models of human space use relative to land cover in studies of the zoonotic malaria *Plasmodium knowlesi* in Malaysian Borneo (172). Locations in which secondary forest and houses are in close proximity have higher rates of transmission. Drones have also been used to assess deforestation and how it increases the risk of transmission (173). Mobile phone data have documented migration patterns (174) and elucidated links between agricultural trends and migration (175), disasters and migration (176), and migration and infectious disease transmission (177, 178). Integrating top-down and bottom-up data collection on health and environmental indicators can provide early warning of disease outbreaks and communities that are reaching the limits of adaptation to multiple environmental challenges (179).

Yet technologies with great promise often have limitations. A simple example is the energy demand of the server farms needed to support ICT and of the billions of objects that make up the Internet of Things. These sources are estimated to account for one per cent of global energy use, and this proportion is increasing (180, 181). Electric vehicles, which may bring substantial reductions in air pollution and associated mortality, require large amounts of electricity when widely used; if the electricity does not come from renewable sources, the benefits are greatly attenuated (182). Promising technologies need to be carefully evaluated using life cycle analysis to gauge the health and environmental benefits accurately, and will often need to be accompanied by equitable demand management.

Even more substantial disbenefits may flow from robotization, automation, 3D printing, and artificial intelligence. While these have the potential to eliminate dangerous or tedious human labour, and to produce goods and services highly efficiently, they may displace many occupations, both manual (such as waiting tables and driving) and professional (such as routine legal services and reading X-rays) (120, 183). This is a controversial prediction; opponents note that successive waves of technological change have led to short-term disruption but not to persisting job loss. However, the pace and extent of AI may signal a change in historical patterns. In any event, careful regulation and well-crafted government policies will be needed to blunt the disruption of large-scale job losses (170).

This portion of the chapter has focused on implementation challenges, in particular on policies and practices that can drive the transition to a healthy, sustainable Anthropocene. Each of these has potential, but each is just a single strategy. We turn now to a broader discussion, of ambitious, multidimensional efforts to transform society. Again, this discussion is not intended to be a complete scan of such efforts; indeed, with so many efforts underway worldwide that would be impossible. Instead we cite selected examples that are particularly instructive regarding human health and well-being. **Box 12.5** outlines an ambitious attempt to harness technology for wider societal goals.

Transition Towns

Another example is the Transition Towns movement, which began in 2006 as a local initiative in Totnes, Devon, UK, a town of under 10,000 people. It aimed to support community-led responses to peak oil and climate change, with community-led strategies for building resilience and quality of life (189, 190). There was a strong emphasis on

Box 12.5. **Lucas Aerospace and the Prospect of Socially Useful Technologies**

While transformation is needed on a large scale, important lessons can be gleaned from localized experience. One such case study is Lucas Aerospace.

In the 1970s Lucas Aerospace was heavily engaged in defence work but also undertook civil aerospace work employing a highly skilled workforce. Lucas epitomized many of the defects of British industry: overdependence on arms expenditure funded by taxpayers' money, and deep, unproductive labour–management antagonism. The trade union movement itself had internal divisions regarding how and whether to engage actively.

The Lucas Aerospace Combine Shop Stewards Committee, a cross-union committee representing workers in 13 unions, began an intensive process in 1975 to develop an alternative to management's 'corporate plan' for the future of Lucas. They realized that merely opposing cuts to jobs was likely to fail and that they could harness the skills and knowledge of the workforce to support the conversion of military production for socially useful civilian purposes. A planning committee was set up at each of the 17 Lucas sites to make a detailed analysis of the design, development, and production capabilities and activities of the plant (184). The Combine committee had a number of external advisors and one of us (AH) had the privilege of working with them.

The Alternative Corporate Plan, announced in January 1976 and updated subsequently, contained 150 new product ideas proposed by the workforce, in six categories: medical equipment, transport vehicles, improved braking systems, energy conservation, oceanics, and telechiric machines, as well as a number of proposals for completely reorganizing Lucas production. Examples included hybrid electric–diesel engines, wind turbines, solar heating components, heat pumps, fuel cells, medical devices such as renal dialysis machines and pacemakers and disability aids. Many offered health benefits, either directly (in medical treatment) or indirectly (by, for example, reducing carbon emissions and air pollution). Many remain relevant almost a half century later. The emphasis on socially useful work was also ground-breaking in its implications for the mental well-being of the workforce and wider society.

The Combine's Alternative Corporate Plan gained considerable attention globally and attracted external charitable funding for academic institutions to support their vision. However, Lucas management rejected the plan, and it was never implemented. Eventually the company withered. Some parts were sold off, and others closed. Lucas Aerospace no longer exists.

One of the leaders (although personally critical of the very concept of leadership) of this movement, Mike Cooley, a design engineer with a vision of how industry could satisfy human needs, wrote a perceptive critique of the human–technology relationship. His commitment was to enhancing human potential rather than subordinating it to technological control (185). Despite the ultimate failure of the Lucas Aerospace workforce to prevail against opposition from management and some trade union officials, their ground-breaking ideas on socially useful production and industrial democracy in action are of renewed relevance (186). The conversion of industries dependent on fossil fuels, armaments or high consumption living to sustainable activities is a task made particularly urgent by the post-COVID economic recession.

This experience also raises a broader issue whether forms of organization emphasizing mutual benefits may be more resilient and conducive to sustainable progress than privately owned companies that concentrate wealth in fewer and fewer hands. In the aftermath of the 2008 financial crisis, cooperative banks, savings and credit cooperatives, and credit unions performed well in many cases and required little help from governments. In contrast, many investor-owned competitors had to be bailed out with large amounts of public funding (187).

Where failures occurred these were largely due to risky speculation by institutions that functioned as the equivalent of central banks to networks of local credit unions or cooperative banks. The poor are particularly well served by savings and credit cooperatives. Membership surveys in Colombia, Kenya, and Rwanda showed that half of members lived below the national poverty lines and nearly half were using financial services for the first time. Importantly, such institutions charge the smallest fees for transfer of remittances from family members working overseas, a major source of financial support for many poor families.

Cases where cooperatives became controlled by politicians, powerful private interests, or their cronies offer a cautionary tale. For example, some agricultural cooperatives in Europe evolved into large multinationals with extensive interests that can collide with those of farmers and the public. They have helped to undermine the reforms of the Common Agricultural Policy that aimed to protect biodiversity, reduce pesticide use, and control pollution (188).

Overall though, with good governance, the various types of cooperative enterprise (financial, worker, producer, and consumer) offer promising prospects for reducing inequities, increasing resilience to shocks, providing decent work, and promoting community cohesion and solidarity.

reducing energy use, on relocalization of economic activity across sectors from agriculture to energy to building to health care, and on public engagement, creativity, and innovation. The movement has grown rapidly; there are over 1000 Transition initiatives in over 50 countries, some in small towns and others in the neighbourhoods of large cities. Strengths include the focus on alternatives to consumer lifestyles, on community-building, and on systems approaches. The Transition movement may also be a strategy for improved health, although evidence on this point is not yet available; a health impact assessment of the Transition Town initiative in Totnes, published six years after the initiative started, described theoretical pathways to improved health, in particular the benefits of social capital, but was not able to document these improvements (191). The movement has been criticized for its small scale, for failing to achieve demographic diversity in many of its localities, and for excessive focus on resilience to energy scarcity rather than offering a 'theory of change' to guide the way to large-scale societal transformation, but nevertheless offers an approach that can be built on (192, 193).

Green Deal and Green New Deal

At the other extreme of spatial scales are proposals to implement economy-wide changes, notably the Green Deal in Europe and the Green New Deal in the USA. Discussion of 'Green New Deal' proposals dates to the decade of the 2000s. In 2007, *New York Times* columnist Thomas Friedman proposed the term (194), and a group of British advocates formed a Green New Deal Group (195), and in 2008, then-Executive Secretary of UNEP, Achim Steiner, proposed a Global Green New Deal to create jobs in green industries and combat climate change (196). In the years that followed, the concept took very different paths in Europe and the USA.

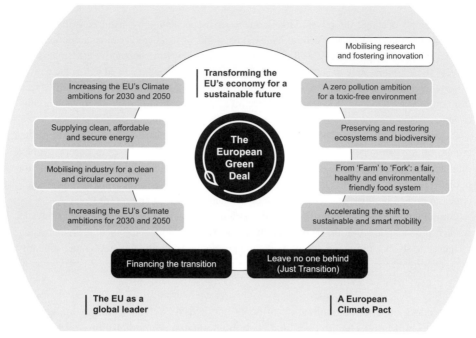

Figure 12.5. Provisions of the European Green Deal.
Source: European Commission. The European Green Deal. Communication from the Commission to the European Parliament, the European Council, the European Economic and Social Committee, and the Committee of the Regions. Brussels: European Commission; 2019.

In Europe, the Green Deal is a set of policy initiatives by the European Commission that aims to reduce GHG emissions by 50–55% by 2030 and to make Europe climate-neutral (with net-zero GHG emissions) by 2050 (**Figure 12.5**) (197). The goals are far-reaching; while energy policy is a central focus, the Green Deal strategy calls for transformation of the EU into 'a fair and prosperous society', and aims to 'protect, conserve and enhance the EU's natural capital, and protect the health and well-being of citizens from environment-related risks'. Accordingly, the European Green Deal contemplates legislation and other actions beyond energy policy – on food (a 'green and healthier agriculture system'), forestry, transport, pollution prevention, the built environment (renovating buildings), biodiversity, and industrial policy (promoting a circular economy). In addition, there is a just transition plan designed to support those regions most dependent on fossil fuels. (In 2015 there were still 128 coal mines across the EU, from Aragón to Silesia, employing more than 238,000 people.)

The prognosis of the European Green Deal is unclear. Each of its policy initiatives would need to be enacted by the complex EU legislative machinery – a formidable but not impenetrable barrier. In fact, in her first State of the Union speech in 2020, EU Commission President Ursula von der Leyen proposed to strengthen the EU's GHG 2030 emission reduction target from the current −40% to at least −55%, suggesting the possibility of increasingly robust action. However, states receiving just transition funding to support decarbonization are not required to commit to a coal phaseout date or to cease investing in natural gas, suggesting that the goals may not be reached throughout the EU. The estimated price of at least €1 trillion over the first decade is substantial but small in relation to Europe's nearly €20 trillion economy (198). While the COVID-19 pandemic may slow progress on the Green Deal, there were many calls to embed Green Deal principles in the pandemic recovery.

In the USA, the Green New Deal had a different origin. It was envisioned by members of civil society and think tanks during the first two decades of the twenty-first century, and was greatly advanced by the activist Sunrise Movement late in 2018–19 in the context of worldwide climate activism. It was inspired by the broad package of economic and social reforms and public works projects launched by US President Franklin D. Roosevelt during the Great Depression, but updated with contemporary social justice, anti-racism, and environmental sustainability ambitions. In 2019, the Green New Deal was presented as a pair of resolutions in the US Congress by Representative Alexandria Ocasio-Cortez and Senator Ed Markey. The resolutions were not detailed legislative roadmaps, in contrast to the European Green Deal. They fell prey to the hyperpartisan environment in the US Congress, and never advanced, although many provisions were incorporated into President Joe Biden's platform.

The Green New Deal called for converting all US electricity generation to renewable and zero-emissions sources within a decade – by 2030. Specific provisions included transition to electric cars, construction of high-speed rail systems, a smart electric grid, energy upgrades to all buildings, and carbon pricing. Alongside these climate-based solutions was a package of social provisions: universal health care, an increased minimum wage, a government jobs programme, family and medical leave, a right to housing, legal action against monopolies. Criticisms of the Green New Deal included the technical, political, and economic feasibility of such rapid decarbonization (199, 200). The future of the US Green New Deal will depend greatly on political developments in the USA.

Green New Deal strategies, in the USA, the UK, and Europe, exemplify the aspirational, multi-sectoral approach to meeting the challenges of the Anthropocene. Some are more explicit about health and well-being than others. However, many positive impacts on health can be identified – through reduced air pollution, increased physical activity, enhanced green space access, reduced global heating, improved housing, and healthier diets (201). As with any large-scale system change, unintended adverse consequences may occur, including through effects on trade with low- and middle-income countries. If and when Green New Deal proposals advance, it will be essential to build explicit health goals and metrics into them, to monitor outcomes, and to make course adjustments as needed.

The Club of Rome Planetary Emergency Plan

The Club of Rome was founded in 1968 and released its influential report, *The Limits to Growth*, in 1972 (202). This report in many ways anticipated Planetary Health thinking. It conceptualized challenges on a global scale, it called for a systems approach, it utilized scenarios (based on early use of computer simulations), and it challenged the notion of perpetual economic growth.

Nearly a half century later, the Club of Rome, in collaboration with the Potsdam Institute for Climate Impact Research, released its *Planetary Emergency Plan* (with an update a year later, addressing COVID-19) (203, 204). These reports identify climate change and eco-logical destruction as a global-scale emergency – a 'dangerous event requiring immediate action to reduce risk of potentially catastrophic results'. It notes, 'Fundamental changes to the environment threaten to undermine the progress we have made in health and life expectancy.' And it identifies opportunities to 'rebuild, improve and regenerate'. The proposed emergency plan includes ten urgent actions under the categories of transforming energy systems, shifting to a circular economy, and creating a just and equitable society founded in human and ecological well-being. Although it also proposes specific commit-ments, such as halting all deforestation, stopping Arctic fossil fuel exploitation, and boosting funding for ecosystem restoration, it stops short of providing a roadmap for global implementation. And although it proposes to 'guarantee the human dimension ... [and] provide for a just transition and well-being economy', it says little about specific mechanisms for promoting health and well-being. Nevertheless, the *Planetary Emergency Plan* is a notable example of a Planetary Health approach that includes specific measures that, if implemented, would advance human health and well-being while reducing our environmental footprint.

Coda: The World after COVID-19

As we write this chapter, the world is battling the COVID-19 pandemic – a battle that will likely continue for many months if not for years. About 115 million cases have been recorded (likely a substantial undercount) and the global death count has surpassed 2.5 million. By the time you are reading this chapter, those numbers will likely seem far in the past, and mercifully low.

Even at this point, in early 2021, there are lessons to be learned from COVID-19; lessons that are highly relevant to Planetary Health. We propose ten lessons, drawn from both successes and failures in confronting the pandemic, informed by the perspectives developed throughout this book.

First, *zoonotic disease spillover is a substantial and growing threat*. This risk has been recognized for a long time, as described in Chapter 3. Pandemics may increasingly be a defining feature of the Anthropocene (205, 206). Action to reduce deforestation, wildlife trafficking, wet markets, and intensive livestock rearing has been grossly inadequate. Such action must be pursued urgently, including through the Convention for Biological Diversity and national legislation. It must also be pursued carefully and with respect for complexity. For example, small-scale animal rearing is a pillar of many low-income rural economies and food systems (207). Abrupt shutdowns of wet markets could have unintended conse-quences including increasing illicit trade; such interventions need to be planned and implemented carefully, in consultation with local voices – as is true for many Planetary Health interventions.

Second, *health system capacity is essential in reducing risk*. Public health systems need to be able rapidly to scale up epidemiological surveillance, diagnostic testing, contact tracing, isolation of infected cases, quarantining of the exposed, protection of high-risk groups, public communication, and vaccination. These strategies work, as demonstrated by initial successes (up to now) in South Korea, Taiwan, Mongolia, New Zealand, Cuba, Rwanda, and Singapore (to name a few) and by failures in the USA, Brazil, the UK, and many other European countries. There needs to be capacity, at the community level, for identifying and supporting those who are ill and the most vulnerable. This should be based on strong systems of community-based primary care. Rapid mobilization of large numbers of community health workers with short training courses and ongoing supervision can help to address epidemics and their wider impacts, as has been shown for cholera and Ebola (208–210). Clinical facilities need surge capacity to accommodate abrupt increases in caseloads, and robust supply chains for health products, food, and other essentials. Initiatives such as the Pandemic Supply Chain Network (211) aim to address bottlenecks in supply and price gouging that can result in rapid increases in supply of vital products in the face of overwhelming demands.

Third, *deprived and marginalized populations are at highest risk in many countries*. This observation is true of nearly all health burdens, and COVID-19 exemplifies it, with higher attack rates and higher case fatality rates in members of ethnic and racial minority groups and among the poor (212–214). The reasons for these disparities are complex, and include employment that entails excess risk of infection, crowded living conditions, disproportionate exposure to environmental risk factors such as air pollution, worse access to personal protective equipment, worse access to medical care, and a disproportionate prevalence of underlying health problems such as cardiovascular disease – traceable, at least in large part, to the environmental and social correlates of racism and poverty. There has also been an epidemic of violence against women accompanying lockdown in many countries, emphasizing the need to scale up resources and services, including in health care, to address this burgeoning threat (215). Steps to protect the public from pandemics need to focus relentlessly on the most vulnerable populations.

Fourth, *political ideology can be a powerful barrier to success, or a powerful predictor of success*. In many parts of the world, populist governments politicized the COVID-19 response, and resisted public health measures. An extreme example is that of President Bolsonaro of Brazil, who actively contradicted science, sought to undermine strategies to contain COVID-19, and propagated misinformation about its impacts (216). In contrast, Norway and Denmark, with high levels of social capital and trust in government, achieved widespread cooperation and were relatively successful in containing the pandemic (217, 218).

Fifth, irrespective of posturing by politicians, *public opinion can strongly support public health measures*. Controlling the pandemic necessitated dramatic changes in behaviour that have had widespread public support in many countries. At least in the short term it is clear that decisive actions to protect public health, backed by strong evidence to justify difficult decisions, can be broadly acceptable, although contested by sometimes vocal minorities. As noted above, in the discussion of neoliberal economics, the pandemic may have reduced fear of the 'nanny state' (219) and driven greater openness to collective solutions. The challenge will be to build on these achievements to ensure that, as economies emerge from COVID-19, the sacrifices made to protect health are not squandered by the rush to re-boot the economy by subsidizing polluters and other businesses whose survival depends on not paying the full economic cost of their activities.

Sixth, *global solidarity is needed to tackle global challenges.* The pandemic exemplified the need for international cooperation and collaboration to develop and test vaccines and potential therapeutic agents. Development Assistance for Health can support 'global health security', including by increasing national capability to implement the International Health Regulations that can reduce the spread of infectious diseases and the risk of repeated waves of infection re-emerging from LMICs to affect HICs (220). The promotion of global health security has sometimes been seen as in competition with universal health coverage, reflecting a concern by high-income nations to minimize the risks of emerging infectious diseases to their own populations. Investments by donor countries, if appropriately designed, can however support both universal health coverage and infectious disease control without creating unnecessary and inefficient vertical programmes, providing mutual benefit for recipient and donor countries (221). Universal access to effective and affordable COVID-19 vaccines is, for example, essential to reduce the risks of the emergence of more transmissible and lethal mutant strains. This recognition of mutual benefit and solidarity is in contrast to the nationalist rhetoric that has become increasingly prevalent in political discourse.

Seventh, *scientific collaboration drives rapid progress.* The pandemic provided many examples of scientists and health professionals overcoming geographical and disciplinary barriers to collaborate for public benefit. More than 100 scientific journals, publishers, and research organizations backed a statement from the Wellcome Trust supporting rapid and open sharing of research on COVID-19. Sharing of preprints, even before peer review, was facilitated, allowing early access to emerging research findings, and funders rapidly scaled up support for collaborative trials and modelling (222). These developments can be models for addressing other Planetary Health challenges. At the same time, rigorous standards of scientific review should be adhered to with the aim of avoiding the premature adoption of unproven and risky interventions (223, 224). Developing coordinated strategies to confront misinformation, including vaccine hesitancy and climate change denial, will be a priority.

Eighth, *the pandemic response may have revealed long-term solutions* useful in meeting Planetary Health challenges. Lockdowns and self-isolation policies greatly reduced economic activity, including local and long-distance travel, with resulting declines in air pollutants such as NO_2 and $PM_{2.5}$ in a range of countries such as China, India, Italy, and the USA, although this was not universal (225–229). In many cities, as vehicular travel fell and demand for outdoor activity grew, streets were reallocated to pedestrian and bicycle use (230, 231). Similarly, shutdowns created a huge burden of cabin fever, anxiety, and depression, stimulating demand for restorative time in natural settings. Many cities closed their parks initially, but re-opened them in response to public demand, and saw usage increase (232, 233). While the long-term impact of the pandemic on commercial property in cities is unclear, at least two possibilities bear mention. First, as firms have learned to accommodate working from home, they may permanently vacate office space to reduce costs. Second, as consumers shift to online shopping, demand for retail space may permanently fall. These trends may provide opportunities to repurpose urban space to affordable housing, a scarce resource in most cities (234, 235). Could these responses portend long-term public demand for cleaner air, and for more active transport infrastructure, parks and green space, and affordable housing? On the other hand, potential negative long-term impacts of the pandemic such as reduced transit use could be addressed, e.g. by incentivizing active travel in cities (236).

Ninth, *the economic impacts of COVID-19 may be severe and long-lasting*. Almost as soon as the pandemic emerged, and economies shut down, unemployment skyrocketed and poverty deepened around the world. The post-COVID economic recession has the potential to exceed the impacts of the 2008 financial crash by perhaps an order of magnitude and could have long-lasting effects akin to the those of the Great Depression in the absence of effective policies to restart the economy (237). Recessions damage both physical and mental health; following the 2008 crash, health indicators worsened considerably, particularly in men, with declines in self-rated health and increased morbidity, psychological distress, and suicide (although traffic fatalities and population-level alcohol consumption declined) (238). Early evidence suggests disproportionate impacts on women, the elderly (239), and people of colour (240). Social safety nets appeared to reduce the adverse effects in a number of European countries. Such safety nets will be necessary to buffer health threats from many disruptions during the Anthropocene.

Tenth, and finally, *the pandemic presents an opportunity for profound system change*. A return to 'normal' – when 'normal' consisted of distributional injustice and unsustainable use of energy and resources – would be a missed opportunity for systemic change. In the absence of well-designed and carefully implemented policies, the post-COVID economic stimulus packages will no doubt cause a rebound in GHG emissions and air pollution. In 2008, for example, recession caused global CO_2 emissions to fall by about 1.9%, which was quickly reversed by a dramatic rebound in 2010 and 2011 (241). This time both the initial fall and the potential rebound could be much larger (242, 243). Moreover, the emphasis on economic recovery without integration of health and environmental sustainability could undermine public support for transition to a green economy (244).

However, the unprecedented infusions of government funds into the world's economies represents an historic opportunity (245). As governments make these investments, they should be guided by criteria regarding which firms should benefit from recapitalization with taxpayers' money. These criteria should not only include firms' economic viability, but should also address environmental, social (including health), and governance (ESG) concerns. Recovery from the crisis could be a great opportunity for rapid decarbonization of the economy and to safeguard health. Companies whose products jeopardize public health and environmental sustainability, whose business model would not be competitive if they paid the full economic costs of their environmental and health externalities, should not generally be bailed out. An exception might be those whose products can be rapidly re-oriented towards health and sustainability, but in these cases ongoing support would depend on achieving these objectives within short timescales.

Existing models provide proof of concept for such investment practices. For example, sovereign wealth funds that manage equity of over US$8 trillion worldwide assess both past ESG performance and recent ESG score improvement when considering whether to purchase ownership stakes in listed companies (246). Similarly, mutual funds actively compete for climate-conscious investment flows, to achieve the 'Low Carbon Designation' created by Morningstar in 2018 (247).

Health criteria might include companies producing vaccines, therapeutic agents, essential diagnostic, protective and other medical equipment, in return for binding agreements to avoid excessive profits in transactions with national health systems. Criteria should also advantage products that improve or protect health. For example, renewable energy technologies yield a double environmental and health benefit, with the potential to prevent millions of premature deaths annually globally from air pollution attributable to burning fossil fuels (248). Similarly, the horticulture and gardening sectors yield health benefits through fruit and vegetable consumption, and the mental and physical benefits of gardening. These sectors also have potential to provide productive employment across a range of professions (249). In order to minimize the health and social impacts of recession, sustainable employment opportunities must also be created as part of the stimulus package. An example is the Obama Administration's Recovery Act following the 2008 recession; this yielded an estimated 192,000 job years of employment in the US renewable energy sector between 2009 and 2012 and markedly reduced the costs of clean renewable energy (250). Large-scale government jobs programmes could provide the labour to insulate millions of buildings, install electric vehicle charging stations, and create parks, cycle paths, and other healthy infrastructure (251).

Finally, the need for a post-COVID economic stimulus is also an opportunity to redirect harmful subsidies from fossil fuels and other damaging products to more productive uses such as renewable energy, health care financing, or reducing the tax burden on the poor. The recovery from the COVID-19 crisis could be a major opportunity to re-orient companies – and thus our societies – towards an economic model that promotes and protects health and the environment.

Conclusions

This book has reported some grim facts: the persistence of inequities, poverty, and suffering in far too many places, the degradation of Earth systems, and the near-certainty that in the absence of decisive action many of these trends will worsen considerably in coming decades. The emergence of the COVID-19 pandemic in 2019–20 has only intensified concern.

But this book has also, we hope, pointed the way to solutions. The path forward builds on the Sustainable Development Goals, aspires to the vision of the lifebuoy economy, and holds health and well-being, equity, and environmental sustainability to be central (**Figure 12.6**). New ways of thinking about our planet – replacing domination with stewardship, overconsumption with moderation, destructive competition with solidarity

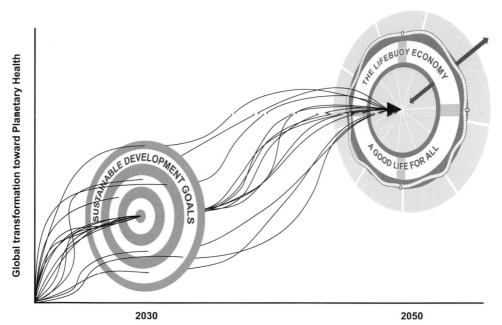

Figure 12.6. A schematic representation of the global transformation toward Planetary Health, in alignment with the Sustainable Development Goals and the principles of the lifebuoy economy.
Credit: Emanuel Santos.

and cooperation, inequity with fairness – will help overcome entrenched conceptual and empathy failures. New ways of investigating and elucidating the workings of our planet and our place on it, the determinants of thriving, and the most effective paths to success, will help overcome long-standing knowledge failures. And new ways of acting on what we know – better governance, better economic arrangements, innovative technologies, changes in attitudes and behaviours – will help overcome implementation challenges. This is an unstable moment, a moment when societies are convulsed by multiple crises – a pandemic, economic ruin, inequalities, racism, the rise of autocrats, accelerating natural disasters. But moments of instability are also moments of opportunity, when with clear vision, strong science, resolve, stamina and courage, we can create a healthier, more equitable and sustainable world.

References

1. Whitmee S, Haines A, Beyrer C, et al. Safeguarding human health in the Anthropocene Epoch: report of The Rockefeller Foundation–*Lancet* Commission on Planetary Health. *The Lancet*. 2015;386(10007):1973–2028. doi: 10.1016/S0140-6736(15)60901-1.
2. UN ESC. The Right to the Highest Attainable Standard of Health. Geneva: United Nations Economic and Social Council Committee on Economic Social and Cultural

Rights; 2000. Available from https://apps.who.int/disasters/repo/13849_files/o/UN_human_rights.htm.

3. Baaki J, Barre A, Bohland P, et al. Gender Just Climate Solutions. Utrecht: Women Engage for a Common Future; 2019. Available from www.wecf.org/gender-just-climate-solutions-5th-edition/.

4. Marmot M, Allen J, Boyce T, Goldblatt P, Morrison J. Health Equity in England: The Marmot Review 10 Years On. UCL Institute of Health Equity; 2020. Available from www.instituteofhealthequity.org/resources-reports/marmot-review-10-years-on.

5. Gugushvili A, Koltai J, Stuckler D, McKee M. Votes, populism, and pandemics. *International Journal of Public Health*. 2020;65(6):721–2. doi: 10.1007/s00038-020-01450-y.

6. Butzer KW. Collapse, environment, and society. *Proceedings of the National Academy of Sciences*. 2012;109(10):3632–9. doi: 10.1073/pnas.1114845109.

7. Raskin P. *Journey to Earthland: The Great Transition to Planetary Civilization*. Boston: Tellus Institute; 2016.

8. Korten DC. *The Great Turning: From Empire to Earth Community*. San Francisco: Berrett-Koehler; 2007.

9. WCED. *Our Common Future. Report of the World Commission on Environment and Development*. New York and Oxford: Oxford University Press; 1987.

10. Naidoo R, Fisher B. Reset Sustainable Development Goals for a pandemic world. *Nature*. 2020;583(7815):198–201. doi: 10.1038/d41586-020-01999-x.

11. UN DESA. The Sustainable Development Goals Report 2020. New York: United Nations Department of Economic and Social Affairs; 2020. Available from https://unstats.un.org/sdgs/report/2020/.

12. Leal Filho W, Brandli LL, Lange Salvia A, Rayman-Bacchus L, Platje J. COVID-19 and the UN Sustainable Development Goals: threat to solidarity or an opportunity? *Sustainability*. 2020;12(13):5343. Available from www.mdpi.com/2071-1050/12/13/5343.

13. Sachs J, Schmidt-Traub G, Lafortune G. Speaking truth to power about the SDGs. *Nature*. 2020;584(7821):344. doi: 10.1038/d41586-020-02373-7.

14. Prescott LS, Logan CA, Albrecht G, et al. The Canmore Declaration: Statement of Principles for Planetary Health. *Challenges*. 2018;9(2). doi: 10.3390/challe9020031.

15. Miles SH, Craddock S. Ethics. In Dellasala DA, Goldstein MI, editors. *Encyclopedia of the Anthropocene*. Elsevier; 2018. pp. 21–7.

16. Benatar S, Upshur R, Gill S. Understanding the relationship between ethics, neoliberalism and power as a step towards improving the health of people and our planet. *The Anthropocene Review*. 2018;5(2):155–76. doi: 10.1177/2053019618760934.

17. Broome J. *Climate Matters: Ethics in a Warming World*. New York: W.W. Norton; 2018.

18. Harris PG. *Global Ethics and Climate Change*. Edinburgh: Edinburgh University Press; 2016.

19. Wettstein HK, French PA. *Ethics and Global Climate Change*. Boston: Wiley; 2016.

20. Tremmel J, Robinson K. *Climate Ethics: Environmental Justice and Climate Change*. London: Taurus; 2014.

21. LaDuke W. Traditional ecological knowledge and environmental futures. *Colorado Journal of Environmental Law and Policy*. 1994;5(1):127–48.

22. Menzies CR, editor. *Traditional Ecological Knowledge and Natural Resource Management*. Lincoln, NE: University of Nebraska Press; 2006.

23. Nelson MK, Shilling D, editors. *Traditional Ecological Knowledge: Learning from Indigenous Practices for Environmental Sustainability*. Cambridge, UK: Cambridge University Press; 2018.

24. Foster A, Cole J, Farlow A, Petrikova I. Planetary Health ethics: beyond first principles. *Challenges*. 2019;10(1):14. Available from www.mdpi.com/2078-1547/10/1/14.

25. Plehwe D, Mirowski P, editors. *The Road from Mont Pèlerin: The Making of the Neoliberal Thought Collective*. Cambridge, MA: Harvard University Press; 2015.

26. Mont Pelerin Society. Statement of Aims 1947. Available from www.montpelerin .org/statement-of-aims/.

27. Maxton GP, Randers J, Suzuki DT, Cameron L, Runde Ø. *Reinventing Prosperity: Managing Economic Growth to Reduce Unemployment, Inequality, and Climate Change*. A report to the Club of Rome. Vancouver and Berkeley: Greystone Books; 2016.

28. Epstein RA. Playing Politics with Coronavirus. *Defining Ideas*. Hoover Institution; 2020. Available from www.hoover.org/research/playing-politics-coronavirus.

29. Blanton D. Fox News poll: Big shift in asking government to 'lend me a hand' amid pandemic, unrest. Fox News. 14 August 2020. Available from www.foxnews.com/ politics/fox-news-poll-big-shift-in-asking-government-to-lend-me-a-hand-amid-pandemic-unrest.

30. Collier P. *The Future of Capitalism*. London: Allen Lane; 2018.

31. Henderson R. *Reimaginging Capitalism in a World on Fire*. New York: Public Affairs; 2020.

32. Klein N. *This Changes Everything: Capitalism vs. the Climate*. New York: Simon & Schuster; 2014.

33. Moore JW, editor. *Anthropocene or Capitalocene? Nature, History, and the Crisis of Capitalism*. Oakland, CA: PM Press/Kairos; 2016.

34. Nogueira A, Ashton WS, Teixeira C. Expanding perceptions of the circular economy through design: eight capitals as innovation lenses. *Resources, Conservation and Recycling*. 2019;149:566–76. https://doi.org/10.1016/j.resconrec.2019.06.021.

35. da Silva J, Fernandes V, Limont M, Rauen WB. Sustainable development assessment from a capitals perspective: analytical structure and indicator selection criteria. *Journal of Environmental Management*. 2020;260:110147. https://doi.org/10.1016/j .jenvman.2020.110147.

36. Guerry AD, Polasky S, Lubchenco J, et al. Natural capital and ecosystem services informing decisions: from promise to practice. *Proceedings of the National Academy of Sciences*. 2015;112(24):7348–55. doi: 10.1073/pnas.1503751112.

37. Kareiva P, Tallis H, Ricketts TH, Daily GC, Polasky S, editors. *Natural Capital: Theory and Practice of Mapping Ecosystem Services*. Oxford: Oxford University Press; 2011.

38. Daily GC, Söderqvist T, Aniyar S, et al. The value of nature and the nature of value. *Science*. 2000;289(5478):395–6. doi: 10.1126/science.289.5478.395.

39. Millennium Ecosystem Assessment Program. *Ecosystems and Human Well-Being: Synthesis*. Washington, DC: Island Press; 2005.

40. Bateman I, Wheeler B. Bringing Health and the Environment into Decision Making: The Natural Capital Approach. Oxford, UK: Rockefeller Foundation Economic Council on Planetary Health at the Oxford Martin School; 2018. Available from www.planetaryhealth.ox.ac.uk/wp-content/uploads/sites/7/2018/06/Health-Env-in-Decision-Making.pdf.

41. Polasky S, Bryant B, Hawthorne P, et al. Inclusive wealth as a metric of sustainable development. *Annual Review of Environment and Resources*. 2015;40(1):445–66. doi: 10.1146/annurev-environ-101813-013253.

42. UNEP. Inclusive Wealth Report. London: UN Environment Programme; 2018. Available from www.unenvironment.org/resources/report/inclusive-wealth-report-2018.

43. Raworth K. *Doughnut Economics: Seven Ways to Think Like a 21st-Century Economist*. London: Random House; 2017.

44. Jackson T. *Prosperity Without Growth: Foundations for the Economy of Tomorrow*, 2nd ed. London: Routledge; 2017.

45. Daly HE. *Beyond Growth: The Economics of Sustainable Development*. Boston: Beacon Press; 1996.

46. Rees WE. Avoiding Collapse: An Agenda for Sustainable Degrowth and Relocalizing the Economy. Vancouver: Canadian Centre for Policy Alternatives; 2014. Available from www.policyalternatives.ca/publications/reports/avoiding-collapse.

47. Kallis G. *Degrowth*. Newcastle upon Tyne: Agenda Publishing; 2018.

48. Bowen A, Hepburn C. Green growth: an assessment. *Oxford Review of Economic Policy*. 2015;30(3):407–22. doi: 10.1093/oxrep/gru029.

49. Hueting R. Why environmental sustainability can most probably not be attained with growing production. *Journal of Cleaner Production*. 2010;18(6):525–30. https://doi.org/10.1016/j.jclepro.2009.04.003.

50. Semieniuk G. Energy in economic growth: is faster growth greener? Department of Economics, SOAS, University of London; 2018. Available from https://EconPapers.repec.org/RePEc:soa:wpaper:208.

51. Trentmann F. *Empire of Things: How We Became a World of Consumers, from the Fifteenth Century to the Twenty-First*. New York: HarperCollins; 2016.

52. de Graaf J, Wann D, Naylor TH. *Affluenza: The All-Consuming Epidemic*. San Francisco: Berrett-Koehler; 2002.

53. Schwartz DT. *Consuming Choices: Ethics in a Global Consumer Age*, 2nd ed. Lanham, MD: Rowman & Littlefield; 2017.

54. Shaw D, Carrington M, Chatzidakis A. *Ethics and Morality in Consumption: Interdisciplinary Perspectives*. Chicago: Taylor and Francis; 2016.

55. Ruvio AA, Belk RW, editors. *The Routledge Companion to Identity and Consumption*. Abingdon, UK and New York: Routledge; 2012.

56. McLaren DP, Childs M. Consumption and Identity. Friends of the Earth; 2013.

57. Schor JB. *The Overspent American: When Buying Becomes You*. New York: Basic Books; 1998.

58. Lander L. The good life versus consumerism and greed: reaching a better understanding of barriers and motivators for sustainability. *Sustainability: The Journal of Record*. 2018;11(2):65–73. doi: 10.1089/sus.2017.0022.

59. Wellcome Trust. Our Planet, Our Health: Responding to a Changing World. London: Wellcome Trust; 2020. Available from wellcome.ac.uk/what-we-do/our-work/our-planet-our-health.

60. ABPI. Worldwide Pharmaceutical Company R&D Expenditure. London: Association of the British Pharmaceutical Industry; 2020. Available from www.abpi.org.uk/facts-and-figures/science-and-innovation/worldwide-pharmaceutical-company-rd-expenditure/.

61. Viergever RF, Hendriks TCC. The 10 largest public and philanthropic funders of health research in the world: what they fund and how they distribute their funds. *Health Research Policy and Systems*. 2016;14(1):12. doi: 10.1186/s12961-015-0074-z.

62. King MD, Howat IM, Candela SG, et al. Dynamic ice loss from the Greenland Ice Sheet driven by sustained glacier retreat. *Communications Earth & Environment*. 2020;1(1):1. doi: 10.1038/s43247-020-0001-2.

63. van Klink R, Bowler DE, Gongalsky KB, et al. Meta-analysis reveals declines in terrestrial but increases in freshwater insect abundances. *Science*. 2020;368(6489):417–20. doi: 10.1126/science.aax9931.

64. FAO. The State of World Fisheries and Aquaculture 2020. Rome: Food and Agriculture Organization of the United Nations; 2020. Available from www.fao.org/documents/card/en/c/ca9229en.

65. Brock PM, Fornace KM, Grigg MJ, et al. Predictive analysis across spatial scales links zoonotic malaria to deforestation. *Proceedings of the Royal Society B: Biological Sciences*. 2019;286(1894):20182351. doi: 10.1098/rspb.2018.2351.

66. Nawaz MO, Henze DK. Premature deaths in Brazil associated with long-term exposure to PM2.5 from Amazon fires between 2016 and 2019. *GeoHealth*. 2020;4(8):e2020GH000268. doi: 10.1029/2020GH000268.

67. Smith MR, Myers SS. Global health implications of nutrient changes in rice under high atmospheric carbon dioxide. *GeoHealth*. 2019;3(7):190–200. doi: 10.1029/2019GH000188.

68. Chang KM, Hess JJ, Balbus JM, et al. Ancillary health effects of climate mitigation scenarios as drivers of policy uptake: a review of air quality, transportation and diet co-benefits modeling studies. *Environmental Research Letters*. 2017;12(11):113001.

69. Hess JJ, Sathish LM, Knowlton K, et al. Building resilience to climate change: pilot evaluation of the impact of India's first heat action plan on all-cause mortality. *Journal of Environmental and Public Health*. 2018;2018:8. doi: 10.1155/2018/7973519.

70. Dhankher OP, Foyer CH. Climate resilient crops for improving global food security and safety. *Plant, Cell & Environment*. 2018;41(5):877–84. doi: 10.1111/pce.13207.

71. Lebowitz AJ. Community collaboration as a disaster mental health competency: a systematic literature review. *Community Mental Health Journal*. 2015;51(2):125–31. doi: 10.1007/s10597-014-9751-6.

72. Stiglitz JE, Fitoussi J-P, Durand M. *For Good Measure: Advancing Research on Well-Being Metrics Beyond GDP*. Paris: OECD Publishing; 2018.

73. Belesova K, Haines A, Ranganathan J, Seddon J, Wilkinson P. Monitoring environmental change and human health: Planetary Health Watch. *The Lancet*. 2020;395(10218):96–8. doi: 10.1016/S0140-6736(19)33042-9.

74. Whitmarsh L, Capstick S, Nash N. Who is reducing their material consumption and why? A cross-cultural analysis of dematerialization behaviours. *Philosophical Transactions A: Mathematical, Physical and Engineering Sciences*. 2017;375(2095). doi: 10.1098/rsta.2016.0376.

75. Oreskes N, Conway EM. *Merchants of Doubt: How a Handful of Scientists Obscured the Truth on Issues from Tobacco Smoke to Global Warming*. London: Bloomsbury Press; 2010.

76. Michaels D. *The Triumph of Doubt: Dark Money and the Science of Deception*. Oxford and New York: Oxford University Press; 2020.

77. Mulvey K, Shulman S. The Climate Deception Dossiers: Internal Fossil Fuel Industry Memos Reveal Decades of Corporate Disinformation. Union of Concerned Scientists; 2015. Available from www.ucsusa.org/global-warming/fight-misinformation/climate-deception-dossiers-fossil-fuel-industry-memos#.XAVrmmiwnD4.

78. Treen KMdI, Williams HTP, O'Neill SJ. Online misinformation about climate change. *WIREs Climate Change*. 2020;11(5):e665. doi: 10.1002/wcc.665.

79. Lees AC, Attwood S, Barlow J, Phalan B. Biodiversity scientists must fight the creeping rise of extinction denial. *Nature Ecology & Evolution*. 2020. doi: 10.1038/s41559-020-01285-z.

80. Garcia D, Galaz V, Daume S. EATLancet vs yes2meat: the digital backlash to the planetary health diet. *The Lancet*. 2019. doi: 10.1016/S0140-6736(19)32526-7.

81. Driscoll DA, Garrard GE, Kusmanoff AM, et al. Consequences of information suppression in ecological and conservation sciences. *Conservation Letters*. 2020; e12757. doi: 10.1111/conl.12757.

82. Bloomfield EF, Tillery D. The circulation of climate change denial online: rhetorical and networking strategies on Facebook. *Environmental Communication*. 2019;13(1): 23–34. doi: 10.1080/17524032.2018.1527378.

83. Vicol D-O. Health Misinformation. London: Full Fact, Africa Check, and Chequeado; 2020. Available from https://fullfact.org/blog/2020/may/health-misinformation/.

84. Lewandowsky S, Ecker UKH, Seifert CM, Schwarz N, Cook J. Misinformation and its correction: continued influence and successful debiasing. *Psychological Science in the Public Interest*. 2012;13(3):106–31. doi: 10.1177/1529100612451018.

85. van der Linden S, Leiserowitz A, Rosenthal S, Maibach E. Inoculating the public against misinformation about climate change. *Global Challenges*. 2017;1(2):1600008. doi: 10.1002/gch2.201600008.

86. Walter N, Murphy ST. How to unring the bell: a meta-analytic approach to correction of misinformation. *Communication Monographs*. 2018;85(3):423–41. doi: 10.1080/03637751.2018.1467564.

87. Kahneman D. *Thinking, Fast and Slow*. New York: Farrar, Straus and Giroux; 2011.

88. Wanger TC, DeClerck F, Garibaldi LA, et al. Integrating agroecological production in a robust post-2020 Global Biodiversity Framework. *Nature Ecology & Evolution*. 2020;4(9):1150–2. doi: 10.1038/s41559-020-1262-y.

89. Chazdon RL, Harvey CA, Komar O, et al. Beyond reserves: a research agenda for conserving biodiversity in human-modified tropical landscapes. *Biotropica*. 2009;41(2):142–53. doi: 10.1111/j.1744-7429.2008.00471.x.

90. Hanefeld J, Vearey J, Lunt N, et al. A global research agenda on migration, mobility, and health. *The Lancet*. 2017;389(10087):2358–9. doi: 10.1016/S0140-6736(17) 31588-X.

91. Frumkin H, Bratman GN, Breslow SJ, et al. Nature contact and human health: a research agenda. *Environmental Health Perspectives*. 2017. doi: 10.1289/EHP1663.

92. Luè A, Bresciani C, Colorni A, et al. Future priorities for a climate-friendly transport: a European strategic research agenda toward 2030. *International Journal of Sustainable Transportation*. 2016;10(3):236–46. doi: 10.1080/15568318.2014.893043.

93. Friel S, Hancock T, Kjellstrom T, et al. Urban health inequities and the added pressure of climate change: an action-oriented research agenda. *Journal of Urban Health: Bulletin of the New York Academy of Medicine*. 2011;88(5):886–95. doi: 10.1007/s11524-011-9607-0.

94. Smit W, Hancock T, Kumaresen J, et al. Toward a research and action agenda on urban planning/design and health equity in cities in low and middle-income countries. *Journal of Urban Health: Bulletin of the New York Academy of Medicine*. 2011;88(5):875–85. doi: 10.1007/s11524-011-9605-2.

95. Hardin G. The Tragedy of the Commons. *Science*. 1968;162(3859):1243–8. doi: 10.1126/science.162.3859.1243.

96. Ostrom E. Beyond markets and states: polycentric governance of complex economic systems. *The American Economic Review*. 2010;100(3):641–72. Available from www.jstor.org/stable/27871226.

97. Wall D. *The Sustainable Economics of Elinor Ostrom: Commons, Contestation and Craft*. London and New York: Routledge; 2014.

98. Thaler RH, Sunstein CR. *Nudge: Improving Decisions about Health, Wealth, and Happiness*. New York: Penguin Books; 2009.

99. Gerasimchuk I, Whitley S, Beaton C, et al. Stories from G20 Countries: Shifting Public Money Out of Fossil Fuels. Winnipeg: International Institute for Sustainable Development; 2018. Available from www.iisd.org/library/stories-g20-countries-shifting-public-money-out-fossil-fuels.

100. Gupta V, Dhillon R, Yates R. Financing universal health coverage by cutting fossil fuel subsidies. *The Lancet Global Health*. 2015;3(6):e306–7. doi: 10.1016/S2214-109X(15)00007-8.

101. Anderson D. $1.7B to clean up orphaned and abandoned wells could create thousands of jobs. CBC News. 17 April 2020.

102. Atansah P, Khandan M, Moss T, Mukherjee A, Richmond J. When Do Subsidy Reforms Stick? Lessons from Iran, Nigeria, and India. Center for Global Development; 2017. Contract No. 111. Available from www.cgdev.org/publication/when-do-subsidy-reforms-stick-lessons-iran-nigeria-and-india.

103. World Bank. Carbon Pricing Dashboard 2020. Available from https://carbonpricingdashboard.worldbank.org/.

104. OECD and World Bank. The FASTER Principles for Successful Carbon Pricing: An Approach Based on Initial Experience; 2015. Contract No. 99570. Available from https://documents.worldbank.org/en/publication/documents-reports/documentdetail/901041467995665361/the-faster-principles-for-successful-carbon-pricing-an-approach-based-on-initial-experience.

105. Stiglitz JE, Stern NH, Duan M, et al. Report of the High-Level Commission on Carbon Prices. Washington, DC: Carbon Pricing Leadership Coalition; 2017. Available from www.carbonpricingleadership.org/report-of-the-highlevel-commission-on-carbon-prices.

106. EU Emissions Trading System reform: Council approves new rules for the period 2021 to 2030 [press release]; 2018. Available from www.consilium.europa.eu/en/press/press-releases/2018/02/27/eu-emissions-trading-system-reform-council-approves-new-rules-for-the-period-2021-to-2030/.

107. Cuevas S, Haines A. Health benefits of a carbon tax. *The Lancet*. 2016;387(10013): 7–9. doi: 10.1016/S0140-6736(15)00994-0.

108. Curtin J, McInerney C, Ó Gallachóir B, et al. Quantifying stranding risk for fossil fuel assets and implications for renewable energy investment: a review of the literature. *Renewable and Sustainable Energy Reviews*. 2019;116:109402. https://doi.org/10.1016/j.rser.2019.109402.

109. Ansar A, Caldecott B, Tilbury J. Stranded Assets and the Fossil Fuel Divestment Campaign: What Does Divestment Mean for the Valuation of Fossil Fuel Assets? Oxford, UK: Stranded Assets Programme, Smith School of Enterprise and the Environment, University of Oxford; 2013. Available from www.smithschool.ox.ac.uk/research/sustainable-finance/publications/SAP-divestment-report-final.pdf.

110. Caldecott B, editor. *Stranded Assets and the Environment: Risk, Resilience, and Opportunity*. Abingdon, UK and New York: Routledge; 2018.

111. van der Ploeg F, Rezai A. Stranded assets in the transition to a carbon-free economy. *Annual Review of Resource Economics*. 2020. doi: 10.1146/annurev-resource-110519-040938.

112. Gold R. PG&E: the first climate-change bankruptcy, probably not the last. *Wall Street Journal*. 18 January 2019.

113. Talberth J, Wysham D. Fossil Fuel Risk Bonds: Safeguarding Public Finances from Product Life Cycle Risks of Oil, Gas, and Coal. West Linn, OR: Center for Sustainable Economy; 2016. Available from https://sustainable-economy.org/wp-content/uploads/2016/06/Fossil-Fuel-Risk-Bonds-May-25.pdf.

114. Goldstein A, Turner WR, Gladstone J, Hole DG. The private sector's climate change risk and adaptation blind spots. *Nature Climate Change*. 2019;9(1):18–25. doi: 10.1038/s41558-018-0340-5.

115. Morana C, Sbrana G. Climate change implications for the catastrophe bonds market: an empirical analysis. *Economic Modelling*. 2019;81:274–94. https://doi.org/10.1016/j.econmod.2019.04.020.

116. TCFD. Final Report: Recommendations of the Task Force on Climate-Related Financial Disclosures. New York: G20 Task Force on Climate-Related Financial Disclosures; 2017. Available from www.fsb-tcfd.org/publications/final-recommendations-report/.

117. Van Parijs P, Vanderborght Y. *Basic Income: A Radical Proposal for a Free Society and a Sane Economy*. Cambridge, MA and London: Harvard University Press; 2017.

118. McKay A. Why a citizens' basic income? A question of gender equality or gender bias. *Work, Employment and Society*. 2007;21(2):337–48. doi: 10.1177/0950017007076643.

119. Ruckert A, Huynh C, Labonté R. Reducing health inequities: is universal basic income the way forward? *Journal of Public Health*. 2018;40(1):3–7. doi: 10.1093/pubmed/fdx006.

120. Ford M. *The Rise of the Robots: Technology and the Threat of Mass Unemployment*. New York: Basic Books; 2015.

121. Gentilini U, Grosh M, Rigolini J, Yemtsov R. *Exploring Universal Basic Income: A Guide to Navigating Concepts, Evidence, and Practices*. Washington, DC: World Bank; 2020.

122. Bidadanure JU. The political theory of universal basic income. *Annual Review of Political Science*. 2019;22(1):481–501. doi: 10.1146/annurev-polisci-050317-070954.

123. Baird S, McKenzie D, Özler B. The effects of cash transfers on adult labor market outcomes. *IZA Journal of Development and Migration*. 2018;8(1):22. doi: 10.1186/s40176-018-0131-9.

124. Kangas O, Floury S, Simanainen M, Chief M. The Basic Income Experiment 2017–2018 in Finland: Preliminary Results. Helsinki: Ministry of Social Affairs and Health; 2019. Available from http://urn.fi/URN:ISBN:978-952-00-4035-2.

125. Forget E. The town with no poverty: the health effects of a Canadian guaranteed annual income field experiment. *Canadian Public Policy*. 2011;37(3):283–305.

126. Forget E. Reconsidering a guaranteed annual income: lessons from Mincome. *Public Sector Digest*. 2015(Fall):18–24.

127. Jones D, Miarnescu I. The Labor Market Impacts of Universal and Permanent Cash Transfers: Evidence from the Alaska Permanent Fund. NBER Working Paper Series. 2020;24312. Available from www.nber.org/papers/w24312.

128. Haarmann C, Haarmann D, Jauch H, et al. Making the Difference: The BIG in Namibia. Basic Income Grant Pilot Project Assessment Report. Windhoek: Namibia NGO Forum, Basic Income Grant Coalition; 2009.

129. Jauch H. The rise and fall of the basic income grant campaign: lessons from Namibia. *Global Labour Journal*. 2015;6(3):336–50. Available from https://mulpress.mcmaster.ca/globallabour/article/view/2367.

130. Levine S, van der Berg S, Yu D. The impact of cash transfers on household welfare in Namibia. *Development Southern Africa*. 2011;28(1):39–59. doi: 10.1080/0376835X.2011.545169.

131. Haushofer J, Shapiro J. The short-term impact of unconditional cash transfers to the poor: experimental evidence from Kenya. *The Quarterly Journal of Economics*. 2016;131(4):1973–2042. doi: 10.1093/qje/qjw025.

132. Beck S, Pulkki-Brännström A-M, San Sebastián M. Basic income – healthy outcome? Effects on health of an Indian basic income pilot project: a cluster randomised trial. *Journal of Development Effectiveness*. 2015;7(1):111–26. doi: 10.1080/19439342.2014.974200.

133. Adeeth CAG, Apoorva S. Universal basic income for India: the way towards right to equality – a review. *Indian Journal of Economics and Development*. 2019;15(1):142–9. doi: 10.5958/2322-0430.2019.00016.7.

134. Davala S, Jhabvala R, Standing G, Mehta SK. *Basic Income: A Transformative Policy for India*. London: Bloomsbury Academic; 2015.

135. Evans DK, Popova A. Cash Transfers and Temptation Goods: A Review of Global Evidence. Washington, DC: World Bank; 2014. Contract No. IE127. Available from http://documents.worldbank.org/curated/en/617631468001808739/Cash-transfers-and-temptation-goods-a-review-of-global-evidence.

136. Arnold C. Pandemic speeds largest test yet of universal basic income. *Nature*. 2020;583(7817):502–3. doi: 10.1038/d41586-020-01993-3.

137. Kenner D. *Carbon Inequality: The Role of the Richest in Tackling Climate Change*. Boca Raton, FL: CRC Press; 2019.

138. Samet JM, Burke TA. Deregulation and the assault on science and the environment. *Annual Review of Public Health*. 2020;41(1):347–61. doi: 10.1146/annurev-publhealth-040119-094056.

139. Brulle RJ. The climate lobby: a sectoral analysis of lobbying spending on climate change in the USA, 2000 to 2016. *Climatic Change*. 2018;149(3):289–303. doi: 10.1007/s10584-018-2241-z.

140. Brulle RJ. Networks of opposition: a structural analysis of U.S. climate change countermovement coalitions 1989–2015. *Sociological Inquiry*. 2019. doi: 10.1111/soin.12333.

141. Rosner D. Webs of denial: climate change and the challenge to public health. *The Milbank Quarterly*. 2016;94(4):733–5. doi: 10.1111/1468-0009.12228.

142. Muñoz Acebes C, Wilkinson D. Rainforest Mafias: How Violence and Impunity Fuel Deforestation in Brazil's Amazon. Human Rights Watch; 2019. Available from www.hrw.org/report/2019/09/17/rainforest-mafias/how-violence-and-impunity-fuel-deforestation-brazils-amazon.

143. Butt N, Lambrick F, Menton M, Renwick A. The supply chain of violence. *Nature Sustainability*. 2019;2(8):742–7. doi: 10.1038/s41893-019-0349-4.

144. Scheidel A, Del Bene D, Liu J, et al. Environmental conflicts and defenders: a global overview. *Global Environmental Change*. 2020;63:102104. https://doi.org/10.1016/j.gloenvcha.2020.102104.

145. Global Witness. Deadly Environment: The Dramatic Rise in Killings of Environmental and Land Defenders. London; 2014. Available from www.globalwitness.org/en/campaigns/environmental-activists/deadly-environment/.

146. Oxfam. Tax on the 'Private' Billions Now Stashed Away in Havens Enough to End Extreme World Poverty Twice Over; 2013. Available from www.oxfam.org/en/press-releases/tax-private-billions-now-stashed-away-havens-enough-end-extreme-world-poverty-twice.

147. Henry JS. Taxing tax havens: how to respond to the Panama Papers. *Foreign Affairs*. 4 April 2016. Available from www.foreignaffairs.com/articles/panama/2016-04-12/taxing-tax-havens.

148. Cobham A, Janský P. Global distribution of revenue loss from corporate tax avoidance: re-estimation and country results. *Journal of International Development*. 2018;30(2):206–32. doi: 10.1002/jid.3348.

149. Shaxson N. Tackling tax havens. *Finance & Development*. 2019;56(3). Available from www.imf.org/external/pubs/ft/fandd/2019/09/tackling-global-tax-havens-shaxon.htm.

150. Cockcroft L. *Global Corruption: Money, Power, and Ethics in the Modern World*. Philadelphia: University of Pennsylvania Press; 2014.

151. Powell M, Wafa D, Mau TA, editors. *Corruption in a Global Context: Restoring Public Trust, Integrity, and Accountability*. London: Routledge; 2019.

152. UNECE. Globally Harmonized System of Classification and Labelling of Chemicals (GHS). Geneva: United Nations Economic Commission for Europe; 2019. Available from www.unece.org/trans/danger/publi/ghs/ghs_welcome_e.html.

153. Gill M, Ebi KL, Smith KR, Whitmarsh L, Haines A. We need health warning labels on points of sale of fossil fuels. *The BMJ Opinion*. 2020. Available from https://blogs.bmj.com/bmj/2020/03/31/we-need-health-warning-labels-on-points-of-sale-of-fossil-fuels/.

154. Leach AM, Emery KA, Gephart J, et al. Environmental impact food labels combining carbon, nitrogen, and water footprints. *Food Policy*. 2016;61:213–23. https://doi.org/10.1016/j.foodpol.2016.03.006.

155. Acquaye AA, Yamoah FA, Feng K. An integrated environmental and fairtrade labelling scheme for product supply chains. *International Journal of Production Economics*. 2015;164:472–83. https://doi.org/10.1016/j.ijpe.2014.12.014.

156. Shangguan S, Afshin A, Shulkin M, et al. A meta-analysis of food labeling effects on consumer diet behaviors and industry practices. *American Journal of Preventive Medicine*. 2019;56(2):300–14. doi: 10.1016/j.amepre.2018.09.024.

157. Anastasiou K, Miller M, Dickinson K. The relationship between food label use and dietary intake in adults: a systematic review. *Appetite*. 2019;138:280–91. doi: 10.1016/j.appet.2019.03.025.

158. Crockett RA, King SE, Marteau TM, et al. Nutritional labelling for healthier food or non-alcoholic drink purchasing and consumption. *Cochrane Database of Systematic Reviews*. 2018;2(2):Cd009315. doi: 10.1002/14651858.CD009315.pub2.

159. Conniff R. Greenwashed timber: how sustainable forest certification has failed. *Yale Environment*. 360. 20 February 2018. Available from https://e360.yale.edu/features/greenwashed-timber-how-sustainable-forest-certification-has-failed.

160. Brown KA, Harris F, Potter C, Knai C. The future of environmental sustainability labelling on food products. *The Lancet Planetary Health*. 2020;4(4):e137–8. doi: 10.1016/S2542-5196(20)30074-7.

161. UNEP. The Status of Climate Change Litigation: A Global Review. Nairobi: United Nations Environment Programme, Law Division; 2017.

162. Setzer J, Byrnes R. Global Trends in Climate Change Litigation: 2020 Snapshot. London: Grantham Research Institute on Climate Change and the Environment and Centre for Climate Change Economics and Policy, London School of Economics and Political Science; 2020.

163. Green M, Volcovici V, Farge E. Climate battles are moving into the courtroom, and lawyers are getting creative. *Reuters Environment*. 2 July 2020.

164. Metzger D, Aidun H. Major Developments in International Climate Litigation in Early 2020. New York: Sabin Center for Climate Change Law, Columbia University; 2020. Available from http://blogs.law.columbia.edu/climatechange/2020/03/12/major-developments-in-international-climate-litigation-in-early-2020/.

165. Cavalcanti MF, Terstegge MJ. The Urgenda case: the Dutch path towards a new climate constitutionalism. *DPCE Online*. 2020;43(2). Available from www.dpceonline.it/index.php/dpceonline/article/view/966.

166. Mintz JA. They threw up their hands: observations on the US Ninth Circuit Court of Appeals' unsatisfying opinion in Juliana v United States. *Journal of Energy & Natural Resources Law*. 2020;38(2):201–4. doi: 10.1080/02646811.2020.1743516.

167. Savaresi A, Auz J. Climate change litigation and human rights: pushing the boundaries. *Climate Law*. 2019;9(3):244–62. https://doi.org/10.1163/18786561-00903006.

168. Wilson G. Envisioning nature's right to a stable climate system. *Sea Grant Law & Policy*. 2020;10(1).

169. Migration Data Portal. Big Data, Migration and Human Mobility. Digital Use Around the World. 2020. Available from https://migrationdataportal.org/themen/big-data-migration-and-human-mobility.

170. Banga K, te Velde DW. Digitalisation and the Future of Manufacturing in Africa. London: Overseas Development Institute (ODI); 2018. Available from www.odi.org/publications/11073-digitalisation-and-future-manufacturing-africa.

171. The World in 2050 (TWI2050). The Digital Revolution and Sustainable Development: Opportunities and Challenges. Laxenburg, Austria: International Institute for Applied Systems Analysis; 2019. Available from http://pure.iiasa.ac.at/id/eprint/15913/.

172. Fornace KM, Alexander N, Abidin TR, et al. Local human movement patterns and land use impact exposure to zoonotic malaria in Malaysian Borneo. *eLife*. 2019;8: e47602. doi: 10.7554/eLife.47602.

173. Hardy A, Makame M, Cross D, Majambere S, Msellem M. Using low-cost drones to map malaria vector habitats. *Parasites & Vectors*. 2017;10(1):29. doi: 10.1186/s13071-017-1973-3.

174. Deville P, Linard C, Martin S, et al. Dynamic population mapping using mobile phone data. *Proceedings of the National Academy of Sciences*. 2014;111(45): 15888–93. doi: 10.1073/pnas.1408439111.

175. Martin-Gutierrez S, Borondo J, Morales AJ, et al. Agricultural activity shapes the communication and migration patterns in Senegal. *Chaos: An Interdisciplinary Journal of Nonlinear Science*. 2016;26(6):065305. doi: 10.1063/1.4952961.

176. Bengtsson L, Lu X, Thorson A, Garfield R, von Schreeb J. Improved response to disasters and outbreaks by tracking population movements with mobile phone network data: a post-earthquake geospatial study in Haiti. *PLoS Medicine*. 2011;8(8): e1001083. doi: 10.1371/journal.pmed.1001083.

177. Bengtsson L, Gaudart J, Lu X, et al. Using mobile phone data to predict the spatial spread of cholera. *Science Reports*. 2015;5:8923. doi: 10.1038/srep08923.

178. Wesolowski A, Metcalf CJE, Eagle N, et al. Quantifying seasonal population fluxes driving rubella transmission dynamics using mobile phone data. *Proceedings of the National Academy of Sciences*. 2015;112(35):11114–9. doi: 10.1073/pnas.1423542112.

179. Haines A, Hanson C, Ranganathan J. Planetary Health Watch: integrated monitoring in the Anthropocene Epoch. *The Lancet Planetary Health*. 2018;2(4):e141–3. doi: 10.1016/s2542-5196(18)30047-0.

180. Masanet E, Shehabi A, Lei N, Smith S, Koomey J. Recalibrating global data center energy-use estimates. *Science*. 2020;367(6481):984–6. doi: 10.1126/science.aba3758.

181. Hittinger E, Jaramillo P. Internet of things: energy boon or bane? *Science*. 2019;364(6438):326. doi: 10.1126/science.aau8825.

182. Peters DR, Schnell JL, Kinney PL, Naik V, Horton DE. Public health and climate benefits and tradeoffs of U.S. vehicle electrification. *GeoHealth*. 2020;4(10): e2020GH000275. doi: 10.1029/2020GH000275.

183. West DM. *The Future of Work: Robots, AI, and Automation*. Washington, DC: Brookings Institution Press; 2018.
184. Wainwright H, Elliot D. *The Lucas Plan: A New Trade Unionism in the Making?* Nottingham, UK: Spokesman; 2018.
185. Cooley M. *Architect or Bee? The Human/Technology Relationship*. Boston: South End Press; 1982.
186. Salisbury B. Story of the Lucas Plan. 2020. Available from https://lucasplan.org.uk/story-of-the-lucas-plan/.
187. Birchall J, Ketilson LH. Resilience of the Cooperative Business Model in Times of Crisis. International Labour Organization, Sustainable Enterprise Programme; 2009. Available from www.ilo.org/empent/Publications/WCMS_108416/lang–en/index.htm.
188. Corporate Europe Observatory. CAP vs Farm to Fork: Will we pay billions to destroy, or to support biodiversity, climate, and farmers? Brussels: Corporate Europe Observatory; 2020. Available from https://corporateeurope.org/en/2020/10/cap-vs-farm-fork.
189. Smith A. The Transition Town Network: a review of current evolutions and renaissance. *Social Movement Studies*. 2011;10(1):99–105. doi: 10.1080/14742837.2011.545229.
190. Hopkins R. *The Transition Handbook: From Oil Dependency to Local Resilience*. White River Junction, VT: Chelsea Green Publishing; 2008.
191. Richardson J, Nichols A, Henry T. Do transition towns have the potential to promote health and well-being? A health impact assessment of a transition town initiative. *Public Health*. 2012;126(11):982–9. doi: 10.1016/j.puhe.2012.07.009.
192. Trainer T. The Transition Towns Movement . . . Going Where? 2018. Available from www.resilience.org/stories/2018-06-07/the-transition-towns-movement-going-where/.
193. Taylor PJ. Transition towns and world cities: towards green networks of cities. *Local Environment*. 2012;17(4):495–508. doi: 10.1080/13549839.2012.678310.
194. Friedman T. A warning from the garden. *New York Times*. 19 January 2007.
195. Green New Deal Group. A Green New Deal: joined-up policies to solve the triple crunch of the credit crisis, climate change and high oil prices. London: New Economics Foundation; 2008. Available from https://greennewdealgroup.org/.
196. Kanter J. The Green New Deal. *New York Times*. 22 October 2008.
197. European Commission. The European Green Deal. Communication from the Commission to the European Parliament, the European Council, the Council, the European Economic and Social Committee and the Committee of the Regions. Brussels: European Commission; 2019.
198. Siddi M. What Are the Prospects for the European Green Deal? London: London School of Economics and Political Science; 2020. Available from https://blogs.lse.ac.uk/europpblog/2020/07/16/what-are-the-prospects-for-the-european-green-deal/.
199. Temple J. Let's keep the Green New Deal grounded in science. *Technology Review*. 18 January 2019. Available from www.technologyreview.com/2019/01/18/137792/lets-keep-the-green-new-deal-grounded-in-science/.
200. Friedman L, Gabriel T. A green new deal is technologically possible. Its political prospects are another question. *New York Times*. 21 February 2019.
201. Haines A, Scheelbeek P. European Green Deal: a major opportunity for health improvement. *The Lancet*. 2020;395(10233):1327–9. doi: 10.1016/S0140-6736(20)30109-4.
202. Meadows DH, Club of Rome. *The Limits to Growth: A Report for the Club of Rome's Project on the Predicament of Mankind*. New York: Universe Books; 1972.

203. Club of Rome. *Planetary Emergency 2.0: Securing a New Deal for People, Nature and Climate*. Winterthur, Switzerland: Club of Rome; 2020.

204. Club of Rome. *Planetary Emergency Plan: Securing a New Deal for People, Nature and Climate*. Winterthur, Switzerland: Club of Rome; 2019.

205. Chin A, Simon GL, Anthamatten P, et al. Pandemics and the future of human–landscape interactions. *Anthropocene*. 2020;31:100256. https://doi.org/10.1016/j.ancene.2020.100256.

206. Rohr JR, Barrett CB, Civitello DJ, et al. Emerging human infectious diseases and the links to global food production. *Nature Sustainability*. 2019;2(6):445–56. doi: 10.1038/s41893-019-0293-3.

207. Petrikova I, Cole J, Farlow A. COVID-19, wet markets, and planetary health. *The Lancet Planetary Health*. 2020;4(6):e213–14. doi: 10.1016/S2542-5196(20)30122-4.

208. Miller NP, Milsom P, Johnson G, et al. Community health workers during the Ebola outbreak in Guinea, Liberia, and Sierra Leone. *Journal of Global Health*. 2018;8(2):020601. doi: 10.7189/jogh-08-020601.

209. Haines A, de Barros EF, Berlin A, Heymann DL, Harris MJ. National UK programme of community health workers for COVID-19 response. *The Lancet*. 2020; 395(10231):1173–5. doi: 10.1016/S0140-6736(20)30735-2.

210. WHO conduct a training programme for community health workers on case management of acute watery diarrhea/cholera in Southcentral Somalia [press release]. Mogadishu: World Health Organization; 2017. Available from https://reliefweb.int/report/somalia/who-conduct-training-program-community-health-workers-case-management-acute-watery.

211. WEF. Pandemic Supply Chain Network (PSCN). Davos: World Economic Forum; 2020. Available from www.weforum.org/covid-action-platform/projects/pandemic-supply-chain-network-pscn.

212. Laurencin CT, McClinton A. The COVID-19 pandemic: a call to action to identify and address racial and ethnic disparities. *Journal of Racial and Ethnic Health Disparities*. 2020;7(3):398–402. doi: 10.1007/s40615-020-00756-0.

213. Chowkwanyun M, Reed AL. Racial health disparities and covid-19 – caution and context. *New England Journal of Medicine*. 2020;383(3):201–3. doi: 10.1056/NEJMp2012910.

214. Selden TM, Berdahl TA. COVID-19 and racial/ethnic disparities in health risk, employment, and household composition. *Health Affairs*. 2020;39(9):1624–32. doi: 10.1377/hlthaff.2020.00897.

215. Roesch E, Amin A, Gupta J, García-Moreno C. Violence against women during covid-19 pandemic restrictions. *BMJ*. 2020;369:m1712. doi: 10.1136/bmj.m1712.

216. Brazil: Bolsonaro sabotages anti-covid-19 efforts. President flouts health authorities' advice, undermines access to information [press release]. New York: Human Rights Watch; 2020. Available from www.hrw.org/news/2020/04/10/brazil-bolsonaro-sabotages-anti-covid-19-efforts.

217. Lakey G. The Nordic secret to battling coronavirus: trust. *Yes! Magazine*. 17 March 2020. Available from www.yesmagazine.org/opinion/2020/03/17/coronavirus-norway/.

218. Rothstein B. Trust is the key to fighting the pandemic. *Scientific American*. 2020. Available from https://blogs.scientificamerican.com/observations/trust-is-the-key-to-fighting-the-pandemic/.

219. Wiley LF, Parmet WE, Jacobson PD. Adventures in nannydom: reclaiming collective action for the public's health. *The Journal of Law, Medicine & Ethics*. 2015;43(s1):73–5. doi: 10.1111/jlme.12221.

220. WHO. *International Health Regulations*. Geneva: World Health Organization; 2005. Available from www.who.int/ihr/publications/9789241596664/en/.

221. Wenham C, Katz R, Birungi C, et al. Global health security and universal health coverage: from a marriage of convenience to a strategic, effective partnership. *BMJ Global Health*. 2019;4(1):e001145. doi: 10.1136/bmjgh-2018-001145.

222. Moorthy V, Henao Restrepo AM, Preziosi MP, Swaminathan S. Data sharing for novel coronavirus (COVID-19). *Bulletin of the World Health Organization*. 2020;98(3):150. doi: 10.2471/blt.20.251561.

223. Avorn J, Kesselheim A. Regulatory decision-making on COVID-19 vaccines during a public health emergency. *JAMA*. 2020;324(13):1284–5. doi: 10.1001/jama.2020.17101.

224. Fihn SD, Perencevich E, Bradley SM. Caution needed on the use of chloroquine and hydroxychloroquine for coronavirus disease 2019. *JAMA Network Open*. 2020;3(4): e209035. doi: 10.1001/jamanetworkopen.2020.9035.

225. Chen K, Wang M, Huang C, Kinney PL, Anastas PT. Air pollution reduction and mortality benefit during the COVID-19 outbreak in China. *The Lancet Planetary Health*. 2020;4(6):e210–12. doi: 10.1016/S2542-5196(20)30107-8.

226. Venter ZS, Aunan K, Chowdhury S, Lelieveld J. COVID-19 lockdowns cause global air pollution declines. *Proceedings of the National Academy of Sciences*. 2020: 202006853. doi: 10.1073/pnas.2006853117.

227. Berman JD, Ebisu K. Changes in U.S. air pollution during the COVID-19 pandemic. *Science of The Total Environment*. 2020;739:139864. doi: https://doi.org/10.1016/j.scitotenv.2020.139864.

228. Shi X, Brasseur GP. The response in air quality to the reduction of Chinese economic activities during the COVID-19 outbreak. *Geophysical Research Letters*. 2020; 47(11):e2020GL088070. doi: 10.1029/2020gl088070.

229. Rodríguez-Urrego D, Rodríguez-Urrego L. Air quality during the COVID-19: PM(2.5) analysis in the 50 most polluted capital cities in the world. *Environmental Pollution*. 2020;266(Pt 1):115042. doi: 10.1016/j.envpol.2020.115042.

230. Caballero S, Rapin P. COVID-19 Made Cities More Bike-Friendly – Here's How to Keep Them That Way. Geneva: World Economic Forum COVID Action Platform; 2020. Available from www.weforum.org/agenda/2020/06/covid-19-made-cities-more-bike-friendly-here-s-how-to-keep-them-that-way/.

231. Goetsch H, Quiros TP. Transport for Development. Washington, DC: World Bank; 2020. Available from https://blogs.worldbank.org/transport/covid-19-creates-new-momentum-cycling-and-walking-we-cant-let-it-go-waste.

232. McCunn LJ. The importance of nature to city living during the COVID-19 pandemic: considerations and goals from environmental psychology. *Cities & Health*. 2020:1–4. doi: 10.1080/23748834.2020.1795385.

233. Day BH. The value of greenspace under pandemic lockdown. *Environmental and Resource Economics*. 2020;76(4);1161–85. doi: 10.1007/s10640-020-00489-y.

234. Kenan Institute of Private Enterprise. How Will COVID-19 Affect Commercial Real Estate? *Kenan Insights*. Chapel Hill, NC: Kenan-Flagler Business School; 2020. Available from https://kenaninstitute.unc.edu/kenan-insight/how-will-covid-19-affect-commercial-real-estate/.

235. Ward JM. The RAND Blog – CalMatters. RAND; 2020. Available from www.rand.org/blog/2020/07/declining-commercial-real-estate-demand-may-provide.html.

236. Gutiérrez A, Miravet D, Domènech A. COVID-19 and urban public transport services: emerging challenges and research agenda. *Cities & Health*. 2020:1–4. doi: 10.1080/23748834.2020.1804291.

237. World Bank Group. *Global Economic Prospects*. Washington, DC: World Bank Group; 2020.
238. Margerison-Zilko C, Goldman-Mellor S, Falconi A, Downing J. Health impacts of the great recession: a critical review. *Current Epidemiology Reports*. 2016;3(1): 81–91. doi: 10.1007/s40471-016-0068-6.
239. Bui TTM, Button P, Picciotti EG. Early Evidence on the Impact of COVID-19 and the Recession on Older Workers. National Bureau of Economic Research; 2020. Available from www.nber.org/papers/w27448.
240. Bernstein J, Jones J. The Impact of the COVID19 Recession on the Jobs and Incomes of Persons of Color. Washington, DC: Center on Budget and Policy Priorities; 2020. Available from www.cbpp.org/research/full-employment/the-impact-of-the-covid19-recession-on-the-jobs-and-incomes-of-persons-of.
241. Peters GP, Marland G, Le Quéré C, et al. Rapid growth in CO_2 emissions after the 2008–2009 global financial crisis. *Nature Climate Change*. 2012;2(1):2–4. doi: 10.1038/nclimate1332.
242. Le Quéré C, Jackson RB, Jones MW, et al. Temporary reduction in daily global CO_2 emissions during the COVID-19 forced confinement. *Nature Climate Change*. 2020;10(7):647–53. doi: 10.1038/s41558-020-0797-x.
243. IEA. Global Energy Review 2020. The Impacts of the Covid-19 Crisis on Global Energy Demand and CO_2 Emissions. Paris: International Energy Agency; 2020. Available from www.iea.org/reports/global-energy-review-2020.
244. Obani PC, Gupta J. The impact of economic recession on climate change: eight trends. *Climate and Development*. 2016;8(3):211–23. doi: 10.1080/17565529.2015.1034226.
245. Guerriero C, Haines A, Pagano M. Health and sustainability in post-pandemic economic policies. *Nature Sustainability*. 2020;3(7):494–6. doi: 10.1038/s41893-020-0563-0.
246. Liang H, Renneboog L. The Global Sustainability Footprint of Sovereign Wealth Funds. European Corporate Governance Institute; 2019. Available from https://ssrn.com/abstract=3516985.
247. Ceccarelli M, Ramelli S, Wagner AF. Low-Carbon Mutual Funds. Swiss Finance Institute; 2020. Research paper 19-13. Available from https://ssrn.com/abstract=3353239.
248. Lelieveld J, Klingmüller K, Pozzer A, et al. Effects of fossil fuel and total anthropogenic emission removal on public health and climate. *Proceedings of the National Academy of Sciences*. 2019;116(15):7192–7. doi: 10.1073/pnas.1819989116.
249. Oxford Economics. The Economic Impact of Ornamental Horticulture and Landscaping in the UK: Report for the Ornamental Horticulture Roundtable Group. Oxford, UK: Oxford Economics; 2018. Available from https://hta.org.uk/resourceLibrary/the-economic-impact-of-ornamental-horticulture-and-landscaping-in-the-uk-pdf.html.
250. Mundaca L, Luth Richter J. Assessing 'green energy economy' stimulus packages: evidence from the U.S. programs targeting renewable energy. *Renewable and Sustainable Energy Reviews*. 2015;42:1174–86. https://doi.org/10.1016/j.rser.2014.10.060.
251. Frumkin H. A new deal for coronavirus recovery. *Medium*. 2020. Available from https://medium.com/@howardfrumkin/a-new-deal-for-coronavirus-recovery-95111ce9d46e.

Index